Simons Symposia

Series Editor

Yuri Tschinkel, Courant Institute of Mathematical Sciences and Simons Foundation, New York University, New York, NY, USA

Working to foster communication and enable interactions between experts, volumes in the Simons Symposia series bring together leading researchers to demonstrate the powerful connection of ideas, methods, and goals shared by mathematicians, theoretical physicists, and theoretical computer scientists.

Symposia volumes feature a blend of original research papers and comprehensive surveys from international teams of leading researchers in thriving fields of study. This blend of approaches helps to ensure each volume will serve not only as an introduction for graduate students and researchers interested in entering the field, but also as the premier reference for experts working on related problems.

The Simons Foundation at its core exists to support basic, discovery-driven research in mathematics and the basic sciences, undertaken in pursuit of understanding the phenomena of our world without specific applications in mind. The foundation seeks to advance the frontiers of research in mathematics and the basic sciences by creating strong collaborations and fostering cross-pollination of ideas between investigators, leading to unexpected breakthroughs and a deeper understanding of the world around us.

More information about this series at http://www.springer.com/series/15045

Werner Müller • Sug Woo Shin • Nicolas Templier
Editors

Relative Trace Formulas

Springer

Editors
Werner Müller
Mathematical Institute
University of Bonn
Bonn, Germany

Sug Woo Shin
Department of Mathematics
University of California
Berkeley, CA, USA

Nicolas Templier
Department of Mathematics, Malott Hall
Cornell University
Ithaca, NY, USA

ISSN 2365-9564 ISSN 2365-9572 (electronic)
Simons Symposia
ISBN 978-3-030-68505-8 ISBN 978-3-030-68506-5 (eBook)
https://doi.org/10.1007/978-3-030-68506-5

Mathematics Subject Classification: 11F66, 11F67, 11F70, 11G18, 11G40, 22E30, 22E50, 22E55

This Springer imprint is published by the registered company Springer Nature Switzerland AG
The registered company address is: Gewerbestrasse 11, 6330 Cham, Switzerland

Preface

The Simons symposium on relative trace formulas was held in Schloss Elmau, Germany, from April 22nd through 28th of 2018. We explored periods and special values of L-functions, integral representations, analytic and arithmetic aspects, theta correspondence, and branching laws. Experts of different specialties discussed these topics together. There were twenty-three participants. We hope that the activities of the symposium and the resulting twelve articles of this proceedings volume will be engaging to newcomers and inspiring to researchers in the field. Each article has been thoroughly refereed. Some of the articles are a synthesis of current knowledge and future directions, while others are research articles that contain original results that have not appeared elsewhere.

This is the third and final volume of a series devoted to the study of the trace formula. The previous two volumes have focused on families of automorphic forms and geometric aspects, respectively. The three symposia were made possible by the endeavor of the Simons Foundation which we would like to thank again for its generous support, as well as Yuri Tschinkel and Meghan Fazzi for their constant assistance in the organizational aspects. We thank the authors for contributing articles to these proceedings and also wish to thank all the anonymous referees. Finally, we thank Springer Nature for their help in publishing these proceedings.

Bonn, Germany — Werner Müller

Berkeley, CA, USA — Sug Woo Shin

Ithaca, NY, USA — Nicolas Templier

Introduction

In the past four decades, the trace formula has developed deep connections with various branches of mathematics in algebra, representation theory, analysis, differential geometry, and algebraic geometry. Nowadays, the body of knowledge in automorphic forms is vast and fascinating. A series of three symposia recently took place on the topic of trace formulas, each with an accompanying proceedings volume. The present volume is the third and final in this series and focuses on relative trace formulas in relation to L-values, arithmetic cycles, and periods of automorphic forms. Relative trace formulas are increasingly important because of their flexibility and wide range of arithmetic applications. The first volume focused on Arthur's trace formula, orbital integrals, Betti numbers, Weyl's law, Sato–Tate equidistribution for families, U_p-eigenvalues, and p-adic L-functions. The second volume focused on methods from algebraic geometry and geometric representation theory and their applications to endoscopy, theta correspondence, Langlands and Arthur packets, local functional equations and γ-factors, graded Hecke algebras, and sheaf-theoretic constructions. In these three proceedings volumes, we have aimed at providing a snapshot of some of the current research and raising interest in pursuing these topics further. Therefore, we opted for a balance between survey articles explaining foundational materials in the style of the 1979 Corvallis proceedings and research articles establishing new results identically to journal publications. We hope that this blend of introductory accounts and research results will stimulate the newcomers and researchers alike.

The collegial format of the symposia allowed a homogeneous set of experts to isolate key difficulties going forward and collectively assess the feasibility of diverse approaches. The main topics of the third symposium included:

- Local and global period integrals, branching laws, integral representations of L-functions, Gan–Gross–Prasad and Ichino–Ikeda conjectures, and automorphic forms on exceptional groups;
- Kuznetsov trace formula for $\mathrm{SL}(2)$, Kloosterman sums, shifted convolution sums, equidistribution, moments of families of central L-values, non-vanishing, and subconvexity;

- Relative trace formula in higher rank, Voronoï summation formula, Arthur classification and distinguished representations of classical groups, bounds towards Ramanujan, theta correspondence, and transfer and relative fundamental lemmas;
- Arithmetic aspects: cohomological automorphic representations, arithmetic cycles on Shimura varieties, algebraicity of critical L-values, and higher Gross–Zagier formula over function fields.

Relative trace formulas are identities that relate period integrals on the spectral side and relative orbital integrals on the geometric side. The most primitive instance is the Petersson trace formula, whose geometric expansion in terms of Kloosterman sums and Bessel functions was first written down by Poincaré in July 1912 in a paper submitted to Ann. Fac. Sci. Toulouse a few days before he died and published posthumously that year. Development started in earnest in the 1980s with the discovery of the Kuznetsov trace formula for SL(2) and with a series of works by Jacquet including one article in 1985 with J. F. Lai with the title "A Relative Trace Formula."

Let G be a reductive group over a global field F and H a subgroup obtained as the fixed-point set of an involution. A **period integral** is obtained when restricting an automorphic function on an adelic quotient $G(F)\backslash G(\mathbb{A}_F)$ to a subquotient $H(F)\backslash H(\mathbb{A}_F)$ and integrating against a character. For simplicity, we shall assume at first that this character is trivial. When we come to discussing Whittaker, Bessel, and Fourier–Jacobi periods, we shall also allow more general spherical subgroups H to be considered.

If we are given an automorphic representation, then the period integrals of its vectors yield a linear functional on its representation space. We say that the representation is **distinguished** by a period integral if this linear functional is non-zero, i.e., if the period integral of some automorphic form in the representation space is non-vanishing. The local counterpart with G and H defined over a local field F involves irreducible smooth representations of $G(F)$ that have a non-zero $H(F)$-invariant vector.

Some of the motivating questions of the field are as follows. What is the relationship between global period integrals and special L-values? In the local situation of representation theory of groups over local fields, can the branching laws be characterized? In many instances where the Langlands functoriality conjecture has been established, it appears that the constructed lift is distinguished by an accompanying period. What are the consequences for functoriality and for periods? What is the relationship between period integrals in the theory of automorphic forms and periods attached to cycles on varieties in arithmetic and algebraic geometry? What is the behavior of period integrals when we vary automorphic forms in families?

All these questions are under current active scrutiny, and the relative trace formula is one of the tools of investigation. A standard strategy is to establish the matching of relative orbital integrals on the geometric sides of two relative trace formulas and deduce from it transfer identities of the relative stable characters associated with tempered L-packets.

As a quick guide for the reader, we give below a context for each of the twelve articles in this volume.

Since the beginning of automorphic forms, their integral on unipotent subgroups has played a major role. Indeed, the theory is structured on parabolic induction, the Jacquet functor, and the analytic continuation of Eisenstein series. The definition of a cusp form is that it integrates to zero on unipotent radicals of all proper parabolic subgroups; that is, all its constant terms vanish. Integrating against non-trivial characters of these unipotent radicals, one generalizes constant terms and obtains **Fourier coefficients** of automorphic forms, also referred to as Whittaker coefficients or Whittaker periods.

In an article in the second volume of these proceedings, D. Jiang and B. Liu have discussed a number of properties of Fourier coefficients for symplectic groups, in relation to the dimensions of nilpotent orbits and Arthur parameters. In the present volume, the article by Pollack gives an introduction to the theory of Fourier coefficients of **modular forms on the group** G_2, which originate from the works of Gross–Wallach in the 1990s. Modular forms on G_2 transform according to quaternionic discrete series representations of $G_2(\mathbb{R})$, and they exhibit similar behavior as holomorphic modular forms on Hermitian symmetric spaces, which transform according to holomorphic discrete series representations, primary instances of which are Siegel modular forms on Sp_{2g}. The article by Pollack finalizes previous works on the subject by establishing the complete Fourier expansion, which includes all the non-generic Fourier coefficients. The article further provides an account of an integral representation of the degree 7 standard L-function developed recently by Gurevich–Segal and gives a new closed expression for its Dirichlet coefficients. A reader interested in automorphic forms on exceptional groups may also consult the article by Finis–Hoffmann–Wakatsuki in the second volume of these proceedings on the geometric side of Arthur's trace formula for G_2.

Integral representations of L**-functions** have existed for GL(1) since the nineteenth century, yet no general theory is known at present, hence the search for new ones is still ongoing. As a step towards a much-desired Lie theoretic criterion that could predict whether a given period integral unfolds to an Eulerian product, the article by Friedberg–Ginzburg proposes a new dimension equation. The equation compares, on the one hand, the sum of the dimensions of the maximal unipotent orbits for which the representations have a non-zero Fourier coefficient, and, on the other hand, the sum of the dimensions of the groups involved. For example, the equation indicates what to expect in unfolding integral kernel constructions arising from the doubling method and theta liftings.

Once an integral representation of an L-function is found, the next set of goals is to study its local and global properties. Consider F a number field, E a quadratic reduced algebra over F, and the **Asai** L**-function** attached to an automorphic representation π of $\mathrm{GL}_n(\mathbb{A}_E)$. If $E = F \times F$ is split, then $\pi = \pi_1 \times \pi_2$ for a pair of automorphic representations π_1, π_2 of $\mathrm{GL}_n(\mathbb{A}_F)$, and this specializes to a Rankin–Selberg L-function whose theory has been developed by Jacquet, Piatetski-Shapiro, and Shalika in the 1980s, where the local L-factor is defined as the greatest common divisor of certain families of **local zeta integrals**, and the local ε-factor is defined through the functional equation satisfied by these zeta integrals.

The article by Beuzart-Plessis partially completes the local theory of Asai L-functions for a general quadratic field extension E/F. The proof treats the Archimedean and non-Archimedean places uniformly by using **global methods**, namely the global functional equation satisfied by the period integrals, and globalization results obtained from Arthur's trace formula that have also featured in the articles by Matz, Müller, and Sarnak–Shin–Templier in the first volume of these proceedings. In the end, the article establishes new results in the Archimedean case and reproves most of the previously known results in the p-adic case. For example, it is established that the γ-factor defined via local zeta integrals coincides with the γ-factor defined via the Langlands–Shahidi method.

The **Gan–Gross–Prasad conjecture** concerns branching laws in the representation theory of classical groups and their L-packets. It has a local component and a global component. In the local case, for classical groups over local fields, the conjectural branching law is in terms of symplectic root numbers of the corresponding Rankin–Selberg product. In the global case, for automorphic representations of classical groups over the adèles, the conjectural branching law is in terms of the non-vanishing of central Rankin–Selberg L-values. The original version of the conjecture pertained to tempered L-parameters, but this has been newly extended to representations of Arthur type in `arXiv:1911.02783` and in the articles by Kobayashi–Speh and Jiang–Zhang in this volume, which we discuss next.

The article by Kobayashi–Speh is a continuation of the article by the same authors in the second volume of these proceedings. It announces some new results, including progress towards the Gan–Gross–Prasad conjecture for **real orthogonal groups**. In more detail, it gives necessary and sufficient conditions for branching laws for smooth irreducible representations of moderate growth of the rank one special orthogonal groups $SO(n+1,1)$ and $SO(n,1)$ in terms of their θ-stable parameters. There are applications to classifying distinguished representations, to non-vanishing results for Archimedean periods, and to invariant bilinear forms on $(\mathfrak{g}, K)$-cohomology. The article fully explores the new territory of non-tempered representations, which for finite-dimensional representations recovers classical branching laws that have been so influential in Lie theory and all adjacent areas of mathematics.

The idea of automorphic descent goes back to the construction by Saito–Kurokawa of non-tempered cuspidal automorphic representations of Sp_4 in the 1970s. It was developed by Ginzburg–Rallis–Soudry in the 1990s with notable applications to global functoriality from GL_n to quasi-split classical groups. The article by D. Jiang and L. Zhang concerns **twisted automorphic descent**, a natural extension that was introduced by the authors and accommodates non-quasi-split groups and general Arthur parameters. It has connections to non-vanishing of certain Rankin–Selberg L-functions, the Jacquet–Langlands correspondence, and the construction of explicit cuspidal automorphic modules by means of their global Arthur parameters. The present article focuses specifically on applications to branching laws and to one direction of the Gan–Gross–Prasad conjecture. One of the main results is that the non-vanishing of the period integral of two automorphic representations on orthogonal groups not only implies the non-vanishing of the central value of a Rankin–Selberg L-function but also constrains the global Arthur parameters of the representations.

The article by Lapid reports on some experimental findings in the representation theory of $\mathrm{GL}_n(F)$ for a p-adic field F, in continuation of his previous article in the second volume of these proceedings, in which he computed all **Kazhdan–Lusztig polynomials** for all pairs of permutations in the symmetric group S_{12}. In the article in the present volume, a possible correlation between deficiency and length is analyzed.

A long-term goal for developing relative trace formulas is to unify the various aspects of relative functoriality, such as theta liftings, matching of relative orbital integrals, distinction of automorphic representations, and the non-vanishing of periods and special values of L-functions. The **Kuznetsov trace formula for** SL(2) is a fundamental instance of relative trace formulas and has been much studied since its discovery in the 1980s. Three articles in this volume touch on its various aspects, which we discuss next.

The article by W. T. Gan and X. Wan uses the theta correspondence to establish the **nonstandard transfer** between the Kuznetsov trace formula for SL(2) and relative trace formulas for orthogonal groups, and also the fundamental lemma for the Hecke algebras. Thereby, they show that known methods for proving functoriality, in this case, theta correspondences, are related to identities at the level of orbital integrals. In turn, these identities between orbital integrals imply global relative character identities by comparison of the geometric sides of two relative trace formulas and by separating out the automorphic representations that contribute to the spectral sides of the respective relative trace formulas. The final step of the article is to deduce local character identities that hold the answer to the question of which representations of the group $\mathrm{SO}_{n+1}(F)$ are $\mathrm{SO}_n(F)$-distinguished, for a p-adic field F.

In the last few years, Munshi has used the Petersson trace formula to develop a GL(2) variant of the **δ-symbol method** introduced by Duke–Friedlander–Iwaniec in the 1990s in the GL(1) context. The idea of the GL(1) version of the δ-symbol method is to equate the Dirac function $\delta(m = n)$ to a bilinear average of additive characters $\exp\left(2i\pi t \frac{m-n}{N}\right)$ over suitably chosen integers t, N, relying on Fourier duality on the multiplicative group $\mathbb{Z}\backslash\mathbb{R}$. The GL(2) variant of the δ-symbol method equates $\delta(m = n)$ with a bilinear average of Hecke eigenvalues $a_f(m)\overline{a_f(n)}$ over a suitable family of modular forms f.

Originally, the GL(1) version of the δ-symbol method has implied subconvexity bounds for character twists of GL(2) principal L-functions, and Munshi has used his GL(2) variant of the δ-symbol method to establish subconvexity bounds for character twists of GL(3) principal L-functions. In his article in the present volume, Munshi returns to this recently solved GL(3) subconvexity problem for character twists, to present an argument that is more transparent and technically simpler. The argument does not rely anymore on the Ramanujan–Selberg conjecture for GL(3) automorphic forms. New unconditional subconvexity results are established, and the article provides the explicit subconvex exponent $1/308$.

Other notable applications of the Kuznetsov trace formula for SL(2) include Linnik's problem on sums of Kloosterman sums, prime geodesic counting, Selberg's $3/16$ bound, equidistribution of special points, primes in arithmetic progressions, and lower bounds for the largest prime factor of $n^2 + 1$. These are discussed in detail in Blomer's article, which surveys the relationship between **relative trace formulas and analytic number theory**.

In higher rank, the spectral and geometric sides of the Kuznetsov trace formula are the subject of current active investigations. Blomer further describes the recent progress made in the last few years, such as Sato–Tate equidistribution for families, average bounds towards Ramanujan, subconvexity for GL(3) principal L-functions in both the eigenvalue and level aspects, and simultaneous non-vanishing for central values of L-functions attached to cusp forms on GSp_4.

Shimura varieties are special kinds of algebraic varieties over number fields with rich symmetries. They can be used by realizing automorphic forms in their cohomology. Moreover, Shimura varieties supply platforms to perform arithmetic and geometric analogues of interesting automorphic constructions like theta series. Shimura varieties also provide an arena where one can test conjectures on general algebraic varieties such as **Deligne's conjecture on special values of L-functions**, the Fontaine–Mazur conjecture on characterizing Galois representations arising from geometry, and Tate's conjecture on algebraic cycles. When the general case is completely out of reach, Shimura varieties can give a series of (still challenging) problems arising from the conjectures, where extra weapons are available, thanks to the rich arithmetic structure and intimate relationship with automorphic forms.

The article by Harris is an example of this, proving a version of Deligne's conjecture for Shimura varieties associated with unitary groups by employing analytic methods in the theory of automorphic forms. In this context, the motivic L-function in Deligne's conjecture is equal (up to a shift) to the automorphic L-function by known instances of the Langlands correspondence and functoriality. The proof is built upon doubling integrals for automorphic L-functions and the geometric method to interpret doubling integrals in terms of cup products in cohomology.

Among several exciting threads of research on Shimura varieties, we choose to focus on Kudla's program in detail in order to stay within the scope of this volume centered on relative trace formulas. His program (developed in collaboration with others) proposes a geometric incarnation of theta series (and related constructions) as a generating function with special algebraic cycles (valued in arithmetic Chow groups) as coefficients, which should be modular in the same way, as classical theta series are modular. Moreover, arithmetic quantities (e.g., intersection numbers, height pairings) attached to these cycles should be related to **central derivatives** of incoherent Eisenstein series and Rankin–Selberg L-functions via precise formulas analogous to the Siegel–Weil formula and the Rallis inner product formula. The Gross–Zagier formula discovered in the 1980s has inspired much subsequent research in this area.

What stands out in these formulas is the appearance of an **incoherent orthogonal space**, meaning an adelic orthogonal space that does not come from an orthogonal space over a number field. Besides its role in the arithmetic Siegel–Weil formula, an incoherent orthogonal space gives rise to Shimura varieties, by switching one Archimedean component to make it coherent. The article by Gross explains this construction precisely and efficiently for both orthogonal and Hermitian spaces and presents conjectures describing the special locus in the mod p fiber of Shimura varieties via another coherent space obtained by switching a non-Archimedean component.

The article by Y. Liu is concerned with carrying over Kudla's program to the case of unitary groups. In the previous works, he handled the case of unitary groups associated with a Hermitian space over a CM quadratic extension of even dimension, and the present article treats the case of a Hermitian space of odd dimension. It is explained how the earlier setup has to be modified to define arithmetic theta liftings by introducing **mixed arithmetic cycles** obtained by tensoring the cohomology of the Shimura variety with the first homology group of a suitable abelian variety. The article further proposes a conjectural arithmetic inner product formula and conjectures that the Beilinson–Bloch height of the mixed arithmetic cycles can be described in terms of central derivatives of L-functions.

Contents

Archimedean Theory and ϵ-Factors for the Asai Rankin-Selberg Integrals

Raphaël Beuzart-Plessis

Abstract In this paper, we partially complete the local Rankin-Selberg theory of Asai L-functions and ϵ-factors as introduced by Flicker and Kable. In particular, we establish the relevant local functional equation at Archimedean places and prove the equality between Rankin-Selberg's and Langlands-Shahidi's ϵ-factors at every place. Our proofs work uniformly for any characteristic zero local field and use as only input the global functional equation and a globalization result for a dense subset of tempered representations that we infer from work of Finis-Lapid-Müller. The results of this paper are used in (R. Beuzart-Plessis: *Plancherel formula for* $\mathrm{GL}_n(F)\backslash \mathrm{GL}_n(E)$ *and applications to the Ichino-Ikeda and formal degree conjectures for unitary groups*, Preprint 2018) to establish an explicit Plancherel decomposition for $\mathrm{GL}_n(F)\backslash \mathrm{GL}_n(E)$, E/F a quadratic extension of local fields, with applications to the Ichino-Ikeda and formal degree conjecture for unitary groups.

1 Introduction

The goal of this paper is to partially complete the local Rankin-Selberg theory of Asai L-functions and ϵ-factors as introduced by Flicker [Fli] and Kable [Kab]. In particular we establish the relevant local functional equation and prove the equality between Rankin-Selberg's and Artin's ϵ-factors (which are the same as Langlands-Shahidi's ϵ-factors by the recent paper [Shank]) in full generality (both for Archimedean and non-Archimedean fields). This will be used in the paper [Beu] to establish an explicit Plancherel decomposition for $\mathrm{GL}_n(F)\backslash \mathrm{GL}_n(E)$, E/F a

The project leading to this publication has received funding from Excellence Initiative of Aix-Marseille University-A*MIDEX, a French "Investissements d'Avenir" programme.

R. Beuzart-Plessis (✉)
Aix-Marseille University, CNRS, Marseille, France

W. Müller et al. (eds.), *Relative Trace Formulas*, Simons Symposia,
https://doi.org/10.1007/978-3-030-68506-5_1

quadratic extension of local fields, with applications to the Ichino-Ikeda and formal degree conjecture for unitary groups.

Recall that the now classical Rankin-Selberg local theory for tensor L-functions and ϵ-factors has been developed by Jacquet-Piatetskii-Shapiro-Shalika, Jacquet-Shalika and Jacquet in [JPSS, JS2] and [Jac]. The main results of those references roughly say that we can define local L-functions of pairs as the "greatest common divisor" (in some loose sense) of certain families of Zeta integrals and local ϵ-factors of pairs through certain functional equations satisfied by those families. Moreover, it is one of the characterizing properties of the local Langlands correspondence of [HT, Hen] and [Sch] that in the p-adic case these so defined local L- and ϵ-factors match the Artin L- and ϵ-factors on the Galois side. In the Archimedean case, the Langlands correspondence is characterized by other means [La] and it is a result of Jacquet and Shalika [JS2, Jac] that Artin L-functions of pairs can indeed be considered as the "greatest common divisor" of the relevant family of Zeta integrals and moreover that those satisfy the correct functional equation with respect to Artin ϵ-factors of pairs.

In [Fli], Flicker has introduced in the non-Archimedean case a family of Zeta integrals that ought to represent the Asai L-function of a given generic irreducible (smooth) representation π of $\mathrm{GL}_n(E)$ where E is a quadratic extension of a non-Archimedean field F (and the Asai L-function is taken with respect to this extension). In particular, he was able to define a Rankin-Selberg type Asai L-function $L^{\mathrm{RS}}(s, \pi, \mathrm{As})$ as the greatest common divisor of his family of Zeta integrals as well as a ϵ-factor $\epsilon^{\mathrm{RS}}(s, \pi, \mathrm{As}, \psi')$ (where $\psi' : F \to \mathbb{S}^1$ is a non-trivial character) through the existence of a functional equation satisfied by the same Zeta integrals. Similar results have been obtained independently by Kable in [Kab]. To be more specific, let ψ be a nontrivial additive character of E which is trivial on F and let $\mathcal{W}(\pi, \psi)$ be the Whittaker model of π with respect to the corresponding standard character of the standard maximal unipotent subgroup $N_n(E)$ of $\mathrm{GL}_n(E)$. The Zeta integrals defined by Flicker and Kable are associated to functions $W \in \mathcal{W}(\pi, \psi)$ and $\phi \in C_c^\infty(F^n)$ and defined by

$$Z(s, W, \phi) = \int_{N_n(F)\backslash \mathrm{GL}_n(F)} W(h)\phi(e_n h)|\det h|_F^s dh$$

where s is a complex parameter, $e_n = (0, \dots, 0, 1)$ and $|.|_F$ the normalized absolute value on F. Flicker and Kable show that this integral converges when the real part of s is large enough and that it is represented by a rational function in q^{-s}, q being the order of the residue field of F. Moreover, they also prove that the vector space spanned by $\{Z(s, W, \phi) \mid W \in \mathcal{W}(\pi, \psi), \phi \in C_c^\infty(F^n)\}$ is a fractional ideal for $\mathbb{C}[q^s, q^{-s}]$ generated by an unique element $L^{\mathrm{RS}}(s, \pi, \mathrm{As})$ of the form $P(q^{-s})^{-1}$ where $P \in \mathbb{C}[T]$ is such that $P(0) = 1$. The next result of [Fli] and [Kab] is the functional equation. For $W \in \mathcal{W}(\pi, \psi)$ set $\widetilde{W}(g) = W(w_n {}^t g^{-1})$ for every

$g \in \mathrm{GL}_n(E)$ where $w_n = \begin{pmatrix} & & 1 \\ & \cdot^{\cdot^{\cdot}} & \\ 1 & & \end{pmatrix}$. Notice that $\widetilde{W} \in \mathcal{W}(\widetilde{\pi}, \psi^{-1})$ where $\widetilde{\pi}$ stands for the contragredient of the representation π. Let $\psi' : F \to \mathbb{S}^1$ be a nontrivial additive character and $\phi \mapsto \widehat{\phi}$ be the usual Fourier transform on $C_c^\infty(F^n)$ defined using the character ψ' and the corresponding autodual Haar measure i.e. for every $\phi \in C_c^\infty(F^n)$ we have

$$\widehat{\phi}(x_1, \ldots, x_n) = \int_{F^n} \phi(y_1, \ldots, y_n)\psi'(x_1 y_1 + \ldots + x_n y_n) dy_1 \ldots dy_n, \quad (x_1, \ldots, x_n) \in F^n,$$

where the measure of integration is chosen such that $\widehat{\widehat{\phi}}(v) = \phi(-v)$. Then the functional equation reads as follows: there exists a unique monomial $\epsilon^{\mathrm{RS}}(s, \pi, \mathrm{As}, \psi')$ in q^{-s} such that

$$\frac{Z(1-s, \widetilde{W}, \widehat{\phi})}{L^{\mathrm{RS}}(1-s, \widetilde{\pi}, \mathrm{As})} = \omega_\pi(\tau)^{n-1} |\tau|_E^{\frac{n(n-1)}{2}(s-1/2)} \lambda_{E/F}(\psi')^{-\frac{n(n-1)}{2}} \epsilon^{\mathrm{RS}}(s, \pi, \mathrm{As}, \psi') \frac{Z(s, W, \phi)}{L^{\mathrm{RS}}(s, \pi, \mathrm{As})}$$

for every $W \in \mathcal{W}(\pi, \psi)$ and $\phi \in C_c^\infty(F^n)$ where ω_π stands for the central character of π, $\tau \in E$ is the unique element so that $\psi(z) = \psi'(\mathrm{Tr}_{E/F}(\tau z))$ for every $z \in E$ (by the assumption made on ψ, we have $\mathrm{Tr}_{E/F}(\tau) = 0$) and $\lambda_{E/F}(\psi')$ is the Langlands constant of the extension E/F (see Sect. 3.2 for a reminder). The attentive reader will have noticed that the ϵ-factor $\epsilon^{\mathrm{RS}}(s, \pi, \mathrm{As}, \psi')$ is normalized differently than in [Fli] and [Kab] (due to the appearance of the factor $\omega_\pi(\tau)^{n-1} |\tau|_E^{\frac{n(n-1)}{2}(s-1/2)} \lambda_{E/F}(\psi')^{-\frac{n(n-1)}{2}}$). The present normalization will be justified a posteriori by the equality between $\epsilon^{\mathrm{RS}}(s, \pi, \mathrm{As}, \psi')$ and the Artin ϵ-factor $\epsilon(s, \pi, \mathrm{As}, \psi')$.

By the work of Anandavardhanan-Rajan [AR] (for π square-integrable) and Matringe [Mat] (for general π) the L-function $L^{\mathrm{RS}}(s, \pi, \mathrm{As})$ matches Shahidi's Asai L-function $L^{\mathrm{Sh}}(s, \pi, \mathrm{As})$ [Sha3, Gold] and hence by [Hen2] also the corresponding Artin L-function $L(s, \pi, \mathrm{As})$ (defined through the local Langlands correspondence). Recently, Anandavardhanan-Kurinczuk-Matringe-Sécherre-Stevens [AKMSS] have also established the equality between $\epsilon^{\mathrm{RS}}(s, \pi, \mathrm{As}, \psi')$ with the Shahidi's ϵ-factor $\epsilon^{\mathrm{Sh}}(s, \pi, \mathrm{As}, \psi')$ when π is supercuspidal. By the recent preprint [Shank] of Shankman this also gives the equality $\epsilon^{\mathrm{RS}}(s, \pi, \mathrm{As}, \psi') = \epsilon(s, \pi, \mathrm{As}, \psi')$ with the Artin ϵ-factor when π is supercuspidal. A similar result has been obtained when $n = 2$ for any representation π in the recent preprint [CCI]. In this paper, we will complete those results in the characteristic zero case by showing that the previous equality between ϵ-factors holds in general and also by working out the Archimedean theory. Moreover, we will also reprove most of the previous results in the p-adic case since our methods can treat uniformly the Archimedean and non-Archimedean case. Our main inputs will be the global functional equation satisfied by the corresponding global Zeta integrals (already established in [Fli, Kab]) as well as a globalization result allowing to realize a dense subset of the tempered dual of $\mathrm{GL}_n(E)$ as local constituents of global cuspidal automorphic representations of GL_n with a control on the ramification. We deduce this globalization result from the recent work of Finis-Lapid-Müller [FLM] on limit multiplicities for cuspidal automorphic representations of GL_n.

We now describe the main result of this paper. Let E/F be a quadratic extension of local fields of characteristic zero. In the Archimedean case, by a smooth representation of $\mathrm{GL}_n(E)$ we will mean a smooth admissible Fréchet representation of moderate growth in the sense of Casselman-Wallach [Cas], [Wall2, Sect. 11] or, which is the same, an admissible SF representation in the sense of [BK]. Let π be a generic irreducible smooth representation of $\mathrm{GL}_n(E)$ and $\mathcal{W}(\pi, \psi)$ be its Whittaker model with respect to a fixed nontrivial additive character $\psi : E \to \mathbb{S}^1$ which we again take to be trivial on F. To $W \in \mathcal{W}(\pi, \psi)$ we associate $\widetilde{W} \in \mathcal{W}(\widetilde{\pi}, \psi^{-1})$ as before. Let $\mathcal{S}(F^n)$ be $C_c^\infty(F^n)$ in the p-adic case and the usual Schwartz space in the Archimedean case. We let $\phi \mapsto \widehat{\phi}$ be the usual Fourier transform on $\mathcal{S}(F^n)$ defined using a nontrivial additive character $\psi' : F \to \mathbb{S}^1$ as before. Let $\tau \in E$ be the unique element so that $\psi(z) = \psi'(\mathrm{Tr}_{E/F}(\tau z))$ for every $z \in E$. Then, for $W \in \mathcal{W}(\pi, \psi)$ and $\phi \in \mathcal{S}(F^n)$ we define as above, whenever convergent, a Zeta integral $Z(s, W, \phi)$. The main result of this paper can now be stated as follows (see Theorems 3.5.1 and 3.5.2):

Theorem 1. *Let $W \in \mathcal{W}(\pi, \psi)$ and $\phi \in \mathcal{S}(F^n)$. Then:*

(i) The integral defining $Z(s, W, \phi)$ is convergent when the real part of s is sufficiently large and moreover it extends to a meromorphic function on $\mathbb{C}$.

(ii) We have the functional equation

$$\frac{Z(1-s, \widetilde{W}, \widehat{\phi})}{L(1-s, \widetilde{\pi}, \mathrm{As})} = \omega_\pi(\tau)^{n-1} |\tau|_E^{\frac{n(n-1)}{2}(s-1/2)} \lambda_{E/F}(\psi')^{-\frac{n(n-1)}{2}} \epsilon(s, \pi, \mathrm{As}, \psi') \frac{Z(s, W, \phi)}{L(s, \pi, \mathrm{As})}$$

where $L(s, \widetilde{\pi}, \mathrm{As})$, $L(s, \pi, \mathrm{As})$ and $\epsilon(s, \pi, \mathrm{As}, \psi')$ stand for the Asai L- and ϵ-factors of Artin.

(iii) The function $s \mapsto \dfrac{Z(s, W, \phi)}{L(s, \pi, \mathrm{As})}$ is holomorphic. Moreover, if π is nearly tempered *(see Sect. 3), for every $s_0 \in \mathbb{C}$ we can choose $W \in \mathcal{W}(\pi, \psi)$ and $\phi \in \mathcal{S}(F^n)$ such that this function does not vanish at s_0.*

We recall here that, in the p-adic case, the theorem above has already been proved by Flicker [Fli, Appendix] and Kable [Kab, Theorem 3] without the restriction on π in the last statement of (iii) but with different definitions of the L and ϵ-factors that we denote here by $L^{\mathrm{RS}}(s,\pi,\mathrm{As})$ and $\epsilon^{\mathrm{RS}}(s,\pi,\mathrm{As},\psi')$. Then, Anandavardhanan-Rajan [AR] (for discrete series) and Matringe [Mat] (extending the result in general) have shown that $L^{\mathrm{RS}}(s,\pi,\mathrm{As}) = L(s,\pi,\mathrm{As})$. Finally, the equality $\epsilon^{\mathrm{RS}}(s,\pi,\mathrm{As},\psi') = \epsilon(s,\pi,\mathrm{As},\psi')$ was already known when π is supercuspidal [AKMSS] or $n = 2$ [CCI]. Therefore, in the p-adic case, the only new aspect of the above theorem is the equality of ϵ-factors $\epsilon^{\mathrm{RS}}(s,\pi,\mathrm{As},\psi') = \epsilon(s,\pi,\mathrm{As},\psi')$ in general. On the other hand, the Archimedean case seems to have remained completely untouched.

One comment is in order: we expect $L(s,\pi,\mathrm{As})$ to be the "greatest common divisor", in a suitable sense, of the family of Zeta integrals $Z(s,W,\phi)$ and therefore the second part of (iii) should be true in general. This is known in the p-adic case by the aforementioned result of Matringe but remains to be seen in the Archimedean case for irreducible generic representations π which are not nearly tempered in the sense of Sect. 3. Actually, we even expect something stronger: there should exist finite families $W_i \in \mathcal{W}(\pi,\psi)$ and $\phi_i \in \mathcal{S}(F^n)$ such that

$$L(s,\pi,\mathrm{As}) = \sum_i Z(s,W_i,\phi_i).$$

Again in the p-adic case, this is known by the very definition of $L^{\mathrm{RS}}(s,\pi,\mathrm{As})$ as a GCD and the result of Matringe. However, in the Archimedean case this result seems harder to establish and in any case unreachable by the method developed in this paper. We refer the reader to [Jac, Theorem 2.7] for a proof of this property in the case of the usual Rankin-Selberg integrals.

We now briefly describe the content of each section of this paper. In Part 2 we gather general results which are not specific to GL_n. In Sect. 2.1, we set up most of our notation for a general connected reductive group. Section 2.2 contains a reminder on properties of topological vector spaces and functions valued in them that we shall use repeatedly in the paper. In Sect. 2.3, we set up some notation and conventions related to representations of local reductive groups. Section 2.4 contains a probably well-known result giving uniform bounds for matrix coefficients of a family of parabolically induced representations. In view of the lack of a proper reference, we provide a proof. In Sect. 2.5, we introduce various "extended" Harish-Chandra Schwartz spaces associated to a generic character on a quasi-split local reductive group. These spaces will be the natural receptacle for the Whittaker models of a family of induced representations. Section 2.6 is devoted to establishing locally uniform bounds for analytic family of Whittaker functions living in the Whittaker models of a family of parabolically induced representations. The proof proceeds by a reduction to the case of usual matrix coefficients by "smoothing" the Whittaker functional and is far more technical in the Archimedean case. Similar (point-wise) bounds have been obtained by Wallach [Wall2, Theorem 15.2.5]. Using the result of Sect. 2.6 and the theory of the Jacquet's functional, we construct in Sect. 2.7 certain "good" sections for Whittaker models of a family of parabolically induced representations. The existence of such families will be a crucial ingredient to extend Theorem 1 from a dense subset of the tempered dual to the set of all generic representations. In Sect. 2.8, we establish a result on the automatic holomorphic continuation of certain functions in several complex variables which are of "finite order" in vertical strips in one of the variable locally uniformly in the remaining ones and satisfying a certain functional equation. This will later be applied to show the meromorphic continuation of the local Asai Rankin-Selberg integrals and control their poles in terms of Asai L-factors. The theory of local Asai Zeta integrals and their functional equations is the object of Part 3. Section 3.2 is a reminder on basic properties of local Asai L- and ϵ-factors of Artin type. In Sect. 3.3, we introduce the relevant Zeta integrals and establish basic convergence results on them. In Sect. 3.4, we recall the meromorphic continuation and functional equation of these integrals in the split case (i.e. when $E = F \times F$) which is due to Jacquet, Piatetski-Shapiro and Shalika [JPSS, JS2, Jac]. In Sect. 3.5, we state the main theorems of this paper pertaining to the same meromorphic continuation and functional equation but in the inert case (i.e. when E/F is a field extension). Section 3.6 is devoted to the unramified computation of these Zeta integrals which is already in the literature in all but one case (i.e. when E/F is a ramified quadratic extension but π, ψ' and ψ are unramified). In Sect. 3.7, we recall the definition of the global Zeta integrals and their functional equation. Section 3.8 contains the aforementioned globalization result due to Finis-Lapid-Müller. The deduction from [FLM] is carefully explained. We also sketch how a similar result (in a slightly weaker form) can be proved for general reductive groups using part of *loc. cit.*. Finally, Sects. 3.9 and 3.10 contain the proof of the main results.

1.1 General Notation

In this paper F will always be a local field of characteristic zero. We will denote by $|.|_F$ the normalized absolute value of F. In the non-Archimedean case, we let $\mathcal{O}_F$ be the ring of integers of F and $q = q_F$ be the cardinality of the residue field of F.

For two positive functions f_1, f_2 on a set X a sentence like

$$f_1(x) \ll f_2(x) \text{ for all } x \in X$$

means that there exists a constant $C > 0$ such that $f_1(x) \leqslant C f_2(x)$ for every $x \in X$. When we want to emphasize that the implicit constant depends on some auxiliary parameters $a, b, c...$ we will write "$f_1(x) \ll_{a,b,c...} f_2(x)$ for all $x \in X$".

For every complex number $z \in \mathbb{C}$, we write $\Re(z)$ and $\Im(z)$ for the real and imaginary parts of z respectively.

2 Preliminaries

2.1 Groups

Let G be a connected reductive group over F. We denote by A_G the maximal split torus in the center of G and by $X^*(G)$, $X^*(A_G)$ the groups of algebraic characters of G and A_G respectively. Set

$$\mathcal{A}_G^* = X^*(G) \otimes \mathbb{R} = X^*(A_G) \otimes \mathbb{R}, \quad \mathcal{A}_{G,\mathbb{C}}^* = X^*(G) \otimes \mathbb{C} = X^*(A_G) \otimes \mathbb{C}$$

and

$$\mathcal{A}_G = \mathrm{Hom}(X^*(G), \mathbb{R}), \quad \mathcal{A}_{G,\mathbb{C}} = \mathrm{Hom}(X^*(G), \mathbb{C})$$

for their duals. We denote by $\langle ., . \rangle$ the natural pairing between $\mathcal{A}_G^*$ and $\mathcal{A}_G$ (resp. between $\mathcal{A}_{G,\mathbb{C}}^*$ and $\mathcal{A}_{G,\mathbb{C}}$). Let $H_G : G(F) \to \mathcal{A}_G$ be the homomorphism characterized by $\langle \chi, H_G(g) \rangle = \log|\chi(g)|_F$ for every $g \in G(F)$ and $\chi \in X^*(G)$. For $\lambda \in \mathcal{A}_{G,\mathbb{C}}^*$ we denote by $g \mapsto g^\lambda$ the unramified character of $G(F)$ given by $g^\lambda = e^{\langle \lambda, H_G(g) \rangle}$. Moreover, we let $\Re(\lambda) \in \mathcal{A}_G^*$ be the *real part* of λ (i.e. its projection to $\mathcal{A}_G^*$ relative to the decomposition $\mathcal{A}_{G,\mathbb{C}}^* = \mathcal{A}_G^* \oplus i\mathcal{A}_G^*$). Let $P = MN$ be a parabolic subgroup of G. Then the restriction map $X^*(A_M) \to X^*(A_G)$ induces surjections $\mathcal{A}_M^* \to \mathcal{A}_G^*$, $\mathcal{A}_{M,\mathbb{C}}^* \to \mathcal{A}_{G,\mathbb{C}}^*$ whose kernels will be denoted by $(\mathcal{A}_M^G)^*$ and $(\mathcal{A}_{M,\mathbb{C}}^G)^*$ respectively.

We fix a minimal parabolic subgroup P_0 of G with a Levi decomposition $P_0 = M_0 N_0$ and we set $A_0 = A_{M_0}$, $\mathcal{A}_0^* = \mathcal{A}_{M_0}^*$, $(\mathcal{A}_0^G)^* = (\mathcal{A}_{M_0}^G)^*$, $\mathcal{A}_0 = \mathcal{A}_{M_0}$, $H_0 = H_{M_0}$. We also choose a maximal compact subgroup K of $G(F)$ which is special in the p-adic case and in good position relative to M_0. We endow K with its unique Haar measure of total mass one. For every parabolic subgroup P of G we have the Iwasawa decomposition $G(F) = P(F)K$. As usual, by a *standard* parabolic subgroup we mean a parabolic subgroup of G containing P_0. If P is such a standard parabolic subgroup we will always write $P = MN$ for its unique Levi decomposition with $M_0 \subseteq M$ and $\overline{P} = M\overline{N}$ for the opposite parabolic subgroup. The restriction map $X^*(M) \to X^*(M_0)$ then induces an embedding $\mathcal{A}_M^* \hookrightarrow \mathcal{A}_0^*$ through which we will always consider $\mathcal{A}_M^*$ as a subspace of $\mathcal{A}_0^*$. We shall also denote by δ_P the modular character of $P(F)$. By the choice of K, for every standard parabolic subgroup $P = MN$ we have $K_P = K_M K_N$ where $K_P = K \cap P(F)$, $K_M = K \cap M(F)$ and $K_N = K \cap N(F)$.

Let $\Delta \subseteq X^*(A_0)$ be the set of simple roots of A_0 in N_0 and $\Delta^\vee \subseteq \mathcal{A}_0$ the corresponding subset of simple coroots. We set

$$\overline{\mathcal{A}_0^+} = \{X \in \mathcal{A}_0 \mid \langle X, \alpha \rangle \leqslant 0 \ \forall \alpha \in \Delta\}$$

$$\overline{(\mathcal{A}_0^*)^+} = \{\lambda \in \mathcal{A}_0^* \mid \langle \lambda, \alpha^\vee \rangle \leqslant 0 \ \forall \alpha \in \Delta\}$$

for the *closed negative Weyl chambers* in $\mathcal{A}_0$ and $\mathcal{A}_0^*$ respectively. We also let

$$M_0^+ = H_0^{-1}(\overline{\mathcal{A}_0^+}) = \{m_0 \in M_0(F) \mid m_0^\alpha \leqslant 1, \ \forall \alpha \in \Delta\}.$$

Let $W^G = \mathrm{Norm}_{G(F)}(M_0)/M_0(F)$ be the Weyl group of M_0. Then, W^G acts on $\mathcal{A}_0^*$ and $\overline{(\mathcal{A}_0^*)^+}$ is a fundamental domain for this action. For every $\lambda \in \mathcal{A}_0^*$ we denote by $|\lambda|$ the unique element in the intersection $W^G\lambda \cap \overline{(\mathcal{A}_0^*)^+}$. We equip $\mathcal{A}_0^*$ with the strict partial order $\prec$ defined by

$$\lambda \prec \mu \text{ if and only if } \mu - \lambda = \sum_{\alpha \in \Delta} x_\alpha \alpha \text{ where } x_\alpha > 0 \text{ for every } \alpha \in \Delta.$$

We fix an algebraic group embedding $\iota : G/A_G \hookrightarrow GL_N$ for some $N \geqslant 1$ and for every $g \in G(F)$ we set

$$\overline{\sigma}(g) = \sup\left(\{1\} \cup \{\log|\iota(g)_{i,j}|_F \mid 1 \leqslant i, j \leqslant N\}\right)$$

where the $\iota(g)_{i,j}$'s denote the entries of the matrix $\iota(g)$.

We denote by $\mathfrak{g}$ the (algebraic) Lie algebra of G. Similar notations will be used for other algebraic groups (i.e. we will denote the Lie algebra of an algebraic group by the corresponding gothic letter). In the Archimedean case, we will also write $\mathfrak{k}$ for the (real) Lie algebra of K and $\mathcal{U}(\mathfrak{k})$, $\mathcal{U}(\mathfrak{g})$ for the enveloping algebras of the complexifications of $\mathfrak{k}$ and $\mathfrak{g}(F)$ (considered as a real Lie algebra) respectively. We identify every element of $\mathcal{U}(\mathfrak{g})$ with the distribution supported at 1 that it defines.

We will assume that all the locally compact topological groups that we encounter have been equipped with Haar measures (bi-invariant Haar measures as we will always integrate, with one exception, over unimodular groups). The precise choices of these Haar measures will always be irrelevant. We denote by $*$ the convolution product on a locally compact topological group H. In the Archimedean case, this convolution product extends to distributions of compact support including elements of $\mathcal{U}(\mathfrak{g})$ and continuous compactly supported functions on closed subgroups (seen as distributions through the choice of a Haar measure on that subgroup).

2.2 *Topological Vector Spaces*

In this paper, by a *topological vector space* (TVS) we always mean a Hausdorff locally convex topological vector space over $\mathbb{C}$. If F is a TVS, we shall denote by F' its continuous dual. The TVS to be considered in this paper will all be LF spaces that is countable direct limit of Fréchet spaces. We will even only encounter *strict* LF spaces i.e. TVS F that can be written as the direct limit of a sequence $(F_n)_n$ of Fréchet spaces where the transition maps $F_n \to F_{n+1}$ are closed embeddings. Strict LF spaces are complete. Moreover, since LF spaces are barreled [Tr, Corollary 3 of Proposition 33.2], they satisfy the uniform boundedness principle (aka Banach-Steinhaus theorem).

We refer the reader to [Bour] for the notions of smooth and holomorphic functions valued in topological vector spaces (TVS) that will be used thoroughly in this paper. Actually, we will only consider smooth and holomorphic functions valued in Fréchet or strict LF spaces for which the following criterion can be applied [Bour, 3.3.1 (v)]:

(2.2.1) Let F be a quasi-complete TVS and M a complex analytic manifold. Then, a function $\varphi : M \to F$ is holomorphic if and only if it is continuous and for some total subspace $H \subset F'$ the scalar-valued functions $m \in M \mapsto \langle \varphi(m), \lambda \rangle$ are holomorphic for every $\lambda \in H$.

Also when we say that a function on a totally disconnected locally compact topological space (e.g. $G(F)$ in the p-adic case) is *smooth* we always mean that it is locally constant.

2.3 Representations

By a *representation* of $G(F)$ we will always mean a smooth representation of finite length with complex coefficients. Here *smooth* has the usual meaning in the p-adic case (i.e. every vector has an open stabilizer) whereas in the Archimedean case it means a smooth admissible Fréchet representation of moderate growth in the sense of Casselman-Wallach [Cas], [Wall2, Sect. 11] or, which is the same, an admissible SF representation in the sense of [BK]. We shall always abuse notation and denote by the same letter a representation and the space on which it acts. In the Archimedean case this space always comes with a topology (it is a Fréchet space) whereas in the p-adic case it will sometimes be convenient, in order to make uniform statements, to equip this space with its finest locally convex topology (it then becomes a strict LF space). For π a representation of $G(F)$ and $\lambda \in \mathcal{A}^*_{G,\mathbb{C}}$, we denote by π_λ the twist of π by the character $g \in G(F) \mapsto g^\lambda$. We let $\mathrm{Irr}(G)$, $\mathrm{Temp}(G)$ and $\Pi_2(G)$ be the sets of isomorphism classes of all irreducible representations, irreducible tempered representations and irreducible square-integrable representations of $G(F)$ respectively. We denote by $\widetilde{\pi}$ the contragredient representation of π (aka smooth dual) and by $\langle .,. \rangle$ the natural pairing between π and $\widetilde{\pi}$. In the Archimedean case, $\widetilde{\pi}$ can be defined as the Casselman-Wallach globalization of the contragredient of the Harish-Chandra module underlying π. An alternative description of $\widetilde{\pi}$, which is more suitable in practice, is as the space of linear forms on π which are continuous with respect *to any* $G(F)$-continuous norm on π together with the natural $G(F)$-action on it (a norm on π is said to be $G(F)$-*continuous* if the action of $G(F)$ on π is continuous for this norm). In the Archimedean case, we denote by π' the topological dual of π (i.e. the space of all continuous linear forms on π). There is a natural action

$g \mapsto \pi'(g)$ of $G(F)$ on π' and if we equip π' with the topology of uniform convergence on compact subsets of π then this action of $G(F)$ on π' is continuous and this allows one to define $\pi'(\varphi)$ for every function $\varphi \in C_c(G(F))$ by integration.

Let $P = MU$ be a parabolic subgroup of G. For σ a representation of $M(F)$, we denote by $i_P^G(\sigma)$ the smooth normalized parabolic induction of σ from $P(F)$ to $G(F)$. The space of $i_P^G(\sigma)$ consists of smooth functions $e : G(F) \to \sigma$ satisfying $e(mug) = \delta_P(m)^{1/2}\sigma(m)e(g)$ for every $(m, u, g) \in M(F) \times U(F) \times G(F)$ (with its natural structure of Fréchet space in the Archimedean case) and the group $G(F)$ acts by right translation. Assume that P is standard. Then, for every $\lambda \in \mathcal{A}_{M,\mathbb{C}}^*$ restriction to K induces a topological isomorphism between $\pi_\lambda = i_P^G(\sigma_\lambda)$ and $\pi_K = i_{K_P}^K(\sigma_{|K_M})$ where $K_P = K \cap P(F)$, $K_M = K \cap M(F)$ and $i_{K_P}^K(\sigma_{|K_M})$ denotes the spaces of smooth functions $e : K \to \sigma$ satisfying $e(muk) = \sigma(m)e(k)$ for every $(m, u, k) \in K_M \times (K \cap U(F)) \times K$. These identifications allows to define a notion of *holomorphic sections* $\lambda \in \mathcal{A}_{M,\mathbb{C}}^* \mapsto e_\lambda \in \pi_\lambda$: we call such a map holomorphic if its composition with the isomorphism $\pi_\lambda \simeq \pi_K$ is holomorphic. It turns out that this notion is actually independent of the choice of K. We also say that an assignment $\lambda \in \mathcal{A}_{M,\mathbb{C}}^* \mapsto T_\lambda \in \pi_\lambda'$ is *holomorphic* if for every holomorphic section $\lambda \mapsto e_\lambda \in \pi_\lambda$ the map $\lambda \in \mathcal{A}_{M,\mathbb{C}}^* \mapsto T_\lambda(e_\lambda)$ is holomorphic. It is equivalent to ask that, identifying T_λ with a linear form on π_K for every $\lambda \in \mathcal{A}_{M,\mathbb{C}}^*$, for every $e \in \pi_K$ the map $\lambda \in \mathcal{A}_{M,\mathbb{C}}^* \mapsto T_\lambda(e)$ be holomorphic: Indeed as π_K is Fréchet hence barreled, if this is so by the Banach-Steinhaus theorem for every compact $\mathcal{K} \subseteq \mathcal{A}_{M,\mathbb{C}}^*$ the subset $\{T_\lambda \mid \lambda \in \mathcal{K}\}$ of π_K' is equicontinuous hence $\mathcal{A}_{M,\mathbb{C}}^* \times \pi_K \to \mathbb{C}$, $(\lambda, e) \mapsto T_\lambda(e)$ is continuous and then we can apply [Jac2, Lemma 1].

Lemma 2.3.1. *Assume that F is Archimedean. Let $\pi \in \mathrm{Irr}(G)$ and let C_K be a Casimir element for K (i.e. the element of $\mathcal{U}(\mathfrak{k})$ associated to a negative definite K-invariant real symmetric bilinear form on the Lie algebra of K). Then,*

(i) $\pi(1 + C_K)$ is a topological isomorphism of π onto itself.

Let p_0 be a $G(F)$-continuous Hilbert norm (i.e. associated to a scalar product) on π which is K-invariant and set $p_\ell = p_0 \circ \pi(1 + C_K)^\ell$ for every integer $\ell \in \mathbb{Z}$. Then,

(ii) The map $\ell \mapsto p_\ell$ is increasing i.e. $p_\ell(v) \leqslant p_{\ell+1}(v)$ for every $\ell \in \mathbb{Z}$ and $v \in \pi$.
(iii) The family of norms $(p_\ell)_{\ell \geqslant 0}$ generates the topology on π.
(iv) Let $\pi^{(\ell)}$ be the completion of π for p_ℓ. Then, the natural pairing $\pi \times \widetilde{\pi} \to \mathbb{C}$ extends to a continuous bilinear form $\pi^{(\ell)} \times \widetilde{\pi} \to \mathbb{C}$ giving an embedding $\pi^{(\ell)} \subset (\widetilde{\pi})'$ for every ℓ and any equicontinuous subset of $(\widetilde{\pi})'$ is contained and bounded in $\pi^{(\ell)}$ for some ℓ.

Proof. For each $\gamma \in \widehat{K}$ the element $1 + C_K$ acts on γ by a scalar $N(\gamma)$ which is greater or equal to 1 (Indeed, C_K being in $\mathcal{U}(\mathfrak{k})^K$ the corresponding operator on γ is scalar and as $C_K = -\sum_i X_i^2$ for a certain basis (X_i) of $\mathfrak{k}$ it is also positive hermitian with respect to any K-invariant scalar product on the space of γ). Let p_0 be a $G(F)$-continuous Hilbert norm on π which is K-invariant (such norm exists as the Harish-Chandra module underlying π admits at least one Hilbert globalization). Set $p_\ell = p_0 \circ \pi(1+C_K)^\ell$ for every integer $\ell \geqslant 0$. Then, by [BK, Proposition 3.9] the family of norms $(p_\ell)_{\ell\geqslant 0}$ generates the topology on π. This gives (iii). For every vector $v \in \pi$ write $v = \sum_{\gamma\in\widehat{K}} v_\gamma$ for its decomposition in K-isotypic components (the corresponding series is absolutely convergent in π). Then, by the K-invariance of p_0 for every $\ell \geqslant 0$ and $v \in \pi$ we have

$$p_{\ell+1}(v) = \left(\sum_{\gamma\in\widehat{K}} p_0(\pi(1+C_K)^{\ell+1}v_\gamma)^2\right)^{1/2} = \left(\sum_{\gamma\in\widehat{K}} N(\gamma)^{2\ell+2} p_0(v_\gamma)\right)^{1/2}$$
$$\geqslant \left(\sum_{\gamma\in\widehat{K}} N(\gamma)^{2\ell} p_0(v_\gamma)\right)^{1/2} = p_\ell(v).$$

As $p_\ell(\pi(1+C_K)v) = p_{\ell+1}(v)$ this shows that $\pi(1+C_K)$ realizes a topological isomorphism between π and one of its closed subspace. Since the subspace of K-finite vectors of π is obviously contained in the image of $\pi(1+C_K)$ and is dense in π this shows (i) and therefore the definition of p_ℓ now also makes sense for $\ell \leqslant 0$. Moreover, the above inequality still holds for every $\ell \in \mathbb{Z}$ (with the same proof). This gives (ii). Finally, let $\widetilde{p}_0$ be the norm on $\widetilde{\pi}$ dual to p_0 i.e.

$$\widetilde{p}_0(v^\vee) = \sup_{\substack{v\in\pi \\ p_0(v)=1}} |\langle v, v^\vee\rangle|, \quad v^\vee \in \widetilde{\pi}.$$

Set $\widetilde{p}_\ell = \widetilde{p}_0 \circ \widetilde{\pi}(1+C_K)^\ell$ for every $\ell \in \mathbb{Z}$. Then, we easily check that $p_{-\ell}$ is the norm dual to $\widetilde{p}_\ell$ for every $\ell \in \mathbb{Z}$. This gives (iv) since the family of norms $(\widetilde{p}_\ell)_{\ell\geqslant 0}$ generates the topology on $\widetilde{\pi}$. ■

2.4 Uniform Bounds for Matrix Coefficients

For every $\lambda \in \mathcal{A}_0^*$, set $\pi_\lambda^0 = i_{P_0}^G(\lambda)$ where we have identified λ with the character it defines on $M_0(F)$. We identify the space of π_λ^0 with $\pi_K^0 := i_{K_{P_0}}^K(1)$ by restriction to K and we equip it with the K-invariant pairing given by

$$\langle e, e'\rangle_0 = \int_K e(k)e'(k)dk, \quad e, e' \in \pi_K^0.$$

By the previous identifications, $\langle .,. \rangle_0$ induces a $G(F)$-invariant continuous pairing between π_λ^0 and $\pi_{-\lambda}^0$ for every $\lambda \in \mathcal{A}_0^*$. Let $e_0 \in \pi_K^0$ be the vector defined by $e_0(k) = 1$ for every $k \in K$. Then, we let

$$\Xi_\lambda^G(g) = \langle \pi_\lambda^0(g)e_0, e_0\rangle_0, \quad \lambda \in \mathcal{A}_0^*, g \in G(F).$$

When $\lambda = 0$, we simply set $\Xi^G = \Xi_0^G$ (it is the usual Harish-Chandra's Xi function see [Var, Sect. II.8.5] and [Wald1, Sect. II.1]). We summarize the basic properties of the functions Ξ_λ^G in the next proposition. In the Archimedean case, most of this is contained in [Wall, Sect. 3.6].

Proposition 2.4.1.

(i) $\Xi_{w\lambda}^G = \Xi_\lambda^G$ *for every* $\lambda \in \mathcal{A}_0^*$ *and every* $w \in W^G$.
(ii) *For every* $\lambda \in \mathcal{A}_0^*$, $\mu \in \mathcal{A}_G^*$ *and* $g \in G(F)$, *we have* $\Xi_{\lambda+\mu}^G(g) = \Xi_\lambda^G(g)g^\mu$.
(iii) *Let* $P = MU$ *be a parabolic subgroup containing* P_0 *(with the Levi component chosen so that* $M \supset M_0$*). Choose for every* $g \in G(F)$ *a decomposition* $g = m_P(g)u_P(g)k_P(g)$ *with* $(m_P(g), u_P(g), k_P(g)) \in M(F) \times U(F) \times K$*. Then, we have*

$$\Xi_\lambda^G(g) = \int_K \delta_P(m_P(kg))^{1/2} \Xi_\lambda^M(m_P(kg))dk$$

for every $\lambda \in \mathcal{A}_0^*$ *and* $g \in G(F)$.
(iv) *There exists* $d > 0$ *such that*

$$\delta_0(m_0)^{1/2} m_0^{|\lambda|} \ll \Xi_\lambda^G(m_0) \ll \delta_0(m_0)^{1/2} m_0^{|\lambda|} \overline{\sigma}(m_0)^d$$

for every $\lambda \in \mathcal{A}_0^*$ *and* $m_0 \in M_0^+$ *where we have set* $\delta_0 = \delta_{P_0}$.

Proof.

(i) This is [Wall, Proposition 3.6.2] in the Archimedean case and [Cas2, Proposition 4.1] in the non-Archimedean case.
(ii) is straightforward.
(iii) follows readily from the isomorphism of "induction by stages" $\pi_\lambda^0 \simeq i_P^G(i_{P_0\cap M}^M(\lambda))$ (see [Wald1, Lemme II.1.6] for the case $\lambda = 0$).
(iv) First we prove the lower bound. By (i), we may assume $\lambda = |\lambda|$. Let $\overline{P}_0 = M_0\overline{N}_0$ be the parabolic subgroup opposite to P_0 and $C \subseteq N_0(F)M_0(F)\overline{N}_0(F)$ a compact neighborhood of 1. We can find a compact subset $C_0 \subseteq M_0(F)$ such that $m_{P_0}(km_0) \in m_0C_0$ for every $k \in C$ and $m_0 \in M_0^+$. By (iii) and (ii), it follows that

$$\begin{aligned}\Xi_\lambda^G(m_0) &= \int_K \delta_0(m_{P_0}(km_0))^{1/2} m_{P_0}(km_0)^\lambda dk \\ &\geqslant \int_{K\cap C} \delta_0(m_{P_0}(km_0))^{1/2} m_{P_0}(km_0)^\lambda dk \\ &\gg \delta_0(m_0)^{1/2} m_0^\lambda = \delta_0(m_0)^{1/2} m_0^{|\lambda|}\end{aligned}$$

for all $m_0 \in M_0^+$. We now prove the upper bound. By [Kos] and [BT, Proposition 4.4.4] (see also [Sil]), for every $k \in K$ and $m_0 \in M_0(F)$, $H_0(m_{P_0}(km_0))$ belongs to the convex hull of $\{wH_0(m_0) \mid w \in W^G\}$. It follows that $m_{P_0}(km_0)^\lambda \leqslant m_0^{|\lambda|}$ for every $k \in K$ and $m_0 \in M_0^+$. Therefore, using (ii) and (iii), we obtain

$$\Xi_\lambda^G(m_0) = \int_K \delta_0(m_{P_0}(km_0))^{1/2} m_{P_0}(km_0)^\lambda dk$$
$$\leqslant m_0^{|\lambda|} \int_K \delta_0(m_{P_0}(km_0))^{1/2} dk = m_0^{|\lambda|} \Xi^G(m_0)$$

for all $m_0 \in M_0^+$. On the other hand, by [Var, Theorem 30 p.339] and [Wald1, Lemme II.1.1] there exists $d > 0$ such that $\Xi^G(m_0) \ll \delta_0(m_0)^{1/2} \overline{\sigma}(m_0)^d$ for all $m_0 \in M_0^+$. The upper bound follows. ∎

Let $P = MU$ be a standard parabolic subgroup and $\sigma \in \mathrm{Temp}(M)$. For every $\lambda \in \mathcal{A}_{M,\mathbb{C}}^*$, we set $\pi_\lambda = i_P^G(\sigma_\lambda)$ and identify the underlying space with $\pi_K := i_{K_P}^K(\sigma_{|K_M})$ where we have set $K_P = K \cap P(F)$ and $K_M = K \cap M(F)$ as before. Similarly, we let $\widetilde{\pi}_\lambda = i_P^G(\widetilde{\sigma}_\lambda)$ and identify its underlying space with $\widetilde{\pi}_K := i_{K_P}^K(\widetilde{\sigma}_{|K_M})$ for every $\lambda \in \mathcal{A}_{M,\mathbb{C}}^*$. We define a bilinear pairing between π_K and $\widetilde{\pi}_K$ by

$$\langle e, e^\vee \rangle = \int_K \langle e(k), e^\vee(k) \rangle_\sigma dk, \quad e \in \pi_K, e^\vee \in \widetilde{\pi}_K$$

where $\langle ., . \rangle_\sigma$ denotes the natural pairing between σ and $\widetilde{\sigma}$. By the previous identifications, $\langle ., . \rangle$ induces a continuous $G(F)$-invariant pairing between π_λ and $\widetilde{\pi}_{-\lambda}$ which identifies the latter with the smooth contragredient of π_λ for every $\lambda \in \mathcal{A}_{M,\mathbb{C}}^*$. The following proposition gives uniform bounds for the matrix coefficients of the π_λ. It is most probably well-known but in view of the lack of a proper reference, we include a proof (see however [Knapp, Proposition 7.14] for the case of K-finite coefficients in the Archimedean case).

Proposition 2.4.2. *There exist continuous semi-norms p and $\widetilde{p}$ on π_K and $\widetilde{\pi}_K$ such that*

$$|\langle \pi_\lambda(g)e, e^\vee \rangle| \leqslant \Xi_{\Re(\lambda)}^G(g) p(e) \widetilde{p}(e^\vee)$$

for every $\lambda \in \mathcal{A}_{M,\mathbb{C}}^*$, $g \in G(F)$ and $(e, e^\vee) \in \pi_K \times \widetilde{\pi}_K$.

Proof. We have

$$\langle \pi_\lambda(g)e, e^\vee \rangle = \int_K \delta_P(m_P(kg))^{1/2} m_P(kg)^\lambda \langle \sigma(m_P(kg)) e(k_P(kg)), e^\vee(k) \rangle_\sigma dk$$

and therefore

$$|\langle \pi_\lambda(g)e, e^\vee\rangle| \leqslant \int_K \delta_P(m_P(kg))^{1/2} m_P(kg)^{\Re(\lambda)} |\langle \sigma(m_P(kg))e(k_P(kg)), e^\vee(k)\rangle_\sigma| dk$$

for every $\lambda \in \mathcal{A}^*_{M,\mathbb{C}}$, $g \in G(F)$ and $(e, e^\vee) \in \pi_K \times \widetilde{\pi}_K$. By [CHH, Theorem 2] and [Sun], there exists continuous semi-norms q, $\widetilde{q}$ on σ and $\widetilde{\sigma}$ such that $|\langle \sigma(m)v, v^\vee\rangle| \leqslant \Xi^M(m)q(v)\widetilde{q}(v^\vee)$ for every $m \in M(F)$ and $(v, v^\vee) \in \sigma \times \widetilde{\sigma}$. (Here we emphasize that *any* semi-norm on σ or $\widetilde{\sigma}$ is continuous in the p-adic case). Consequently,

$$|\langle \pi_\lambda(g)e, e^\vee\rangle| \leqslant p(e)\widetilde{p}(e^\vee) \int_K \delta_P(m_P(kg))^{1/2} m_P(kg)^{\Re(\lambda)} \Xi^M(m_P(kg)) dk$$

for every $\lambda \in \mathcal{A}^*_{M,\mathbb{C}}$, $g \in G(F)$, $(e, e^\vee) \in \pi_K \times \widetilde{\pi}_K$ where we have set $p(e) = \sup_{k\in K} q(e(k))$ and $\widetilde{p}(e^\vee) = \sup_{k\in K} \widetilde{q}(e^\vee(k))$. By Proposition 2.4.1, the above integral is equal to $\Xi^G_{\Re(\lambda)}(g)$ whereas p and $\widetilde{p}$ clearly define continuous semi-norms on π_K and $\widetilde{\pi}_K$ respectively. ■

2.5 Harish-Chandra Schwartz Spaces of Whittaker Functions

From this section and until the end of Sect. 2.7, we assume that G is quasi-split. Thus $P_0 = B$ is a Borel subgroup, $T = M_0$ is a maximal torus and A_0 is the maximal split torus in T. Let $\xi : N_0(F) \to \mathbb{S}^1$ be a generic character. For every $\lambda \in \mathcal{A}^*_0$, we denote by $\mathcal{C}_\lambda(N_0(F)\backslash G(F), \xi)$ the space of functions $W : G(F) \to \mathbb{C}$ satisfying:

- $W(ug) = \xi(u)W(g)$ for every $(u, g) \in N_0(F) \times G(F)$;
- If F is p-adic: W is right-invariant by a compact-open subgroup $J \subseteq G(F)$ and there exists $R > 0$ such that for every $d > 0$ we have an inequality

$$|W(tk)| \ll \left(\prod_{\alpha\in\Delta}(1 + t^\alpha)^R\right) \Xi^G_\lambda(t)\overline{\sigma}(t)^{-d}, \quad t \in T(F), k \in K;$$

- If F is Archimedean: W is smooth and there exists $R > 0$ such that for every $u \in \mathcal{U}(\mathfrak{g})$ and $d > 0$ we have an inequality

$$|(R(u)W)(tk)| \ll \left(\prod_{\alpha\in\Delta}(1 + t^\alpha)^R\right) \Xi^G_\lambda(t)\overline{\sigma}(t)^{-d}, \quad t \in T(F), k \in K.$$

Then, $\mathcal{C}_\lambda(N_0(F)\backslash G(F), \xi)$ has a natural locally convex topology making it into a LF space. Notice that we have $\mathcal{C}_\lambda(N_0(F)\backslash G(F), \xi) = \mathcal{C}_{w\lambda}(N_0(F)\backslash G(F), \xi)$ for every $w \in W^G$ and $\lambda \in \mathcal{A}^*_0$ (by Proposition 2.4.1(i)). The next lemma shows that $\mathcal{C}_\lambda(N_0(F)\backslash G(F),\xi)$ is actually a Fréchet space in the Archimedean case and a strict LF space in the p-adic case (in particular it is complete). Although it is probably well-known (see [Jac, Proposition 3.1] for a similar result) for the sake of completeness we provide a full proof.

Lemma 2.5.1. *Let $\lambda \in \mathcal{A}_0^*$. Then, for every $R > 0$ and $d > 0$ there exists a continuous semi-norm $p_{R,d}$ on $\mathcal{C}_\lambda(N_0(F)\backslash G(F),\xi)$ such that*

$$|W(tk)| \leqslant p_{R,d}(W)\left(\prod_{\alpha\in\Delta}(1+t^\alpha)^{-R}\right)\delta_0(t)^{1/2}t^{|\lambda|}\overline{\sigma}(t)^{-d}$$

for every $W \in \mathcal{C}_\lambda(N_0(F)\backslash G(F),\xi)$, $t \in T(F)$ and $k \in K$.

Proof. By the uniform boundedness principle it suffices to show that for every $R > 0$, $d > 0$ and $W \in \mathcal{C}_\lambda(N_0(F)\backslash G(F),\xi)$ we have

$$|W(tk)| \ll \left(\prod_{\alpha\in\Delta}(1+t^\alpha)^{-R}\right)\delta_0(t)^{1/2}t^{|\lambda|}\overline{\sigma}(t)^{-d}, \quad t\in T(F), k\in K.$$

Notice that by Proposition 2.4.1(iv), there exist $R_0 > 0$ and $d_0 > 0$ such that

$$\Xi_\lambda^G(t) \ll \left(\prod_{\alpha\in\Delta}(1+t^\alpha)^{R_0}\right)\delta_0(t)^{1/2}t^{|\lambda|}\overline{\sigma}(t)^{d_0}, \quad t\in T(F).$$

Indeed, for every $t \in T(F)$ there exists $w \in W^G$ such that $wtw^{-1} \in T^+$ and since w admits a lift in K, by Proposition 2.4.1(iv) there exists $d_0 > 0$ such that

$$\Xi_\lambda^G(t) = \Xi_\lambda^G(wtw^{-1}) \ll \delta_0(t)^{1/2}(wtw^{-1})^{|\lambda|}\overline{\sigma}(t)^{d_0} = t^{w^{-1}|\lambda|-|\lambda|}\delta_0(t)^{1/2}t^{|\lambda|}\overline{\sigma}(t)^{d_0}.$$

As $|\lambda| \in \overline{(\mathcal{A}_0^*)^+}$, $w^{-1}|\lambda| - |\lambda|$ is a nonnegative linear combination of simple roots. Hence, there exists $R_0 > 0$ (which we can of course choose independently of w) so that

$$t^{w^{-1}|\lambda|-|\lambda|} \leqslant \prod_{\alpha\in\Delta}(1+t^\alpha)^{R_0}.$$

Therefore, we are left with showing that for every $R > 0$, $d > 0$ and $W \in \mathcal{C}_\lambda(N_0(F)\backslash G(F),\xi)$ we have

$$|W(tk)| \ll \left(\prod_{\alpha\in\Delta}(1+t^\alpha)^{-R}\right)\Xi_\lambda^G(t)\overline{\sigma}(t)^{-d}, \quad t\in T(F), k\in K.$$

In the p-adic case we can prove the following stronger inequality by essentially the same argument as in the proof of [Wald2, Lemme 3.7]: for every compact-open subgroup J of $G(F)$, there exists $c = c_J > 0$ such that for every $W \in \mathcal{C}_\lambda(N_0(F)\backslash G(F), \xi)^J$ and $d > 0$ we have

$$|W(tk)| \ll \left(\prod_{\alpha\in\Delta} \mathbf{1}_{]0,c]}(t^\alpha)\right) \Xi_\lambda^G(t)\overline{\sigma}(t)^{-d}, \quad t \in T(F), k \in K$$

where $\mathbf{1}_{]0,c]}$ stands for the characteristic function of the interval $]0, c]$. Thus, we only consider the Archimedean case. Let $W \in \mathcal{C}_\lambda(N_0(F)\backslash G(F), \xi)$ and $d > 0$. Clearly, it suffices to show the existence of $R > 0$ such that for every Δ-tuple $\underline{N} = (N_\alpha)_{\alpha\in\Delta}$ of nonnegative integers we have

$$\left(\prod_{\alpha\in\Delta} t^{N_\alpha \alpha}\right) |W(tk)| \ll \left(\prod_{\alpha\in\Delta} (1+t^\alpha)^R\right) \Xi_\lambda^G(t)\overline{\sigma}(t)^{-d}, \quad t \in T(F), k \in K. \tag{2.5.1}$$

Let $d\xi : \mathfrak{n}_0(F) \to \mathbb{C}$ denote the differential of ξ at the origin. Since ξ is generic for every $\alpha \in \Delta$ there exists $X_\alpha \in \mathfrak{n}_0(F)$ such that $d\xi(X_\alpha) = 1$ and $\mathrm{Ad}(a)X_\alpha = \alpha(a)X_\alpha$ for every $a \in A_0(F)$. We make such a choice. Fix a norm $\|.\|$ on $\mathfrak{n}_0(F)$. As $T(F)/A_0(F)$ is compact, there exists $c > 0$ such that $\|\mathrm{Ad}(tk)^{-1}X_\alpha\| \leqslant ct^{-\alpha}$ for every $t \in T(F)$, $k \in K$ and $\alpha \in \Delta$. Then, for every Δ-tuple $\underline{N} = (N_\alpha)_{\alpha\in\Delta}$ of nonnegative integers, setting $u_{\underline{N}} = \prod_{\alpha\in\Delta} X_\alpha^{N_\alpha} \in \mathcal{U}(\mathfrak{n}_0)$ (the product being taken in some fixed order), we have

$$\begin{aligned} |W(tk)| &= \left|\prod_{\alpha\in\Delta} d\xi(X_\alpha)^{N_\alpha} W(tk)\right| \\ &= \left|(L(-\epsilon u_{\underline{N}})W)(tk)\right| = \left|\left(R(\epsilon\, \mathrm{Ad}(tk)^{-1}u_{\underline{N}})W\right)(tk)\right| \\ &\leqslant \left(\prod_{\alpha\in\Delta} t^{-N_\alpha\alpha}\right) \sup_{u\in\mathcal{K}_N} |(R(u)W)(tk)| \end{aligned}$$

for every $t \in T(F)$, $k \in K$ where ϵ is a certain sign and $\mathcal{K}_N$ is the compact subset of $\mathcal{U}(\mathfrak{n}_0)$ consisting of products of $N = \sum_{\alpha\in\Delta} N_\alpha$ elements of $\mathfrak{n}_0(F)$ of norm smaller than c. By definition of $\mathcal{C}_\lambda(N_0(F)\backslash G(F), \xi)$, there exists $R > 0$ such that for every integer $N \geqslant 0$ we have

$$\sup_{u\in\mathcal{K}_N} |(R(u)W)(tk)| \ll \left(\prod_{\alpha\in\Delta} (1+t^\alpha)^R\right) \Xi_\lambda^G(t)\overline{\sigma}(t)^{-d}, \quad t \in T(F), k \in K.$$

This shows 2.5.1 and ends the proof of the lemma. ■

2.6 Uniform Bounds for Families of Whittaker Functions

We continue with the setting of the previous sections, still assuming that G is quasi-split and fixing a generic character $\xi : N_0(F) \to \mathbb{S}^1$. Let $P = MN$ be a standard parabolic subgroup, $\sigma \in \mathrm{Temp}(M)$ and set $\pi_\lambda = i_P^G(\sigma_\lambda)$ for every $\lambda \in \mathcal{A}^*_{M,\mathbb{C}}$. We identify as before the space of π_λ with $\pi_K := i_{K_P}^K(\sigma_{|K_M})$. We assume given a family $J_\lambda \in \mathrm{Hom}_{N_0}(\pi_\lambda, \xi)$ of Whittaker functionals on $\pi_\lambda = i_P^G(\sigma_\lambda)$ for $\lambda \in \mathcal{A}^*_{M,\mathbb{C}}$ i.e. a family of continuous linear forms $J_\lambda : \pi_\lambda \to \mathbb{C}$ satisfying $J_\lambda \circ \pi_\lambda(u) = \xi(u) J_\lambda$ for every $u \in N_0(F)$ and $\lambda \in \mathcal{A}^*_{M,\mathbb{C}}$. We moreover suppose that the family $\lambda \mapsto J_\lambda \in (\pi_\lambda)'$ is holomorphic in the sense of Sect. 2.3. By Frobenius reciprocity, for every $\lambda \in \mathcal{A}^*_{M,\mathbb{C}}$, J_λ induces a continuous $G(F)$-equivariant linear map

$$\widetilde{J}_\lambda : \pi_\lambda \to C^\infty(N_0(F)\backslash G(F), \xi)$$

where $C^\infty(N_0(F)\backslash G(F), \xi)$ stands for the space of all smooth functions $W : G(F) \to \mathbb{C}$ such that $W(ug) = \xi(u)W(g)$ for every $(u, g) \in N_0(F) \times G(F)$. Recall that in Sect. 2.1 we have defined a strict partial order $\prec$ on $\mathcal{A}_0^*$.

Proposition 2.6.1.

(i) *For every $\lambda \in \mathcal{A}^*_{M,\mathbb{C}}$ and $\mu \in \mathcal{A}_0^*$ such that $|\Re(\lambda)| \prec \mu$, the image of $\widetilde{J}_\lambda$ is included in $\mathcal{C}_\mu(N_0(F)\backslash G(F), \xi)$ and the resulting linear map (that we will still denote by $\widetilde{J}_\lambda$)*

$$\pi_\lambda \to \mathcal{C}_\mu(N_0(F)\backslash G(F), \xi)$$

is continuous.

(ii) *Let $\mu \in (\mathcal{A}_0^G)^*$ and set $\mathcal{U}[\prec \mu] = \{\lambda \in (\mathcal{A}^G_{M,\mathbb{C}})^* \mid |\Re(\lambda)| \prec \mu\}$ (an open subset of $(\mathcal{A}^G_{M,\mathbb{C}})^*$). Then, the family of continuous linear maps*

$$\lambda \in \mathcal{U}[\prec \mu] \mapsto \widetilde{J}_\lambda \in \mathrm{Hom}_{G(F)}(\pi_\lambda, \mathcal{C}_\mu(N_0(F)\backslash G(F), \xi))$$

is analytic in the sense that for every analytic section $\lambda \mapsto e_\lambda \in \pi_\lambda$ the resulting map $\lambda \in \mathcal{U}[\prec \mu] \mapsto \widetilde{J}_\lambda(e_\lambda) \in \mathcal{C}_\mu(N_0(F)\backslash G(F), \xi)$ is analytic.

Proof. We will show the following:

(2.6.1) For every compact subset $\mathcal{K} \subset \mathcal{A}^*_{M,\mathbb{C}}$, there exists $R > 0$ and a continuous semi-norm p on π_K such that

$$|J_\lambda(\pi_\lambda(t)e)| \leqslant p(e)\,\Xi_{\Re(\lambda)}(t) \left(\prod_{\alpha \in \Delta} (1 + t^\alpha)^R \right)$$

for every $\lambda \in \mathcal{K}$, $t \in T(F)$ and $e \in \pi_K$.

Before proving 2.6.1, we explain how the proposition can be deduced from this. The first part can readily be inferred from 2.6.1 together with the following inequality which is a consequence of Proposition 2.4.1 (iv) and the Cartan decomposition $G(F) = KT^+K$: If $|\Re(\lambda)| \prec \mu$ then for every $d > 0$ we have

$$\Xi_{\Re(\lambda)}(g) \ll \Xi_\mu(g)\overline{\sigma}(g)^{-d}, \quad g \in G(F). \tag{2.6.2}$$

For (ii), by the criterion 2.2.1 and since for every $e \in \pi_K$ and $g \in G(F)$ the map $\lambda \in \mathcal{A}^*_{M,\mathbb{C}} \mapsto \widetilde{J}_\lambda(e)(g) = J_\lambda(\pi_\lambda(g)e)$ is analytic, it suffices to check that for every $e \in \pi_K$ the map $\lambda \in \mathcal{U}[\prec \mu] \mapsto \widetilde{J}_\lambda(e) \in \mathcal{C}_\mu(N_0(F)\backslash G(F), \xi)$ is continuous. Let $\lambda_0 \in \mathcal{U}[\prec \mu]$ and $e \in \pi_K$. Then, by definition of the topology on $\mathcal{C}_\mu(N_0(F)\backslash G(F), \xi)$, we need to show the following: there exists $R > 0$ such that for every $\epsilon > 0$ and $d > 0$ if $\lambda \in \mathcal{U}[\prec \mu]$ is sufficiently close to λ_0 then

$$\sup_{\substack{t\in T(F)\\ k\in K}} \Xi_\mu(t)^{-1}\left(\prod_{\alpha\in\Delta}(1+t^\alpha)^{-R}\right)\overline{\sigma}(t)^d \left|J_\lambda(\pi_\lambda(tk)e) - J_{\lambda_0}(\pi_{\lambda_0}(tk)e)\right| < \epsilon. \tag{2.6.3}$$

Let $R > 0$ and p be a continuous semi-norm on π_K so that 2.6.1 is satisfied on a compact neighborhood $\mathcal{K} \subseteq \mathcal{U}[\prec \mu]$ of λ_0. Then, there exists $C > 0$ such that

$$\sup_{\substack{t\in T(F)\\ k\in K}} \Xi_\mu(t)^{-1}\left(\prod_{\alpha\in\Delta}(1+t^\alpha)^{-R}\right)\overline{\sigma}(t)^{d+1} \left|J_\lambda(\pi_\lambda(tk)e)\right| < C$$

for every $\lambda \in \mathcal{K}$. Indeed, this follows from the (easily checked) fact that the inequality 2.6.2 can be made uniform on $\mathcal{K}$. Set $M = 2\epsilon^{-1}C$. Let $T[> M]$ denote the subset of elements $t \in T(F)$ such that $\overline{\sigma}(t) > M$ and $T[\leqslant M]$ be its complement. Then, the above inequality easily implies

$$\sup_{\substack{t\in T[>M]\\ k\in K}} \Xi_\mu(t)^{-1}\left(\prod_{\alpha\in\Delta}(1+t^\alpha)^{-R}\right)\overline{\sigma}(t)^d \left|J_\lambda(\pi_\lambda(tk)e)\right| < \frac{\epsilon}{2} \tag{2.6.4}$$

for every $\lambda \in \mathcal{K}$. On the other hand, $T[\leqslant M]$ is compact modulo $A_G(F)$ so that by continuity of the map $(\lambda, g) \mapsto J_\lambda(\pi_\lambda(g)e)$ for $\lambda \in (\mathcal{A}^G_{M,\mathbb{C}})^*$ sufficiently close to λ_0 we have

$$\sup_{\substack{t\in T[\leqslant M]\\ k\in K}} \Xi_\mu(t)^{-1}\left(\prod_{\alpha\in\Delta}(1+t^\alpha)^{-R}\right)\overline{\sigma}(t)^d \left|J_\lambda(\pi_\lambda(tk)e) - J_{\lambda_0}(\pi_{\lambda_0}(tk)e)\right| < \epsilon.$$

Together with 2.6.4 this readily implies 2.6.3 thus showing (ii) provided 2.6.1 is satisfied.

We now prove 2.6.1. Set $\widetilde{\pi}_\lambda = i_P^G(\widetilde{\sigma}_\lambda)$ that we identify with $\widetilde{\pi}_K := i_{K_P}^K(\widetilde{\sigma}_{|K_M})$ for every $\lambda \in \mathcal{A}^*_{M,\mathbb{C}}$. We fix a pairing $\langle .,. \rangle$ between π_K and $\widetilde{\pi}_K$ as in Sect. 2.4 that identifies $\widetilde{\pi}_{-\lambda}$ with the smooth contragredient of π_λ for every $\lambda \in \mathcal{A}^*_{M,\mathbb{C}}$. This pairing induces an embedding $\widetilde{\pi}_K \subseteq \pi'_K$.

First, we treat the p-adic case. Since ξ is generic, we can choose for every $\alpha \in \Delta$ a vector $X_\alpha \in \mathfrak{n}_0(F)$ so that $\xi(\exp(X_\alpha)) \neq 1$ and $\mathrm{Ad}(a)X_\alpha = \alpha(a)X_\alpha$ for all $a \in A_0(F)$. Let $J \subseteq K$ be a compact-open subgroup. Then, there exists $C = C_J > 0$ such that for every $\alpha \in \Delta$ and $t \in T(F)$ with $t^\alpha \geqslant C$ we have $t^{-1}\exp(X_\alpha)t \in J$. Let $e \in (\pi_K)^J$, $\lambda \in \mathcal{A}^*_{M,\mathbb{C}}$ and $t \in T(F)$. Then, if there exists $\alpha \in \Delta$ such that $t^\alpha \geqslant C$ we have

$$\xi(\exp(X_\alpha))J_\lambda(\pi_\lambda(t)e) = J_\lambda(\pi_\lambda(t)\pi_\lambda(t^{-1}\exp(X_\alpha)t)e) = J_\lambda(\pi_\lambda(t)e)$$

hence (as $\xi(\exp(X_\alpha)) \neq 1$) $J_\lambda(\pi_\lambda(t)e) = 0$. Thus,

$$J_\lambda(\pi_\lambda(t)e) = 0$$

unless $t^\alpha < C$ for every $\alpha \in \Delta$. On the other hand, there exists a compact-open subgroup $J' \subseteq K$ such that $J' \subseteq \mathrm{Ker}(\xi)tJt^{-1}$ as soon as $t^\alpha < C$ for every $\alpha \in \Delta$ where we have denoted by $\mathrm{Ker}(\xi)$ the kernel of ξ. Hence, if $t^\alpha < C$ for every $\alpha \in \Delta$ we have

$$J_\lambda(\pi_\lambda(t)e) = J_\lambda(\pi(e_{J'})\pi_\lambda(t)e)$$

where $e_{J'} := \mathrm{vol}(J')^{-1}\mathbf{1}_{J'}$. The family of functionals $\lambda \mapsto J_\lambda \circ \pi(e_{J'})$ is represented by an analytic section $\lambda \in \mathcal{A}^*_{M,\mathbb{C}} \mapsto e_\lambda^\vee \in \widetilde{\pi}_\lambda = i_P^G(\widetilde{\sigma}_\lambda)$ and by what we just saw for every $t \in T(F)$ we have

$$J_\lambda(\pi_\lambda(t)e) = \begin{cases} \langle \pi_\lambda(t)e, e_\lambda^\vee \rangle & \text{if } t^\alpha < C \text{ for every } \alpha \in \Delta, \\ 0 & \text{otherwise.} \end{cases}$$

The inequality 2.6.1 (with $R = 0$) is now a consequence of Proposition 2.4.2.

Next, we treat the Archimedean case. For this, we need to introduce another set of notation. Fix a $M(F)$-continuous Hilbert norm $\widetilde{q}_0$ on $\widetilde{\sigma}$ which is K_M-invariant and let $\widetilde{p}_0$ be the K-invariant Hilbert norm on $\widetilde{\pi}_K$ defined by

$$\widetilde{p}_0(e^\vee) = \left(\int_K \widetilde{q}_0(e^\vee(k))^2 \right)^{1/2}, \quad e^\vee \in \widetilde{\pi}_K.$$

Then, for every $\lambda \in \mathcal{A}^*_{M,\mathbb{C}}$, $\widetilde{p}_0$ induces a $G(F)$-continuous norm on $\widetilde{\pi}_\lambda$ through the identification $\widetilde{\pi}_\lambda \simeq \widetilde{\pi}_K$. Let C_K be a Casimir operator for K. Set, as in Lemma 2.3.1, $\widetilde{p}_\ell = \widetilde{p}_0 \circ \widetilde{\pi}_K(1 + C_K)^\ell$ for every integer $\ell \in \mathbb{Z}$ and $\widetilde{\pi}_K^{(\ell)}$ for the completion of $\widetilde{\pi}_K$ with respect to $\widetilde{p}_\ell$ (a Banach space). Then, by Lemma 2.3.1, we have natural inclusions $\widetilde{\pi}_K^{(\ell)} \subseteq \pi'_K$ for every ℓ, every equicontinuous subset of π'_K is contained and bounded in some $\widetilde{\pi}_K^{(\ell)}$ where $\ell \leqslant 0$ and the family of norms $(\widetilde{p}_\ell)_{\ell\geqslant 0}$ generates the topology on $\widetilde{\pi}_K$. Let $\ell \leqslant 0$ and $m \geqslant -\ell$ be integers and $\varphi \in C_c^{2m}(G(F))$ then we have

$$\pi'_\lambda(\varphi)\Lambda \in \widetilde{\pi}_K^{(\ell+m)} \tag{2.6.5}$$

for every $\lambda \in \mathcal{A}^*_{M,\mathbb{C}}$ and $\Lambda \in \widetilde{\pi}_K^{(\ell)}$. Indeed, we have

$$\begin{aligned}\widetilde{p}_{\ell+m}(\widetilde{\pi}_\lambda(\varphi)e^\vee) &= \widetilde{p}_0\left(\widetilde{\pi}_\lambda(L(1+C_K)^{\ell+m}R(1+C_K)^{-\ell}\varphi)\widetilde{\pi}_K(1+C_K)^\ell e^\vee\right)\\ &\leqslant \left\|L(1+C_K)^{\ell+m}R(1+C_K)^{-\ell}\varphi\right\|_{L^1}\times \sup_{g\in\mathrm{Supp}(\varphi)}\|\widetilde{\pi}_\lambda(g)\|_{\widetilde{p}_0}\times\widetilde{p}_\ell(e^\vee)\end{aligned}$$

for every $\lambda \in \mathcal{A}^*_{M,\mathbb{C}}$ and $e^\vee \in \widetilde{\pi}_K$ where $\left\|L(1+C_K)^{\ell+m}R(1+C_K)^{-\ell}\varphi\right\|_{L^1}$ denotes the L^1-norm of $L(1+C_K)^{\ell+m}R(1+C_K)^{-\ell}\varphi \in C_c(G(F))$ and for every $g \in G(F)$ we have denoted by $\|\widetilde{\pi}_\lambda(g)\|_{\widetilde{p}_0}$ the operator norm of $\widetilde{\pi}_\lambda(g)$ with respect to the norm $\widetilde{p}_0$. This proves 2.6.5 by density of $\widetilde{\pi}_K$ in $\widetilde{\pi}_K^{(\ell')}$ for every ℓ'. Moreover, since the function $\lambda \in \mathcal{A}^*_{M,\mathbb{C}} \mapsto \sup_{g\in\mathrm{Supp}(\varphi)}\|\widetilde{\pi}_\lambda(g)\|_{\widetilde{p}_0}$ is easily seen to be locally bounded, we have that the operator norm of $\pi'_\lambda(\varphi)$ seen as a continuous linear map from $\widetilde{\pi}_K^{(\ell)}$ to $\widetilde{\pi}_K^{(\ell+m)}$ is locally bounded in λ.

Let $\mathcal{K} \subseteq \mathcal{A}^*_{M,\mathbb{C}}$ be a compact subset. By continuity of $\lambda \mapsto J_\lambda \in \pi'_K$ for the weak-$\star$ topology and the uniform boundedness principle, the family $\{J_\lambda \mid \lambda \in \mathcal{K}\}$ is equicontinuous. Hence, there exists $\ell \leqslant 0$ such that this family is included and bounded in $\widetilde{\pi}_K^{(\ell)}$ and, by what we have just seen, for every $m \geqslant -\ell$ and $\varphi \in C_c^{2m}(G(F))$ the family

$$\{J_\lambda \circ \pi_\lambda(\varphi) \mid \lambda \in \mathcal{K}\}$$

is included and bounded in $\widetilde{\pi}_K^{(\ell+m)}$. Let p and $\widetilde{p}$ be continuous semi-norms on π_K and $\widetilde{\pi}_K$ as in Proposition 2.4.2. Then, for m sufficiently large $\widetilde{p}$ extends (uniquely) to a continuous semi-norm on $\widetilde{\pi}_K^{(m+\ell)}$ and the inequality of Proposition 2.4.2 still holds for every $(e, e^\vee) \in \pi_K \times \widetilde{\pi}_K^{(m+\ell)}$ (by density of $\widetilde{\pi}_K$ in $\widetilde{\pi}_K^{(m+\ell)}$). Therefore, we have obtained:

(2.6.6) For m sufficiently large and every $\varphi \in C_c^{2m}(G(F))$ there exists a constant $C > 0$ such that

$$|J_\lambda(\pi_\lambda(\varphi)\pi_\lambda(g)e)| \leqslant Cp(e)\,\Xi^G_{\Re(\lambda)}(g)$$

for all $\lambda \in \mathcal{K}$, $g \in G(F)$ and $e \in \pi_K$.

Let $\overline{B} = T\overline{N}_0$ be the Borel subgroup opposite to B and let $Y_1, \ldots, Y_b$ be an $\mathbb{R}$-basis of $\bar{\mathfrak{b}}(F)$. Set $\Delta_{\overline{B}} = Y_1^2+\ldots+Y_b^2 \in \mathcal{U}(\bar{\mathfrak{b}})$ and let m be a positive integer that we assume sufficiently large in what follows. By elliptic regularity [BK, Lemma 3.7] there exist $\varphi^1_{\overline{B}} \in C_c^{m'}(\overline{B}(F))$ and $\varphi^2_{\overline{B}} \in C_c^\infty(\overline{B}(F))$ where $m' = 2m - \dim(\overline{B}) - 1$ such that

$$\varphi^1_{\overline{B}} * \Delta^m_{\overline{B}} + \varphi^2_{\overline{B}} = \delta_1^{\overline{B}}$$

in the sense of distributions where $\delta_1^{\overline{B}}$ denotes the Dirac distribution at 1 on $\overline{B}(F)$. Let $\varphi_N \in C_c^\infty(N_0(F))$ be such that

$$\int_{N_0(F)} \varphi_N(u)\xi(u)du = 1.$$

Then, setting $\varphi^i = \varphi_N * \varphi^i_{\overline{B}}$ for $i = 1, 2$, for every $\lambda \in \mathcal{A}^*_{M,\mathbb{C}}$, $e \in \pi_K$ and $t \in T(F)$, we have

$$\begin{aligned} J_\lambda(\pi_\lambda(t)e) &= J_\lambda(\pi_\lambda(\varphi^1_{\overline{B}})\pi_\lambda(\Delta^m_{\overline{B}})\pi_\lambda(t)e) + J_\lambda(\pi_\lambda(\varphi^2_{\overline{B}})\pi_\lambda(t)e) \\ &= J_\lambda(\pi_\lambda(\varphi^1)\pi_\lambda(\Delta^m_{\overline{B}})\pi_\lambda(t)e) + J_\lambda(\pi_\lambda(\varphi^2)\pi_\lambda(t)e). \end{aligned} \tag{2.6.7}$$

Noticing that $\varphi^i \in C_c^{m'}(G(F))$ for $i = 1, 2$, from 2.6.6 we deduce that for m sufficiently large there exists $C > 0$ such that

$$|J_\lambda(\pi_\lambda(t)e)| \leqslant C(p(\pi_\lambda(\mathrm{Ad}(t)^{-1}\Delta^m_{\overline{B}})e) + p(e))\Xi^G_{\Re(\lambda)}(t) \tag{2.6.8}$$

for every $\lambda \in \mathcal{K}$, $t \in T(F)$ and $e \in \pi_K$. To get 2.6.1 it only remains to notice that there exist $R > 0$ and a continuous semi-norm q on π_K such that

$$p(\pi_\lambda(\mathrm{Ad}(t)^{-1}\Delta^m_{\overline{B}})e) \leqslant \left(\prod_{\alpha\in\Delta}(1+t^\alpha)^R\right) q(e)$$

for every $\lambda \in \mathcal{K}$, $t \in T(F)$ and $e \in \pi_K$. This follows from the fact that the function $(\lambda, t) \in \mathcal{A}^*_{M,\mathbb{C}} \times T(F) \mapsto \pi_\lambda(\mathrm{Ad}(t)^{-1}\Delta^m_{\overline{B}})$ is polynomial (see [Del, Proposition 1]). ■

2.7 *Application to the Existence of Good Sections for Whittaker Models*

We continue with the setting of the previous section: G is quasi-split, $\xi : N_0(F) \to \mathbb{S}^1$ is a generic character, $P = MN$ is a standard parabolic subgroup, $\sigma \in \mathrm{Temp}(M)$ and we set $\pi_\lambda = i_P^G(\sigma_\lambda)$ for every $\lambda \in \mathcal{A}^*_{M,\mathbb{C}}$ whose space is identified as before with $\pi_K := i^K_{K_P}(\sigma_{|K_M})$. Let $w_0 \in W^G$ be the longest element of the Weyl group and choose a lift $\widetilde{w_0} \in G(F)$. Let ξ^- be the generic character of

$\overline{N}_0(F) = w_0 N_0(F) w_0^{-1}$ defined by $\xi^-(\overline{u}) = \xi(\widetilde{w_0}\overline{u}\widetilde{w_0}^{-1})$, $\overline{u} \in \overline{N}_0(F)$. Assume that σ is generic with respect to the restriction of ξ^- to $\overline{N}_0(F) \cap M(F)$ i.e. there exists a continuous nonzero linear form $\ell : \sigma \to \mathbb{C}$ such that $\ell \circ \sigma(\overline{u}) = \xi^-(\overline{u})\ell$ for every $\overline{u} \in \overline{N}_0(F) \cap M(F)$. Then, the construction of Jacquet's integral [Wall2, Sect. 15.4], [CS] provides us with a holomorphic family of Whittaker functionals

$$\lambda \in \mathcal{A}^*_{M,\mathbb{C}} \mapsto J_\lambda \in \mathrm{Hom}_{N_0}(\pi_\lambda, \xi)$$

as in the previous section which is everywhere non-vanishing. For every $\lambda \in \mathcal{A}^*_{M,\mathbb{C}}$ we denote by $\mathcal{W}(\pi_\lambda, \xi)$ the corresponding Whittaker model i.e. the space of functions of the form $g \in G(F) \mapsto J_\lambda(\pi_\lambda(g)e)$ for $e \in \pi_K$.

Corollary 2.7.1. *For every $\lambda_0 \in (\mathcal{A}^G_{M,\mathbb{C}})^*$ and $W_0 \in \mathcal{W}(\pi_{\lambda_0}, \xi)$ there exists a map*

$$\lambda \in (\mathcal{A}^G_{M,\mathbb{C}})^* \mapsto W_\lambda \in \mathcal{W}(\pi_\lambda, \xi)$$

such that:

- *for every $\mu \in (\mathcal{A}^G_0)^*$ and $\lambda \in \mathcal{U}[\prec \mu] = \{\lambda \in (\mathcal{A}^G_{M,\mathbb{C}})^* \mid |\Re(\lambda)| \prec \mu\}$ we have $W_\lambda \in \mathcal{C}_\mu(N_0(F)\backslash G(F), \xi)$ and the resulting map*

$$\lambda \in \mathcal{U}[\prec \mu] \mapsto W_\lambda \in \mathcal{C}_\mu(N_0(F)\backslash G(F), \xi)$$

  is analytic;
- *$W_{\lambda_0} = W_0$.*

Proof. Indeed, there exists $e \in \pi_K$ such that (with the notation of Sect. 2.6), $W_0 = \widetilde{J}_{\lambda_0}(e)$ and then it suffices to set $W_\lambda = \widetilde{J}_\lambda(e)$ for every $\lambda \in (\mathcal{A}^G_{M,\mathbb{C}})^*$: the required properties immediately follows from Proposition 2.6.1. ■

2.8 Holomorphic Continuation of Certain Functions

For every $C, D \in \mathbb{R} \cup \{-\infty\}$ with $D > C$ we set $\mathcal{H}_{>C} = \{s \in \mathbb{C} \mid \Re(s) > C\}$ and $\mathcal{H}_{]C,D[} = \{s \in \mathbb{C} \mid C < \Re(s) < D\}$. A *vertical strip* is a subset of $\mathbb{C}$ which is the closure of $\mathcal{H}_{]C,D[}$ for some $C, D \in \mathbb{R}$ with $D > C$. Let M be a complex analytic manifold and $C \in \mathbb{R} \cup \{-\infty\}$. Then, we say that a holomorphic function $Z : \mathcal{H}_{>C} \times M \to \mathbb{C}$ is *of finite order in vertical strips in the first variable locally uniformly in the second variable* if there exists $d \in \mathbb{R}$ such that for every vertical strip $V \subseteq \mathcal{H}_{>C}$ and every compact subset $\mathcal{K}_M \subseteq M$, we have

$$\sup_{(s,t) \in V \times \mathcal{K}_M} e^{-|s|^d} |Z(s,t)| < \infty.$$

If the above inequality holds with $d = 0$ for every vertical strip $V \subseteq \mathcal{H}_{>C}$ and every compact subset $\mathcal{K}_M \subseteq M$, we also say that Z is *bounded in vertical strips in the first variable locally uniformly in the second variable*. Similarly, we say that Z is *rapidly decreasing in vertical strips in the first variable locally uniformly in the second variable* if for every vertical strip $V \subseteq \mathcal{H}_{>C}$, every $N > 0$ and every compact subset $\mathcal{K}_M \subseteq M$ we have

$$\sup_{(s,t)\in V\times\mathcal{K}_M} (1+|s|)^N |Z(s,t)| < \infty.$$

When M is a point, i.e. Z is just a function of one complex variable, we will just say "of finite order in vertical strips", resp. "bounded in vertical strips", resp. "rapidly decreasing in vertical strips". The following proposition will be needed to extend Theorem 1 from nearly tempered representations to all (generic) representations.

Proposition 2.8.1. *Let M be a connected complex analytic variety, $U \subseteq M$ a nonempty connected relatively compact open subset and*

$$Z_+, Z_- : \mathbb{C} \times U \to \mathbb{C}$$

be two holomorphic functions satisfying

$$Z_+(s,t) = Z_-(-s,t) \tag{2.8.1}$$

for every $(s,t) \in \mathbb{C} \times U$. Moreover, we assume that for each $\epsilon \in \{\pm\}$ the following two conditions are satisfied

(1) Z_ϵ is of finite order in vertical strips in the first variable locally uniformly in the second variable;
(2) For every connected relatively compact open subset $U' \subseteq M$ containing U, there exists $C > 0$ such that Z_ϵ admits a (necessarily unique) holomorphic continuation to $\mathcal{H}_{>C} \times U'$ which is again of finite order in vertical strips in the first variable locally uniformly in the second variable.

Then, Z_+ and Z_- extend to holomorphic functions on all of $\mathbb{C} \times M$ which are of finite order in vertical strips in the first variable locally uniformly in the second variable and still satisfy 2.8.1.

Proof. Let $D > 0$ and $U' \subseteq M$ be a connected relatively compact open subset containing U. Clearly, we just need to show that Z_+, Z_- extend to holomorphic functions on $\mathcal{H}_{]-D,D[} \times U'$ of finite order in vertical strips in the first variable locally uniformly in the second variable (the functional equation 2.8.1 will then hold automatically for these extensions by connectedness). Choosing another connected relatively compact open subset $U'' \subseteq M$ containing the closure of U' and using condition (2) for U'', we see that there exists $C > 0$ and a nonnegative integer n such that $(s,t) \mapsto e^{s^n} Z_\pm(s,t)$ extend to holomorphic functions on $\mathcal{H}_{>C} \times U' \cup \mathbb{C} \times U$ which are rapidly decreasing in vertical strips in the first variable locally uniformly in the second variable. Moreover, up to multiplying Z_+ and Z_- by $(s,t)\mapsto e^{s^n}$, we may assume that $n=0$. Then, for $D>C$ and $\epsilon\in\{\pm\}$, we set

$$\mathcal{Z}_\epsilon^{<D}(s,t)=\frac{1}{2\pi}\left(\int_{-\infty}^{+\infty}\frac{Z_\epsilon(D+ix,t)}{D+ix-s}dx-\int_{-\infty}^{+\infty}\frac{Z_{-\epsilon}(D+ix,t)}{D+ix+s}dx\right)$$

for every $(s,t)\in\mathcal{H}_{]-D,D[}\times U'$. The above expression converges absolutely and defines a holomorphic function on $\mathcal{H}_{]-D,D[}\times U'$ which is bounded in vertical strips in the first variable locally uniformly in the first variable. Moreover, by the functional equation 2.8.1 and since Z_+, Z_- are holomorphic functions on $\mathbb{C}\times U$ that are rapidly decreasing in vertical strips in the first variable, by Cauchy's integration formula $\mathcal{Z}_\epsilon^{<D}$ and Z_ϵ coincide on $\mathcal{H}_{]-D,D[}\times U$ for $\epsilon\in\{\pm\}$. Therefore, $\mathcal{Z}_\epsilon^{<D}$ is the sought-after extension. ■

3 Zeta Integrals and Statement of the Main Theorems

3.1 Notation

Let F be a local field of characteristic zero and E be either a quadratic extension of F (inert case) or $F\times F$ (split case). We write $|.|_F$ and $|.|_E$ for the normalized absolute values of F and E respectively. Thus, in the split case we have $|(\lambda,\mu)|_E=|\lambda|_F|\mu|_F$ for every $(\lambda,\mu)\in E$ and in both cases we have $|x|_E=|x|_F^2$ for every $x\in F$. In the non-Archimedean case, we let $\mathcal{O}_F$ and $\mathcal{O}_E$ be the rings of integers of F and E respectively. We fix non-trivial additive characters $\psi':F\to\mathbb{S}^1$ and $\psi:E\to\mathbb{S}^1$ and we assume that ψ *is trivial on* F. We denote by τ the unique element of E such that $\psi(z)=\psi'(\mathrm{Tr}_{E/F}(\tau z))$ for every $z\in E$ where $\mathrm{Tr}_{E/F}$ stands for the trace of the extension E/F.

Let $n\geqslant 1$ be an integer. We write G_n for the algebraic group GL_n, $\mathfrak{gl}_n$ for its Lie algebra and we denote by Z_n, B_n, A_n and N_n the subgroups of scalar, resp. upper triangular, resp. diagonal, resp. unipotent upper triangular matrices in G_n. By a *standard* parabolic subgroup P of G_n we mean a parabolic subgroup containing B_n. We then write $P=MU$ for its unique Levi decomposition with $A_n\subseteq M$. We denote by δ_n and $\delta_{n,E}$ the modular characters of $B_n(F)$ and $B_n(E)$ respectively. We will denote by K_n the standard maximal compact subgroup of $G_n(F)$ that is $K_n=G_n(\mathcal{O}_F)$ in the p-adic case and $K_n=O(n)$ or $U(n)$ in the cases where $F=\mathbb{R}$ or $\mathbb{C}$ respectively. For $g\in G_n$, we write ${}^tg$ for its transpose and $g_{i,j}$, $1\leqslant i,j\leqslant n$, for its entries. Also, for $a\in A_n$ we simply write a_i for $a_{i,i}$ $(1\leqslant i\leqslant n)$. Let $\mathcal{A}^*=\mathcal{A}^*_{A_n}$, $\mathcal{A}^*_{\mathbb{C}}=\mathcal{A}^*_{A_n,\mathbb{C}}$ and $(\mathcal{A}^{G_n})^*=(\mathcal{A}^{G_n}_{A_n})^*$. We identify $\mathcal{A}^*$ with $\mathbb{R}^n$ through the choice of the basis $\chi_1,\ldots,\chi_n$ of $X^*(A_n)$ where χ_i $(1\leqslant i\leqslant n)$ is defined by $\chi_i(a)=a_i$ for every $a\in A_n$. Then, the closed negative Weyl chamber $\overline{(\mathcal{A}^*)^+}$ with respect to B_n is the cone of n-tuples $\lambda=(\lambda_1,\ldots,\lambda_n)\in\mathbb{R}^n$ such that $\lambda_1\leqslant\ldots\leqslant\lambda_n$ and the subspace $(\mathcal{A}^{G_n})^*$ consists of vectors $\lambda\in\mathbb{R}^n$ with

$\lambda_1 + \ldots + \lambda_n = 0$. Let $\lambda = (\lambda_1, \ldots, \lambda_n) \in \mathcal{A}^*$. Then, defining $|\lambda|$ as in Sect. 2.1, we have $|\lambda| = (\lambda_{w(1)}, \ldots, \lambda_{w(n)})$ where w is any permutation of $\{1, \ldots, n\}$ such that $\lambda_{w(1)} \leqslant \ldots \leqslant \lambda_{w(n)}$. Moreover, the character $a \mapsto a^\lambda$ of $A_n(E)$ is given by

$$a^\lambda = |a_1|_E^{\lambda_1} \ldots |a_n|_E^{\lambda_n}.$$

We set $\min(\lambda) := \min(\lambda_1, \ldots, \lambda_n)$. Notice that if $\lambda \in \overline{(\mathcal{A}^*)^+}$ then $\min(\lambda) = \lambda_1$.

For $\pi \in \mathrm{Irr}(G_n(E))$ we will denote by ω_π its central character seen as a character on $E^\times$ through the natural identification $Z_n(E) \simeq E^\times$. Let $P = MU$ be a standard parabolic subgroup of G_n. Then, M is of the form

$$M = G_{n_1} \times \ldots \times G_{n_k}$$

for some integers $n_1, \ldots, n_k$ with $n_1 + \ldots + n_k = n$. Let $\tau_i \in \mathrm{Irr}(G_{n_i}(E))$, for every $1 \leqslant i \leqslant k$, be an irreducible representation of $G_{n_i}(E)$ so that $\sigma = \tau_1 \boxtimes \ldots \boxtimes \tau_k$ is an irreducible representation of $M(E)$. We will denote by

$$\tau_1 \times \ldots \times \tau_k$$

the (normalized) induced representation $i_{P(E)}^{G_n(E)}(\sigma)$. A representation $\pi \in \mathrm{Irr}(G_n(E))$ is *generic* if it admits a nonzero Whittaker functional with respect to any (or equivalently one) generic character of $N_n(E)$. We will denote by $\mathrm{Irr}_{\mathrm{gen}}(G_n(E))$ the subset of generic representations in $\mathrm{Irr}(G_n(E))$. By [Zel, Theorem 9.7] and [Vog, Theorem 6.2f], every $\pi \in \mathrm{Irr}_{\mathrm{gen}}(G_n(E))$ is isomorphic to a representation of the form $\tau_1 \times \ldots \tau_k$ where for each $1 \leqslant i \leqslant k$, τ_i is an essentially square-integrable (i.e. an unramified twist of a square-integrable) representation of some $G_{n_i}(E)$. We will say that a representation π of $G_n(E)$ is *nearly tempered* if it is isomorphic to an induced representation of the form

$$(\tau_1 \otimes |\det|_E^{\lambda_1}) \times \ldots \times (\tau_k \otimes |\det|_E^{\lambda_k})$$

where for each $1 \leqslant i \leqslant k$, $\tau_i \in \Pi_2(G_{n_i}(E))$ for some $n_i \geqslant 1$ and λ_i is a real number with $|\lambda_i| < \frac{1}{4}$. Again by [Zel, Theorem 9.7] and also [VS], every nearly tempered representation is irreducible and generic. We will denote by $\mathrm{Irr}_{\mathrm{ntemp}}(G_n(E)) \subseteq \mathrm{Irr}_{\mathrm{gen}}(G_n(E))$ the subset of nearly tempered representations.

We equip $\mathrm{Temp}(G_n(E))$ with a topology as follows. For each standard parabolic subgroup $P = MU$ of G_n and $\sigma \in \Pi_2(M(E))$ the map $\lambda \in i\mathcal{A}_M^* \mapsto i_{P(E)}^{G_n(E)}(\sigma_\lambda) \in \mathrm{Temp}(G_n(E))$ identifies a certain quotient of $i\mathcal{A}_M^*$ with a subset of $\mathrm{Temp}(G_n(E))$. Then, we endow $\mathrm{Temp}(G_n(E))$ with the unique topology whose connected components are precisely these subsets equipped with the quotient topology from $i\mathcal{A}_M^*$.

We define generic characters $\psi_n : N_n(E) \to \mathbb{S}^1$, $\psi_n' : N_n(F) \to \mathbb{S}^1$ by

$$\psi_n'(u) = \psi'\left((-1)^n \sum_{i=1}^{n-1} u_{i,i+1}\right) \text{ and } \psi_n(u) = \psi\left((-1)^n \sum_{i=1}^{n-1} u_{i,i+1}\right).$$

If π is an irreducible generic representation of $G_n(E)$, we write $\mathcal{W}(\pi, \psi_n)$ for its Whittaker model (with respect to ψ_n). For every Whittaker function $W \in C^\infty(N_n(E)\backslash G_n(E), \psi_n)$, we define $\widetilde{W} \in C^\infty(N_n(E)\backslash G_n(E), \psi_n^{-1})$ by

$$\widetilde{W}(g) = W(w_n{}^t g^{-1}) \quad (g \in G_n(E))$$

where $w_n = \begin{pmatrix} & & 1 \\ & \cdot^{\cdot^{\cdot}} & \\ 1 & & \end{pmatrix}$. Then, if π is an irreducible generic representation of $G_n(E)$ the map $W \mapsto \widetilde{W}$ induces a (topological) isomorphism $\mathcal{W}(\pi, \psi_n) \simeq \mathcal{W}(\widetilde{\pi}, \psi_n^{-1})$.

We let $\mathcal{S}(F^n) = C_c^\infty(F^n)$ be the space of all locally constant and compactly supported complex-valued functions on F^n in the p-adic case and $\mathcal{S}(F^n)$ be the usual Schwartz space on F^n in the Archimedean case. We denote by $\phi \mapsto \widehat{\phi}$ be the usual Fourier transform on F^n defined using the additive character ψ' and the corresponding autodual measure i.e. for every $\phi \in \mathcal{S}(F^n)$ we have

$$\widehat{\phi}(x_1, \ldots, x_n) = \int_{F^n} \phi(y_1, \ldots, y_n)\psi'(x_1y_1 + \ldots + x_ny_n)dy_1 \ldots dy_n, \quad (x_1, \ldots, x_n) \in F^n,$$

where the measure of integration is chosen such that $\widehat{\widehat{\phi}}(v) = \phi(-v)$. Finally, we write e_n for the vector $e_n = (0, \ldots, 0, 1) \in F^n$.

3.2 Asai L-Functions and Epsilon Factors

Let W_F be the Weil group of F and set

$$W_F' = \begin{cases} W_F \times SL_2(\mathbb{C}) & \text{if } F \text{ is } p\text{-adic}, \\ W_F & \text{if } F \text{ is Archimedean}, \end{cases}$$

for the Weil-Deligne group of F. In the inert case, we define similarly the Weil-Deligne group W_E' of E. An *admissible* complex representation of W_F' (resp. W_E') is by definition a continuous morphism $\phi : W_F' \to \mathrm{GL}(M)$ (resp. $\phi : W_E' \to GL(M)$), where M is a finite dimensional complex vector space, which is semi-simple and algebraic when restricted to $SL_2(\mathbb{C})$ (in the p-adic case). To any admissible complex representation $\phi : W_F' \to \mathrm{GL}(M)$, we associate a local L-factor $L(s, \phi)$ and a local ϵ-factor $\epsilon(s, \phi, \psi')$ as in [Ta, §3] and [GR, §2.2]. In the p-adic case, $L(s, \phi)$ is of the form $P(q^{-s})$ where $P \in \mathbb{C}[T]$ is such that $P(0) = 1$ whereas $\epsilon(s, \phi, \psi')$ is of the form $cq^{n(s-1/2)}$ where $n \in \mathbb{Z}$ and $c = \epsilon(1/2, \phi, \psi') \in \mathbb{C}^\times$. In the Archimedean case, $L(s, \phi)$ is a product of $\dim(M)$ factors of the form $\pi^{-(s+s_0)/2}\Gamma((s + s_0)/2)$ for some $s_0 \in \mathbb{C}$ where Γ denotes the usual gamma function and $\epsilon(s, \varphi, \psi')$ is of the form $cQ^{s-1/2}$ where $Q \in \mathbb{R}_+^*$ and $c = \epsilon(1/2, \varphi, \psi') \in \mathbb{C}^\times$. Using these invariants, we define the local γ-factor of ϕ as

$$\gamma(s, \phi, \psi') = \epsilon(s, \phi, \psi')\frac{L(1 - s, \widetilde{\phi})}{L(s, \phi)}$$

where $\widetilde{\phi}$ stands for the contragredient of ϕ. Setting $\psi'_\lambda(x) = \psi'(\lambda x)$ for every $(\lambda, x) \in F^\times \times F$, we have

$$\epsilon(s, \phi, \psi'_\lambda) = (\det \phi)(\lambda)|\lambda|_F^{\dim(M)(s-\frac{1}{2})}\epsilon(s, \phi, \psi')$$

for every $\lambda \in F^\times$ where we have identified $\det \phi$ with a character of $F^\times$ through class field theory. Moreover, if we identify the absolute value $|.|_F$ to a character $W'_F \to \mathbb{R}_+^*$ (again through class field theory), we have $L(s, \phi \otimes |.|_F^{s_0}) = L(s+s_0, \phi)$ and $\epsilon(s, \phi \otimes |.|_F^{s_0}, \psi') = \epsilon(s+s_0, \phi, \psi')$ for every $s_0 \in \mathbb{C}$. Let $\eta_{E/F}$ be the quadratic character of W'_F associated to the extension E/F and set

$$\lambda_{E/F}(\psi') = \epsilon(\frac{1}{2}, \eta_{E/F}, \psi').$$

This is sometimes called the *Langlands constant* of the extension E/F. It is a fourth root of unity which is trivial in the split case. Moreover, in the inert case if ϕ is an admissible complex representation of W'_E of dimension n, by the inductivity of ϵ-factors in degree 0 [Ta, Theorem 3.4.1], we have

$$\epsilon(s, \phi, \psi'_E) = \lambda_{E/F}(\psi')^{-n}\epsilon(s, \mathrm{Ind}_{W'_E}^{W'_F}(\phi), \psi') \tag{3.2.1}$$

where we have set $\psi'_E := \psi' \circ \mathrm{Tr}_{E/F}$ and $\mathrm{Ind}_{W'_E}^{W'_F}$ denotes the functor of induction from W'_E to W'_F (which sends admissible representations to admissible representations).

Assume that we are in the inert case. Fix $s \in W'_F \setminus W'_E$. Let $\phi : W'_E \to \mathrm{GL}(M)$ be an admissible representation. We defined $\mathrm{As}(\phi) : W'_F \to \mathrm{GL}(M \otimes M)$ by $\mathrm{As}(\phi)(w) = \phi(w) \otimes \phi(sws^{-1})$ for $w \in W'_E$ and $\mathrm{As}(\phi)(s) = (\mathrm{Id}_M \otimes \phi(s^2)) \circ \iota$ where ι is the linear automorphism of $M \otimes M$ sending $u \otimes v$ to $v \otimes u$. Then, $\mathrm{As}(\phi)$ is an admissible representation of W'_F and we set $L(s, \phi, \mathrm{As}) = L(s, \mathrm{As}(\phi))$, $\epsilon(s, \phi, \mathrm{As}, \psi') = \epsilon(s, \mathrm{As}(\phi), \psi')$, $\gamma(s, \phi, \mathrm{As}, \psi') = \gamma(s, \mathrm{As}(\phi), \psi')$ and call them the Asai L-function, ϵ-factor and γ-factor of ϕ respectively. If ϕ decomposes as a direct sum

$$\phi = \phi_1 \oplus \ldots \oplus \phi_k$$

of admissible representations of W'_E then we have

$$\mathrm{As}(\phi) = \bigoplus_{i=1}^{k} \mathrm{As}(\phi_i) \oplus \bigoplus_{1 \leqslant i < j \leqslant k} \mathrm{Ind}_{W'_E}^{W'_F}(\phi_i \otimes \phi_j^c) \tag{3.2.2}$$

where ϕ_j^c stands for the representation $w \mapsto \phi_j(sws^{-1})$. We have $\mathrm{As}(\phi \otimes |.|_E^{s_0}) = \mathrm{As}(\phi) \otimes |.|_F^{2s_0}$ for every $s_0 \in \mathbb{C}$. Moreover, if $\phi = \chi$ is a character of W'_E, that we identify to a character of $E^\times$ through class field theory, then $\mathrm{As}(\chi)$ is the character of W'_F corresponding to the restriction of χ to $F^\times$.

Let $\pi \in \mathrm{Irr}(G_n(E))$. Then, the local Langlands correspondence for GL_n [La, HT, Hen, Sch] associates to π an admissible complex representation $\phi_\pi : W'_E \to \mathrm{GL}(M)$ of dimension n. We set $L(s, \pi, \mathrm{As}) = L(s, \phi_\pi, \mathrm{As})$, $\epsilon(s, \pi, \mathrm{As}, \psi') = \epsilon(s, \phi_\pi, \mathrm{As}, \psi')$ and $\gamma(s, \pi, \mathrm{As}, \psi') = \gamma(s, \phi_\pi, \mathrm{As}, \psi')$ and call them the Asai L-function, ϵ-factor and γ-factor of π respectively. We have $\det \mathrm{As}(\phi_\pi) = (\omega_\pi)^n \eta_{E/F}^{n(n-1)/2}$ and therefore

$$\epsilon(s, \pi, \mathrm{As}, \psi'_\lambda) = \omega_\pi(\lambda)^n |\lambda|_F^{n^2(s-\frac{1}{2})} \eta_{E/F}(\lambda)^{\frac{n(n-1)}{2}} \epsilon(s, \pi, \mathrm{As}, \psi') \tag{3.2.3}$$

for every $\lambda \in F^\times$. Let $z \in \mathbb{R}_+^*$. We also set

$$L(s, \pi, \mathrm{As}) = L(s, \mathrm{As}(\phi_\pi)).$$

Assume now that we are in the split case. Let $\pi = \pi_1 \boxtimes \pi_2$ be an irreducible representation of $G_n(E) = G_n(F) \times G_n(F)$. Let $\phi_1, \phi_2 : W'_F \to \mathrm{GL}_n(\mathbb{C})$ be the admissible complex representations associated to π_1 and π_2 respectively by the local Langlands correspondence. Then, we set

$$L(s, \pi, \mathrm{As}) = L(s, \pi_1 \times \pi_2) = L(s, \phi_1 \otimes \phi_2),$$
$$\epsilon(s, \pi, \mathrm{As}, \psi') = \epsilon(s, \pi_1 \times \pi_2, \psi') = \epsilon(s, \phi_1 \otimes \phi_2, \psi')$$

and

$$\gamma(s, \pi, \mathrm{As}, \psi') = \gamma(s, \pi_1 \times \pi_2, \psi') = \gamma(s, \phi_1 \otimes \phi_2, \psi').$$

We now return to the inert case. Let $P = MU$ be a standard parabolic subgroup of G_n and $n_1, \ldots, n_k \in \mathbb{N}^*$ so that

$$M(E) = G_{n_1}(E) \times \ldots \times G_{n_k}(E).$$

For each $1 \leqslant i \leqslant k$, let $\sigma_i \in \Pi_2(G_{n_i}(E))$ and set $\sigma = \sigma_1 \boxtimes \ldots \boxtimes \sigma_k \in \Pi_2(M(E))$. We identify $\mathcal{A}^*_{M,\mathbb{C}}$ with $\mathbb{C}^k$ so that for every $(\lambda_1, \ldots, \lambda_k) \in \mathbb{C}^k$, we have

$$\sigma_\lambda = \sigma_1 |\det|_E^{\lambda_1} \boxtimes \ldots \boxtimes \sigma_k |\det|_E^{\lambda_k}.$$

We set $\pi_\lambda = i_{P(E)}^{G_n(E)}(\sigma_\lambda)$ for every $\lambda \in \mathcal{A}^*_{M,\mathbb{C}}$. Then, π_λ is irreducible for almost every λ. For such λ, $L(s, \pi_\lambda, \mathrm{As})$, $\epsilon(s, \pi_\lambda, \mathrm{As}, \psi')$ and $\gamma(s, \pi_\lambda, \mathrm{As}, \psi')$ are defined as above. For the remaining λ's, we define these factors as the one associated to the unique irreducible subquotient π_λ^0 of π_λ which is the Langlands quotient of $i_{Q(E)}^{G_n(E)}(\sigma_\lambda)$ for any parabolic Q with Levi M for which λ is positive (in the large sense).

Lemma 3.2.1.

(i) For every $\lambda = (\lambda_1, \ldots, \lambda_k) \in \mathcal{A}^*_{M,\mathbb{C}}$ *we have*

$$L(s, \pi_\lambda, \mathrm{As}) = \prod_{i=1}^{k} L(s + 2\lambda_i, \sigma_i, \mathrm{As}) \prod_{1 \leqslant i < j \leqslant k} L(s + \lambda_i + \lambda_j, \sigma_i \times \sigma_j)$$

and

$$\epsilon(s, \pi_\lambda, \mathrm{As}, \psi') = \lambda_{E/F}(\psi')^{\frac{k(k-1)}{2}} \prod_{i=1}^{k} \epsilon(s + 2\lambda_i, \sigma_i, \mathrm{As}, \psi')$$
$$\prod_{1 \leqslant i < j \leqslant k} \epsilon(s + \lambda_i + \lambda_j, \sigma_i \times \sigma_j, \psi'_E).$$

(ii) Set $U = \{(\lambda_1, \ldots, \lambda_k) \in \mathcal{A}^*_{M,\mathbb{C}} \mid |\Re(\lambda_i)| < \frac{1}{4}\}$. *Then, for every* $s \in \mathbb{C}$ *with* $\Re(s) = \frac{1}{2}$ *the function* $\lambda \in U \mapsto \gamma(s, \pi_\lambda, \mathrm{As}, \psi')$ *is holomorphic. Moreover, for* π *nearly tempered,* $L(s, \pi, \mathrm{As})$ *has no poles in* $\{\Re(s) \geqslant \frac{1}{2}\}$.

(iii) For any $z \in \mathbb{R}_+^*$ *the function* $(s, \lambda) \mapsto L(s, \pi_\lambda, \mathrm{As})^{-1}$ *is holomorphic. Moreover, it is of finite order in vertical strips in the first variable locally uniformly in the second variable (see Sect. 2.8).*

Proof.

(i) By compatibility of the Langlands correspondence with parabolic induction and unramified twists, for every $\lambda = (\lambda_1, \ldots, \lambda_k) \in \mathcal{A}^*_{M,\mathbb{C}}$ the Langlands parameter of π_λ (or rather π_λ^0) is given by

$$\phi_{\pi_\lambda} = \phi_{\sigma_1} \otimes |.|_E^{\lambda_1} \oplus \ldots \oplus \phi_{\sigma_k} \otimes |.|_E^{\lambda_k}.$$

Therefore, the identities of the lemma follow directly from 3.2.1 and 3.2.2.

(ii) is a consequence of (i) and the fact that for π_1, π_2 tempered the L-functions $L(s, \pi_1, \mathrm{As})$ and $L(s, \pi_1 \times \pi_2)$ have no poles in $\{\Re(s) > 0\}$.
(iii) This is a consequence of (i) and the fact that the Gamma function decays exponentially in vertical strips. ■

3.3 Definition and Convergence of the Local Zeta Integrals

Let $\pi \in \mathrm{Irr}_{\mathrm{gen}}(G_n(E))$. For every $W \in \mathcal{W}(\pi, \psi_n)$, $\phi \in \mathcal{S}(F^n)$ and $s \in \mathbb{C}$ we define, whenever convergent, a Zeta integral

$$Z(s, W, \phi) = \int_{N_n(F)\backslash G_n(F)} W(h)\phi(e_n h)|\det h|_F^s dh.$$

Recall that for every $C \in \mathbb{R}$ we have set $\mathcal{H}_{>C} = \{z \in \mathbb{C} \mid \Re(z) > C\}$ (see Sect. 2.8). By Corollary 2.7.1, there exists $\mu \in \mathcal{A}^*$ such that $\mathcal{W}(\pi, \psi_n) \subseteq \mathcal{C}_\mu(N_n(E)\backslash G_n(E), \psi_n)$. Thus, the next lemma will show that the above Zeta integrals are at least convergent in some right half-plane.

Lemma 3.3.1. *Let $\mu \in \mathcal{A}^*$ and $\phi \in \mathcal{S}(F^n)$. Then, for every $W \in \mathcal{C}_\mu(N_n(E)\backslash G_n(E), \psi_n)$ the integral*

$$Z(s, W, \phi) = \int_{N_n(F)\backslash G_n(F)} W(h)\phi(e_n h)|\det h|_F^s dh$$

is absolutely convergent for all $s \in \mathcal{H}_{>-2\min(\mu)}$. Moreover, the function $s \mapsto Z(s, W, \phi)$ is holomorphic and bounded in vertical strips. More precisely, for every vertical strip $V \subseteq \mathcal{H}_{>-2\min(\mu)}$ there exists a continuous semi-norm $p_{V,\phi}$ on $\mathcal{C}_\mu(N_n(E)\backslash G_n(E), \psi_n)$ such that

$$|Z(s, W, \phi)| \leqslant p_{V,\phi}(W)$$

for every $W \in \mathcal{C}_\mu(N_n(E)\backslash G_n(E), \psi_n)$ and $s \in V$.

Proof. Let $W \in \mathcal{C}_\mu(N_n(E)\backslash G_n(E), \psi_n)$. By the Iwasawa decomposition $G_n(F) = N_n(F)A_n(F)K_n$, we need to show the convergence of

$$\int_{K_n}\int_{A_n(F)} |W(ak)||\phi(e_n ak)||\det a|_F^t \delta_n(a)^{-1} da dk$$

for $t > -2\min(\mu)$. Since for each $R > 0$ we have $|\phi(e_n ak)| \ll (1+|a_n|)^{-R}$ for all $a \in A_n(F)$ and $k \in K_n$, by Lemma 2.5.1 we are reduced to show the convergence of

$$\int_{A_n(F)} \prod_{i=1}^{n-1}(1+|\frac{a_i}{a_{i+1}}|)^{-R}(1+|a_n|)^{-R}\delta_{n,E}(a)^{1/2}a^{|\mu|}|\det a|_F^t\delta_n(a)^{-1}da =$$

$$\int_{A_n(F)} \prod_{i=1}^{n-1}(1+|\frac{a_i}{a_{i+1}}|)^{-R}(1+|a_n|)^{-R}\prod_{i=1}^{n}|a_i|^{t+2|\mu|_i}da$$

locally uniformly for $\frac{R}{n} - 2\min(\mu) > t > -2\min(\mu)$ where $|\mu|_1, \ldots, |\mu|_n$ denote the coordinates of $|\mu|$. This last fact follows from the elementary inequality

$$\prod_{i=1}^{n-1}(1+|\frac{a_i}{a_{i+1}}|)^{-R}(1+|a_n|)^{-R} \leqslant \prod_{i=1}^{n}(1+|a_i|)^{-R/n}$$

together with the convergence of the integral

$$\int_{F^\times}(1+|x|)^{-R/n}|x|^{t+r}d^\times x$$

locally uniformly for $\frac{R}{n} - r > t > -r$ for every $r \in \mathbb{R}$. ■

We recall that we have introduced in Sect. 3.1 the notion of nearly tempered representation of $G_n(E)$.

Lemma 3.3.2. *Assume that π is a generic irreducible representation of $G_n(E)$ which is nearly tempered. Then, there exists $\epsilon > 0$ such that for every $W \in \mathcal{W}(\pi, \psi_n)$ and $\phi \in \mathcal{S}(F^n)$ the integral defining $Z(s, W, \phi)$ converges absolutely on $\mathcal{H}_{\frac{1}{2}-\epsilon}$ and defines a holomorphic function bounded in vertical strips there.*

Proof. Let $W \in \mathcal{W}(\pi, \psi_n)$ and $\phi \in \mathcal{S}(F^n)$. By assumption, there exists a parabolic subgroup $P = MU$ of G_n, a discrete series σ of $M(E)$ and $\lambda = (\lambda_1, \ldots, \lambda_n) \in \mathcal{A}^*_{M,\mathbb{C}} \subseteq \mathcal{A}^*_{\mathbb{C}} = \mathbb{C}^n$ satisfying $|\Re(\lambda_i)| < \frac{1}{4}$ for every $1 \leqslant i \leqslant n$ such that $\pi \simeq i_{P(E)}^{G_n(E)}(\sigma_\lambda)$. Let $\rho = (\frac{n-1}{2}, \ldots, \frac{1-n}{2}) \in \mathcal{A}^*$ be half the sum of the roots of A_n in B_n. Then, for every $\eta > 0$ we have $|\Re(\lambda)| \prec |\Re(\lambda)| + \eta\rho$. Therefore, by Proposition 2.6.1(i) (and standard properties of the Jacquet's functional [Wall2, Sect. 15.4]) we have

$$\mathcal{W}(\pi, \psi_n) \subseteq \mathcal{C}_{|\Re(\lambda)|+\eta\rho}(N_n(E)\backslash G_n(E), \psi_n)$$

for any $\eta > 0$. By the previous lemma it follows that for any $\eta > 0$, $Z(s, W, \phi)$ converges absolutely on $\mathcal{H}_{-2\min(|\Re(\lambda)|+\eta\rho)}$ and defines a holomorphic function there. Since

$$\mathcal{H}_{-2\min(|\Re(\lambda)|)} = \bigcup_{\eta>0}\mathcal{H}_{-2\min(|\Re(\lambda)|+\eta\rho)}$$

we deduce that $Z(s, W, \phi)$ converges absolutely on $\mathcal{H}_{-2\min(|\Re(\lambda)|)}$ and defines a holomorphic function there. Finally, the inequalities satisfied by λ imply that $\mathcal{H}_{-2\min(|\Re(\lambda)|)} = \mathcal{H}_{\frac{1}{2}-\epsilon}$ for some $\epsilon > 0$ and the lemma follows. ■

We end this section with the following non-vanishing result.

Lemma 3.3.3. *For every $s_0 \in \mathbb{C}$, there exist finite families $W_i \in \mathcal{W}(\pi, \psi_n)$ and $\phi_i \in \mathcal{S}(F^n)$ indexed by $i \in I$ such that the function $s \mapsto \sum_{i\in I} Z(s, W_i, \phi_i)$ (which is defined on some right-half plane) admits a holomorphic continuation to $\mathbb{C}$ which is non-vanishing at s_0.*

Proof. Let P_n be the mirabolic subgroup of G_n (i.e. the subgroup of elements $g \in G_n$ with last row $(0, \dots, 0, 1)$), U_n be the unipotent radical of P_n and $\overline{U}_n = {}^tU_n$. Then, we have the decomposition $G_n(F) = P_n(F)Z_n(F)\overline{U}_n(F)$ and correspondingly for Haar measures $dh = |\det p|_F^{-1} d_r p dz d\overline{u}$ where $d_r p$ denotes a right Haar measure on $P_n(F)$. Thus, by Lemma 3.3.1, for $\Re(s) \gg 1$ the expression defining $Z(s, W, \phi)$ is absolutely convergent and we have

$$Z(s, W, \phi)$$
$$= \int_{Z_n(F)\times \overline{U}_n(F)} \int_{N_n(F)\backslash P_n(F)} W(pz\overline{u})|\det p|_F^{s-1} d_r p \phi(e_n z\overline{u})|\det z|_F^s dz d\overline{u}$$
$$= \int_{Z_n(F)\times \overline{U}_n(F)} \int_{N_n(F)\backslash P_n(F)} W(p\overline{u})|\det p|_F^{s-1} d_r p \phi(e_n z\overline{u})\omega_\pi(z)|\det z|_F^s dz d\overline{u}$$

for every $W \in \mathcal{W}(\pi, \psi_n)$ and $\phi \in \mathcal{S}(F^n)$. Let $\varphi_Z \in C_c^\infty(Z_n(F))$ and $\varphi_{\overline{U}} \in C_c^\infty(\overline{U}_n(F))$. Then, there exists a unique $\phi = \phi_{\varphi_Z, \varphi_{\overline{U}}} \in C_c^\infty(F^n)$ such that $\phi(e_n z\overline{u}) = \varphi_Z(z)\varphi_{\overline{U}}(\overline{u})$ for every $(z, u) \in Z_n(F) \times \overline{U}_n(F)$. For such a ϕ, the above identity becomes

$$Z(s, W, \phi) = \int_{N_n(F)\backslash P_n(F)} \left(R(\varphi_{\overline{U}})W\right)(p)|\det p|_F^{s-1} d_r p$$
$$\int_{Z_n(F)} \varphi_Z(z)\omega_\pi(z)|\det z|^s dz, \quad \Re(s) \gg 1$$

for every $W \in \mathcal{W}(\pi, \psi_n)$. Let $f \in C_c^\infty(N_n(E)\backslash P_n(E), \psi_n)$. By [GK], [Jac3, Proposition 5] and [Kem], there exists $W_0 \in \mathcal{W}(\pi, \psi_n)$ whose restriction to $P_n(E)$ coincides with f. By Dixmier-Malliavin [DM] (in the Archimedean case), there exist finite families $(W_i)_{i\in I}$ and $(\varphi_{\overline{U},i})_{i\in I}$ of elements in $\mathcal{W}(\pi, \psi_n)$ and $C_c^\infty(\overline{U}_n(F))$ respectively such that

$$W_0 = \sum_{i\in I} R(\varphi_{\overline{U},i})W_i.$$

Choose $\varphi_Z \in C_c^\infty(Z_n(F))$ such that

$$\int_{Z_n(F)} \varphi_Z(z)\omega_\pi(z)|\det z|^{s_0}dz \neq 0$$

(Notice that the above integral is absolutely convergent for any complex value of s_0). Set $\phi_i = \phi_{\varphi_Z,\varphi_{\overline{U},i}}$ for every $i \in I$. Then, by the above, for $\Re(s) \gg 1$ we have

$$\begin{aligned}
&\sum_{i\in I} Z(s, W_i, \phi_i) \\
&\quad = \sum_{i\in I} \int_{N_n(F)\backslash P_n(F)} \left(R(\varphi_{\overline{U},i})W_i\right)(p)|\det p|_F^{s-1}d_r p \int_{Z_n(F)} \varphi_Z(z)\omega_\pi(z)|\det z|^s dz \\
&\quad = \int_{N_n(F)\backslash P_n(F)} W_0(p)|\det p|_F^{s-1}d_r p \int_{Z_n(F)} \varphi_Z(z)\omega_\pi(z)|\det z|^s dz \\
&\quad = \int_{N_n(F)\backslash P_n(F)} f(p)|\det p|_F^{s-1}d_r p \int_{Z_n(F)} \varphi_Z(z)\omega_\pi(z)|\det z|^s dz.
\end{aligned}$$

The above integrals are convergent for any $s \in \mathbb{C}$ uniformly on compacta and therefore the resulting expression defines a holomorphic function on $\mathbb{C}$. Moreover, we can certainly choose f so that

$$\int_{N_n(F)\backslash P_n(F)} f(p)|\det p|_F^{s_0-1}d_r p \neq 0.$$

By our choice of φ_Z this implies that the holomorphic continuation of $\sum_{i\in I} Z(s, W_i, \phi_i)$ does not vanish at s_0. ■

3.4 Local Functional Equation: The Split Case

In this section we assume that we are in the split case i.e. $E = F \times F$. Let $\pi \in \mathrm{Irr}_{\mathrm{gen}}(G_n(E))$. Then $\pi = \pi_1 \boxtimes \pi_2$ for some $\pi_1, \pi_2 \in \mathrm{Irr}(G_n(F))$. In the case where $\psi(x, y) = \psi'(x)\psi'(-y)$ for $(x, y) \in E$ (i.e. $\tau = (1, -1)$) and $W = W_1 \otimes W_2$ for some $W_1 \in \mathcal{W}(\pi_1, \psi'_n)$, $W_2 \in \mathcal{W}(\pi_2, \psi_n'^{-1})$, for $\phi \in \mathcal{S}(F^n)$ the Zeta integral $Z(s, W, \phi)$ belongs to a family of expressions studied by Jacquet-Piatetskii-Shapiro-Shalika and Jacquet in [JPSS] and [Jac]. By the main results of those references together with some of the characterizing properties of the local Langlands correspondence for GL_n [HT, Hen, Sch], in this situation $Z(s, W, \phi)$ admits a meromorphic continuation to $\mathbb{C}$ satisfying the functional equation

$$\begin{aligned}
Z(1-s, \widetilde{W}, \widehat{\phi}) &= \omega_{\pi_2}(-1)^{n-1}\gamma(s, \pi_1 \times \pi_2, \psi')Z(s, W, \phi) \\
&= \omega_\pi(\tau)^{n-1}|\tau|_E^{\frac{n(n-1)}{2}(s-1/2)}\lambda_{E/F}(\psi')^{-\frac{n(n-1)}{2}}\gamma(s, \pi, \mathrm{As}, \psi')Z(s, W, \phi).
\end{aligned}$$

Let us remark here that our conventions are slightly different to the ones in [Jac]: the Fourier transform $\phi \mapsto \widehat{\phi}$ in *loc. cit.* is normalized using the character ψ'^{-1} rather than ψ' (see [Jac, §2]) which results in composing the one used in this paper with the involution $x \in F^n \mapsto -x$. On the other hand, in the functional equation of [Jac, Theorem 2.1] the term $\omega_{\pi_2}(-1)^{n-1}$ is replaced by $\omega_{\pi_1}(-1)^{n-1}$. Finally, in [Jac] the local γ-factors are also normalized differently: although in this paper we have followed the normalization of [Ta] and [GR], in *loc. cit.* the convention is opposite and what we denote here by $\gamma(s, \pi_1 \times \pi_2, \psi')$ corresponds to the factor $\gamma(s, \pi_1 \times \pi_2, \psi'^{-1})$ in [Jac] (see in particular the appendix [Jac, Section 16]). Taking all of these into account, it is a straightforward exercise to see that the above functional equation is equivalent to the one given in [Jac, Theorem 2.1].

In the p-adic case the meromorphic continuation of $s \mapsto Z(s, W, \phi)$ and the above equality extend to any $W \in \mathcal{W}(\pi, \psi)$, by linearity, whereas in the Archimedean case such an extension follows from the remark after Theorem 2.3 of [Jac]. We record this as a theorem by removing the assumption on ψ.

Theorem 3.4.1. *Assume that $E = F \times F$. Let $\pi \in \mathrm{Irr}_{\mathrm{gen}}(G_n(E))$. Then, for every $W \in \mathcal{W}(\pi, \psi_n)$ and $\phi \in \mathcal{S}(F^n)$ the function $s \mapsto Z(s, W, \phi)$ has a meromorphic extension to $\mathbb{C}$ satisfying the functional equation*

$$Z(1-s, \widetilde{W}, \widehat{\phi}) = \omega_\pi(\tau)^{n-1} |\tau|_E^{\frac{n(n-1)}{2}(s-1/2)} \lambda_{E/F}(\psi')^{-\frac{n(n-1)}{2}} \gamma(s, \pi, \mathrm{As}, \psi') Z(s, W, \phi). \tag{3.4.1}$$

<u>Proof</u>. By the above discussion, the theorem holds when $\psi(x, y) = \psi'(x)\psi'(-y)$ for every $(x, y) \in E$. We therefore just need to study the effect of replacing ψ by ψ_λ for some $\lambda \in F^\times$ where $\psi_\lambda(z) = \psi(\lambda z)$ for every $z \in E$. Doing so amounts to replacing τ by $\lambda\tau$. Define the generic character $\psi_{n,\lambda}$ as ψ_n using ψ_λ instead of ψ. Then, there is an isomorphism $\mathcal{W}(\pi, \psi_n) \to \mathcal{W}(\pi, \psi_{n,\lambda})$, $W \mapsto W_\lambda$ given by

$$W_\lambda(g) = W(a(\lambda)g), \quad g \in G_n(E)$$

where $a(\lambda) = \begin{pmatrix} \lambda^{n-1} & & \\ & \ddots & \\ & & 1 \end{pmatrix}$. By the change of variable $h \mapsto a(\lambda)^{-1}h$ we have

$$Z(s, W_\lambda, \phi) = \delta_n(a(\lambda)) |\det a(\lambda)|_F^{-s} Z(s, W, \phi), \quad \Re(s) \gg 1. \tag{3.4.2}$$

This already shows that $Z(s, W_\lambda, \phi)$ has a meromorphic continuation if and only if $Z(s, W, \phi)$ has one. On the other hand, for all $g \in G_n(E)$ we have

$$\begin{aligned} \widetilde{(W_\lambda)}(g) &= W(a(\lambda) w_n {}^t g^{-1}) = W(w_n {}^t (w_n a(\lambda)^{-1} w_n^{-1} g)^{-1}) \\ &= \widetilde{W}(w_n a(\lambda)^{-1} w_n^{-1} g) = \omega_\pi(\lambda)^{n-1} (\widetilde{W})_\lambda(g) \end{aligned}$$

so that $\widetilde{W_\lambda} = \omega_\pi(\lambda)^{n-1}(\widetilde{W})_\lambda$ and finally (by 3.4.2)

$$Z(1-s, \widetilde{W_\lambda}, \widehat{\phi}) = \omega_\pi(\lambda)^{n-1}\delta_n(a(\lambda))|\det a(\lambda)|_F^{s-1} Z(1-s, \widetilde{W}, \widehat{\phi}). \tag{3.4.3}$$

From 3.4.2 and 3.4.3 it follows that if $Z(s, W, \phi)$ and $Z(s, \widetilde{W}, \widehat{\phi})$ satisfy 3.4.1 then

$$\begin{aligned} &Z(1-s, \widetilde{W_\lambda}, \widehat{\phi}) \\ &= \omega_\pi(\lambda)^{n-1}|\det a(\lambda)|_F^{2s-1}\omega_\pi(\tau)^{n-1}|\tau|_E^{\frac{n(n-1)}{2}(s-1/2)}\lambda_{E/F}(\psi')^{-\frac{n(n-1)}{2}}\gamma(s, \pi_1 \times \pi_2, \psi') \\ &\times Z(s, W_\lambda, \phi) \\ &= \omega_\pi(\lambda\tau)^{n-1}|\lambda\tau|_E^{\frac{n(n-1)}{2}(s-1/2)}\lambda_{E/F}(\psi')^{-\frac{n(n-1)}{2}}\gamma(s, \pi_1 \times \pi_2, \psi')Z(s, W_\lambda, \phi) \end{aligned}$$

which is precisely the functional equation for ψ replaced by ψ_λ. ■

3.5 Local Functional Equation: The Inert Case

In this Section we assume that E/F is a quadratic field extension. We are going to state two theorems which are the main results of this paper. These theorems will be proved in Sects. 3.9 and 3.10. The first result is the exact analog of Theorem 3.4.1 in the inert case:

Theorem 3.5.1. *Assume that E/F is a quadratic field extension. Let $\pi \in \mathrm{Irr}_{\mathrm{gen}}(G_n(E))$. Then, for every $W \in \mathcal{W}(\pi, \psi_n)$ and $\phi \in \mathcal{S}(F^n)$ the function $s \mapsto Z(s, W, \phi)$ has a meromorphic extension to $\mathbb{C}$ and satisfies the functional equation*

$$Z(1-s, \widetilde{W}, \widehat{\phi}) = \omega_\pi(\tau)^{n-1}|\tau|_E^{\frac{n(n-1)}{2}(s-1/2)}\lambda_{E/F}(\psi')^{-\frac{n(n-1)}{2}}\gamma(s, \pi, \mathrm{As}, \psi')Z(s, W, \phi). \tag{3.5.1}$$

Theorem 3.5.2. *Assume that E/F is a quadratic field extension. Let $\pi \in \mathrm{Irr}_{\mathrm{gen}}(G_n(E))$. Then, for every $W \in \mathcal{W}(\pi, \psi_n)$ and $\phi \in \mathcal{S}(F^n)$ the function*

$$s \mapsto \frac{Z(s, W, \phi)}{L(s, \pi, \mathrm{As})}$$

is holomorphic and of finite order in vertical strips. Moreover, if π is nearly tempered for every $s_0 \in \mathbb{C}$ there exist $W \in \mathcal{W}(\pi, \psi_n)$ and $\phi \in \mathcal{S}(F^n)$ for which the above function does not vanish at s_0.

As in the introduction, we recall here that a big part of Theorems 3.5.1 and 3.5.2 was already known in the p-adic case. More precisely, in this case the functional equation of Theorem 3.5.1 was established by Flicker [Fli, Appendix] and Kable [Kab, Theorem 3] with a possibly different γ-factor $\gamma^{\mathrm{RS}}(s,\pi,\mathrm{As},\psi') = \epsilon^{\mathrm{RS}}(s,\pi,\mathrm{As},\psi')\frac{L^{\mathrm{RS}}(1-s,\pi^\vee,\mathrm{As})}{L^{\mathrm{RS}}(s,\pi,\mathrm{As})}$. Here, by its very definition, $L^{\mathrm{RS}}(s,\pi,\mathrm{As})$ is the inverse of a polynomial in q^{-s} with 1 as constant term such that $\frac{Z(s,W,\phi)}{L^{\mathrm{RS}}(s,\pi,\mathrm{As})}$ is always holomorphic and non-vanishing at any given point $s_0 \in \mathbb{C}$ for suitable choice of the pair $(W,\phi) \in \mathcal{W}(\pi,\psi_n) \times \mathcal{S}(F^n)$. The equality $L^{\mathrm{RS}}(s,\pi,\mathrm{As}) = L(s,\pi,\mathrm{As})$ was then proved first by Anandavardhanan-Rajan [AR] for π square-integrable and then by Matringe [Mat] in general. On the other hand, the identity of ϵ-factors $\epsilon^{\mathrm{RS}}(s,\pi,\mathrm{As},\psi') = \epsilon(s,\pi,\mathrm{As},\psi')$ was previously only known when π is supercuspidal [AKMSS] or $n=2$ [CCI]. Therefore, this identity of ϵ-factors is the only truly new result when E and F are p-adic. However, we won't rely on those previous works in the proof, except for the global functional equation of Flicker and Kable (see Sect. 3.7), and moreover we will be able to treat mostly uniformly non-Archimedean and Archimedean fields.

The aim of the next lemma is to check that Theorems 3.5.1 and 3.5.2 do not depend on the choices of ψ and ψ'.

Lemma 3.5.3. *If Theorems 3.5.1 and 3.5.2 hold for one pair (ψ,ψ') of nontrivial additive characters of E, F respectively with ψ trivial on F then they hold for any such pair.*

Proof. The independence on the choice of ψ can be proved exactly the same way as in the proof of Theorem 3.4.1. Clearly Theorem 3.5.2 and the meromorphic extension of $s \mapsto Z(s,W,\phi)$ for every $W \in \mathcal{W}(\pi,\psi_n)$ and $\phi \in \mathcal{S}(F^n)$ do not depend on the choice of ψ'. Thus, it only remains to show the independence of the functional equation 3.5.1 on the choice of ψ'. If we replace ψ' by ψ'_λ defined by $\psi'_\lambda(x) = \psi'(\lambda x)$ for every $x \in F$ for some $\lambda \in F^\times$ but keep ψ fixed then we have to replace τ by $\lambda^{-1}\tau$. Therefore, the functional equation 3.5.1 for ψ' replaced by ψ'_λ reads

$$Z(1-s,\widetilde{W},\widehat{\phi}^{\psi'_\lambda}) = \omega_\pi(\lambda^{-1}\tau)^{n-1}|\lambda^{-1}\tau|_E^{\frac{n(n-1)}{2}(s-1/2)}$$
$$\lambda_{E/F}(\psi'_\lambda)^{-\frac{n(n-1)}{2}}\gamma(s,\pi,\mathrm{As},\psi'_\lambda)Z(s,W,\phi) \qquad (3.5.2)$$

for every $W \in \mathcal{W}(\pi,\psi_n)$ and $\phi \in \mathcal{S}(F^n)$ where $\widehat{\phi}^{\psi'_\lambda}$ stands for the Fourier transform of ϕ with respect to ψ'_λ (rather than ψ') and the corresponding autodual Haar measure on F^n. We have the relation $\widehat{\phi}^{\psi'_\lambda}(v) = |\lambda|_F^{n/2}\widehat{\phi}(\lambda v)$ $(v \in F^n)$. Therefore, by the change of variable $h \mapsto \lambda^{-1}h$, we have

$$Z(1-s,\widetilde{W},\widehat{\phi}^{\psi'_\lambda}) = |\lambda|_F^{n(s-1/2)}\omega_\pi(\lambda)Z(1-s,\widetilde{W},\widehat{\phi}). \qquad (3.5.3)$$

On the other hand,

$$\lambda_{E/F}(\psi'_\lambda) = \eta_{E/F}(\lambda)\lambda_{E/F}(\psi') \qquad (3.5.4)$$

whereas by 3.2.3,

$$\gamma(s,\pi,\mathrm{As},\psi'_\lambda)=\omega_\pi(\lambda)^n|\lambda|_F^{n^2(s-1/2)}\eta_{E/F}(\lambda)^{n(n-1)/2}\gamma(s,\pi,\mathrm{As},\psi'). \quad (3.5.5)$$

Combining 3.5.3, 3.5.4 and 3.5.5 we see that the functional equation 3.5.2 for (ψ,ψ'_λ) reduces to the one for (ψ,ψ') (that is 3.5.1). ■

The next lemma is straightforward and allows to restrict to representations with unitary central character.

Lemma 3.5.4. *Let π be a generic irreducible representation of $G_n(E)$ and $\lambda\in\mathcal{A}^*_{G_n,\mathbb{C}}$. Then if Theorems 3.5.1 and 3.5.2 hold for π_λ they also hold for π.*

3.6 Unramified Computation

In this section, we consider the case where F is non-Archimedean and the extension E/F is either inert or split. Recall that an irreducible representation π of $G_n(E)$ is said to be *unramified* if it admits a nonzero $G_n(\mathcal{O}_E)$-fixed vector in which case $\pi^{G_n(\mathcal{O}_E)}$ is a line. We also say that the characters ψ' and ψ are *unramified* if the maximal fractional ideals on which they are trivial are $\mathcal{O}_F$ and $\mathcal{O}_E$ respectively. If π and ψ are unramified there exists a unique $W\in\mathcal{W}(\pi,\psi_n)^{G_n(\mathcal{O}_E)}$ such that $W(1)=1$. The following unramified computation is standard and already in the literature in all but one case. We provide a proof in this missing case.

Lemma 3.6.1. *Assume that F is non-Archimedean and that $\pi\in\mathrm{Irr}_{\mathrm{gen}}(G_n(E))$, ψ' and ψ are all unramified. Let $W\in\mathcal{W}(\pi,\psi_n)^{G_n(\mathcal{O}_E)}$ be normalized by $W(1)=1$ and $\phi\in\mathcal{S}(F^n)$ be the characteristic function of $\mathcal{O}_F^n$. Then, for $\Re(s)\gg 1$ we have*

$$Z(s,W,\phi)=\mathrm{vol}(N_n(\mathcal{O}_F)\backslash G_n(\mathcal{O}_F))L(s,\pi,\mathrm{As}).$$

Proof. In the split case this is [JS, Proposition 2.3] and in the inert case when the extension E/F is unramified this is [Fli2, Proposition 3]. Thus, it only remains to deal with the case where E/F is a ramified field extension. Let ϖ_F be an uniformizer of F. For any n-tuple $\lambda=(\lambda_1,\ldots,\lambda_n)$ of integers set

$$a(\lambda)=\begin{pmatrix}\varpi_F^{\lambda_1}&&\\&\ddots&\\&&\varpi_F^{\lambda_n}\end{pmatrix}.$$

Then, by the Iwasawa decomposition we have (for $\Re(s)\gg 1$)

$$Z(s,W,\phi)=\mathrm{vol}(N_n(\mathcal{O}_F)\backslash G_n(\mathcal{O}_F))\sum_{\lambda\in\mathbb{Z}^n}W(a(\lambda))\phi(\varpi_F^{\lambda_n}e_n)|\det a(\lambda)|_F^s\delta_n(a(\lambda))^{-1} \tag{3.6.1}$$

For every n-tuple $\lambda = (\lambda_1,\ldots,\lambda_n)$ of decreasing integers let s_λ be the Schur function as defined in [Fli2, §3]. Let $t = (t_1,\ldots,t_n) \in (\mathbb{C}^\times)^n/\mathfrak{S}_n$ be the Satake parameter of π. Then, by [CS, Shin], $W(a(\lambda))$ is zero unless $\lambda_1 \geqslant \ldots \geqslant \lambda_n$ in which case it equals $\delta_{n,E}(a(\lambda))^{1/2}s_{2\lambda}(t)$. Moreover as $\phi = \mathbf{1}_{\mathcal{O}_F^n}$ we have $\phi(\varpi_F^{\lambda_n}e_n) = 0$ if $\lambda_n < 0$ and 1 otherwise. Therefore, by 3.6.1 and since $\delta_{n,E} = \delta_n^2$ on $A_n(F)$ we get

$$\begin{aligned}\mathrm{vol}(N_n(\mathcal{O}_F)\backslash G_n(\mathcal{O}_F))^{-1}Z(s,W,\phi) &= \sum_{\substack{\lambda\in\mathbb{Z}^n\\ \lambda_1\geqslant\ldots\geqslant\lambda_n\geqslant 0}} s_{2\lambda}(t)|\det a(\lambda)|_F^s\\ &= \sum_{\substack{\lambda\in\mathbb{Z}^n\\ \lambda_1\geqslant\ldots\geqslant\lambda_n\geqslant 0}} s_{2\lambda}(t)q_F^{-s(\lambda_1+\ldots+\lambda_n)}\\ &= \sum_{\substack{\lambda\in\mathbb{Z}^n\\ \lambda_1\geqslant\ldots\geqslant\lambda_n\geqslant 0}} s_{2\lambda}(q_F^{-s/2}t_1,\ldots,q_F^{-s/2}t_n)\end{aligned}$$

But by [M] equality 5.(a) p.77 we have

$$\sum_{\substack{\lambda\in\mathbb{Z}^n\\ \lambda_1\geqslant\ldots\geqslant\lambda_n\geqslant 0}} s_{2\lambda}(q_F^{-s/2}t_1,\ldots,q_F^{-s/2}t_n)=\prod_{1\leqslant i\leqslant n}(1-q_F^{-s}t_i^2)^{-1}\prod_{1\leqslant i<j\leqslant n}(1-q_F^{-s}t_it_j)^{-1}.$$

On the other hand, since π is unramified it is of the form $\chi_1\times\ldots\times\chi_n$ where the χ_i are unramified character of $E^\times$. Thus, by Lemma 3.2.1(i), we have

$$\begin{aligned}L(s,\pi,\mathrm{As}) &= \prod_{i=1}^n L(s,\chi_i,\mathrm{As})\prod_{1\leqslant i<j\leqslant n}L(s,\chi_i\chi_j)=\prod_{i=1}^n L(s,\chi_{i|F^\times})\prod_{1\leqslant i<j\leqslant n}L(s,\chi_i\chi_j)\\ &= \prod_{1\leqslant i\leqslant n}(1-q_F^{-s}t_i^2)^{-1}\prod_{1\leqslant i<j\leqslant n}(1-q_F^{-s}t_it_j)^{-1}\end{aligned}$$

and this ends the proof of the lemma. ■

Corollary 3.6.2. *Assume that E/F is a quadratic extension of p-adic fields and that $\pi \in \mathrm{Irr}_{\mathrm{gen}}(G_n(E))$ is an unramified generic representation of $G_n(E)$. Then, there exist $W \in \mathcal{W}(\pi,\psi_n)$ and $\phi \in \mathcal{S}(F^n)$ such that $s \mapsto Z(s,W,\phi)$ and $s \mapsto Z(s,\widetilde{W},\widehat{\phi})$ are not identically zero and admit meromorphic continuation to $\mathbb{C}$ satisfying*

$$Z(1-s,\widetilde{W},\widehat{\phi})=\omega_\pi(\tau)^{n-1}|\tau|_E^{\frac{n(n-1)}{2}(s-1/2)}\lambda_{E/F}(\psi')^{-\frac{n(n-1)}{2}}\gamma(s,\pi,\mathrm{As},\psi')Z(s,W,\phi).$$

Proof. By Lemma 3.5.3 (or rather its proof) the statement of the corollary does not depend on the choice of (ψ, ψ'). Therefore, we may assume that both ψ' and ψ are unramified. Then, taking for W the unique element of $\mathcal{W}(\pi, \psi_n)^{G_n(\mathcal{O}_E)}$ normalized by $W(1) = 1$ and for $\phi \in \mathcal{S}(F^n)$ the characteristic function of $\mathcal{O}_F^n$, by the previous lemma we have

$$Z(s, W, \phi) = \mathrm{vol}(N_n(\mathcal{O}_F)\backslash G_n(\mathcal{O}_F))L(s, \pi, \mathrm{As}),$$

$$Z(s, \widetilde{W}, \widehat{\phi}) = \mathrm{vol}(N_n(\mathcal{O}_F)\backslash G_n(\mathcal{O}_F))L(s, \widetilde{\pi}, \mathrm{As})$$

for $\Re(s) \gg 1$. This shows that both $s \mapsto Z(s, W, \phi)$ and $s \mapsto Z(s, \widetilde{W}, \widehat{\phi})$ are not identically zero and admit meromorphic continuation to $\mathbb{C}$ satisfying

$$Z(1-s, \widetilde{W}, \widehat{\phi}) = \frac{L(1-s, \widetilde{\pi}, \mathrm{As})}{L(s, \pi, \mathrm{As})} Z(s, W, \phi).$$

Thus, it only remains to check that

$$\epsilon(s, \pi, \mathrm{As}, \psi') = \omega_\pi(\tau)^{1-n} |\tau|_E^{\frac{n(n-1)}{2}(1/2-s)} \lambda_{E/F}(\psi')^{\frac{n(n-1)}{2}}.$$

Since π is unramified, it is an induced representation of the form $\chi_1 \times \ldots \times \chi_n$ where the χ_i's are unramified characters of $E^\times$. By Lemma 3.2.1(i) we have

$$\epsilon(s, \pi, \mathrm{As}, \psi') = \lambda_{E/F}(\psi')^{\frac{n(n-1)}{2}} \prod_{1 \leqslant i < j \leqslant n} \epsilon(s, \chi_i \chi_j, \psi'_E) \times \prod_{i=1}^{n} \epsilon(s, \chi_{i|F^\times}, \psi') \tag{3.6.2}$$

Since $\chi_{i|F^\times}$ and ψ' are unramified we have $\epsilon(s, \chi_{i|F^\times}, \psi') = 1$ for all $1 \leqslant i \leqslant n$. On the other hand, since χ_i, χ_j and $\psi = \psi'_{E,\tau}$ are unramified we have

$$\epsilon(s, \chi_i \chi_j, \psi'_E) = |\tau|_E^{1/2-s} \chi_i(\tau)^{-1} \chi_j(\tau)^{-1} \epsilon(s, \chi_i \chi_j, \psi) = |\tau|_E^{1/2-s} \chi_i(\tau)^{-1} \chi_j(\tau)^{-1}$$

for every $1 \leqslant i < j \leqslant n$. Therefore 3.6.2 becomes

$$\begin{aligned} \epsilon(s, \pi, \mathrm{As}, \psi') &= \lambda_{E/F}(\psi')^{\frac{n(n-1)}{2}} |\tau|_E^{\frac{n(n-1)}{2}(1/2-s)} \prod_{1 \leqslant i < j \leqslant n} \chi_i(\tau)^{-1} \chi_j(\tau)^{-1} \\ &= \lambda_{E/F}(\psi')^{\frac{n(n-1)}{2}} |\tau|_E^{\frac{n(n-1)}{2}(1/2-s)} \omega_\pi(\tau)^{1-n} \end{aligned}$$

which is exactly what we wanted. ■

3.7 Global Zeta Integrals and Their Functional Equation

In this section we let k'/k be a quadratic extension of number fields. For every place v of k we denote by k_v the corresponding completion of k and set $k'_v = k_v \otimes_k k'$. Also if v is non-Archimedean, we let $\mathcal{O}_v$, $\mathcal{O}_{k',v}$ be the ring of integers of k_v and k'_v respectively. Let $\mathbb{A} = \prod'_v k_v$ and $\mathbb{A}_{k'} = \mathbb{A} \otimes_k k' = \prod'_v k'_v$ be the adèle rings of k and k' respectively and $|.|$ be the normalized absolute value on $\mathbb{A}$. We will also denote by $\mathbb{A}_{k',f} = \prod'_{v\nmid\infty} k'_v$ the finite adèles of k' and by $k'_\infty = \prod_{v|\infty} k'_v$ the product of the Archimedean completions of k' (so that $\mathbb{A}_{k'} = \mathbb{A}_{k',f} \times k'_\infty$). Let Ψ' and Ψ be nontrivial additive characters of $k\backslash\mathbb{A}$ and $k'\backslash\mathbb{A}_{k'}$ respectively with Ψ *trivial* on $k\backslash\mathbb{A}$. For every place v we let Ψ_v be the local component of Ψ at v to which we associate a generic character $\Psi_{n,v} : N_n(k'_v) \to \mathbb{S}^1$ as before. Then $\Psi_n = \prod_v \Psi_{n,v}$ define a character of $N_n(\mathbb{A}_{k'})$ which is at the same time trivial on $N_n(k')$ and $N_n(\mathbb{A})$. Let Z_∞ be the connected component (for the usual topology) of $Z_n(\mathbb{R})$. By the diagonal embedding $\mathbb{R} \hookrightarrow k'_\infty$, we consider Z_∞ as a central subgroup of $G_n(k'_\infty)$. For ω a continuous unitary character of Z_∞ we denote by $\mathcal{A}_{\mathrm{cusp}}(Z_\infty G_n(k')\backslash G_n(\mathbb{A}_{k'}), \omega)$ the space of smooth functions $\varphi : G_n(k')\backslash G_n(\mathbb{A}_{k'}) \to \mathbb{C}$ (here smooth means that φ is right invariant by an open subgroup of $G_n(\mathbb{A}_{k',f})$ and C^∞ in the $G(k'_\infty)$ component) with all their derivatives of moderate growth in the sense of [MW, §I.2.3], having central character ω (i.e. $\varphi(zg) = \omega(z)\varphi(g)$ for every $(z, g) \in Z_\infty \times G_n(\mathbb{A}_{k'})$) and satisfying

$$\int_{N(k')\backslash N(\mathbb{A}_{k'})} \varphi(ug)du = 0, \quad g \in G_n(\mathbb{A}_{k'})$$

for every parabolic subgroup $P = MN$ of G_n. By a cuspidal automorphic representation of $G_n(\mathbb{A}_{k'})$ we mean a closed topologically irreducible subrepresentation of the space $\mathcal{A}_{\mathrm{cusp}}(Z_\infty G_n(k')\backslash G_n(\mathbb{A}_{k'}), \omega)$ for some character ω. If Π is such a cuspidal automorphic representation of $G_n(\mathbb{A}_{k'})$ then there exists an isomorphism

$$\Pi \simeq \widehat{\bigotimes_{v|\infty}} \Pi_v \otimes {\bigotimes_{v\nmid\infty}}' \Pi_v \tag{3.7.1}$$

where for every place v of k, Π_v is an irreducible representation of $G_n(k'_v)$ in the sense of Sect. 2.3. Moreover, the representation Π_v is generic for every place v, unramified for almost all places v and the restricted tensor product above is taken with respect to a family of (almost everywhere defined) $G_n(\mathcal{O}_{k',v})$-fixed vectors.

Let Π be a cuspidal automorphic representation of $G_n(\mathbb{A}_{k'})$ and let, for every place v of k, W_v be a Whittaker function in $\mathcal{W}(\Pi_v, \Psi_{n,v})$ such that W_v is $G_n(\mathcal{O}_{k',v})$-invariant and satisfies $W_v(1) = 1$ for almost all places v. Set

$$W = \prod_v W_v.$$

Then, W is a well-defined function on $G_n(\mathbb{A}_{k'})$ satisfying $W(ug) = \Psi_n(u)W(g)$ for every $(u, g) \in N_n(\mathbb{A}_{k'}) \times G_n(\mathbb{A}_{k'})$. We define similarly $\widetilde{W} = \prod_v \widetilde{W}_v$. For every place v, choose a function $\phi_v \in \mathcal{S}(k_v'^n)$ such that $\phi_v = \mathbf{1}_{G_n(\mathcal{O}_{k',v})}$ for almost all places v and set $\phi = \prod_v \phi_v$ (a function on $\mathbb{A}_{k'}^n$). We define similarly $\widehat{\phi} = \prod_v \widehat{\phi_v}$ where the local Fourier transforms are defined with respect to the local components Ψ_v' of Ψ'. For $s \in \mathbb{C}$, we define, whenever convergent, the following global analog of the Zeta integrals discussed in the previous sections:

$$Z(s, W, \phi) = \int_{N_n(\mathbb{A})\backslash G_n(\mathbb{A})} W(h)\phi(e_n h)|\det h|^s dh.$$

Then, the following result is a consequence of [Kab, Propositions 5 & 6] (although, strictly speaking, the discussion of *loc. cit.* assumes that k'/k splits at all Archimedean places, this assumption is not used in the proof of the result below).

Theorem 3.7.1. *When $\Re(s)$ is sufficiently large, the integral defining $Z(s, W, \phi)$ is absolutely convergent and we have*

$$Z(s, W, \phi) = \prod_v Z(s, W_v, \phi_v).$$

Moreover, the function $s \mapsto Z(s, W, \phi)$ has a meromorphic continuation to $\mathbb{C}$ that satisfies the functional equation

$$Z(s, W, \phi) = Z(1 - s, \widetilde{W}, \widehat{\phi}).$$

3.8 A Globalization Result

We continue with the setting fixed in Sect. 3.7. Let v_0, v_1 be two distinct places of k with v_1 non-Archimedean. Recall that in Sect. 2.3 we have equipped the set $\mathrm{Temp}(G_n(k'_{v_0}))$ of tempered representations of $G_n(k'_{v_0})$ with a topology. We have the following globalization result which is a simple consequence of the main result of [FLM]. For convenience, we explain the deduction.

Theorem 3.8.1. *Let U be an open subset of* $\mathrm{Temp}(G_n(k'_{v_0}))$. *Then, there exists a cuspidal automorphic representation Π of $G_n(\mathbb{A}_{k'})$ such that $\Pi_{v_0} \in U$ and Π_v is unramified for every non-Archimedean place $v \notin \{v_0, v_1\}$.*

Proof. Let S_∞ be the set of all Archimedean places of k and set $S = S_\infty \cup \{v_0\}$, $k'_S = \prod_{v|S} k'_v$. Denote by $\mathrm{Temp}(G_n(k'_S))$ the set of all (isomorphism classes of) irreducible tempered representations of $G_n(k'_S)$ i.e. representations which can be written as (completed) tensor products of tempered irreducible representations of $G_n(k'_v)$ for every $v \in S$. Of course, we have an identification $\mathrm{Temp}(G_n(k'_S)) =$

$\prod_{v\in S}\mathrm{Temp}(G_n(k'_v))$ and we equip this set with the product topology. Clearly, it suffices to show that for every open U of $\mathrm{Temp}(G_n(k'_S))$ there exists a cuspidal automorphic representation Π of $G_n(\mathbb{A}_{k'})$ such that $\Pi_S := \widehat{\bigotimes_{v\in S}}\Pi_v$ belongs to U and Π_v is unramified for every $v \notin S\cup\{v_1\}$. Let $G_n(\mathbb{A}_{k'})^1$ and $G_n(k'_S)^1$ be the subgroups of elements g in $G_n(\mathbb{A}_{k'})$ and $G(k'_S)$ respectively satisfying $|\det g|_{\mathbb{A}_{k'}} = 1$ where $|.|_{\mathbb{A}_{k'}}$ denotes the normalized absolute value on $\mathbb{A}_{k'}$. Then, we have decompositions $G_n(\mathbb{A}_{k'}) = G_n(\mathbb{A}_{k'})^1 \times Z_\infty$ and $G_n(k'_S) = G_n(k'_S)^1 \times Z_\infty$. Let $\mathrm{Temp}(G_n(k'_S)^1)$ be the set of isomorphism classes of tempered representations of $G_n(k'_S)^1$ i.e. representations obtained by restriction of representations in $\mathrm{Temp}(G_n(k'_S))$ that we equip with the quotient topology. Then, we have an isomorphism $\mathrm{Temp}(G_n(k'_S)) \simeq \mathrm{Temp}(G_n(k'_S)^1) \times \widehat{Z_\infty}$. Let U be an open subset of $\mathrm{Temp}(G_n(k'_S))$ that we may assume to be of the form $U_1 \times V$ where $U_1 \subset \mathrm{Temp}(G_n(k'_S)^1)$ and $V \subset \widehat{Z_\infty}$ are open. Let $\mathfrak{p}$ be the prime ideal in $\mathcal{O}_{k'}$ corresponding to v_1 and for all $m \geqslant 1$ let $K^S(\mathfrak{p}^m)$ be the open subgroup $K_{v_1}(m) \times \prod_{v\notin S\cup\{v_1\}} G_n(\mathcal{O}_{k',v})$ of $\prod'_{v\notin S} G_n(k'_v)$ where $K_{v_1}(m) := \{k \in G_n(\mathcal{O}_{k',v_1}) \mid k \equiv I_n \bmod \mathfrak{p}^m\}$. Let μ_{Pl} be the Plancherel measure for $G_n(k'_S)^1$. Since $\mu_{\mathrm{Pl}}(U_1)$ is nonzero, by [FLM, Theorem 2] there exists an irreducible representation Π' of $G_n(\mathbb{A}_{k'})$ such that $\Pi' \mid_{G_n(\mathbb{A}_{k'})^1} \hookrightarrow L^2(G_n(k')\backslash G_n(\mathbb{A}_{k'})^1)$, $(\Pi'_S) \mid_{G_n(k'_S)^1} \in U_1$ and $(\Pi')^{K^S(\mathfrak{p}^m)} \neq 0$ for some $m \geqslant 1$. Let $\omega \in V$ and denote by Π the unique irreducible representation of $G_n(\mathbb{A}_{k'})$ whose restriction to $G_n(\mathbb{A}_{k'})^1$ coincides with $\Pi' \mid_{G_n(\mathbb{A}_{k'})^1}$ and whose central character equals ω on Z_∞. Then $\Pi \hookrightarrow L^2(Z_\infty G_n(k')\backslash G_n(\mathbb{A}_{k'}), \omega)$, $\Pi_S \in U$ and $\Pi^{K^S(\mathfrak{p}^m)} \neq 0$. Moreover, since Π is tempered at every Archimedean place by [Wall3] we have that Π is actually a cuspidal (rather than just square-integrable) automorphic representation of $G_n(\mathbb{A}_{k'})$. Obviously, Π has all of the desired properties. ■

Remark 1. The proof of the above result uses the full strength of [FLM] but in fact we can get a slightly weaker result which is however sufficient for the application we have in mind without appealing to the analysis of the continuous part of the spectral side of Arthur's weighted trace formula which is the hard part of [FLM]. More precisely, the following ought to be provable (although the author hasn't checked all the details) using the limit property [FLM, Proposition 3] of the geometric side of Arthur's weighted trace formula together with Sauvageot's density principle [Sau]:

> Let G be any connected reductive group over k. Let v_0, v_1, v_2 be three distinct places of k with v_1, v_2 non-Archimedean. Let U be an open subset of $\mathrm{Temp}(G(k_{v_0}))$ and π_1 be a supercuspidal representation of $G(k_{v_1})$. Then, there exists a cuspidal automorphic representation Π of $G(\mathbb{A})$ such that $\Pi_{v_0} \in U$, Π_{v_1} is isomorphic to an unramified twist of π_1 and Π_v is unramified for every non-Archimedean place $v \notin \{v_0, v_1, v_2\}$.

3.9 Proof of Theorems 3.5.1 and 3.5.2 in the Nearly Tempered Case

Let the setting be as in Sect. 3.5. We choose a quadratic extension k'/k of number fields together with a place v_0 of k such that:

- there exists an isomorphism $k'_{v_0}/k_{v_0} \simeq E/F$ that we fix from now on;
- every Archimedean place of k different from v_0 splits in k'.

Given a place v of k, a nontrivial additive character ψ'_v of k_v and a generic irreducible representation π_v of $G_n(k'_v)$ we let $\gamma^{Sh}(s, \pi_v, \mathrm{As}, \psi'_v)$ be the local Asai γ-factor defined by Shahidi (see [Sha3, Gold]). The goal of this section is to show the following:

Theorem 3.9.1. *For every $\pi \in \mathrm{Irr}_{\mathrm{ntemp}}(G_n(E))$, $W \in \mathcal{W}(\pi, \psi_n)$, $\phi \in \mathcal{S}(F^n)$ and $s \in \mathbb{C}$ with $\Re(s) = \frac{1}{2}$ we have*

$$Z(1-s, \widetilde{W}, \widehat{\phi}) = \omega_\pi(\tau)^{n-1}|\tau|_E^{\frac{n(n-1)}{2}(s-1/2)}\lambda_{E/F}(\psi')^{-\frac{n(n-1)}{2}}\gamma^{Sh}(s, \pi, \mathrm{As}, \psi')Z(s, W, \phi).$$

First we explain how to deduce from it Theorems 3.5.1 and 3.5.2 in the nearly tempered case. Let $\pi \in \mathrm{Irr}_{\mathrm{ntemp}}(G_n(E))$, $W \in \mathcal{W}(\pi, \psi_n)$ and $\phi \in \mathcal{S}(F^n)$. By the main results of [Sha2] (in the Archimedean case) and [Shank] (in the p-adic case) we have

$$\gamma^{Sh}(s, \pi, \mathrm{As}, \psi') = \gamma(s, \pi, \mathrm{As}, \psi'). \tag{3.9.1}$$

Therefore, as $s \mapsto \gamma(s, \pi, \mathrm{As}, \psi')$ is a meromorphic function on $\mathbb{C}$, the functional equation of Theorem 3.9.1 together with Lemma 3.3.2 implies that $s \mapsto Z(s, W, \phi)$ admits a meromorphic continuation to $\mathbb{C}$ and that the functional equation is satisfied for every $s \in \mathbb{C}$. This already shows Theorem 3.5.1. For Theorem 3.5.2, we rewrite the functional equation as

$$\frac{Z(1-s, \widetilde{W}, \widehat{\phi})}{L(1-s, \widetilde{\pi}, \mathrm{As})} = \omega_\pi(\tau)^{n-1}|\tau|_E^{\frac{n(n-1)}{2}(s-1/2)}\lambda_{E/F}(\psi')^{-\frac{n(n-1)}{2}}\epsilon(s, \pi, \mathrm{As}, \psi')\frac{Z(s, W, \phi)}{L(s, \pi, \mathrm{As})}. \tag{3.9.2}$$

By Lemma 3.3.2 and since the functions $L(s, \pi, \mathrm{As})$ and $L(1-s, \widetilde{\pi}, \mathrm{As})$ have no zeros, the functions $s \mapsto \frac{Z(s,W,\phi)}{L(s,\pi,\mathrm{As})}$ and $s \mapsto \frac{Z(1-s,\widetilde{W},\widehat{\phi})}{L(1-s,\widetilde{\pi},\mathrm{As})}$ are holomorphic on $\{\Re(s) \geqslant \frac{1}{2}\}$ and $\{\Re(s) \leqslant \frac{1}{2}\}$ respectively. Since the epsilon factor $\epsilon(s, \pi, \mathrm{As}, \psi')$ has no zeros either, this implies together with 3.9.2 that $s \mapsto \frac{Z(s,W,\phi)}{L(s,\pi,\mathrm{As})}$ is holomorphic on $\{\Re(s) \geqslant \frac{1}{2}\} \cup \{\Re(s) \leqslant \frac{1}{2}\} = \mathbb{C}$. Moreover, as π is nearly tempered the L-functions $L(s, \pi, \mathrm{As})$ and $L(s, \widetilde{\pi}, \mathrm{As})$ are holomorphic in $\{\Re(s) \geqslant \frac{1}{2}\}$ (Lemma 3.2.1(ii)). Hence, by Lemma 3.3.3 for every $s_0 \in \mathbb{C}$ with $\Re(s_0) \geqslant 1/2$

(resp. $\Re(s_0) \leqslant 1/2$) there exist $W \in \mathcal{W}(\pi, \psi_n)$ and $\phi \in \mathcal{S}(F^n)$ with $\frac{Z(s_0,W,\phi)}{L(s_0,\pi,\mathrm{As})} \neq 0$ (resp. $\frac{Z(1-s_0,\widetilde{W},\widehat{\phi})}{L(1-s_0,\widetilde{\pi},\mathrm{As})} \neq 0$). By 3.9.2 this implies that for every $s_0 \in \mathbb{C}$ there exist $W \in \mathcal{W}(\pi, \psi_n)$ and $\phi \in \mathcal{S}(F^n)$ with $\frac{Z(s_0,W,\phi)}{L(s_0,\pi,\mathrm{As})} \neq 0$. Finally, by Lemma 3.2.1 (iii) and Lemma 3.3.2, there exists $\epsilon > 0$ such that both $s \mapsto \frac{Z(s,W,\phi)}{L(s,\pi,\mathrm{As})}$ and $s \mapsto \frac{Z(s,\widetilde{W},\widehat{\phi})}{L(s,\widetilde{\pi},\mathrm{As})}$ are of finite order in vertical strips of $\mathcal{H}_{>\frac{1}{2}-\epsilon}$. Together with the functional equation 3.9.2, this shows that $s \mapsto \frac{Z(s,W,\phi)}{L(s,\pi,\mathrm{As})}$ is of finite order in vertical strips on all of $\mathbb{C}$. This ends the proof of Theorem 3.5.2.

We now proceed to the proof of Theorem 3.9.1. Let Ψ and Ψ' be nontrivial characters of $k\backslash\mathbb{A}$ and $k'\backslash\mathbb{A}_{k'}$ with Ψ trivial on $\mathbb{A}$. By Lemma 3.5.3 and the identity 3.9.1 we may assume that $\psi' = \Psi'_{v_0}$ and $\psi = \Psi_{v_0}$. We now show that it suffices to establish Theorem 3.9.1 for a dense subset $D \subseteq \mathrm{Temp}(G_n(E))$. Indeed, let $\pi \in \mathrm{Irr}_{\mathrm{ntemp}}(G_n(E))$ that we write as $\pi = i_{P(E)}^{G_n(E)}(\sigma_{\lambda_0})$ where $P = MU$ is a standard parabolic subgroup of G_n, $\sigma \in \Pi_2(M(E))$ and $\lambda_0 = (\lambda_{0,1}, \dots, \lambda_{0,n}) \in \mathcal{A}^*_{M,\mathbb{C}} \subseteq \mathcal{A}^*_{\mathbb{C}} = \mathbb{C}^n$ satisfies

$$|\Re(\lambda_{0,i})| < 1/4, \quad 1 \leqslant i \leqslant n. \tag{3.9.3}$$

By Lemma 3.5.4, we may assume that $\lambda_0 \in (\mathcal{A}^G_{M,\mathbb{C}})^*$. Set $\pi_\lambda = i_{P(E)}^{G_n(E)}(\sigma_\lambda)$ for every $\lambda \in \mathcal{A}^*_{M,\mathbb{C}}$ (so that $\pi = \pi_{\lambda_0}$). Let $W \in \mathcal{W}(\pi, \psi_n)$ and $\phi \in \mathcal{S}(F^n)$. Let $\rho = (\frac{n-1}{2}, \dots, \frac{1-n}{2}) \in (\mathcal{A}^{G_n})^*$ be half the sum of the roots of A_n in B_n. Then $|\Re(\lambda_0)| \prec |\Re(\lambda_0)| + \epsilon\rho$ for any $\epsilon > 0$ and by 3.9.3 we can choose $\epsilon > 0$ such that $-2\min(|\Re(\lambda_0)| + \epsilon\rho) < \frac{1}{2}$. Set $\mu = |\Re(\lambda_0)| + \epsilon\rho$ and $\mathcal{U}[\prec \mu] = \{\lambda \in (\mathcal{A}^{G_n}_{M,\mathbb{C}})^* \mid |\Re(\lambda)| \prec \mu\}$. By Corollary 2.7.1, there exists a map

$$\lambda \in (\mathcal{A}^G_{M,\mathbb{C}})^* \mapsto W_\lambda \in \mathcal{W}(\pi_\lambda, \psi_n)$$

such that $W_{\lambda_0} = W$, $W_\lambda \in \mathcal{C}_\mu(N_n(E)\backslash G_n(E), \psi_n)$ for every $\lambda \in \mathcal{U}[\prec \mu]$ and the induced map

$$\lambda \in \mathcal{U}[\prec \mu] \mapsto W_\lambda \in \mathcal{C}_\mu(N_n(E)\backslash G_n(E), \psi_n)$$

is analytic. Assume that Theorem 3.9.1 holds for a dense subset $D \subseteq \mathrm{Temp}(G_n(E))$. Then, since this theorem is insensitive to unramified twists (Lemma 3.5.4), there exists a dense subset $D_\sigma \subseteq i(\mathcal{A}^{G_n}_M)^*$ such that Theorem 3.9.1 holds for π_λ for every $\lambda \in D_\sigma$. In particular, we have

$$Z(1-s, \widetilde{W}_\lambda, \widehat{\phi}) = \omega_\pi(\tau)^{n-1}|\tau|_E^{\frac{n(n-1)}{2}(s-1/2)}\lambda_{E/F}(\psi')^{-\frac{n(n-1)}{2}}\gamma^{Sh}(s, \pi_\lambda, \mathrm{As}, \psi')Z(s, W_\lambda, \phi) \tag{3.9.4}$$

for every $\lambda \in D_\sigma$ and $s \in \mathbb{C}$ with $\Re(s) = \frac{1}{2}$. By the choice of μ and Lemma 3.3.1 for any $s \in \mathbb{C}$ with $\Re(s) = \frac{1}{2}$ the maps $\lambda \in \mathcal{U}[\prec \mu] \mapsto Z(s, W_\lambda, \phi)$ and $\lambda \in \mathcal{U}[\prec \mu] \mapsto Z(1-s, \widetilde{W}_\lambda, \widehat{\phi})$ are holomorphic. By 3.9.1 and Lemma 3.2.1(ii), the map

$\lambda \in \mathcal{U}[\prec \mu] \mapsto \gamma^{Sh}(s, \pi_\lambda, \mathrm{As}, \psi')$ is also holomorphic. Therefore, by density of D_σ in $i(\mathcal{A}_M^G)^*$, 3.9.4 is also satisfied for every $\lambda \in i(\mathcal{A}_M^G)^*$ (by continuity) hence for every $\lambda \in \mathcal{U}[\prec \mu]$ (by analyticity) and in particular for $\lambda = \lambda_0$. This shows that Theorem 3.9.1 holds for π and ends the proof of the reduction.

Let $U \subset \mathrm{Temp}(G_n(E))$ be an open subset. We are now going to show that there exists $\pi \in U$ for which Theorem 3.9.1 holds. Let v_1 be a non-Archimedean place of k which splits in k'. By Theorem 3.8.1, there exists a cuspidal automorphic representation Π of $G_n(\mathbb{A}_{k'})$ such that $\Pi_{v_0} \in U$ and Π_v is unramified for every non-Archimedean place $v \notin \{v_0, v_1\}$. Let $S = S_\infty \cup \{v_0, v_1\}$ where S_∞ stands for the set of all Archimedean places of k and let T be the (finite) set of places v of k outside S which ramify in k' or where Ψ'_v or Ψ_v is not unramified. Then, by [Sha3, Theorem 3.5(4)] the product

$$L^{S\cup T}(s, \Pi, \mathrm{As}) = \prod_{v \notin S\cup T} L(s, \Pi_v, \mathrm{As}),$$

which converges for $\Re(s)$ sufficiently large, admits a meromorphic continuation to $\mathbb{C}$ and we have

$$L^{S\cup T}(s, \Pi, \mathrm{As}) = \prod_{v\in S\cup T} \gamma^{Sh}(s, \Pi_v, \mathrm{As}, \Psi'_v) L^{S\cup T}(1-s, \Pi^\vee, \mathrm{As}). \tag{3.9.5}$$

On the other hand, combining Theorem 3.7.1 with Lemma 3.6.1, we find that

$$\prod_{v\in S\cup T} Z(s, W_v, \phi_v) L^{S\cup T}(s, \Pi, \mathrm{As}) = \prod_{v\in S\cup T} Z(1-s, \widetilde{W}_v, \widehat{\phi}_v) L^{S\cup T}(1-s, \Pi^\vee, \mathrm{As}) \tag{3.9.6}$$

for every $(W_v)_v \in \prod_{v\in S\cup T} \mathcal{W}(\Pi_v, \Psi_{n,v})$ and $(\phi_v)_v \in \prod_{v\in S\cup T} \mathcal{S}(k_v^n)$. Notice that we haven't proven the meromorphic continuation of $Z(s, W_v, \phi_v)$ yet but, by Theorem 3.7.1 and the aforementioned result of Shahidi, we at least know that the products

$$\prod_{v\in S\cup T} Z(s, W_v, \phi_v) \text{ and } \prod_{v\in S\cup T} Z(1-s, \widetilde{W}_v, \widehat{\phi}_v)$$

admit meromorphic continuations. For every place $v \in (S\cup T)\setminus\{v_0\}$, either v splits in k' or v is non-Archimedean and Π_v is unramified. Therefore, by Theorem 3.4.1, Corollary 3.6.2 and Lemma 3.3.3 we can choose the local data so that the functions $s \mapsto Z(s, W_v, \phi_v)$ and $s \mapsto Z(1-s, \widetilde{W}_v, \widehat{\phi}_v)$ admit meromorphic continuation to $\mathbb{C}$ which are not identically zero and satisfying the functional equation

$$Z(1-s, \widetilde{W}_v, \widehat{\phi}_v) = \omega_{\Pi_v}(\tau)^{n-1} |\tau|_{k'_v}^{\frac{n(n-1)}{2}(s-1/2)} \lambda_{k'_v/k_v}(\Psi'_v)^{-\frac{n(n-1)}{2}} \gamma(s, \Pi_v, \mathrm{As}, \Psi'_v) Z(s, W_v, \phi_v)$$

for every $v \in (S \cup T) \setminus \{v_0\}$. Here $\tau \in k'$ is the only trace zero element such that $\Psi(z) = \Psi'(\mathrm{Tr}_{\mathbb{A}_{k'}/\mathbb{A}}(\tau z))$ for all $z \in \mathbb{A}_{k'}$. This already implies that the local Zeta integrals $Z(s, W, \phi)$ and $Z(1-s, \widetilde{W}, \widehat{\phi})$ admit meromorphic continuation for every $W \in \mathcal{W}(\Pi_{v_0}, \psi_n)$ and $\phi \in \mathcal{S}(F^n)$. Moreover, combining the above identities with the product formula

$$\prod_{v \in S \cup T} \omega_{\Pi_v}(\tau)^{n-1} |\tau|_{k'_v}^{\frac{n(n-1)}{2}(s-1/2)} \lambda_{k'_v/k_v}(\Psi'_v)^{-\frac{n(n-1)}{2}}$$
$$= \prod_v \omega_{\Pi_v}(\tau)^{n-1} |\tau|_{k'_v}^{\frac{n(n-1)}{2}(s-1/2)} \lambda_{k'_v/k_v}(\Psi'_v)^{-\frac{n(n-1)}{2}} = 1$$

(which is a consequence of the fact that Π_v, k'_v/k_v, Ψ_v and Ψ'_v are all unramified outside $S \cup T$) identity 3.9.6 becomes

$$\omega_{\Pi_{v_0}}(\tau)^{n-1} |\tau|_E^{\frac{n(n-1)}{2}(s-1/2)} \lambda_{E/F}(\psi')^{-\frac{n(n-1)}{2}} Z(s, W, \phi) L^{S \cup T}(s, \Pi, \mathrm{As})$$
$$= Z(1-s, \widetilde{W}, \widehat{\phi}) \prod_{v \in S \cup T \setminus \{v_0\}} \gamma(s, \Pi_v, \mathrm{As}, \Psi'_v) L^{S \cup T}(1-s, \Pi^\vee, \mathrm{As})$$

for every $W \in \mathcal{W}(\Pi_{v_0}, \psi_n)$ and $\phi \in \mathcal{S}(F^n)$. Taking the quotient with 3.9.5 and using 3.9.1, we obtain

$$\omega_{\pi_{v_0}}(\tau)^{n-1} |\tau|_E^{\frac{n(n-1)}{2}(s-1/2)} \lambda_{E/F}(\psi')^{-\frac{n(n-1)}{2}} \gamma^{Sh}(s, \Pi_{v_0}, \mathrm{As}, \psi') Z(s, W, \phi)$$
$$= Z(1-s, \widetilde{W}, \widehat{\phi}) \prod_{v \in S \cup T \setminus \{v_0\}} \frac{\gamma(s, \Pi_v, \mathrm{As}, \Psi'_v)}{\gamma^{Sh}(s, \Pi_v, \mathrm{As}, \Psi'_v)} = Z(1-s, \widetilde{W}, \widehat{\phi})$$

for every $W \in \mathcal{W}(\Pi_{v_0}, \psi_n)$ and $\phi \in \mathcal{S}(F^n)$. This shows that Theorem 3.9.1 holds for Π_{v_0} hence, as U was arbitrary, for a dense subset of $\mathrm{Temp}(G_n(E))$. By the previous reduction, this proves Theorem 3.9.1 in all cases. ■

3.10 Proof of Theorems 3.5.1 and 3.5.2 in the General Case

Let π be a generic irreducible representation of $G_n(E)$. There exists a standard parabolic subgroup $P = MU$ of G_n, $\sigma \in \Pi_2(M(E))$ and $\lambda_0 \in \mathcal{A}^*_{M,\mathbb{C}}$ such that $\pi = i_{P(E)}^{G_n(E)}(\sigma_{\lambda_0})$. By Lemma 3.5.4, we may assume that $\lambda_0 \in (\mathcal{A}^{G_n}_{M,\mathbb{C}})^*$. Set $\pi_\lambda = i_{P(E)}^{G_n(E)}(\sigma_\lambda)$ for every $\lambda \in \mathcal{A}^*_{M,\mathbb{C}}$. Let $W \in \mathcal{W}(\pi, \psi_n)$ and $\phi \in \mathcal{S}(F^n)$. Choose a section

$$\lambda \in (\mathcal{A}^G_{M,\mathbb{C}})^* \mapsto W_\lambda \in \mathcal{W}(\pi_\lambda, \psi_n)$$

as in Corollary 2.7.1 with $W_{\lambda_0} = W$. For $\mu \in (\mathcal{A}^{G_n})^*$ we define the open subset $\mathcal{U}[\prec \mu]$ of $(\mathcal{A}_{M,\mathbb{C}}^G)^*$ as in Corollary 2.7.1. In particular, for every $\mu \in (\mathcal{A}^{G_n})^*$ the above section induces an analytic map

$$\lambda \in \mathcal{U}[\prec \mu] \mapsto W_\lambda \in \mathcal{C}_\mu(N_n(E)\backslash G_n(E), \psi_n).$$

Decomposing the ϵ-factor $\epsilon(s, \pi_\lambda, \mathrm{As}, \psi')$ according to Lemma 3.2.1(i) we see that there exists $C > 0$, a linear form L on $(\mathcal{A}_{M,\mathbb{C}}^G)^*$ and an element $u \in \mathbb{C}$ of norm 1 such that

$$\omega_\pi(\tau)^{n-1}|\tau|_E^{\frac{n(n-1)}{2}(s-1/2)}\lambda_{E/F}(\psi')^{-\frac{n(n-1)}{2}}\epsilon(s, \pi_\lambda, \mathrm{As}, \psi') = uC^{L(\lambda)+s-\frac{1}{2}}$$

for every $\lambda \in (\mathcal{A}_{M,\mathbb{C}}^G)^*$ and $s \in \mathbb{C}$. Let v be a square root of u and set

$$\epsilon_{1/2}(s, \pi_\lambda, \psi') := v(\sqrt{C})^{L(\lambda)+s-\frac{1}{2}}, \quad \lambda \in (\mathcal{A}_{M,\mathbb{C}}^G)^*, s \in \mathbb{C}$$

so that by the above

$$\omega_\pi(\tau)^{n-1}|\tau|_E^{\frac{n(n-1)}{2}(s-1/2)}\lambda_{E/F}(\psi')^{-\frac{n(n-1)}{2}}\epsilon(s, \pi_\lambda, \mathrm{As}, \psi') = \epsilon_{1/2}(s, \pi_\lambda, \psi')^2 \tag{3.10.1}$$

for every $\lambda \in (\mathcal{A}_{M,\mathbb{C}}^G)^*$ and $s \in \mathbb{C}$. We are now going to apply Proposition 2.8.1 to the two following (for the moment partially defined) functions on $\mathbb{C} \times (\mathcal{A}_{M,\mathbb{C}}^G)^*$:

$$Z_+(s, \lambda) = \epsilon_{1/2}(\frac{1}{2} + s, \pi_\lambda, \psi')\frac{Z(\frac{1}{2} + s, W_\lambda, \phi)}{L(\frac{1}{2} + s, \pi_\lambda, \mathrm{As})}$$

and

$$Z_-(s, \lambda) = \epsilon_{1/2}(\frac{1}{2} + s, \pi_\lambda, \psi')^{-1}\frac{Z(\frac{1}{2} + s, \widetilde{W}_\lambda, \widehat{\phi})}{L(\frac{1}{2} + s, \widetilde{\pi}_\lambda, \mathrm{As})}.$$

Let $U \subseteq (\mathcal{A}_{M,\mathbb{C}}^G)^*$ be the nonempty relatively compact connected open subset of vectors $\lambda = (\lambda_1, \ldots, \lambda_n) \in (\mathcal{A}_{M,\mathbb{C}}^G)^* \subseteq \mathcal{A}_\mathbb{C}^*$ satisfying $|\lambda_i| < \frac{1}{4}$ for every $1 \leqslant i \leqslant n$. Then, for every $\lambda \in U$, the representation π_λ is nearly tempered. Set as usual $\rho = (\frac{n-1}{2}, \ldots, \frac{1-n}{2}) \in (\mathcal{A}^{G_n})^*$ for half the sum of the roots of A_n in B_n. By Lemma 3.3.2 the functions Z_+, Z_- are at least well-defined on $\{\Re(s) \geqslant \frac{1}{2}\} \times U$. Moreover, by the discussion of the previous section, Z_+ and Z_- admit holomorphic extensions to $\mathbb{C} \times U$ which are of finite order in vertical strips in the first variable locally uniformly in the second variable and satisfying the functional equation

$$Z_+(s, \lambda) = Z_-(-s, \lambda), \quad s \in \mathbb{C}, \lambda \in U. \tag{3.10.2}$$

Let $U' \subseteq (\mathcal{A}^G_{M,\mathbb{C}})^*$ be a relatively compact connected open subset containing U. Then, there exists $\mu \in (\mathcal{A}^{G_n})^*$ such that $U' \subseteq \mathcal{U}[\prec \mu]$. Hence, by Lemmas 3.3.1 and 3.2.1(iii), there exists C such that Z_+ and Z_- admit holomorphic extensions to $\mathcal{H}_{>C} \times U'$ which are of finite order in vertical strips in the first variable locally uniformly in the second variable. Therefore, Z_+ and Z_- satisfy the hypothesis of Proposition 2.8.1 and they extend to holomorphic functions on all of $\mathbb{C} \times (\mathcal{A}^G_{M,\mathbb{C}})^*$ of finite order in vertical strips in the first variable locally uniformly in the second variable satisfying 3.10.2 for every $(s, \lambda) \in \mathbb{C} \times (\mathcal{A}^G_{M,\mathbb{C}})^*$. Specializing to $\lambda = \lambda_0$, by definition of Z_+, Z_- and the relation 3.10.1, we deduce that $Z(s, W, \phi)$ and $Z(s, \widetilde{W}, \widehat{\phi})$ admit meromorphic continuation to $\mathbb{C}$ satisfying the functional equation

$$Z(1-s, \widetilde{W}, \widehat{\phi}) = \omega_\pi(\tau)^{n-1} |\tau|_E^{\frac{n(n-1)}{2}(s-1/2)} \lambda_{E/F}(\psi')^{-\frac{n(n-1)}{2}} \gamma(s, \pi, \mathrm{As}, \psi') Z(s, W, \phi).$$

This already proves Theorem 3.5.1 for π. Moreover, we also obtain that $\dfrac{Z(s, W_\lambda, \phi)}{L(s, \pi_\lambda, \mathrm{As})}$ is holomorphic of finite order in vertical strips thus implying Theorem 3.5.2 for π.

References

[AKMSS] U.K. Anandavardhanan, R. Kurinczuk, N. Matringe, V. Sécherre, S. Stevens, *Galois self-dual cuspidal types and Asai local factors*. Preprint. arXiv:1807.07755

[AR] U.K. Anandavardhanan, C.S. Rajan, Distinguished representations, base change, and reductibility for unitary groups. Int. Math. Res. Not. **14**(14), 841–854 (2005)

[BK] J.N. Bernstein, B. Krötz, Smooth Fréchet globalizations of Harish-Chandra modules. Isr. J. Math. **199**(1), 45–111 (2014)

[Beu] R. Beuzart-Plessis, *Plancherel formula for* $\mathrm{GL}_n(F)\backslash\mathrm{GL}_n(E)$ *and applications to the Ichino-Ikeda and formal degree conjectures for unitary groups*. Preprint (2018)

[Bour] N. Bourbaki, Éléments de mathématique. Fasc. XXXIII. Variétés différentielles et analytiques. Fascicule de résultats (Paragraphes 1 á 7). Actualités Scientifiques et Industrielles, No. 1333 (Hermann, Paris, 1967), 97 pp.

[BT] F. Bruhat, J. Tits, Groupes réductifs sur un corps local. Inst. Hautes Études Sci. Publ. Math. **41**, 5–251 (1972)

[Cas] W. Casselman, Canonical extensions of Harish-Chandra modules to representations of G. Can. J. Math. **41**(3), 385–438 (1989)

[Cas2] W. Casselman, The unramified principal series of p-adic groups. I. The spherical function. Compositio Math. **40**(3), 387–406 (1980)

[CCI] S.-Y. Chen, Y. Cheng, I. Ishikawa, *Gamma factors for Asai representations of* GL_2. Preprint (2018). arXiv:1810.12561

[CHH] M. Cowling, U. Haagerup, R. Howe, Almost L^2 matrix coefficients. J. für die reine und angewandte Math. **387**, 97–110 (1988)

[CS] W. Casselman, J.A. Shalika, The unramified principal series of p-adic groups II; the Whittaker function. Compositio Math. **41**, 207–231 (1980)

[Del] P. Delorme, Homomorphismes de Harish-Chandra liés aux K-types minimaux des séries principales généralisées des groupes de Lie réductifs connexes. Ann. Sci. École Norm. Sup. (4) **17**(1), 117–156 (1984)

[DM] J. Dixmier, P. Malliavin, Factorisations de fonctions et de vecteurs indefiniment differentiables. Bull. Sci. Math. (2) **102**(4), 307–330 (1978)

[FLM] T. Finis, E. Lapid, W. Müller, Limit multiplicities for principal congruence subgroups of $GL(n)$ and $SL(n)$. J. Inst. Math. Jussieu **14**(3), 589–638 (2015)

[Fli] Y.Z. Flicker, On zeroes of the twisted tensor L-function. Math. Ann. **297**, 199–219 (1993)

[Fli2] Y.Z. Flicker, Twisted tensors and Euler products. Bull. Soc. Math. France **116**(3), 295–313 (1988)

[GK] I.M. Gelfand, D. Kazhdan, Representations of the group $GL(n, K)$ where K is a local field, in *Lie Groups and Their Representations* (Proc. Summer School, Bolyai János Math. Soc., Budapest, 1971), pp. 95–118. Halsted, New York (1975)

[Gold] D. Goldberg, Some results on reducibility for unitary groups and local Asai L-functions. J. Reine Angew. Math. **448**, 65–95 (1994)

[GR] B.H. Gross, M. Reeder, Arithmetic invariants of discrete Langlands parameters. Duke Math. J. **154**(3), 431–508 (2010)

[HT] M. Harris, R. Taylor, *The Geometry and Cohomology of Some Simple Shimura Varieties*, With an appendix by Vladimir G. Berkovich. Annals of Mathematics Studies, vol. 151 (Princeton University Press, Princeton, NJ, 2001), viii+276 pp. ISBN:0-691-09090-4

[Hen] G. Henniart, Une preuve simple des conjectures de Langlands pour $GL(n)$ sur un corps p-adique. Invent. Math. **139**(2), 439–455 (2000)

[Hen2] G. Henniart, Correspondance de Langlands et fonctions L des carrés extérieur et symétrique. IMRN **4**, 633–673 (2010)

[Jac] H. Jacquet, *Archimedean Rankin-Selberg Integrals*, in *Automorphic Forms and L-Functions II. Local Aspects*. Contemp. Math., vol. 489 (Amer. Math. Soc., Providence, RI, 2009), pp. 57–172

[Jac2] H. Jacquet, *Integral Representation of Whittaker Functions*, in *Contributions to Automorphic Forms, Geometry, and Number Theory* (Johns Hopkins Univ. Press, Baltimore, MD, 2004), pp. 373–419

[Jac3] H. Jacquet, Distinction by the quasi-split unitary group. Isr. J. Math. **178**(1), 269–324 (2010)

[JPSS] H. Jacquet, I.I. Piatetski-Shapiro, J.A. Shalika, Rankin-Selberg convolutions. Am. J. Math. **105**(2), 367–464 (1983)

[JS] H. Jacquet, J.A. Shalika, On Euler products and the classification of automorphic representations I. Am. J. Math. **103**, 449–558 (1981)

[JS2] H. Jacquet, J.A. Shalika, *Rankin-Selberg Convolutions: Archimedean Theory*, in *Festschrift in Honor of I. I. Piatetski-Shapiro on the Occasion of His Sixtieth Birthday, Part I* (Ramat Aviv, 1989). Israel Math. Conf. Proc., vol. 2 (Weizmann, Jerusalem, 1990), pp. 125–207

[Kab] A.C. Kable, Asai L-function and Jacquet's conjecture. Am. J. Math. **126**, 789–820 (2004)

[Kem] A. Kemarsky, A note on Kirillov model for representations of $GL_n(\mathbb{C})$. C. R. Math. Acad. Sci. Paris **353**(7), 579–582 (2015)

[Knapp] A.W. Knapp, *Representation Theory of Semisimple Groups. An Overview Based on Examples*. Princeton Mathematical Series, vol. 36 (Princeton University Press, Princeton, NJ, 1986), xviii+774 pp. ISBN:0-691-08401-7

[Kos] B. Kostant, On convexity, the Weyl group and the Iwasawa decomposition. Ann. Sci. École Norm. Sup. (4) **6**, 413–455 (1973)

[La] R.P. Langlands, *On the Classification of Irreducible Representations of Real Algebraic Groups*, in *Representation Theory and Harmonic Analysis on Semisimple Lie Groups*. Math. Surveys Monogr., vol. 31 (Amer. Math. Soc., Providence, RI, 1989), pp. 101–170

[M] I. MacDonald, *Symmetric Functions and Hall Polynomials*, 2nd edn. (Clarendon Press, Oxford, 1995)

[Mat] N. Matringe, Distinction and Asai L-functions for generic representations of general linear groups over p-adic fields. IMRN **2011**(1), 74–95 (2011)

[MW] C. Mœglin, J.-L. Waldspurger, *Spectral Decomposition and Eisenstein Series. Une paraphrase de l'Écriture*. Cambridge Tracts in Mathematics, vol. 113 (Cambridge University Press, Cambridge, 1995), xxviii+338 pp.

[Sau] F. Sauvageot, Principe de densité pour les groupes réductifs. Compositio Math., **108**(2), 151–184 (1997)

[Sch] P. Scholze, The local Langlands correspondence for GL_n over p-adic fields. Invent. Math. **192**(3), 663–715 (2013)

[Shin] T. Shintani, On an explicit formula for class 1 "Whittaker functions" on GL_n over p-adic fields. Proc. Jpn. Acad. **52**, 180–182 (1976)

[Sil] A.J. Silberger, Convexity for a simply connected p-adic group. Bull. Am. Math. Soc. **81**(5), 910–912 (1975)

[Sha2] F. Shahidi, Local coefficients as Artin factors for real groups. Duke Math. J. **52**(4), 973–1007 (1985)

[Sha3] F. Shahidi, A proof of Langlands conjecture for Plancherel measure; complementary series of $\mathfrak{p}$-adic groups. Ann. Math. (2), **132**, 273–330 (1990)

[Shank] D. Shankman, *Local Langlands correspondence for Asai L-functions and epsilon factors*. Prepublication (2018). arXiv:1810.11852

[Sun] B. Sun, Bounding matrix coefficients for smooth vectors of tempered representations. Proc. AMS **137**(1), 353–357 (Jan 2009)

[Ta] J. Tate, Number theoretic background, in automorphic forms, representations, and L-functions. Proc. Symposia Pure Math. AMS **33**, 3–26 (1979)

[Tr] F. Trèves, *Topological Vector Spaces, Distributions and Kernels* (Academic Press, New York, London, 1967), xvi+624 pp.

[Var] V.S. Varadarajan, *Harmonic Analysis on Real Reductive Groups*. Lecture Notes in Mathematics, Vol. 576 (Springer, Berlin, New York, 1977), v+521 pp.

[Vog] D. Vogan, Gel'fand-Kirillov dimension for Harish-Chandra modules. Invent. Math. **48**, 75–98 (1978)

[VS] D. Vogan, B. Speh, Reducibility of generalized principal series representations. Acta Math. **145**(3–4), 227–299 (1980)

[Wald1] J.-L. Waldspurger, La formule de Plancherel pour les groupes p-adiques (d'après Harish-Chandra). J. Inst. Math. Jussieu **2**(2), 235–333 (2003)

[Wald2] J.-L. Waldspurger, Une formule intégrale reliée à la conjecture locale de Gross-Prasad, 2e partie: extension aux représentations tempérées, in Sur les conjectures de Gross et Prasad. I. Astérisque **346**, 171–312 (2012)

[Wall] N.R. Wallach, *Real Reductive Groups I*. Pure and Applied Mathematics, vol. 132 (Academic Press, Boston, MA, 1988), xx+412 pp.

[Wall2] N.R. Wallach, *Real Reductive Groups II*. Pure and Applied Mathematics, vol. 132-II (Academic Press, Boston, MA, 1992), xiv+454 pp. ISBN:0-12-732961-7

[Wall3] N. Wallach, *On the Constant Term of an L^2-Automorphic Form*. Operator Algebras and Group Representations, vol. II (Monog. Stud. Math., Vol. 18) (Pittman, London, 1984), pp. 227–237

[Zel] A.V. Zelevinsky, Induced representations of reductive p-adic groups. II. On irreducible representations of GL(n). Ann. Sci. École Norm. Sup. (4) **13**(2), 165–210 (1980)

The Relative Trace Formula in Analytic Number Theory

Valentin Blomer

Abstract We discuss a variety of applications of the relative trace formula of Kuznetsov type in analytic number theory and the theory of automorphic forms.

2010 Mathematics Subject Classification Primary 11F72

1 The Poisson Summation Formula

The mother of all trace formulae is the Poisson summation formula. If $f : \mathbb{R} \to \mathbb{C}$ is a sufficiently nice function, say Schwartz class for simplicity, then

$$\sum_{n\in\mathbb{Z}} f(n) = \sum_{n\in\mathbb{Z}} \widehat{f}(n) \tag{1}$$

where

$$\widehat{f}(y) = \int_{-\infty}^{\infty} f(x)e(-xy)\,\mathrm{d}x, \quad e(x) = e^{2\pi i x}$$

is the Fourier transform. Applying (1) formally with $f(x) = x^{-s}$, one obtains the functional equation of the Riemann zeta function, and this heuristic argument can easily be made rigorous. More generally, the Poisson summation formula implies the functional equations for all Dirichlet L-functions [IK, Section 4.6], and is in fact essentially a re-statement of those.

Author partially supported by DFG grant BL 915/2-2.

V. Blomer (✉)
Mathematisches Institut, Bonn, Germany
e-mail: blomer@math.uni-bonn.de

W. Müller et al. (eds.), *Relative Trace Formulas*, Simons Symposia,
https://doi.org/10.1007/978-3-030-68506-5_2

Why is the Poisson summation formula a trace formula? Given a function f as above, we consider the convolution operator

$$L_f : L^2(\mathbb{R}/\mathbb{Z}) \to L^2(\mathbb{R}/\mathbb{Z})$$

given by

$$L_f(g)(x) = \int_{\mathbb{R}} f(x-y)g(y)\,dy = \int_{\mathbb{R}/\mathbb{Z}} k(x,y)g(y)\,dy$$

where

$$k(x,y) = \sum_{n\in\mathbb{Z}} f(x-y+n). \tag{2}$$

The functions $e_n(x) = e(nx)$ are an orthonormal basis of Eigenfunctions: $L_f(e_n) = \widehat{f}(n)e_n$. Computing the trace of L_f in two ways we obtain

$$\sum_{n\in\mathbb{Z}} \widehat{f}(n) = \int_{\mathbb{R}/\mathbb{Z}} k(x,x)\,dx = \sum_{n\in\mathbb{Z}} f(n).$$

An analogous formula can be established for the multi-dimensional torus $(\mathbb{R}/\mathbb{Z})^n$. The Poisson summation formula is one of the most frequently used tools in analytic number theory, and we give two standard applications:

(a) *The Pólya-Vinogradov inequality* ([Po, Vi], see [IK, Section 12.4] for a modern treatment). If χ is a primitive Dirichlet character modulo q, then

$$\sum_{n\leq x} \chi(n) \ll q^{1/2}\log q$$

for all $x \geq 1$. This shows cancellation in character sums of length slightly larger than the square-root of the modulus. The proof is simple: we split the sum over n into residue classes modulo q, smooth the characteristic function on $[1, x]$ a little bit and apply the Poisson summation formula. This leads to the bound

$$\sum_{n\leq x} \chi(n) \ll 1 + \frac{1}{\sqrt{q}}\Big|\sum_{n\in\mathbb{Z}} \bar{\chi}(n)w(n)\Big|$$

where $w(n) \ll \min(q/|n|, q^2/|n|^2)$. The zero frequency $n = 0$ vanishes, and a trivial estimate on the non-zero frequencies gives the desired bound.

(b) *The number of lattice points in growing discs* ([Sz], see [IK, Section 4.4] or [Co, Section 10.2.6] for a modern treatment with complete details). Given a disc of radius $\sqrt{R}$ centered at the origin, it contains about πR lattice points as $R \to \infty$. The error term can easily be bounded $O(R^{1/2})$ (the circumference of the disc), but an application of the two-dimensional Poisson summation formula improves this error to $O(R^{1/3})$. Here the integral transform involves a Bessel function J_0 as an integral kernel. This can also be interpreted as an instance of Weyl's law on the flat torus.

The experienced reader will observe that both applications are essentially nothing but applications of functional equations of suitable zeta functions.

2 Harmonic Analysis on Upper Half Plane

The group $(\mathbb{R}/\mathbb{Z})^n$ is abelian, and for the rest of this note we will consider spectral summation formulae for non-abelian groups. In this section the underlying group is $G = \mathrm{SL}_2(\mathbb{R})$ and for simplicity we fix the lattice $\Gamma = \mathrm{SL}_2(\mathbb{Z})$. For $K = \mathrm{SO}(2)$, the quotient is $X := \Gamma\backslash G/K$ is in natural bijection with the quotient of the upper half plane $\mathbb{H} = \{z \in \mathbb{C} \mid \Im z > 0\}$ by Γ. The space $L^2(X)$ is acted on by various self-adjoint and pairwise commuting operators: the hyperbolic Laplace operator

$$\Delta = -y^2(\partial_x^2 + \partial_y^2),$$

its non-Archimedean counterparts, the Hecke operators

$$T_n(\phi)(z) = \frac{1}{\sqrt{n}} \sum_{ad=n} \sum_{b \,(\mathrm{mod}\, d)} \phi\left(\frac{az+b}{d}\right)$$

for $n \in \mathbb{N}$, and the reflection operator $T_{-1}(\phi)(z) = \phi(-\bar{z})$. Since X is not compact, we have a spectral decomposition

$$L^2(X) = L^2_{\mathrm{disc}}(X) \oplus L^2_{\mathrm{cont}}(X) = L^2_{\mathrm{cusp}}(X) \oplus L^2_{\mathrm{res}}(X) \oplus L^2_{\mathrm{cont}}(X)$$

where $L^2_{\mathrm{cusp}}(X)$ is the subspace of $\phi \in L^2(X)$ satisfying $\int_0^1 \phi(x+iy)\,dx = 0$ for all $y > 0$, $L^2_{\mathrm{cont}}(X)$ is the continuous spectrum spanned by incomplete Eisenstein series, and in our case $L^2_{\mathrm{disc}}(X)$ consists only of the one-dimensional space of constant functions. A non-constant joint Eigenfunction of Δ, all T_n ($n \in \mathbb{N}$) and T_{-1} with respective eigenvalues $\lambda_\phi = t_\phi^2 + 1/4$, $\lambda_\phi(n)$ with $n \in \mathbb{N}$, and $\epsilon_\phi \in \{\pm 1\}$ has a Fourier expansion

$$\phi(x+iy) = \text{constant term } + \sum_{n \neq 0} \frac{\lambda_\phi(n)}{\sqrt{|n|}} e(nx) W_{it_\phi}(4\pi |n| y)$$

where $W_{it}(y) = W_{0,it}(y)$ is the standard Whittaker function and for $n < 0$ we define $\lambda_\phi(n) = (-1)^{(1-\epsilon_\phi)/2}\lambda_\phi(-n)$. By definition, the constant term vanishes if and only if $\phi \in L^2_{\mathrm{cusp}}(X)$. The fact that the Hecke eigenvalues are essentially the

Fourier coefficients of ϕ is a specialty of the group GL(2), this is not the case for general groups.

We proceed to derive a sort of analogue of the Poisson summation formula. Given a sufficiently nice function $f : K\backslash G/K \to \mathbb{C}$, say smooth and compactly supported, we construct a kernel function $K : X \times X \to \mathbb{C}$ by

$$K(x_1, x_2) = \sum_{\gamma \in \Gamma} f(x_1^{-1}\gamma x_2), \tag{3}$$

which is is the analogue of (2) and gives rise to a convolution operator $L_f(g)(x_1) = \int_X K(x_1, x_2)g(x_2)\,\mathrm{d}x_2$. Again we try to compute the trace of L_f in two ways: as the sum over the eigenvalues and as the integral of the kernel over the diagonal. As our space X is non-compact, there is a convergence problem that is usually solved by applying a suitable truncation operator. The integral of the kernel is computed explicitly by splitting the sum over $\gamma \in \Gamma$ into conjugacy classes. In this way one obtains (the most basic form of) the Selberg trace formula as a sum over the spectrum on the one side and a sum over conjugacy classes of matrices $\gamma \in \Gamma$ on the other, which features in particular class numbers and regulators of quadratic number fields in its hyperbolic terms. That the arithmetic of quadratic number fields arises is not surprising as 2-by-2 matrices have quadratic minimal polynomials. See e.g. [Str, Proposition 2] for a general and completely explicit version of this formula.

For the purpose of analytic number theory, the following variation of this procedure is very useful and leads to the Bruggeman-Kuznetsov formula [Br, Ku]. Instead of integrating $K(x_1, x_2)$ over the diagonal $x_1 = x_2 \in X$, one can integrate it over $N(\mathbb{Z})\backslash N(\mathbb{R}) \times N(\mathbb{Z})\backslash N(\mathbb{R})$ (with $N = \{(\begin{smallmatrix} 1 & * \\ & 1 \end{smallmatrix})\}$) against a character of this group. More precisely, we consider

$$\int_0^\infty \int_0^1 \int_0^1 K\left(\xi_1 + i\frac{y}{|n|}, \xi_2 + i\frac{y}{|m|}\right) e(-\xi_1 n + \xi_2 m)\,\mathrm{d}\xi_1\,\mathrm{d}\xi_2 \frac{\mathrm{d}y}{y}$$

for $n, m \in \mathbb{Z}\backslash\{0\}$. Using the spectral expansion of K, the unipotent integrals feature the n-th and m-the Fourier coefficient of members of an orthonormal basis of $L^2(X)$. On the other hand, one can insert the definition (3) of K and compute the integral using the Bruhat decomposition of $\Gamma = \mathrm{SL}_2(\mathbb{Z})$, which in the present case can be stated very elementarily as

$$\mathrm{SL}_2(\mathbb{Z}) = N(\mathbb{Z}) \cup \bigcup_{c \in \mathbb{N}} \bigcup_{\substack{d \,(\mathrm{mod}\, c) \\ (d,c)=1}} N(\mathbb{Z}) \begin{pmatrix} * & * \\ c & d \end{pmatrix} N(\mathbb{Z}). \tag{4}$$

The details, due to Zagier, can be found in [Jo, Section 1]. This identifies the Kuznetsov formula as a relative trace formula in the sense of Jacquet. The final formula reads as follows:

$$2\pi \sum_{\phi\in\mathcal{B}} \frac{\lambda_\phi(n)\lambda_\phi(m)}{L(1,\mathrm{Ad}^2\phi)} h(t_\phi) + \int_{\mathbb{R}} \frac{\sigma_{it}(n)\sigma_{it}(m)}{|\zeta(1+2it)|^2} h(t)\,\mathrm{d}t$$
$$= \delta_{n,m} \int_{\mathbb{R}} h(t)\tanh(\pi t)t\,\mathrm{d}t + \sum_c \frac{1}{c} S(n,m,c)\widehat{h}_{\pm}\left(\frac{nm}{c^2}\right) \tag{5}$$

where $n, m \in \mathbb{Z} \setminus \{0\}$, h is a sufficiently nice test function, $\mathcal{B}$ is an orthonormal basis of joint Eigenfunction of $L^2_{\mathrm{cusp}}(X)$, $\sigma_{it}(n) = \sum_{ad=|n|}(a/d)^{it}$ is the Hecke eigenvalue of the corresponding Eisenstein series,

$$S(n,m,c) = \sum_{\substack{d \pmod c\\ (d,c)=1}} e\left(\frac{nd+m\bar{d}}{c}\right) \tag{6}$$

is the standard Kloosterman sum, and $\widehat{h}_{\pm}$ is a certain integral transform of h given by a Bessel kernel depending on the sign $\pm = \mathrm{sgn}(nm)$. The right hand side of (5) reflects directly the decomposition (4). The appearance of a Bessel function is not a coincidence, it is an Archimedean analogue of the Kloosterman sum, as can be seen, for instance by the formula [GR, 8.432.7]

$$K_0\left(\frac{\sqrt{nm}}{c}\right) = \frac{1}{2}\int_{\mathbb{R}} \exp\left(-\frac{1}{c}\left(nx+\frac{m}{x}\right)\right)\frac{\mathrm{d}x}{x}.$$

The adjoint L-function at 1 is proportional to the square of the L^2-norm of a Hecke-normalized cusp form, and $|\zeta(1+2it)|^2$ is the corresponding regularized version for Eisenstein series.

There is a different way of proving the formula (5), which is essentially Kuznetsov's original argument and analogous to the proof of the related Petersson formula for holomorphic cusp forms [IK, Proposition 14.5]. The linear form $L^2(X) \to \mathbb{C}$, $\phi \mapsto \lambda_\phi(n)$ (say for $n > 0$) has a kernel that is essentially given by a Poincaré series

$$P_n(z) = \sum_{\gamma\in N(\mathbb{Z})\backslash\Gamma} f(n\cdot \Im\gamma z)e(n\cdot\Re\gamma z)$$

for a suitable function f. Indeed, by unfolding we have

$$\langle\phi, P_n\rangle = \lambda_\phi(n)\sqrt{n}\int_0^\infty W_{it_\phi}(4\pi y)\overline{f(y)}\frac{\mathrm{d}y}{y^2}.$$

Again by the Bruhat decomposition, one can compute the Fourier coefficients of P_n in terms of Kloosterman sums. Computing now the inner product $\langle P_n, P_m\rangle$ by Parseval on the one hand and by unfolding on the other one derives (5). It requires some analytic virtuosity with integral transforms and Bessel functions to play with the test function f so that the left hand side of (5) features the given function h and to compute the Bessel transform $\widehat{h}$.

The shape of the Kuznetsov formula and the Selberg trace formula is similar in some respects, but there are also some important differences:

- The spectral side of the Kuznetsov formula does not contain the residual spectrum, in particular the discrete spectrum is cupsidal.
- The spectral side of the Kuznetsov formula is weighted by an L-value at the edge of the critical strip. For the group GL(2), such L-values can be removed and included with relatively little effort, see e.g. [IK, Proposition 26.15] for a prototype. For applications to L-functions involving period formulae it is often desirable to have an additional factor $1/L(1, \mathrm{Ad}^2\phi)$ in the cuspidal spectrum, but in other situations one may prefer a summation formula without an extra L-value.
- The Kloosterman sums in the Kuznetsov formula are easier to handle from the point of view of analytic number theory than the class numbers that appear in the Selberg trace formula. This is an important advantage of the Kuznetsov formula in practice (see the applications below). Theoretically, it is possible to switch directly between class numbers and sums of Kloosterman sums (cf. [SY, Theorem 1.3] or [Al]). This is ultimately "just" Poisson summation, yet the passage is rather subtle.

Unlike the Poisson summation formula, the Selberg trace formula and the Kuznetsov formula are asymmetric in the sense that the spectral side and the geometric side have a very different shape. It is an important feature that the Kuznetsov formula can nevertheless be inverted. For a sufficiently nice test function H and integers $n, m \in \mathbb{Z} \setminus \{0\}$, the sum

$$\sum_c \frac{1}{c} S(n, m, c) H\left(\frac{nm}{c^2}\right)$$

can be expressed in terms of Fourier coefficients of automorphic forms of the form

$$\sum_{\phi \in \mathcal{B}} \frac{\lambda_\phi(n)\lambda_\phi(m)}{L(1, \mathrm{Ad}^2\phi)} \check{H}(t_\phi)$$

for a suitable transform $\check{H}$ of H (essentially the inverse of $h \mapsto \hat{h}_\pm$), along with two similar sums involving the holomorphic spectrum and the Eisenstein series. It is the subtle inversion of the Sears-Titchmarsh transform $h \mapsto \widehat{h}_+$ (see [ST]) that requires the *entire* GL(2) spectrum including holomorphic cusp forms, see [IK, Section 16.4] for details. For the transform $h \mapsto \widehat{h}_-$, one needs the simpler Kontorovich-Lebedev inversion formula [IK, (16.46)], where no holomorphic contribution shows up (which is consistent with the fact that holomorphic forms have no negative Fourier coefficients). As we will see in the next section, the fact that (5) can also be read meaningfully from right to left is absolutely crucial for applications in analytic number theory and can hardly be underestimated.

3 Applications

In this section we present a variety of applications of the Kuznetsov formula (5) and generalizations thereof to congruence subgroups, number fields and half-integral weight automorphic forms. They are roughly ordered in a thematic way, although there is no clear separation. Some applications use (5) from left to right, some from right to left and some in both directions usually with a Voronoi step or a Cauchy-Schwarz step in between. Some of the problems are seemingly completely unrelated to the theory of automorphic forms and it is the sometimes accidental appearance of Kloosterman sums that opens the door. Needless to say that Kuznetsov formula is only one of many tools to solve many of the below mentioned problems, but it is a crucial one. Each of these problems deserves a survey article of its own, and I can only sketch a few ideas. I hope that the interested reader will consult the referenced literature for more details.

3.1 *Statistics of eigenvalues*

(a) *Vertical Sato-Tate laws.* Given a Hecke Eigenform ϕ, the statistical distribution of the eigenvalues $\lambda_\phi(p)$ as p varies over primes has received a lot of attention. For holomorphic cusp forms it is known to follow the Sato-Tate distribution [BGHT]

$$\frac{\log P}{P} \sum_{\substack{p \leq P \\ p \text{ prime}}} f(\lambda(p)) \to \frac{1}{2\pi} \int_{-2}^{2} f(x)\sqrt{4 - x^2}\, \mathrm{d}x, \qquad P \to \infty$$

(f a compactly supported continuous function), but for Maaß forms this is wide open. One can modify the question, fix the prime p and ask instead for the statistical distribution as one varies over the spectrum. This is known as a "vertical Sato-Tate law". It was first addressed by Sarnak [Sa1] who showed using the Selberg trace formula that

$$\frac{12}{T^2} \sum_{\lambda_\phi \leq T} f(\lambda_\phi(p)) \to \frac{1}{2\pi} \int_{-2}^{2} f(x)\sqrt{4 - x^2} \frac{p+1}{p + 2 + \frac{1}{p} - x^2}\, \mathrm{d}x, \qquad T \to \infty,$$

for a fixed prime p and f as above. Serre [Se] showed similar results for holomorphic forms using the Eichler trace formula. Note that if p tends to infinity, this approaches the semicircle distribution. This question can also be addressed with the Kuznetsov formula (see e.g. [BBR, Proposition 2]), in which case the spectral sum is naturally weighted by an L-value:

$$\frac{12}{T^2}\sum_{\lambda_\phi \le T} f(\lambda_\phi(p))\frac{\zeta(2)}{L(1,\mathrm{Ad}^2\phi)} \to \frac{1}{2\pi}\int_{-2}^{2} f(x)\sqrt{4-x^2}\,\mathrm{d}x, \quad T\to\infty.$$

Interestingly, this slightly different counting procedure produces the semicircle distribution "on the nose". For the proof, we assume by a standard approximation argument that $f(x) = x^k$ for some $k \in \mathbb{N}$, apply the Hecke relations and estimate the Kloosterman term in the Kuznetsov formula for instance by the Weil bound (8) below.

(b) *Density results for Maaß forms violating the Ramanujan conjecture.* The previous two asymptotics show in particular that the Ramanujan conjecture at p is satisfied "with probability one". By a small variation of the argument, this can be made quantitative as follows ([BBR, Proposition 1]): for a prime $p \le T$, $\alpha > 2$ and $\varepsilon > 0$ we have

$$\#\{\lambda_\phi \le T^2 : |\lambda_\phi(p)| \ge \alpha\} \ll_\varepsilon T^{2-\frac{8\log(\alpha/2)}{\log p}+\varepsilon}. \tag{7}$$

Upon choosing $\alpha = p^{1/4+\varepsilon}$, this recovers in particular the Selberg-type bound $\lambda_\phi(p) \ll p^{1/4+\varepsilon}$ (since the right hand side of (7) is less than 1, so the set on the left hand side must be empty). It is at this point where the Kuznetsov formula is advantageous compared to the Selberg trace formula [Sa1, Theorem 1]. The latter would replace the constant 8 in the exponent only by 4,[1] and the corresponding version of (7) would only interpolate between the trivial representation (with $\lambda_{\text{triv}}(p) = p^{1/2} + p^{-1/2}$) and the tempered spectrum, that is, the bound is trivial on the tempered spectrum $\alpha = 2$ and excludes the possibility of $\lambda_\phi(p)| \ge p^{1/2+\varepsilon}$. Had we used trivial bounds on the Kloosterman sum, the Selberg trace formula and Kuznetsov formula would produce the same result, but the option of getting extra cancellation from the Weil bound (8) makes the Kuznetsov formula advantageous in this situation.

This type of argument works also at the infinite place. Using the test function $h(t) = (X^{it} + X^{-it})^2(t^2+1)^{-2}$ that blows up at the exceptional spectrum, one obtains Selberg's $3/16$ bound as a lower bound for non-trivial eigenvalues of the Laplacian on $\Gamma\backslash\mathbb{H}$, see [IK, (16.58)] as well as density results for exceptional eigenvalues of large level.

(c) *Large sieve estimates.* A different kind of statistical result is provided by large sieve estimates that show in some sense orthogonality of Hecke eigenvalues against arbitrary families of sequences. More precisely, if a_n is any sequence of complex numbers, then [DI1, Theorem 2]

$$\sum_{t_\phi \le T}\Big|\sum_{n\le N} a_n\lambda_\phi(n)\Big|^2 \ll_\varepsilon (T^2+N)^{1+\varepsilon}\sum_{n\le N}|a_n|^2$$

[1]Which follows from Sarnak's argument [Sa1] with small modifications.

for all $\varepsilon > 0$. This is essentially best possible, since the first term on the right hand side is achieved when all n-sums have square-root cancellation and the second term is achieved if a_n is chosen to be $\overline{\lambda_{\phi_0}(n)}$ for some fixed ϕ_0. Note that a simple application of Cauchy's inequality would replace $T^2 + N$ with $T^2 N$. Such general estimates are very useful in analytic number theory. One of the most common applications is a bound for moments of L-functions of a family $\mathcal{F}$ that recovers in many cases the Lindöf hypothesis on average over the family if the conductor of the L-functions is at most $|\mathcal{F}|^4$ (this follows from an approximate functional equation). The proof of the large sieve inequality is a double application of the Kuznetsov formula. Opening the square, one applies the Kuznetsov formula to the t_ϕ-sum. After an application of Cauchy's inequality, one can re-arrange the Kloosterman term to make it amenable to a second application of the Kuznetsov formula in the other direction. This is not involutory because of Cauchy's inequality in between.

3.2 *Arithmetic Applications: Kloosterman Sums, Primes and Arithmetic Functions*

(a) *Linnik's problem: cancellation in sums of Kloosterman sums.* Kloosterman sums as in (6) are ubiquitous in number theory. They first came up in Kloosterman's investigation [Klo] of quaternary quadratic forms by the circle method. Weil's bound [We]

$$|S(m, n, c)| \leq \tau(c)(m, n, c)^{1/2} c^{1/2} \tag{8}$$

(where τ denotes the divisor function) is best possible for individual sums, but quite often in number theory *sums of Kloosterman sums*

$$\mathcal{S}_{m,n}(X; W) := \sum_{c} S(n, m, c) W(c/X)$$

over the modulus c come up for some fixed smooth function W and some large parameter X. Weil's bound yields immediately $\mathcal{S}_{m,n}(X; W) \ll_{W,m,n} X^{3/2} \log X$, but the Kuznetsov formula (along with the fact that $\mathrm{SL}_2(\mathbb{Z})$ has no exceptional eigenvalues, see [DI1, p. 261] for a quick proof) provides substantial cancellation in this sum:

$$\mathcal{S}_{m,n}(X; W) \ll_{W,m,n,\varepsilon} X^{1+\varepsilon} \tag{9}$$

for all $\varepsilon > 0$. This is very remarkable because it goes far beyond what algebraic geometry can achieve and it is one of the very few examples where exponential sums can be averaged non-trivially over the modulus. The bound (9) and generalizations thereof, essentially due to Kuznetsov [Ku] and developed in

[DI1], were historically the first applications of (5) and the cornerstone for all applications in this subsection.

(b) *Shifted convolution sums.* A typical question in number theory concerns the correlation of additive shifts of multiplicative function. A standard example is an asymptotic evaluation of

$$\sum_{n\leq x} \tau(n)\tau(n+1) = \sum_{a\leq x} \sum_{\substack{m\leq x\\ m\equiv 1 \,(\mathrm{mod}\, a)}} \tau(m)$$

(as can be seen by opening the first divisor function and writing $n = ab$). The inner sum can be transformed by the Voronoi summation formula into a sum over Kloosterman sums $S(m, 1, a)$. It is then the Kuznetsov formula that can evaluate the a-sum over Kloosterman sums with respect to the modulus very precisely and yields the strongest error terms.[2] The (somewhat involved) details can for instance be found in [Meu]. If the sum over x is smooth, one obtains square root cancellation in the error term. Such estimates and generalizations thereof play an important role in the theory of the Riemann zeta-function (see below).

(c) *Primes in long arithmetic progressions.* The distribution of prime numbers

$$\pi(x; q, a) := \#\{p \leq x \mid p \equiv a \,(\mathrm{mod}\, q),\, p \text{ prime}\}$$

in arithmetic progressions with $(a, q) = 1$ is a very delicate topic due to possible Siegel zeros of Dirichlet L-functions. For many applications one can get away with a few arithmetic progressions in which primes are potentially ill-distributed if the majority of progressions is well-behaved. The standard result of this type is the Bombieri-Vinogradov theorem [IK, Theorem 17.1] which essentially states that $\pi(x; q, a) \sim \phi(q)^{-1}\pi(x; 1, 0)$ for most pairs a, q with $(a, q) = 1$ as long as $q \leq x^c$ for some $c < 1/2$. Going beyond $c > 1/2$ can be crucial in applications. In a sequence of papers culminating in [BFI], Bombieri, Friedlander, Fouvry and Iwaniec managed to achieve this at least in certain situations, a crucial ingredient being non-trivial bounds for sums over (incomplete) Kloosterman sums as they follow from the Kuznetsov formula. Although these extended versions of the Bombieri-Vinogradov theorem have some restrictive conditions (e.g. a has to be fixed), it is remarkable that they go beyond the ranges that would be covered by the Riemann hypothesis. This can hardly be underestimated and is the basis of the following application.

(d) *The first case of Fermat's last theorem.* If p is a prime, then the first case of Fermat's last theorem gives insolubility of the equation $x^p + y^p + z^p = 0$ in integers x, y, z such that $p \nmid xyz$. Adleman and Heath-Brown [AHB] showed

---

[2] It is also possible to detect the condition additive shift by some variant of the circle method. After some elementary Fourier analysis this also leads to Kloosterman sums. Although relatively similar, in many cases the above approach has slight advantages.

that this follows for infinitely many primes (in a strong quantitative sense) if a certain estimate on primes in arithmetic progression holds; this was proved by Fourvy [Fo] based (among other things) on the work of Bombieri-Friedlander-Iwaniec [BFI].

(e) *Proportion of zeta zeros on the critical line.* Selberg proved in 1942 that a positive proportion of zeros lie on the critical line $\Re s = 1/2$. About 30 years later, Levinson showed by a different method that in fact at least 1/3 of all zeros lie on the critical line (see [Iv, Section 10.1] for a brief sketch of the method). It ultimately depends on computing the mean square of a linear combination of $\zeta(s)$ and its derivatives multiplied by a carefully chosen Dirichlet polynomial whose length is responsible for the numerical value of the proportion. Based on mean values for (incomplete) Kloosterman sums following from (5), Conrey [Con] improved the proportion of zeros on the critical line to slightly over 40%.[3]

(f) *The largest prime factor of n^2+1.* It is an old conjecture that there are infinitely many primes of the form $n^2 + 1$, but a proof is unfortunately far out of reach by current methods. As an approximation, one can consider the largest prime factor of $n^2 + 1$. It is trivial that this can be at least as big as n infinitely often.[4] Chebychev indicated a method to show that it is bigger than Cn infinitely often for every given constant $C > 1$. This was refined by Hooley. The key point is that the distribution of $n^2 + 1$ in residue classes leads to Kloosterman sums (cf. [DI2, Lemma 2]): one has

$$\sum_{n^2+1\equiv 0 \,(\mathrm{mod}\, m)} e\left(\frac{nh}{m}\right) \approx \sum_{\substack{r^2+s^2=m\\ r,s>0,\,(r,s)=1}} e\left(\frac{h\bar{r}}{s}\right)$$

which after Poisson summation in r gives Kloosterman sums. Using bounds on sums of Kloosterman sums, Deshouillers and Iwaniec showed [DI2] that the largest prime factor of $n^2 + 1$ is infinitely often at least $n^{6/5}$. At the time of writing, the current record is $n^{1.279}$ [Mer].

(g) *Equidistribution of roots of quadratic congruences.* The previous example is ultimately concerned with the equidistribution of roots of quadratic congruences, a topics that goes back (in principle to Gauß and) Hooley, Hejhal, Bykovskiĭ, Zavorotny and others with the strongest results based on the spectral theory of automorphic forms. Duke, Friedlander and Iwaniec [DFI] went one step further and coupled this with a sieve to show that the equidistribution property remains true among *prime* moduli: let f be an integral quadratic polynomial without real roots and $h \neq 0$. Then

[3] At the time of writing, the current record is slightly over 41%.

[4] For any prime $p \equiv 1 \pmod 4$ choose $1 \leq n < p$ with $n^2 \equiv -1 \pmod p$.

$$\sum_{\substack{p\le x\\ p \text{ prime}}} \sum_{\substack{\nu \,(\mathrm{mod}\, p)\\ p\mid f(\nu)}} e\left(\frac{h\nu}{p}\right) = o\Big(\sum_{p\le x} 1\Big)$$

as $x \to \infty$. A beautiful set of lecture notes proving this result *ab ovo* is [Ko]. The case of polynomials with real roots is treated in [To].

(h) *Prime geodesic theorem.* The prime geodesics on the upper half plane play a similar role for the Selberg zeta functions as the primes for the Riemann zeta function. In particular they satisfy a "prime number theorem", which follows from a standard application of the Selberg trace formula: if p runs through the lengths of prime geodesics on $\Gamma\backslash\mathbb{H}$, then

$$\sum_{p\le X} \log p = X + O(X^{3/4}).$$

Henryk Iwaniec observed that an *additional* application of the Kuznetsov formula can improve the error term. These ideas have been refined, and the current record [SY] is an error term of $O(X^{25/36+\varepsilon})$ with $25/36 = 0.69\ldots$ which uses the Kuznetsov formula again, but rather indirectly through the work of Conrey-Iwaniec [CI].

3.3 Applications to L-Functions I

L-function occur naturally in families. While their individual behaviour is often hard to understand, their statistical properties in families are more amenable to the tools of analytic number theory. If the family is given by spectral properties, spectral summation formulae such as the Kuznetsov formula belong to the key tools to study analytic properties of L-functions in various average senses. Therefore all of the following applications of the Kuznetsov formula are concerned with moments of L-functions in one way or another. Usually the L-functions are encoded into the Kuznetsov formula by means of an approximate functional equation, but in principle it is also possible to use a suitable integral representation. This area is a huge industry, and the following discussion can highlight, inevitably, only a very small selection of results.

(a) *Symmetry types of L-functions.* In [KS], Katz and Sarnak introduced the notion of symmetry type for a family of L-functions. A common approach to this determination is to analyze the density of low-lying zeros in the specified family which should mimic the appropriate random matrix model. With current technology this is only possible for test functions whose Fourier transforms have restricted support. The most frequently used quantity is the so-called one-level density which essentially counts the number of zeros within $1/\log C$ of the origin (where C is the conductor of the L-function). The key tools here are a spectral summation formula (like the Kuznetsov formula or the Petersson formula for holomorphic forms) and Weil's explicit formula (see e.g. [IK, Theorem 5.11]) to convert zeros into primes, and the results typically assume the Riemann hypothesis. One of the most influential papers in this direction is [ILS], covering many families of L-functions. The one-level density is not unrelated to the analytic rank of the L-function (i.e. its order at $s = 1$). Applications to ranks of elliptic curves are given for instance in [Yo1].

(b) *Spectral expansion of the fourth moment of the Riemann zeta function.* Motohashi observed that for a Schwartz class function w, the fourth moment

$$\int_{-\infty}^{\infty} |\zeta(1/2+it)|^4 w(t)\, \mathrm{d}t$$

of the Riemann zeta function has a beautiful expansion as a spectral sum of third powers of central L-values of automorphic forms for $\mathrm{SL}(2,\mathbb{Z})$, see [Mo] for a detailed and self-contained proof and extensive discussion. This formula is quite remarkable from an aesthetic point of view, relating two rather different families of L-functions in an exact identity, but it is also very useful theoretically and gives the strongest bounds for the fourth moment in short intervals (by choosing w appropriately and analyzing the dual side of the formula). The Kuznetsov formula enters after opening the fourth moment: using an approximate functional equation, we obtain an expression very roughly of the form

$$\int_{-\infty}^{\infty} \sum_{n,m \ll (1+|t|)^{1/2}} \frac{\tau(n)\tau(m)}{n^{1/2+it} m^{1/2-it}} w(t)\, \mathrm{d}t.$$

Here it is convenient to sum the diagonals and write $m = n + h$, so that shifted convolution sums of the type $\sum_n \tau(n)\tau(n+h)$ (with some weights) emerge.

(c) *Beyond endoscopy.* Given a reductive algebraic group G, the cuspidal automorphic representations π of $G(\mathbb{A})$ are expected to be functorial transfers from other groups if $L(s,\pi,\rho)$ has a pole for some finite dimensional representation $\rho : {}^LG \to \mathrm{GL}_n(\mathbb{C})$. Langlands [La] put forward the idea to study this by the trace formula. The idea is to consider a trace formula whose spectral side is weighted by $\mathrm{ord}_{s=1} L(s,\pi,\rho)$ and to try to match the geometric side with the geometric side of the corresponding other group. Sarnak suggested to study this with the Kuznetsov formula instead. In [Ve], Venkatesh carried out this programme with the Kuznetsov formula for the case $G = \mathrm{GL}(2)$ and ρ the symmetric square representation, which indeed classifies precisely the dihedral representations. It would be very interesting to study higher rank groups or higher dimensional representations ρ, and we refer the reader also to [Al] where Langlands' ideas of "Beyond endoscopy" are investigated with the Selberg trace formula.

(d) *Non-vanishing of L-functions.* Evaluating asymptotically a moment of a family of L-functions can prove that one, infinitely many, or even a large proportion of members in this family are non-zero. Whether or not an L-function vanishes at a certain point in the critical strip (often the central point $s = 1/2$) is an important question, sometimes with unexpected applications.

If $\Gamma \subseteq \mathrm{SL}_2(\mathbb{Z})$ is a finite-volume, but non-cocompact subgroup, all but finitely many eigenvalues are embedded in the continuous spectrum, and it is in general unclear how to prove a Weyl law for the discrete spectrum alone. In fact, it may well be that there is no discrete spectrum at all except for the constant function. Philipps and Sarnak [PhS] considered deformations given by curves in the Teichmüller space of Γ, and for congruence subgroups Γ they gave a criterion in terms of non-vanishing of L-functions which tells us (conditionally under the assumption of the simplicity of the spectrum of $L^2(\mathrm{SL}_2(\mathbb{Z})\backslash\mathbb{H})$) that such deformations destroy cusp forms, i.e. the eigenvalue becomes a pole of the scattering matrix.[5] Using the Kuznetsov formula, Luo [Lu], evaluated a mollified second moment of the respective Rankin-Selberg L-functions to show that indeed a positive proportion of these L-values doesn't vanish.

Also higher rank examples can be treated with Kuznetsov formula (5): in [BLM] it is shown that for any self-dual cuspidal automorphic representation Π on $\mathrm{GL}(4, \mathbb{Q})\backslash\mathrm{GL}(4, \mathbb{A})$ which is unramified at all finite places, there exist infinitely cusp forms π for $\mathrm{SL}(2, \mathbb{Z})$ such that $L(1/2, \Pi \times \pi) \neq 0$. This is the consequence of a certain reciprocity formula relating different families of L-functions in an exact identity.

3.4 Applications to L-Functions II: Subconvexity and Equidistribution

The Generalized Riemann Hypothesis implies best possible bounds for L-functions on the critical line (cf. [IK, Corollary 5.20]). In absence of a proof of GRH, it is a very interesting and very hard problem to obtain bounds for L-functions that are better than the generic convexity bound that is implied by the functional equation. Such bounds can often imply that "something or other is equidistributed" [Frl, p. 373], and we give a number of examples where the corresponding subconvexity bound is a consequence of the Kuznetsov formula. From a technical and conceptual point of view, these are often the hardest applications of the Kuznetsov formula, since the relative trace formula is coupled with various other transformation formulae (Voronoi, Poisson, or even another application of the Kuznetsov formula), so that extremely precise knowledge on the exact shape of the various terms in the formula is required.

(a) *Sums of three squares.* An asymptotic formula for the number of solutions to $x^2 + y^2 + z^2 = n$ should in principle follow from Siegel's mass formula:

[5] This can be phrased as an analogue of Fermi's golden rule.

the ternary theta series is decomposed into an Eisenstein part that gives the main term as a product of local densities, and a cuspidal part, whose Fourier coefficients should be substantially dominated by the main term and therefore constitute an error term. There are many subtleties for positive ternary quadratic forms (see e.g. [Bl1] for a survey), but a major problem is our limited knowledge towards the Ramanujan conjecture for half-integral weight cusp forms. The key breakthrough here is due to Iwaniec and Duke [Iw, Du], based on analyzing the Salié-type sums in the half-integral weight Kuznetsov formula. There is an alternative route based on Waldspurger's theorem that translates the problem into a problem of obtaining subconvexity for twisted automorphic L-functions. This can also be achieved by Kuznetsov formula (applied in a very different way) and gives the currently strongest results [BH]. Once an asymptotic formula with a power saving error is available, one can ask the finer question of equidistribution of the solutions in shrinking regions on the sphere of radius $\sqrt{n}$ which was first obtained in [DSP]. A striking result is an analysis of the variance in small annuli [HR], which also uses the Kuznetsov formula.

(b) *Equidistribution of Heegner points.* A similar problem asks for equidstribution of Heegner points and generalizations thereof which can again be rephrased as a subconvexity problem of twisted L-functions. Here the quadratic form $x^2 + y^2 + z^2$ is replaced with the discriminant form $q(x, y, z) = y^2 - 4xz$. For negative d, one associates to each $\mathrm{SL}_2(\mathbb{Z})$-orbit of a primitive solution to $q(x, y, z) = d$ the $\mathrm{SL}_2(\mathbb{Z})$-orbit of the Heegner point $(-y + i\sqrt{|d|})/(2x) \in \mathbb{H}$. After a conditional proof of Linnik, Duke showed in his seminal paper [Du] that these points equidistribute on the modular surface. This has initiated a lot of work, and we refer the reader for instance to [Mi, HM, Coh, Zha, LMY, Yo3] for generalizations and extensions. See [MV] for an excellent survey.

Duke's theorem has been generalized to higher rank and in particular to GL(3) in [ELMV] where the authors obtain analogous equidistribution result for equivalence classes of maximal compact flats. As in Duke's original theorem, this can be rephrased in algebraic terms and gives statements about the distribution of orders in totally real cubic fields and of integer points on certain homogeneous varieties. For instance, if P_i runs through a sequence of integral, cubic, monic, irreducible polynomials with three real roots and increasing discriminant, then (a projection of) the set of integral 3×3 matrices M with $P_i(M) = 0$ becomes equidistributed with respect to a natural $\mathrm{PGL}_3(\mathbb{R})$-invariant measure. One (of many) inputs is a subconvexity bound for automorphic L-functions with nebentypus established in [BHM] using the GL(2) Kuznetsov formula.

(c) *Mass distribution problems.* A different class of problems is concerned with the equidistribution of the mass of a sequence of automorphic forms whose spectral parameter (or potentially also its level) is tending to infinity. The random wave conjecture of Berry [Be] states that Eigenfunctions of a classically ergodic system behaves like Gaussian random variables in the large eigenvalue limit. Although some exceptions are known [RS], the mass of (a sequence of) automorphic forms ϕ_j, $j = 1, 2, \ldots \to \infty$, is generically expected to become equidistributed. This can be measured in at least three interesting ways:

- in the weak-$*$ limit: $|\phi_j|^2 d\mu \to c\, d\mu$ for a constant $c > 0$. This is related to the QUE conjecture of Rudnick and Sarnak [RS]. For Hecke Eigenfunctions on $SL_2(\mathbb{Z})\backslash\mathbb{H}$ this is related to subconvexity of certain degree 8 L-functions (degree 4 L-functions for Eisenstein series). For Eisenstein series and cuspidal CM-forms (dihedral forms), much progress has been made for instance in [PeS, Sa2, LS] long before the breakthrough results of Lindenstrauss and Soundararajan. More refined quantitative recent results can be found for instance in [Hu, BK2, Yo2]. All of them are based on certain mean value bounds of L-functions by the Kuznetsov formula.
- by considering $\|\phi_j\|_p/\|\phi_j\|_2$ for some $2 < p \leq \infty$. Bounds for this quantity show that large peaks cannot occur, or at least can occur only rarely. Particularly interesting is the case $p = 4$, because by Parseval and the formula of Watson-Ichino [Ic] this can be expressed as a mean value of central L-values, see for instance [Bl3, BKY, BK1, Hu].
- by considering restriction norm problems to interesting submanifolds $Y \subseteq X$. In some interesting cases the restriction $\|\phi|_Y\|_{2,Y}^2/\|\phi\|_{2,X}^2$ can be expressed in terms of central values of L-functions. Typical situations are Gross-Prasad pairs (see [II]) of subgroups, but also the period integral for $GL_n \times GL_{n-1}$ Rankin-Selberg convolutions belongs into this class. By Parseval this can often be translated into a mean value of L-functions that in favourable cases can be attacked with the Kuznetsov formula, see e.g. [BKY, LY, LLY] and [Sa3] for a general overview.

Finally it is worth mentioning that subconvexity of L-functions has become an important topic in the theory of L-functions in its own sake. On the one hand it is an excellent test case of the strength of available methods, on the other hand the search for subconvexity has also led to a better understanding of available tools and in particular the Kuznetsov formula. Even in absence of arithmetic applications it broadens our understanding of L-functions. This is huge area, and we can only mention here a very small set of selected examples focusing on higher rank L-functions, for instance [Bl2, JM, Li, Su].

4 Other Groups

After the great success of the Kuznetsov formula for the group GL(2), it is natural to investigate what it has to offer in other situations. The generalization to other rank 1 groups does not pose essential difficulties, and the shape of the formula is relatively similar. In fact, the formalism presented in Sect. 2 works for any reductive algebraic group. General versions for GL(n) can be found in [Go, Section 11] and [Frb, Theorem D]. Applications in analytic number theory require very precise and explicit information on both sides of the formula, and this is not at all easy to obtain for higher rank groups. In particular, some of the key open questions involve

- a good understanding of the various types of Kloosterman sums as certain kinds of exponential sums over finite fields and rings;
- a good understanding of the growth and oscillatory properties of the relevant integral transforms, and also the problem to what extent they can be inverted.

Nevertheless, the Kuznetsov formula is a versatile tool for statistical investigations of automorphic forms and L-functions potentially for arbitrary groups and perhaps analytically easier to handle than the Selberg trace formula. A thorough study of its scope and limitations in higher rank has just begun, but is a rapidly growing body of work. The following discussion highlights a few specific examples.

4.1 The Group GL(3)

The Weyl group of GL(3) has 6 elements, so the right hand side of the Kuznetsov formula consists (a priori) of 6 terms. Two of them vanish, one is the identity, and we are left with two hyper-Kloosterman terms and the long Weyl element Kloosterman term. The latter is usually dominant in applications. The first explicit computations go back to the late 1980's in [BFG] where all relevant Kloosterman sums were completely determined, and their algebro-geometric properties were studied in [Frb, Ste]. To get a feeling for the complexity, an explicit version of the long Weyl element Kloosterman sum for GL(3) looks as follows: let $n, m \in (\mathbb{Z} \setminus \{0\})^2$ and ψ_n, ψ_m be the corresponding characters on unipotent upper-triangular 3×3 matrices, explicitly

$$\psi_n \begin{pmatrix} 1 & x_2 & * \\ & 1 & x_1 \\ & & 1 \end{pmatrix} = e(n_1 x_1 + n_2 x_2).$$

Let w the long Weyl element, $c = \mathrm{diag}(1/c_2, c_2/c_1, c_1)$ for some $c_1, c_2 \in \mathbb{N}$ and U the integral unipotent upper-triangular 3×3 matrices. Then by the explicit Bruhat decomposition the long Weyl element Kloosterman sum corresponding to the "modulus" c and the characters ψ_n, ψ_m is given by

$$S(\psi_n, \psi_m, c) = \sum_{\gamma \in bcwb' \in U \backslash \mathrm{SL}_3(\mathbb{Z})/U} \psi_n(b)\psi_m(b').$$

Unfolding the definition, this equals

$$\sum_{\substack{B_1, C_1 \,(\mathrm{mod}\, c_1) \\ B_2, C_2 \,(\mathrm{mod}\, c_2) \\ c_1 C_2 + B_1 B_2 + c_2 C_1 \equiv 0 \,(\mathrm{mod}\, c_1 c_2) \\ (B_j, C_j, c_j) = 1}} e\left(\frac{m_2 B_1 + n_1(Y_1 c_2 - Z_1 B_2)}{c_1} - \frac{m_1 B_2 + n_2(Y_2 c_1 - Z_2 B_1)}{c_2}\right) \tag{10}$$

where Y_j, Z_j are chosen such that $Y_jB_j + Z_jC_j \equiv 1 \pmod{c_j}$ for $j = 1, 2$. If $(c_1, c_2) = 1$, this factorizes into a product of two ordinary Kloosterman sums to modulus c_1 and c_2. The Archimedean Whittaker transform is even more complicated and only recently has been made explicit by Buttcane [Bu1].

With these developments in place, many of the applications in Sect. 3.1 can be performed in a similar way (although sometimes quantitatively weaker): vertical Sato-Tate laws and density results for exceptional eigenvalues can be found along these lines in [BBR, Bl4, BBM, BZ]. A large sieve has been established in [BB1].

The invertibility of the Kuznetsov formula was the key to all applications in Sect. 3.2. First important steps in this direction were established in [Ye, Bu2], the most complete solution is contained in [Bu3], where a general test function can be put on the long Weyl element term. As one may expect, one of the difficulties lies in the fact that the spectral side contains not only the spherical spectrum, but all weights simultaneously. In the case of a non-abelian maximal compact subgroup $K = \mathrm{SO}(3)$, this is a non-trivial obstacle. However, it remains an open question where (and if) the long Weyl element Kloosterman sums (10) come up "in nature", i.e. in number theoretic problems. Applications of arithmetic significance are still a desideratum.

Some work has been done on L-functions corresponding to the applications presented in Sects. 3.3 and 3.4. Symmetry types of L-function using the one-level density have been investigated in [GK]. The underlying analysis of the Kuznetsov formula was recently generalized to GL(4) in [GSW].

The first subconvexity result for genuine GL(3) L-functions (by any method) has been obtained in [BB2] using the Kuznetsov formula: let $\pi \subseteq L^2(\mathrm{SL}_3(\mathbb{Z})\backslash\mathrm{SL}_3(\mathbb{R}))$ be an irreducible, cuspidal, spherical representation with Langlands parameter μ in "generic position", i.e. satisfying

$$c \le \frac{|\mu_j|}{\|\mu\|} \le C \quad (1 \le j \le 3) \quad \text{and} \quad c \le \frac{|\mu_i - \mu_j|}{\|\mu\|} \le C \quad (1 \le i < j \le 3)$$

for two constants $C > c > 0$. This covers 99% of all Maaß forms (choosing c, C appropriately), but misses, for instance, self-dual forms. Then

$$L(1/2, \pi) \ll \|\mu\|^{\frac{3}{4} - \frac{1}{120000}},$$

where the convexity bound is $L(1/2, \pi) \ll \|\mu\|^{3/4+\varepsilon}$.

4.2 *The Group* GSp(4)

Another interesting group with real rank 2 is the group Sp(4). As the Siegel upper half plane admits a complex structure, we have the notion of holomorphic cusp forms, for which an analogue of the Petersson formula has been worked out in detail by Kitaoka [Ki]. This formula has recently been used in analytic number theory.

For Siegel modular forms, there is a big difference between Fourier coefficients and Hecke eigenvalues. The unipotent periods produce Fourier coefficients in the relative trace formula, and it is a remarkable fact that by Böcherer's conjecture [Bo, FuMo] they are related to central L-values.

Using the Petersson-Kitaoka relative trace formula, local spectral equidistribution problems in the sense of Sect. 3.1 have been considered in [KST]. The papers [Bl5, Wa] compute mean values of spinor L-functions with applications to non-vanishing. This uses the full force of the formula. The analysis of the proof manipulates the Kloosterman terms in a rather suitable way, as they also contribute to the main term in the asymptotic formula.

4.3 Groups of Unbounded Rank

The asymptotic distribution of Satake parameters has been obtained in a rather general context. Important works in this direction using the currently most advanced analysis of the Selberg trace formula for the group $\mathrm{GL}(n)$ and even arbitrary classical groups are [MT, FiMa]. Using a version of the Kuznetsov formula, strong density results for Satake parameters are obtained in [Bl4, Bl6] for the group $\Gamma_0(q) \subseteq \mathrm{SL}_n(\mathbb{Z})$ of matrices whose lowest row is congruent to $(0, \dots, 0, *)$ modulo q. This is based, among other things, on a detailed analysis of certain $\mathrm{GL}(n)$ Kloosterman sums. For a discussion of these results in particular in connection with Sarnak's density conjecture we refer the reader to the introduction of [Bl6]. A Petersson type formula is used in [KL] to obtain asymptotic distribution results for Satake parameters on $\mathrm{GSp}(2n)$.

Acknowledgement The author was partially supported by DFG grant BL 915/2-2.

References

[AHB] L. Adleman, D.R. Heath-Brown, The first case of Fermat's last theorem. Invent. Math. **79**, 409–416 (1985)

[Al] S.A. Altuğ, Beyond endoscopy via the trace formula, II: asymptotic expansions of Fourier transforms and bounds towards the Ramanujan conjecture. Am. J. Math. **139**, 863–913 (2017)

[BGHT] T. Barnet-Lamb, D. Geraghty, M. Harris, R. Taylor, A family of Calabi-Yau varieties and potential automorphy II. Publ. Res. Inst. Math. Sci. **47**, 29–98 (2011)

[Be] M. Berry, Regular and irregular semiclassical wavefunctions. J. Phys. A **10**, 2083–2091 (1977)

[Bl1] V. Blomer, Ternary quadratic forms and sums of three squares with restricted variables, in *The Anatomy of Integers*, ed. by J.-M. de Koninck, A. Granville, F. Luca, CRM Proceedings & Lecture Notes, vol. 46 (2008), pp. 1–17

[Bl2] V. Blomer, Subconvexity for twisted *L-functions on* GL(3). Am. J. Math. **134**, 1385–1421 (2012)

[Bl3] V. Blomer, On the 4-norm of an automorphic form. J. Eur. Math. Soc. **15**, 1825–1852 (2013)

[Bl4] V. Blomer, Applications of the Kuznetsov formula on GL(3). Invent. Math. **194**, 673–729 (2013)

[Bl5] V. Blomer, Spectral summation formula for GSp(4) and moments of spinor L-functions. J. Eur. Math. Soc. **21**, 1751–1774 (2019)

[Bl6] V. Blomer, Density theorems for GL(n) (2019). arxiv:1906.07459

[BB1] V. Blomer, J. Buttcane, Global decomposition of GL(3) Kloosterman sums and the spectral large sieve. J. Reine Angew. Math. **757**, 51–88 (2019)

[BB2] V. Blomer, J. Buttcane, On the subconvexity problem for L-functions on GL(3). Ann. Sci. Ecole Norm. Sup. **53**, 1441–1500 (2020)

[BHM] V. Blomer, G. Harcos, P. Michel, Bounds for modular L-functions in the level aspect. Ann. Sci. Ecole Norm. Sup. **40**, 697–740 (2007)

[BKY] V. Blomer, R. Khan, M. Young, Distribution of mass of holomorphic cusp forms. Duke Math. J. **162**, 2609–2644 (2013)

[BBR] V. Blomer, J. Buttcane, N. Raulf, A Sato-Tate law for GL(3). Comm. Math. Helv. **89**, 895–919 (2014)

[BBM] V. Blomer, J. Buttcane, P. Maga, Applications of the Kuznetsov formula on GL(3): the level aspect. Math. Ann. **369**, 723–759 (2017)

[BH] V. Blomer, G. Harcos, Twisted L-functions over number fields and Hilbert's eleventh problem. Geom. Funct. Anal. **20**, 1–52 (2010)

[BLM] V. Blomer, X. Li, S. Miller, A spectral reciprocity formula and non-vanishing of L-functions on GL(4) $\times$ GL(2). J. Number Theory Prime **205**, 1–43 (2019)

[Bo] S. Böcherer, Bemerkungen über die Dirichletreihen von Koecher und Maaß. Math. Gottingensis **68** (1986)

[BFI] E. Bombieri, J. Friedlander, H. Iwaniec, Primes in arithmetic progressions to large moduli. Acta Math. **156**, 203–251 (1986)

[Br] R. Bruggeman, Fourier coefficients of cusp forms. Invent. Math. **45**, 1–18 (1978)

[BFG] D. Bump, S. Friedberg, D. Goldfeld, Poincaré series and Kloosterman sums for SL(3, $\mathbb{Z}$). Acta Arith. **50**, 31–89 (1988)

[Bu1] J. Buttcane, The spectral Kuznetsov formula on SL(3). Trans. AMS **368**, 6683–6714 (2016)

[Bu2] J. Buttcane, The arithmetic Kuznetsov formula on GL(3), I: the Whittaker case. Rev. Mat. Iberoamericana. To appear

[Bu3] J. Buttcane, The arithmetic Kuznetsov formula on GL(3), II: the general case (2019). arxiv:1909.09232

[BK1] V. Buttcane, R. Khan, L^4-norms of Hecke newforms of large level. Math. Ann. **362**, 699–715 (2015)

[BK2] J. Buttcane, R. Khan, On the fourth moment of Hecke–Maass forms and the random wave conjecture. Compos. Math. **153**, 1479–1511 (2017)

[BZ] J. Buttcane, F. Zhou, Plancherel distribution of Satake parameters of Maass cusp forms on GL_3. IMRN (5), 1417–1444 (2020)

[Coh] P. Cohen, Hyperbolic equidistribution problems on Siegel 3-folds and Hilbert modular varieties. Duke Math. J. **129**, 87–127 (2005)

[Co] H. Cohen, *Number Theory Volum II: Analytic and Modern Tools*. Graduate Texts in Mathematics, vol. 240 (Springer, Berlin, 2007)

[Con] B. Conrey, More than two fifths of the zeros of the Riemann zeta function are on the critical line. J. Reine Angew. Math. **399**, 1–26 (1989)

[CI] B. Conrey, H. Iwaniec, The cubic moment of central values of automorphic L-functions. Ann. Math. **151**, 1175–1216 (2000)

[DI1] J.-M. Deshouillers, H. Iwaniec, Kloosterman sums and Fourier coefficients of cusp forms. Invent. Math. **70**, 219–288 (1982)

[DI2] J.-M. Deshouillers, H. Iwaniec, On the greatest prime factor of $n^2 + 1$. Ann. Inst. Fourier (Grenoble) **32**, 1–11 (1982)

[Du] W. Duke, Hyperbolic distribution problems and half-integral weight Maass forms. Invent. Math. **92**, 73–90 (1988)

[DSP] W. Duke, R. Schulze-Pillot, Representation of integers by positive ternary quadratic forms and equidistribution of lattice points on ellipsoids. Invent. Math. **99**, 49–57 (1990)

[DFI] W. Duke, J. Friedlander, H. Iwaniec, Equidistribution of roots of a quadratic congruence to prime moduli. Ann. Math. **141**, 423–441 (1995)

[ELMV] M. Einsiedler, E. Lindenstrauss, P. Michel, A. Venkatesh, Distribution of periodic torus orbits and Duke's theorem for cubic fields. Ann. Math. **173**, 815–885 (2011)

[FiMa] T. Finis, J. Matz, On the asymptotics of Hecke operators for reductive groups. arxiv:1905.09078

[Fo] E. Fouvry, Théorème de Brun-Titchmarsh: application au théorème de Fermat. Invent. Math. **79**, 383–407 (1985)

[Frb] S. Friedberg, Poincaré series for $\mathrm{GL}(n)$: Fourier expansion, Kloosterman sums, and algebreo-geometric estimates. Math. Z. **196**, 165–188 (1987)

[Frl] J. Friedlander, Bounds for L-functions, in *Proceedings of the International Congress of Mathematicians, ICM '94* (Birkhäuser, Basel, 1995), pp. 363–373

[FuMo] M. Furusawa, K. Morimoto, Refined global Gross-Prasad conjecture on special Bessel periods and Böcherer's conjecture. J. EMS. To appear. (2021)

[Go] D. Goldfeld, *Automorphic forms and L-functions for the group* $\mathrm{GL}(n, \mathbb{R})$. Cambridge Studies in Advanced Mathematics, vol. 99 (2006)

[GK] D. Goldfeld, A. Kontorovich, On the GL(3) Kuznetsov formula with applications to symmetry types of families of L-functions, in *Automorphic Representations and L-Functions*. Tata Institute of Fundamental Research Studies in Mathematics, vol. 22 (2013), pp. 263–310

[GSW] D. Goldfeld, E. Stade, M. Woodbury, An orthogonality relation for $\mathrm{GL}(4, \mathbb{R})$ (2019). arxiv:1910.13586

[GR] I.S. Gradshteyn, I.M. Ryzhik, *Tables of Integrals, Series, and Products*, 7th edn. (Academic Press, New York, 2007)

[HM] G. Harcos, P. Michel, The subconvexity problem for Rankin-Selberg L-functions and equidistribution of Heegner points. II. Invent. Math. **163**, 581–655 (2006)

[Hu] P. Humphries, Equidistribution in shrinking sets and L^4-norm bounds for automorphic forms. Math. Ann. **371**, 1497–1543 (2018)

[HR] P. Humphries, M. Radziwiłł, Optimal small scale equidistribution of lattice points on the sphere, Heegner points, an closed geodesics. arxiv:1910.01360

[Ic] A. Ichino, Trilinear forms and the central values of triple product L-functions. Duke Math. J. **145**, 281–307 (2008)

[II] A. Ichino, T. Ikeda, On the periods of automorphic forms on special orthogonal groups and the Gross-Prasad conjecture. Geom. Funct. Anal. **19**, 1378–1425 (2010)

[Iv] A. Ivić, *The Riemann Zeta-Function: Theory and Applications*. Reprint of the 1985 original (Dover Publications, New York, 2003)

[Iw] H. Iwaniec, Fourier coefficients of modular forms of half-integral weight. Invent. Math. **87**, 385–401 (1987)

[IK] H. Iwaniec, E. Kowalski, *Analytic Number Theory*. AMS Colloquium Publications, vol. 53 (American Mathematical Society, Providence, 2004)

[ILS] I. Iwaniec, W. Luo, P. Sarnak, Low lying zeros of families of L-functions. Inst. Hautes Études Sci. Publ. Math. **91**, 55–131 (2000)

[Jo] D. Joyner, On the Kuznetsov-Bruggeman formula for a Hilbert modular surface having one cusp. Math. Z. **203**, 59–104 (1990)

[JM] M. Jutila, Y. Motohashi, Uniform bound for Hecke L-functions. Acta Math. **195**, 61–115 (2005)

[KS] N. Katz, P. Sarnak. *Random Matrices, Frobenius eigenvalues, and Monodromy*. AMS Colloquium Publications, Providence, vol. 45 (1999)

[Ki] Y. Kitaoka, Fourier coefficients of Siegel cusp forms of degree 2. Nagoya Math. J. **93**, 149–171 (1984)

[Klo] H.D. Kloosterman, On the representation of numbers in the form $ax^2 + by^2 + cz^2 + dt^2$. Acta Math. **49**, 407–464 (1926)

[KL] A. Knightly, C. Li, On the distribution of Satake parameters for Siegel modular forms. Doc. Math. **24**, 677–747 (2019)

[Ko] E. Kowalski, *Un cours de théorie analytique des nombres*. Cours Spécialisés. Société mathématique de France, vol. 13 (2004)

[KST] E. Kowalski, A. Saha, J. Tsimerman, Local spectral equidistribution for Siegel modular forms and applications. Compos. Math. **148**, 335–384 (2012)

[Ku] N.V. Kuznetsov, The Petersson conjecture for cusp forms of weight zero and the Linnik conjecture. Sums of Kloosterman sums. Math. USSR-Sb **39**, 299–342 (1981)

[La] R. Langlands, Beyond endoscopy, in *Contributions to Automorphic Forms, Geometry and Number Theory*, pp. 611–697 (Johns Hopkins University Press, Baltimore, 2004)

[Li] X. Li, Bounds for GL(3) $\times$ GL(2) L-functions and GL(3) L-functions. Ann. Math. **173**, 301–336 (2011)

[LY] X. Li, M. Young, The L^2 restriction norm of a GL_3 Maass form. Compos. Math. **148**, 675–717 (2012)

[LLY] X. Li, S.-C. Liu, M. Young, The L^2 restriction norm of a Maass form on $SL_{n+1}(\mathbb{Z})$. Math. Ann. **371**, 1301–1335 (2018)

[LMY] S.-C. Liu, R. Masri, M. Young, Subconvexity and equidistribution of Heegner points in the level aspect. Compos. Math. **149**, 1150–1174 (2013)

[Lu] W. Luo, Nonvanishing of L-values and the Weyl law. Ann. Math. **154**, 477–502 (2001)

[LS] W. Luo, P. Sarnak, Quantum ergodicity of Eigenfunctions on $PSL_2(\mathbb{Z})\backslash\mathbb{H}_2$. Inst. Hautes Études Sci. Publ. Math. **81**, 207–237 (1995)

[MT] J. Matz, N. Templier, Sato-Tate equidistribution for families of Hecke–Maass forms on $SL(n, \mathbb{R})/SO(n)$. arxiv:1505.07285

[Mer] J. Merikoski, Largest prime factor of $n^2 + 1$. arxiv:1908.08816

[Meu] T. Meurman, *On the Binary Additive Divisor Problem*. Number Theory (Turku, 1999) (de Gruyter, Berlin, 2001), pp. 223–246

[Mi] P. Michel, The subconvexity problem for Rankin-Selberg L-functions and equidistribution of Heegner points. Ann. Math. **160**, 185–236 (2004)

[MV] P. Michel, A. Venkatesh, Equidistribution, L-functions and Ergodic theory: on some problems of Yu. Linnik, in *Proceedings ICM Zürich* (2006), pp. 421–457

[Mo] Y. Motohashi, *Spectral Theory of the Riemann Zeta-Function*. Cambridge Tracts in Mathematics, Cambridge, vol. **127** (1997)

[PeS] Y. Petridis, P. Sarnak, Quantum unique ergodicity for $SL_2(\mathcal{O})\backslash\mathbb{H}_3$ and estimates for L-functions. J. Evol. Equ. **1**, 277–290 (2001)

[PhS] R. Philipps, P. Sarnak, On cusp forms for co-finite subgroups of PSL(2, $\mathbb{R}$). Invent. Math. **80**, 339–364 (1985)

[Po] G. Pólya, *Über die Verteilung der quadratischen Reste und Nichtreste*. Gött. Nachr. 21–29 (1918)

[RS] Z. Rudnick, P. Sarnak, The behavior of eigenstates of arithmetic hyperbolic manifolds. Comm. Math. Phys. **161**, 195–213 (1994)

[Sa1] P. Sarnak, Statistical properties of eigenvalues of the Hecke operators, in *Analytic Number Theory and Diophantine Problems (Stillwater, 1984)*. Progress in Mathematics, vol. **70** (Birkhäuser, Boston, 1987), pp. 321–331

[Sa2] P. Sarnak, Estimates for Rankin-Selberg L-functions and quantum unique ergodicity. J. Funct. Anal. **184**, 419–453 (2001)

[Sa3] P. Sarnak, Letter to Reznikov. https://publications.ias.edu/node/498

[ST] D. Sears, E. Titchmarsh, Some Eigenfunction formulae. Quart. J. Math. **1**, 165–175 (1950)

[Se] J.-P. Serre, Répartition asymptotique des valeurs propres de l'operateur de Hecke T_p. J. Amer. Math. Soc. **10**, 75–102 (1997)

[Ste] G. Stevens, Poincaré series on GL(r) and Kloostermann sums. Math. Ann. **277**, 25–51 (1987)

[Str] A. Strömbergsson, Some remarks on a spectral correspondence for maass waveforms. Int. Math. Res. Not. **2001**(10), 505–517
[SY] K. Soundararajan, M. Young, The prime geodesic theorem. J. Reine Angew. Math. **676**, 105–120 (2013)
[Su] E. Suvitie, On a short spectral sum involving inner products of a holomorphic cusp form and Maass forms. Acta Arith. **144**, 395–418 (2010)
[Sz] W. Szerpiński, Über ein Problem aus der analytischen Zahlentheorie. Prace mat.-fiz. **17**, 77–118 (1906)
[To] A. Toth, Roots of quadratic congruences. Int. Math. Res. Not. 719–739 (2000)
[Ve] A. Venkatesh, Beyond endoscopy and special forms on GL(2). J. Reine Angew. Math. **577**, 23–80 (2004)
[Vi] I.M. Vinogradov, Über eine asymptotische Formel aus der Theorie der binären quadratischen Formen. J. Soc. Phys. Math. Permǐ **1**, 18–28 (1918)
[Wa] F. Waibel, Moments of spinor L-functions and symplectic Kloosterman sums. Quart. J. Math. **70**, 1411–1436 (2019)
[We] A. Weil, On some exponential sums. Proc. Nat. Acad. Sci. U. S. A. **34**, 204–207 (1948)
[Ye] Y. Ye, A Kuznetsov Formula for Kloosterman Sums on GL_n. Ramanujan J. **4**, 385–395 (2000)
[Yo1] M. Young, Low-lying zeros of families of elliptic curves. J. Amer. Math. Soc. **19**, 205–250 (2006)
[Yo2] M. Young, The quantum unique ergodicity conjecture for thin sets. Adv. Math. **286**, 958–1016 (2016)
[Yo3] M. Young, Weyl-type hybrid subconvexity bounds for twisted L-functions and Heegner points on shrinking sets. J. Eur. Math. Soc. **17**, 1545–1576 (2017)
[Zha] S.-W. Zhang, Equidistribution of CM-points on quaternion Shimura varieties. Int. Math. Res. Not., 3657–3689 (2005)

Dimensions of Automorphic Representations, L-Functions and Liftings

Solomon Friedberg and David Ginzburg

Abstract There are many Rankin-Selberg integrals representing Langlands L-functions, and it is not apparent what the limits of the Rankin-Selberg method are. The Dimension Equation is an equality satisfied by many such integrals that suggests a priority for further investigations. However there are also Rankin-Selberg integrals that do not satisfy this equation. Here we propose an extension and reformulation of the dimension equation that includes many additional cases. We explain some of these cases, including the new doubling integrals of the authors, Cai and Kaplan. We then show how this same equation can be used to understand theta liftings, and how doubling integrals fit into a lifting framework. We give an example of a new type of lift that is natural from this point of view.

2010 Mathematics Subject Classification Primary 11F66; Secondary 11F27, 11F70, 17B08, 22E50, 22E55

1 Introduction

Two broad classes of integrals appear frequently in the theory of automorphic forms. Let G be a reductive group defined over a number field F, ρ be a complex analytic representation of the L-group of G, and π be an irreducible automorphic representation of $G(\mathbb{A})$. First, one may sometimes represent the Langlands L-function $L(s, \pi, \rho)$ (for $\Re(s) \gg 0$) as an integral, and the desired analytic properties of this L-function may then be deduced from the integral representation. Such constructions, often called Rankin-Selberg integrals, have a long history with many examples (Google Scholar lists 3370 results for the phrase "Rankin-Selberg

S. Friedberg (✉)
Department of Mathematics, Boston College, Chestnut Hill, MA, USA
e-mail: solomon.friedberg@bc.edu

D. Ginzburg
School of Mathematical Sciences, Tel Aviv University, Tel Aviv, Israel
e-mail: ginzburg@post.tau.ac.il

W. Müller et al. (eds.), *Relative Trace Formulas*, Simons Symposia,
https://doi.org/10.1007/978-3-030-68506-5_3

integral”), with Eisenstein series and their Fourier coefficients appearing in many of the integrals. See Bump [2] for an engaging survey. Second, given two reductive groups G, H there is sometimes a function $\theta(g,h)$ on $G(\mathbb{A}) \times H(\mathbb{A})$ that is $G(F) \times H(F)$ invariant and that may be used as an integral kernel to transport automorphic representations on $G(\mathbb{A})$ to $H(\mathbb{A})$. One considers the functions

$$f_\varphi(h) := \int_{G(F)\backslash G(\mathbb{A})} \varphi(g)\,\theta(g,h)\,dg$$

as φ runs over the functions in π and lets σ be the representation of $H(F)\backslash H(\mathbb{A})$ generated by the functions f_φ. The most familiar example is the classical theta correspondence, where the integral kernel θ is constructed from the Weil representation (see Gan [15] for recent progress), while other classes of such integrals are given by the authors and Bump [7] and by Leslie [34].

We pose two natural questions. First, the construction of Rankin-Selberg integrals representing L-functions is often quite involved, and it may take a great deal of work to see that a specific integral is Eulerian. Is there any commonality among the known Rankin-Selberg integrals that can be used to decide whether a specific integral is a worthy candidate for investigation, or to say it differently, to rule out integrals as being unlikely to represent an L-function? Second, is there any way to know whether an automorphic function of two variables $\theta(g,h)$ is likely to be a useful kernel function, and if so, can one predict the properties of the representation σ from those of π and θ? For example, given π and θ, when is it reasonable to think that σ might be generic?

An answer to the first question was proposed by the second named author in [18]: the *dimension equation* describes an equality between the dimensions of groups and the dimensions of the representations that is satisfied by many integrals that represent L-functions. Below we develop the dimension question in some detail and illustrate it in many cases, including cases of integrals that represent the product of two or more distinct Langlands L-functions in separate complex variables. In fact a refinement of the dimension equation expands its applicability to additional cases. We shall explain this refinement, and show how this allows us to include doubling integrals including the new doubling integrals of the authors, Cai and Kaplan [11]. Then we shall use the dimension equation to discuss integral kernels, showing how the equation gives an indication of what to expect in integral kernel constructions, and explain how doubling integrals may be used to bridge these two classes of constructions. Last, we shall pursue the dimension equation farther in specific cases, providing new examples for further study.

To conclude our introduction, we describe the contents by section. In Sect. 2 we introduce unipotent orbits and state the dimension equation, following [18]. Though this paper may be read independently from [18], it is a natural continuation of that work. Then, in Sect. 3 we give many examples of Rankin-Selberg integrals that satisfy the dimension equation. We conclude the section by presenting an exotic example—an integral of Rankin-Selberg type that is Eulerian by two different choices of Eisenstein series, one satisfying the dimension equation but the other not. This motivates the need to extend the dimension equation. This extension is described in Sect. 4, and examples of the extended dimension equation are presented in Sect. 5. Next, in Sect. 6 integral kernels are connected to the dimension equation, and the use of the dimension equation to predict aspects of the resulting correspondence is illustrated. Section 7 revisits doubling integrals from the perspective of integral kernels, and formulates a general classification question. Then in Sect. 8 a new low rank example is presented and its analysis is described in brief. The global theory has a local counterpart, and the final Section illustrates the local properties that appear in this context.

Part of the preparation of this paper occurred when the first-named author was a visitor at the Simons Center for Geometry and Physics, and he expresses his appreciation for this opportunity. While the paper was in the final stages of preparation, Aaron Pollack communicated to us that Shrenik Shah has independently suggested an extension of the dimension equation. We also thank Yuanqing Cai for helpful comments.

2 The Dimension Equation

Fix a number field F and let $\mathbb{A}$ be its ring of adeles. If G is an algebraic group defined over F, then we write $[G]$ for the quotient $G(F)\backslash G(\mathbb{A})$.

We begin by recalling some facts about unipotent orbits. References for this material are Collingwood and McGovern [12] and Ginzburg [17]. Let G be a reductive group defined over F, $\overline{F}$ be an algebraic closure of F, and let $\mathcal{O}$ be a unipotent orbit of $G(\overline{F})$. If $G \subset GL(V)$ is a classical group then these orbits are indexed by certain partitions of $\dim(V)$. For example, if $G = GL_n$ then they are indexed by all partitions of n, while if $G = Sp_{2n}$ then they are indexed by the partitions of $2n$ such that each odd part occurs an even number of times. When $\mathcal{O}$ is indexed by a partition $\mathcal{P}$, we simply write $\mathcal{O} = \mathcal{P}$.

For convenience suppose that G is a split classical group. Fix a Borel subgroup $B = TN$ with unipotent radical N, let Φ denote the positive roots with respect to B, and for $\alpha \in \Phi$ let $x_\alpha(t)$ denote the corresponding one-parameter subgroup of N. Then one attaches a unipotent subgroup $U_\mathcal{O} \subseteq N$ to $\mathcal{O}$. This group may be described as follows. If $\mathcal{O}$ is given in partition form by $\mathcal{O} = (p_1^{e_1} \dots p_k^{e_k})$ (we show repeated terms in a partition using exponential notation) let $h_\mathcal{O}(t)$ be the diagonal matrix whose entries are $\{t^{p_i-2j-1} \mid 0 \le j \le p_i - 1\}$ repeated e_i times, with the entries arranged in non-increasing order in terms of power of t. This gives rise to a filtration $N \supset N_1 \supset N_2 \supset \cdots$ of N, where $N_i = N_{i,\mathcal{O}}$ is given by

$$N_{i,\mathcal{O}}(F) = \{x_\alpha(r) \in N(F) \mid \alpha \in \Phi \text{ and } h_\mathcal{O}(t)x_\alpha(r)h_\mathcal{O}(t)^{-1} = x_\alpha(t^j r) \text{ for some } j \text{ with } j \ge i\}. \tag{1}$$

Also, let $G_\mathcal{O}$ be the stabilizer of $h_\mathcal{O}(t)$ in G.

Fix a nontrivial additive character ψ of $F\backslash\mathbb{A}$, and let $L_{2,\mathcal{O}} = N_{2,\mathcal{O}}/[N_{2,\mathcal{O}}, N_{2,\mathcal{O}}]$. Then the characters of the abelian group $L_{2,\mathcal{O}}(\mathbb{A})$ may be identified with $L_{2,\mathcal{O}}(F)$. Also $G_{\mathcal{O}}(F)$ acts on the characters of $L_{2,\mathcal{O}}(\mathbb{A})$ and hence on $L_{2,\mathcal{O}}(F)$ by conjugation. Over the algebraically closed field $\overline{F}$, the action of $G_{\mathcal{O}}(\overline{F})$ on $L_{2,\mathcal{O}}(\overline{F})$ has an open orbit, and the stabilizer in $G_{\mathcal{O}}(\overline{F})$ of a representative for this orbit is a reductive group whose Cartan type is uniquely determined by the orbit. A character $\psi_{\mathcal{O}}$ of $L_{2,\mathcal{O}}(F)$ will be called a generic character associated to $\mathcal{O}$ if the connected component of its stabilizer in $G_{\mathcal{O}}(F)$ has, after base change to $\overline{F}$, the same Cartan type. We caution the reader that there may be infinitely many $G_{\mathcal{O}}(F)$-orbits of generic characters associated to a single $\mathcal{O}$. (For an example, see [18, p. 162].)

Suppose that G is a reductive group, $\mathcal{O}$ is a unipotent orbit of G, and $\psi_{\mathcal{O}}$ is a generic character. Let $U_{\mathcal{O}} = N_{2,\mathcal{O}}$ and regard $\psi_{\mathcal{O}}$ as a character of $U_{\mathcal{O}}(\mathbb{A})$ in the canonical way. If φ is an automorphic form on $G(\mathbb{A})$, its *Fourier coefficient with respect to* $\mathcal{O}$ is defined to be

$$\varphi^{U_{\mathcal{O}},\psi_{\mathcal{O}}}(g) = \int_{[U_{\mathcal{O}}]} \varphi(ug)\psi_{\mathcal{O}}(u)\,du, \qquad g \in G(\mathbb{A}).$$

If π is an automorphic representation of $G(\mathbb{A})$, we say that π has *nonzero Fourier coefficients with respect to* $\mathcal{O}$ if the set of functions $\varphi^{U_{\mathcal{O}},\psi_{\mathcal{O}}}(g)$ is not identically zero as φ runs over the space of π and $\psi_{\mathcal{O}}$ runs over set of generic characters associated to $\mathcal{O}$.

We recall that the set of unipotent orbits has a natural partial ordering, which corresponds to inclusion of Zariski closures and corresponds to the dominance order for the associated partitions. There is a unique maximal unipotent orbit $\mathcal{O}_{\max}$ and for this orbit the group $U_{\mathcal{O}_{\max}}$ is N, the unipotent radical of the Borel subgroup. If π is an automorphic representation of $G(\mathbb{A})$ we let $\mathcal{O}(\pi)$ denote the set of maximal unipotent orbits for which π has nonzero Fourier coefficients. For example, the automorphic representation π is generic if and only if $\mathcal{O}(\pi) = \{\mathcal{O}_{\max}\}$.

We remark that these same definitions apply without change in the covering group case. Indeed, if $\widetilde{G}$ is a cover of $G(\mathbb{A})$, then $N(\mathbb{A})$ embeds canonically in $\widetilde{G}(\mathbb{A})$ and so the same definitions apply. For example, in [6] the authors construct a representation Θ_{2n+1} of the double cover of $SO_{2n+1}(\mathbb{A})$ for $n \geq 4$ such that $\mathcal{O}(\Theta_{2n+1}) = (2^{2m}1^{2j+1})$, where $n = 2m + j$ with $j = 0$ or 1. Here and below we omit the set notation when the set $\mathcal{O}(\pi)$ is a singleton and identify $\mathcal{O}(\pi)$ with the partition.

Recall that each nilpotent orbit in a Lie algebra has a dimension. In fact, these dimensions are computed for classical groups in terms of partitions in [12, Corollary 6.1.4]. We use this to discuss and work with the dimensions of the unipotent orbits under consideration here. To illustrate, on the symplectic group Sp_{2n}, let $\mathcal{O}$ be the unipotent orbit associated to the partition $(n_1n_2\ldots n_r)$ of $2n$. For such a partition to be a symplectic partition it is required that all odd numbers in the partition occur with even multiplicity. Then

$$\dim \mathcal{O} = 2n^2 + n - \frac{1}{2}\sum_{i=1}^{r}(2i-1)n_i - \frac{1}{2}a, \tag{2}$$

where a is the number of odd integers in the partition $(n_1 n_2 \dots n_r)$. Importantly, there is also a connection between the dimension of an orbit $\mathcal{O}$ and the Fourier coefficients above. That is:

$$\tfrac{1}{2}\dim \mathcal{O} = \dim N_{2,\mathcal{O}} + \tfrac{1}{2}\dim N_{1,\mathcal{O}}/N_{2,\mathcal{O}}. \tag{3}$$

As this suggests, when $N_{1,\mathcal{O}} \neq N_{2,\mathcal{O}}$, there is another natural coefficient of Fourier-Jacobi type (that is, involving a theta series); see [18, equation (7)], for details, and the discussion of (16) at the end of Sect. 4 below.

Throughout the rest of this paper we make the following assumption:

Assumption 1. *For all automorphic representations π under consideration, the dimension of each orbit in the set $\mathcal{O}(\pi)$ is the same.*

We know of no examples where Assumption 1 does not hold, and it is expected that it is always true [17, Conjecture 5.10]. In fact, we know of no examples where $\mathcal{O}(\pi)$ is not a singleton. Under Assumption 1, we write $\dim \mathcal{O}(\pi)$ for the dimension of any orbit in the set $\mathcal{O}(\pi)$. Given a representation for which Assumption 1 holds, define (following [17, Definition 5.15]) the Gelfand-Kirillov dimension of π:

$$\dim(\pi) = \tfrac{1}{2}\dim \mathcal{O}(\pi),$$

(compare [1, (3.2.6)] and [32, (2.4.2), Remark (iii)]). By definition, if χ is an idele class character then $\dim(\chi) = 0$. Similarly if ψ is an additive character of $[U]$ where U is a unipotent group then $\dim(\psi) = 0$.

It is expected that it is possible to compute the unipotent orbit attached to an Eisenstein series from the unipotent orbits of its inducing data. See [18, Section 4.3]. This has been confirmed in a number of cases [10], [35]. In general, suppose that P is a parabolic subgroup of G with Levi decomposition $P = MN$, and τ is an automorphic representation of $M(\mathbb{A})$ such that Assumption 1 holds for τ, and consider the Eisenstein series $E_\tau(g,s)$ on $G(\mathbb{A})$ corresponding to the induced representation $\mathrm{Ind}_{P(\mathbb{A})}^{G(\mathbb{A})}\tau\delta_P^s$. Then one expects that

$$\dim E_\tau(g,s) = \dim \tau + \dim U. \tag{4}$$

To give one example, if τ is generic then E_τ is too, and so (4) holds. In discussing specific integrals, we shall assume that (4) always holds below.

We now have the information needed to explain the Dimension Equation of [18, Definition 3]. Suppose one has a Rankin-Selberg integral over the various groups G_j and involving various automorphic representations π_i, and the integral unfolds to unique functionals which are factorizable. Here both the adelic modulo rational quotients in the integral and the automorphic representations are with respect to the same field F and ring of adeles $\mathbb{A}$. Then the expectation formulated in [18] is that the sum of the dimensions of the groups is (generally) equal to the sum of the dimensions of the automorphic representations.

Definition 1. *The Dimension Equation is the equality*

$$\sum_j \dim G_j = \sum_i \dim \pi_i.$$

A number of examples of this equation in the context of Rankin-Selberg integrals are presented in [18]. To clarify the extent of applicability of this equation, we will give additional examples in the next Section.

3 Examples of the Dimension Equation

We begin with classical examples of the dimension equation. Riemann's second proof of the analytic continuation of the zeta function represents it as a Mellin transform of the Jacobi theta function, and this integral satisfies the dimension equation (both sides are 1). Similarly, the Hecke integral

$$\int_{[\mathbb{G}_m]} \varphi\begin{pmatrix} a & \\ & 1 \end{pmatrix} |a|^s \, d^\times a \tag{5}$$

satisfies the equation (both sides are again 1), and the classical Rankin-Selberg integral

$$\int_{[PGL_2]} \varphi_1(g)\varphi_2(g)E(g,s)\, dg \tag{6}$$

has group of dimension 3 and three representations that are each of dimension 1.

More generally, the Rankin-Selberg integrals of Jacquet, Piatetski-Shapiro and Shalika [29, 30] on $GL_n \times GL_k$ satisfy the dimension equation. Suppose that π_1, π_2 are irreducible cuspidal automorphic representations of $GL_n(\mathbb{A})$, $GL_k(\mathbb{A})$ resp. Since all cuspidal automorphic representations of general linear groups are generic, we have $\dim(\pi_1) = n(n-1)/2$ and $\dim(\pi_2) = k(k-1)/2$. Let $\phi_i \in \pi_i$ for $i = 1, 2$.

If $k = n$ then the Ranklin-Selberg integral representing $L(s, \pi_1 \times \pi_2)$, generalizing (6), is given by

$$\int_{[PGL_n]} \varphi_1(g)\varphi_2(g)E(g,s)\, dg \tag{7}$$

where E is the ("mirabolic") Eisenstein series induced from the standard parabolic subgroup P of type $(n-1, 1)$ and character $\delta_P^s \eta^{-1}$, where δ_P is the modular character of P and η is the product of the central characters of π_1 and π_2. In this integral the dimension of the group is $n^2 - 1$. As for the representations, the orbit of the Eisenstein series E is (21^{n-2}), so it is of dimension $n - 1$ (compare (4)). The dimension equation is the equality

$$n^2 - 1 = 2(n(n-1)/2) + n - 1.$$

If $k \neq n$, suppose $k < n$ without loss. When $k = n - 1$ the integral is a direct generalization of (5):

$$\int_{[GL_{n-1}]} \varphi_1 \begin{pmatrix} g & \\ & 1 \end{pmatrix} \varphi_2(g) |\det g|^s \, dg.$$

Here the group is of dimension $(n-1)^2$ and the representations π_1, π_2 are of dimension $n(n-1)/2$ and $(n-1)(n-2)/2$ resp. Since

$$(n-1)^2 = n(n-1)/2 + (n-1)(n-2)/2$$

the dimension equation holds. However, if $k < n - 1$ then a similar integral

$$\int_{[GL_k]} \varphi_1 \begin{pmatrix} g & \\ & I_{n-k} \end{pmatrix} \varphi_2(g) |\det g|^s \, dg$$

does not satisfy the dimension equation: the group is of dimension k^2 while the representations are of total dimension $n(n-1)/2 + k(k-1)/2 > k^2$. One way to satisfy the equation is to introduce an additional integration over a group of dimension $(n^2 - n - k^2 - k)/2$. And indeed the integral of Jacquet, Piatetski-Shapiro and Shalika

$$\int_{[GL_k]} \int_{[Y_{n,k}]} \varphi_1 \left(y \begin{pmatrix} g & \\ & I_{n-k} \end{pmatrix} \right) \varphi_2(g) \psi(y) |\det g|^s \, dg \tag{8}$$

includes an integration over the group $Y_{n,k}$ consisting of upper triangular $n \times n$ unipotent matrices whose upper left $(k+1) \times (k+1)$ corner is the identity matrix, a group which has dimension $n(n-1)/2 - k(k+1)/2$. To be sure, the integral of [29, 30] requires an additive character ψ of this group (the restriction of the standard generic character to $Y_{n,k}$), and this finer level of detail is not seen by the dimension equation, but the dimension equation already makes the introduction of the group $Y_{n,k}$ natural.

Another example of the dimension equation is practically tautological. If $E(g, s)$ is an Eisenstein series on a reductive group G formed from generic inducing data, then a straightforward calculation shows that it too is generic. If $E(g, s)$ is generic with respect to the unipotent subgroup N and character ψ_N, then its Whittaker coefficient

$$\int_{[N]} E(n,s)\psi_N(n)\,dn$$

satisfies the dimension equation, since both the group and the representation have dimension $\dim(N)$. The Langlands-Shahidi method studies such coefficients systematically and uses them to get information about L-functions. See Shahidi [39] and the references there.

Our next example of the dimension equation is given by the integral representing the Asai L-function for GL_n (Flicker [13]). Suppose K/F is a quadratic extension, and for this example let $\mathbb{A}$ denote the adeles of F and $\mathbb{A}_K$ the adeles of K. Suppose that Π is an irreducible cuspidal automorphic representation on $GL_n(\mathbb{A}_K)$, and $\varphi \in \Pi$. Then this integral is of the form

$$\int_{Z(\mathbb{A})GL_n(F)\backslash GL_n(\mathbb{A})} \varphi(g)E(g,s)\,dg,$$

where $E(g,s)$ is again a mirabolic Eisenstein series on $GL_n(\mathbb{A})$ (the central character of Π is built into E so the product is invariant under the center $Z(\mathbb{A})$ of $GL_n(\mathbb{A})$). Here the group has dimension n^2-1. This time the automorphic form is on the $\mathbb{A}$-points of the restriction of scalars $\mathrm{Res}_{K/F}GL_n$. Thus the dimension of π viewed as an automorphic representation over F is twice its dimension over K. That is, in this context $\dim(\Pi) = 2 \times \frac{n(n-1)}{2}$. The dimension equation is satisfied since

$$n^2 - 1 = 2 \times \frac{n(n-1)}{2} + n - 1.$$

There are also integrals that represent the product of two different Langlands L-functions *in two separate complex variables.* These also satisfy the dimension equation. We give a number of examples. The first instance of such an integral, representing the product of the standard and exterior square L-functions, was provided by Bump and Friedberg [3]. Suppose that π is a cuspidal automorphic representation of GL_n. If $n = 2k$ is even, then the integral is of the form

$$\int_{[GL_1\backslash(GL_k\times GL_k)]} \varphi(\iota(g_1,g_2))E(g_1,s_1)\left|\frac{\det g_2}{\det g_1}\right|^{s_2-1/2} dg_1\,dg_2,$$

where $\iota : GL_k \times GL_k \to GL_{2k}$ is a certain embedding, φ is in π, and E is the mirabolic Eisenstein series on GL_k. In the quotient $GL_1\backslash(GL_k \times GL_k)$, the group GL_1 acts diagonally. This integral satisfies the dimension equation since the group is of dimension $2k^2-1$ and the representations are of dimensions $(2k)(2k-1)/2$ and $k-1$. If $n = 2k+1$ is odd, then the integral is similar, but is now over $[GL_1\backslash(GL_{k+1} \times GL_k)]$ with the Eisenstein series on GL_{k+1}. The dimension equation is satisfied in the case that n is odd since

$$(k+1)^2 + k^2 - 1 = (2k+1)(2k)/2 + ((k+1) - 1).$$

Three additional examples of Rankin-Selberg integrals representing the product of Langlands L-functions attached to the standard and spin representations in separate complex variables are found in Bump, Friedberg and Ginzburg [5]. Suppose π on GSp_4 is generic with trivial central character and φ is in the space of π. Let P, Q be the two non-conjugate standard maximal parabolic subgroups of GSp_4 and let E_P, E_Q be the Eisenstein series induced from from the modular functions $\delta_P^{s_1}$, $\delta_Q^{s_2}$ resp. Then the integral is of the form

$$\int_{[GL_1\backslash GSp_4]} \varphi(g) E_P(g, s_1) E_Q(g, s_2)\, dg.$$

Here the group is of dimension 10 and the representations are of dimensions 4, 3, 3 resp., so the dimension equation is satisfied.

Suppose next that π is on GSp_6, generic, and has trivial central character, and let H be the subgroup of $GL_2 \times GSp_4$ of pairs of group elements (h_1, h_2) with equal similitude factors. Then, for a certain embedding $\iota : H \to GSp_6$, the integral given there is of the form

$$\int_{[GL_1\backslash H]} \varphi(\iota(h_1, h_2)) E_1(h_1, s_1) E_2(h_2, s_2)\, dh_1\, dh_2$$

where E_1 (resp. E_2) is a Borel (resp. Siegel) Eisenstein series on GL_2 (resp. GSp_4). Here the group is of dimension 13 and the three representations in the integral are of dimensions 9, 1, 3 (resp.), as the Siegel Eisenstein series has unipotent orbit (2^2).

Third suppose that π is on GSp_8 and is once again generic. Introduce the subgroup H of $GL_2 \times GSp_6$ of pairs with equal similitude factors and let $\iota : H \to GSp_8$ be a certain embedding given in [5]. Then the authors and Bump prove that the integral

$$\int_{[GL_1\backslash H]} \varphi(\iota(h_1, h_2)) E(h_2, s_1, s_2)\, dh_1\, dh_2$$

is once again a product of two different Langlands L-functions, where E is the two-variable Eisenstein series on GSp_6 induced from the standard parabolic with Levi factor $GL_1 \times GL_2$. The group has dimension 24, π has dimension 16, and indeed E has dimension 8.

To present one further example, Pollack and Shah [38] give an integral representing the product of three Langlands L-functions in three distinct complex variables. If π is on PGL_4, then this is of the form

$$\int_{[PGL_4]} \varphi(g) E_1(g, w) E_2(g, s_1, s_2)\, dg$$

where E_1 and E_2 are the Eisenstein series with Levi factors $GL_2 \times GL_2$ and $GL_1^2 \times GL_2$, resp. Here the dimension of the group in the integral is 15 and the representations are of dimensions 6,4,5 resp. Those authors also present a related integral on $GU(2,2)$ and the same dimension count holds.

The dimension equation holds as well when covering groups are involved. For example, the symmetric square integrals of Bump-Ginzburg [9] and Takeda [40] may be checked to satisfy this.

Nonetheless, not all Rankin-Selberg integrals satisfy the dimension equation in the form presented in [18, 19]. Here is a simple, striking example. Suppose π is a cuspidal automorphic representation of $PGL_4(\mathbb{A})$, φ is in the space of π, and $E(g,s)$ is an Eisenstein series on the group $PGSp_4(\mathbb{A})$. Consider the integral

$$\int_{[PGSp_4]} \varphi(g)E(g,s)\,dg. \tag{9}$$

The basic dimension equation states that

$$10 = \dim(PGSp_4) = \dim(\pi) + \dim(E).$$

Since π is generic, it is of dimension 6, so this equation requires an Eisenstein series of dimension 4, that is, a generic Eisenstein series. In fact there is an Eulerian integral with this data. Using the GSp_4 Eisenstein series with trivial central character induced from an automorphic representation τ on GL_2 via the parabolic with Levi $GL_1 \times GL_2$ (the Klingen parabolic), the integral unfolds to the degree 12 L-function $L(s, \pi \times \tau, \wedge^2 \times \text{standard})$. Indeed, after using low rank isogenies one sees that this integral is essentially the same as the integral for the $SO_n \times GL_k$ standard L-function obtained by Ginzburg [16] in the case $n = 6$, $k = 2$. For these parameters, the L-function is realized as an integral of an SO_6 automorphic form against an Eisenstein series on SO_5.

However, there is *another* Rankin-Selberg integral of the form (9) that is also Eulerian! That is the case that E is an Eisenstein series of dimension 3 induced from the modular function δ_P^s of the Siegel parabolic P of Sp_4 (one may also twist by a character of GL_1). Indeed, after using low rank isogenies to again regard this as an integral of an SO_6 automorphic form against an Eisenstein series on SO_5, this is the integral treated by the authors and Bump [4] with $n = 2, m = 1$ (see (1.4) there). The integral is zero unless the automorphic representation π is a lift from Sp_4, and in that case it represents a degree 5 L-function. However the dimension equation does not hold, since the Siegel Eisenstein series is of dimension 3, not 4.

There are other important examples of integrals that represent L-functions but that do not satisfy the dimension equation of [18]. One class of such integrals are the doubling integrals, a class first constructed by Piatetski-Shapiro and Rallis [36], which represent the standard L-function of an automorphic representation on a classical group G, twisted by GL_1. The original doubling integral was for split classical groups but the construction has been extended or used by a number of authors including Yamana [41], Gan [14], Lapid and Rallis [33], and Ginzburg and Hundley [21]. The doubling construction was extended to the tensor product L-function for $G \times GL_k$ by Cai, Friedberg, Kaplan and Ginzburg [11]. (This extension is particularly helpful as it allows one to use the converse theorem to prove lifting results.) Another class of integrals representing L-functions that does not satisfy the original dimension equation is the "new way" integrals, a class also initiated by Piatetski-Shapiro and Rallis [37]. Once again this class has been extended, in particular there are the integrals of Bump, Furusawa and Ginzburg [8] and of Gurevich-Segal [28]. A third type of integral that is outside the scope of the dimension equation is the Godement-Jacquet integral [27]. Finally, the WO-model integrals of [4] do not satisfy the dimension equation.

These last examples all raise the question: is it possible to modify the dimension equation so that it encompasses these examples? In fact, the answer is yes. We explain this in the next Section below.

4 Extending the Dimension Equation

Let us begin with the general form of a Rankin-Selberg integral, following [19]. Let G_i, $1 \leq i \leq l$ be reductive groups defined over F, π_i be irreducible automorphic representations of $G_i(\mathbb{A})$, and let $\varphi_i \in \pi_i$ be automorphic forms. Suppose that at least one automorphic form, say $\varphi_1 \in \pi_1$, is cuspidal so that the integral to be considered converges. Let $U_i \subset G_i$ be subgroups attached to *some* unipotent orbits of G_i, and let ψ_i be characters of $[U_i]$. Here the groups U_i may be trivial, and the characters ψ_i need not be in 'general position' for U_i; in particular we do not assume that U_i is attached to the unipotent orbit of π_i. We suppose that there is a reductive group G such that for each i, the stabilizer of ψ_i inside a suitable Levi subgroup of G_i contains G (up to isomorphism). In this case if φ_i is an automorphic form on G_i then the Fourier coefficient $\varphi_i^{U_i,\psi_i}$ is automorphic as a function of $G(\mathbb{A})$. Then we consider the integral

$$\int_{[Z\backslash G]}\int_{[U_1]}\int_{[U_2]}\cdots\int_{[U_l]}\varphi_1(u_1g)$$
$$\varphi_2(u_2g)\ldots\varphi_l(u_lg)\,\psi_1(u_1)\psi_2(u_2)\ldots\psi_l(u_l)\,du_l\ldots du_2\,du_1\,dg$$
$$=\int_{[Z\backslash G]}\varphi_1^{U_1,\psi_1}(g)\varphi_2^{U_2,\psi_2}(g)\ldots\varphi_l^{U_l,\psi_l}(g)\,dg, \tag{10}$$

where Z is the center of G and the central characters are chosen compatibly so that the integrand is $Z(\mathbb{A})$-invariant.

We are concerned with the case that one automorphic form, say φ_l, is an Eisenstein series induced from a Levi subgroup $P = MN$ and an automorphic representation τ of $M(\mathbb{A})$. Write φ_l as $\varphi_l(g,s)$ and write the associated section $f_\tau(g,s)$, so for $\Re(s) \gg 0$ we have

$$\varphi_l(g,s) = \sum_{\gamma \in P(F)\backslash G(F)} f_\tau(\gamma g, s).$$

Then the integral (10) is a function of a complex variable s, defined for $\Re(s) \gg 0$ and with analytic continuation (possibly with poles) and functional equation by virtue of the corresponding properties for the Eisenstein series $\varphi_l(g,s)$.

Such integrals are expected to represent L-functions when they can be shown to be equal to adelic integrals of some factorizable coefficients of the functions in the integrand. This requires an unfolding process. We begin with the unfolding described in [19], and then explain how the assumptions there may be weakened to arrive at a more general case. To describe the unfolding considered in [19], we introduce the following notation.

First, suppose that for each i, $1 \le i \le l-1$, the automorphic representation π_i attached to φ_i has the property that $\mathcal{O}(\pi_i)$ consists of a single unipotent orbit $\mathcal{O}_i$ with unipotent group V_i, and that φ_i has nonzero Fourier coefficients with respect to the generic character ψ_{V_i} of $[V_i]$. Write the associated Fourier coefficient

$$L_i(\varphi_i, g) = \int_{[V_i]} \varphi_i(vg)\psi_{V_i}(v)\,dv, \qquad g \in G_i(\mathbb{A}). \tag{11}$$

We emphasize that this Fourier coefficients is not a number, but a function of $g \in G_i(\mathbb{A})$. Also, since the first step in analyzing an integral of the form (10) is to unfold the Eisenstein series $\varphi_l(g,s)$, let $V_l = V_\tau N$ where V_τ is the unipotent subgroup of $M \subseteq G_l$ corresponding to the maximal unipotent orbit $\mathcal{O}_\tau$ attached to τ (again assumed unique), and let ψ_{V_l} be a corresponding character of the unipotent subgroup V_τ extended to a character of V_l which is trivial on N. Then we write

$$L_l(f_\tau, g, s) = \int_{[V_l]} f_\tau(vg,s)\psi_{V_l}(v)\,dv, \qquad g \in G_l(\mathbb{A}). \tag{12}$$

We suppose that after an unfolding process, the integral (10) may be shown to equal an integral involving the Fourier coefficients (11), (12). To write the unfolded integral, for $1 \le i \le l$, let R_i be a *unipotent* subgroup of G_i, and ψ_{R_i} be a character of $R_i(\mathbb{A})$. For $1 \le i \le l-1$ write

$$\varphi_i^{R_i}(g) = \int_{R_i(\mathbb{A})} L_i(\varphi_i, rg)\psi_{R_i}(r)\,dr, \qquad g \in G_i(\mathbb{A})$$

and similarly let

$$f_\tau^{R_l}(g,s) = \int_{R_l(\mathbb{A})} L_l(f_\tau, rg, s)\psi_{R_l}(r)\,dr, \qquad g \in G_l(\mathbb{A}).$$

Then, following [19], we suppose that for $\Re(s) \gg 0$ the integral (10) unfolds to

$$\int_{Z(\mathbb{A})M(\mathbb{A})\backslash G(\mathbb{A})} \varphi_1^{R_1}(g)\varphi_2^{R_2}(g)\dots\varphi_{l-1}^{R_{l-1}}(g)f_\tau^{R_l}(w_0 g, s)\,dg, \tag{13}$$

where M is a subgroup of G and w_0 is a Weyl group element. Such an integral is called a *unipotent global integral.*

For the convenience of the reader, we give an example. The integral (7) is of the form (10) with $l = 3$, groups $G_1 = G_2 = G_3 = GL_n$, π_1, π_2 cuspidal, and π_3 the automorphic representation corresponding to the mirabolic Eisenstein series on $GL_n(\mathbb{A})$. The groups U_1, U_2, U_3 and their characters are trivial, and G in (10) is again GL_n. Since π_1, π_2 are cuspidal on GL_n they are generic, so V_1 and V_2 are each the standard maximal unipotent subgroup of GL_n and L_1 and L_2 are global Whittaker functionals. Since π_3 is induced from a character, V_τ is trivial and $L_3(f_\tau, g, s) = f_\tau(g, s)$. The unfolding in (13) is given in this case in [30, Sections (4.3)–(4.5)]. One arrives at (13) with M equal to the maximal unipotent subgroup of GL_n and each of the groups R_i trivial. For an additional example, see [19, p. 173].

Returning to the general case of (13), when the functionals

$$\mathcal{L}_i(\varphi_i) = \varphi_i^{R_i}(e),\ 1 \le i \le l-1; \qquad \mathcal{L}_{l,s}(f_\tau) = f_\tau^{R_l}(e, s)$$

are each factorizable, then the unfolded integral (13) is Eulerian. Such an integral is called a *Eulerian unipotent integral* in [19]. In fact, this broad class of integrals includes all the Rankin-Selberg integrals presented above which satisfy the original form of the dimension equation. More generally, the dimension equation is expected to hold for all Eulerian unipotent integrals, and a classification of this class of integrals is initiated in [19].

Our goal now is to extend the dimension equation to include many of the examples noted at the last section, that is, Rankin-Selberg integrals that do not satisfy the dimension equation. To do so, we begin with the same integral (10) but we extend the notation so that each φ_i is now either a single automorphic form in π_i or a pair of automorphic forms, one in π_i and the other in its contragredient $\widetilde{\pi}_i$. We also relax the description of the unfolding above in two ways. We suppose once again that the integral unfolds to an Eulerian expression (13). However, in this expression we no longer assume that L_i is a Fourier coefficient given by an integral of the form (11) or (12) over a *unipotent* subgroup V_i against a character of the maximal unipotent orbit attached to π_i or τ. Instead we allow the integrals in (11), (12) to be over arbitrary subgroups of G_i. For example, L_i might be an integral realizing a unique functional such as the Shalika functional or the spherical functional. In this case, we do not use $\dim \pi_i$ in the dimension equation. Instead, we replace this term by the dimension of the full group that realizes the unique functional.

To be specific, for $1 \leq i \leq l-1$ suppose that there is an algebraic group $X_i \subset G_i$, not necessarily unipotent, such that

$$L_i(\varphi_i)(g) = \int_{[X_i]} \varphi_i(xg)\, \psi_{X_i}(x)\, dx,$$

where ψ_{X_i} is a character of $[X_i]$, and let $\mathcal{L}_i$ be the associated functional on π_i. In this case we define the dimension of $\mathcal{L}_i$ to be the dimension of the algebraic group X_i. For example, if $\varphi_i = (\phi, \phi')$ is a pair of automorphic functions with $\phi \in \pi_i$, $\phi' \in \widetilde{\pi}_i$, we may consider $\mathcal{L}_i$ to be the functional that assigns to φ_i the global matrix coefficient

$$\mathcal{L}_i(\varphi_i) = \int_{[G_i]} \phi(g)\phi'(g)\, dg. \tag{14}$$

In this case we define $\dim(\mathcal{L}_i) = \dim G_i$. Similarly, for the Eisenstein series induced from P, we consider

$$L_l(f_\tau, g, s) = \int_{[X_l]} f_\tau(xg, s)\psi_{X_l}(v)\, dv, \qquad g \in G_l(\mathbb{A}),$$

and we define $\dim(\mathcal{L}_l) = \dim X_l + \dim N$. Note that we include the dimension of N in the dimension of the functional $\mathcal{L}_l$. This gives, by definition, an extension of (4).

This definition of the dimension of a functional requires a coda. To explain why, suppose that $E(g, s)$ is the mirabolic Eisenstein series on $GL_3(\mathbb{A})$. This function has unipotent orbit (21) (in fact, it generates the automorphic minimal representation for $GL_3(\mathbb{A})$), so it has dimension 2. However, if $e_{i,j}$ denotes the matrix with 1 in position (i, j) and 0 elsewhere, then it is easy to prove that

$$\int_{(F\backslash\mathbb{A})^2} E(I_3 + re_{1,2} + me_{1,3}, s)\, \psi(r)\, dr\, dm$$
$$= \int_{(F\backslash\mathbb{A})^3} E(I_3 + re_{1,2} + me_{1,3} + ne_{2,3}, s)\, \psi(r)\, dr\, dm\, dn$$

and in fact both corresponding functionals are nonzero and unique. More generally, when an orbit $\mathcal{O}$ is small the integral defining the Fourier coefficient with respect to that orbit will have additional invariance properties (there is a nontrivial group that normalizes $U_{\mathcal{O}}$ and stabilizes $\psi_{\mathcal{O}}$) and so can be enlarged in a similar way. Hence to define the dimension of a functional $\mathcal{L}$ we must specify that if a functional over a smaller group also realizes $\mathcal{L}$ then we use that smaller group in defining the dimension of $\mathcal{L}$. Note that this may depend on the representation under consideration.

We broaden the dimension equation to the following *extended dimension equation*.

Definition 2. *The Extended Dimension Equation is the equality*

$$\dim(G) - \dim(Z) + \sum_{i=1}^{l} \dim(U_i) = \sum_{i=1}^{l} \dim(\mathcal{L}_i). \tag{15}$$

That is, the sum of the dimensions of the groups in the Rankin-Selberg integral (10) is equal to the sum of dimensions of the *functionals* that are obtained after unfolding. The key point is that we are using the dimensions of functionals in place of the Gelfand-Kirillov dimensions of the representations π_i.

In practice, to use the extended dimension equation we aim at specific factorizable functionals (Whittaker functionals, Shalika functionals, etc.) or matrix coefficients (14) for the $\mathcal{L}_i$. Then the equation prescribes a relation on the groups. It turns out that this relation is often highly useful for predicting integrals that represent L-functions.

One may make minor modifications in the above formalism keeping the same principle. For example, the initial integral may be over G rather than $Z\backslash G$, as in (8). Also, if there is more than one Eisenstein series that is unfolded in the integral, then the set up and equation may be modified to reflect this.

One may ask why such an equation ought to hold when the integral unfolds (indeed, one may ask this for the original dimension equation as well). In some cases when we perform the unfolding of a global integral the generalized dimension equation is preserved. A first example of this invariance is indicated on p. 3 of [20]. However, we regard this equation basically as an experimental tool that allows us to sort among many possible integrals those that are most likely to be of interest.

We conclude this section by comparing the extended dimension equation with the original form of the dimension equation, where $\dim(\mathcal{L}_i)$ is replaced by $\dim(\pi_i)$, a quantity which is computed by (3). Suppose first that π is an automorphic representation such that $\mathcal{O}(\pi)$ is a single unipotent orbit $\mathcal{O}$. Recall that there is an associated filtration $N_{i,\mathcal{O}}$ of N given by (1). Let φ_π be an automorphic form in the space of π. If $N_{1,\mathcal{O}} = N_{2,\mathcal{O}}$, and if an integral involving φ_π unfolds to the Fourier coefficient $\varphi_\pi^{U_\mathcal{O},\psi_\mathcal{O}}$ i.e. to the functional (11) given by integration over $[N_{2,\mathcal{O}}]$, then since $\dim(\pi) = \dim(N_{2,\mathcal{O}})$ (by (3)), in this situation $\dim \mathcal{L} = \dim \pi$. If instead $N_{1,\mathcal{O}} \neq N_{2,\mathcal{O}}$, then $N_{1,\mathcal{O}}/N_{2,\mathcal{O}}$ has the structure of a Heisenberg group. In this situation, it is often the case that an integral involving φ_π unfolds to a Fourier-Jacobi coefficient of the form

$$L(\varphi_\pi)(h) := \int\limits_{[N_{1,\mathcal{O}}]} \theta_{Sp}(l(v)h)\varphi_\pi(vh)\,\psi_{N_1}(v)\,dv, \tag{16}$$

where θ_{Sp} is a certain theta function obtained via the Weil representation. (Here we might need to involve covering groups.) See for example [24, Section 3.2]; the notation is given in detail there. If $\mathcal{L}$ is the functional obtained by composing L with evaluation at the identity, then in this case we would have $\dim \mathcal{L} = \dim N_{1,\mathcal{O}}$.

However, if Θ_{Sp} is the representation corresponding to θ_{Sp}, then it is known that $\dim \Theta_{Sp} = \frac{1}{2}\dim(N_{1,\mathcal{O}}/N_{2,\mathcal{O}})$. Since $\dim \pi = \dim N_{2,\mathcal{O}} + \frac{1}{2}\dim(N_{1,\mathcal{O}}/N_{2,\mathcal{O}})$ (see (3)), we obtain the equality $\dim \Theta_{Sp} + \dim \pi = \dim N_{1,\mathcal{O}}$. Thus the definition of $\dim \mathcal{L}$ is in fact consistent with original dimension equation in this case. We conclude that when Fourier-Jacobi coefficients are used to construct Eulerian integrals (see for example [22]), the different versions of the dimension equation we have presented are consistent, and it is accurate to label (15) an extension of the original dimension equation. Last, for the Eisenstein series, if the integral in the inducing data τ unfolds to the Whittaker functional then the definition of $\dim(\mathcal{L}_l)$ is consistent with (4) and with the original Dimension Equation.

5 Examples of the Extended Dimension Equation

In this section we offer examples of the extended dimension equation, and explain how other classes of integrals representing L-functions fit into the picture. To begin, *both* integrals of the form (9) satisfy this extended dimension equation. Indeed, if E is the Siegel Eisenstein series whose unipotent orbit has dimension 3, then this integral unfolds to a WO model for π in the sense of [4]. This model involves an integration over a 3-dimensional reductive group (a form of SO_3) as well as a 4-dimensional unipotent group so in this case the dimension of the functional applied to π is 7. The modified dimension equation does indeed hold, in the form $10 = 7+3$ (in contrast to the integral involving the Klingen Eisenstein series, where the contributions from the two functions in the integrand were 6 and 4).

Next we discuss doubling integrals. This is a class of integrals introduced by Piatetski-Shapiro and Rallis [36], of the form

$$\int\limits_{G(F)\times G(F)\backslash G(\mathbb{A})\times G(\mathbb{A})} \varphi_\pi(g)\varphi_\sigma(h)E(\iota(g,h),s)\,dg\,dh \qquad (17)$$

where G is a symplectic or orthogonal group, π and σ are two irreducible cuspidal automorphic representations of $G(\mathbb{A})$, and φ_π, φ_σ are in the corresponding spaces of automorphic forms. The Eisenstein series is defined on an auxilliary group $H(\mathbb{A})$ and $\iota : G \times G \to H$ is an injection. They show that after unfolding the Eisenstein series, the open orbit representative involves the inner product

$$< \varphi_\pi, \varphi_\sigma >= \int\limits_{G(F)\backslash G(\mathbb{A})} \varphi_\pi(g)\varphi_\sigma(g)\,dg \qquad (18)$$

as inner integration. This integral is nonzero unless σ is the contragredient of π, and in that case, the integral involves the functional (14), that is, the matrix coefficient, which is factorizable. It is readily checked that the original form of the dimension equation does not hold.

The dimension of the functional (14) is equal to $\dim(G)$. Thus the extended dimension equation attached to the integral (17) is

$$2\dim(G) = \dim(G) + \dim(E),$$

that is,

$$\dim(G) = \dim(E).$$

Moreover, this equation is satisfied by the integrals in [36]. For example, consider integral (17) with $G = Sp_{2n}$. In this case the Eisenstein series $E(\cdot, s)$ is defined on the group $H(\mathbb{A})$ with $H = Sp_{4n}$. It is the maximal parabolic Eisenstein series attached to the parabolic P with Levi factor GL_{2n} obtained by inducing the modular character δ_P^s. Thus $\dim(E) = \dim(U(P))$, where $U(P)$ is the unipotent radical of P. This is given by

$$\dim(U(P)) = \frac{1}{2}(1 + 2 + \cdots + 2n) = n(2n+1) = \dim(Sp_{2n}).$$

Thus the extended dimension equation indeed holds for the doubling integral (17). We remark that $E(\cdot, s)$ is attached to the unipotent orbit (2^{2n}).

It may similarly be confirmed that the dimension equation holds for the generalized doubling integrals of Cai et al. [11], that represent the Rankin-Selberg L-function on $G \times GL_k$, where G is a classical group, attached to the tensor product of the standard representations. These integrals are of the form

$$\int\limits_{G(F)\times G(F)\backslash G(\mathbb{A})\times G(\mathbb{A})} \varphi_\pi(g)\,\varphi_\sigma(h)\,E_\tau^{U,\psi_U}(\iota(g,h), s)\,dg\,dh \qquad (19)$$

where now E_τ is an Eisenstein series on a larger group H whose construction depends on τ and the superscripts on E denote a Fourier coefficient with respect to a unipotent group U and character ψ_U such that $\iota(G \times G)$ is contained in the stabilizer of ψ_U inside the normalizer of U in H. Suppose that $G = Sp_{2n}$. Then $H = Sp_{4kn}$, the Eisenstein series is induced from a generalized Speh represntation on GL_{2kn} which has unipotent orbit (k^{2n}), and the (U, ψ_U) coefficient is one corresponding to the unipotent orbit $((2k-1)^{2n}1^{2n})$ in H. The integral once again unfolds to an integral involving matrix coefficients. Due to the Fourier coefficient with respect to (U, ψ_U), the extended dimension equation in this case becomes

$$2\dim(Sp_{2n}) + \dim(U) = \dim(Sp_{2n}) + \dim(E_\tau). \qquad (20)$$

To show that this is true, it follows from [17] that

$$\dim(E_\tau) = \frac{1}{2}\dim((k)^{2n}) + \dim(U(P)),$$

where $\dim((k)^{2n})$ is the dimension of the unipotent orbit of the inducing data, and $U(P)$ is the unipotent radical of the maximal parabolic P inside H whose Levi factor is GL_{2kn}. (That is, (4) holds in this situation.) The number $\frac{1}{2}\dim((k)^{2n}))$ is equal to the dimension of the unipotent radical of the parabolic subgroup of GL_{2nk} whose Levi part is GL_{2n}^k, that is, $2n^2k(k-1)$. It is then easy to check that (20) holds.

There are other classes of Eulerian integrals. The 'new way' integrals of [37] unfold to functionals that are not unique. They satisfy the extended dimension equation, albeit tautologically. The integrals of Godement-Jacquet type may be obtained from doubling integrals after unfolding. Hence they should be regarded as belonging to this paradigm. Whether or not this is helpful for efforts to extend the method (the Braverman-Kazhdan-Ngô program) remains to be seen.

6 Integral Kernels and the Dimension Equation

Integral kernels appear often in the theory of automorphic forms as a way to relate automorphic forms on one group to automorphic forms on a different group. In this brief Section, we explain the connection between such constructions and the dimension equation and illustrate the use of this equation to detect properties of such a correspondence. We follow [18] and provide an additional example of interest. Then in the next Section we turn to a similar analysis using the extended dimension equation.

Suppose that G, H, L are reductive groups and there is an embedding $\iota : G \times H \to L$ such that the images of G and H in L commute. Let U be a unipotent subgroup of L and ψ_U be a character of $[U]$ whose stabilizer in L contains $\iota(G(\mathbb{A}), H(\mathbb{A}))$. Let Θ denote an automorphic representation of $L(\mathbb{A})$. Then one may seek to construct a lifting from automorphic representations of $G(\mathbb{A})$ to $H(\mathbb{A})$ as follows.

Let π denote an irreducible cuspidal representation of $G(\mathbb{A})$. Let σ be the representation of $H(\mathbb{A})$ generated by all functions of the form

$$f(h) = \int\limits_{[G]}\int\limits_{[U]} \varphi_\pi(g)\,\theta(u(g,h))\,\psi_U(u)\,du\,dg, \tag{21}$$

with θ in the space of Θ and φ_π in the space of π. Notice that from the properties of Θ, the functions $f(h)$ are $H(F)$-invariant functions on $H(\mathbb{A})$. As a first case, suppose that σ is an irreducible automorphic representation of $H(\mathbb{A})$. In that case, the dimension equation attached to this construction is

$$\dim(G) + \dim(U) + \dim(\sigma) = \dim(\pi) + \dim(\Theta). \tag{22}$$

That is, since the integral (21) gives a representation on $H(\mathbb{A})$ instead of an L-function, we include the dimension of the lift, σ, with the dimensions of the groups. See [18, Section 6]. More generally, suppose that σ is in the discrete part of the space $L^2(H(F)Z_H(\mathbb{A})\backslash H(\mathbb{A}), \omega)$, where Z_H is the center of H and ω is a central character (this is true, for example, when σ is cuspidal). In that case we expect that at least one of the summands of σ will satisfy Eq. (22).

This simple equation turns out to be quite powerful. We illustrate with an example. Suppose that $G = Sp_{2n}$, $H = SO_{2k}$, $L = Sp_{4nk}$, and Θ is the classical theta representation. Note that the unipotent orbit attached to Θ is (21^{4nk-2}), and $\dim(\Theta) = 2nk$. Suppose also that π is generic, so $\dim(\pi) = n^2$. Since $\dim(G) = 2n^2 + n$, the dimension equation (22) becomes

$$\dim(\sigma) = 2nk - n - n^2.$$

As a first consequence, if $k < \frac{n+1}{2}$ then the equation would assert that $\dim(\sigma) < 0$. So we expect that the lift must be zero. Second, let us ask for which k the lift σ can be generic. In that case, we would have $\dim(\sigma) = k^2 - k$. We conclude that a necessary condition for σ to be generic is the condition

$$k^2 - k = 2nk - n - n^2.$$

For a fixed n, there are two solutions to this equation, namely $k = n + 1$ and $k = n$. And indeed, both these consequences of the dimension equation are true. The lift does vanish if $k < \frac{n+1}{2}$. And the lift to SO_{2k} with $k = n + 1$ is always generic while the lift with $k = n$ is sometimes generic, and these are the only cases where the lift of a generic cuspidal automorphic representation is generic. See [23, Cor. 2.3] and the last two paragraphs in Section 2; for the analogous local result see Proposition 2.4 there.

7 Doubling Integrals and Integral Kernels

In this Section we connect doubling integrals and integral kernels. First let us describe such doubling integrals in general. Suppose that H is a group, $U \subseteq H$ is a unipotent subgroup, and ψ_U is a character of $[U]$. Suppose that there is an embedding $\iota : G \times G \to N_H(U)$ whose image fixes ψ_U (under conjugation). Let π, σ be automorphic representations of a group $G(\mathbb{A})$. Then we consider integrals of the form

$$\int_{[G\times G]} \int_{[U]} \varphi_\pi(g)\, \varphi_\sigma(h)\, E(u\iota(g,h), s, f_s)\psi_U(u)\, du\, dg\, dh. \tag{23}$$

Note that this includes the integrals (17) considered by Piatetski-Shapiro and Rallis but that we allow an extra unipotent integration, as in (19). We say that the integral (23) is a doubling integral if for $\Re(s)$ large it is equal to

$$\int_{G(\mathbb{A})} \int_{U_0(\mathbb{A})} < \varphi_\pi, \sigma(g)\varphi_\sigma > f_W(\delta u_0(g, 1), s)\psi_U(u_0)\, du_0\, dg, \tag{24}$$

for some subgroup U_0 of U, some Fourier coefficient f_W of the section f_s, and some $\delta \in H(F)$. Due to the matrix coefficient, this integral is zero unless σ is the contragredient of π, so we suppose this from now on.

The extended dimension equation attached to the integral (23) is

$$\dim(G) + \dim(U) = \dim(E). \tag{25}$$

We emphasize that if a certain integral satisfies the extended dimension equation (25) it does not necessarily means that the integral will be non-zero or Eulerian. This can only be determined after the unfolding process. The advantage of the Eq. (25) is that it eliminates unlikely candidates.

Suppose that (23) is a doubling integral. Then we may make an integral kernel as follows. Fix a cuspidal automorphic representation π of $G(\mathbb{A})$, and consider the functions

$$f(h) = \int_{[G]} \int_{[U]} \varphi_\pi(g) E(u(g, h), s)\psi_U(u)\, du\, dg. \tag{26}$$

Arguing as in Theorem 1 of Ginzburg and Soudry [25] we deduce that the representation σ generated by these functions is a certain twist of the representation π. In particular we have $\dim(\sigma) = \dim(\pi)$. Hence the dimension equation (22) is the same as the extended dimension equation (25). Accortdingly, one can view the construction given by (21) as a generalization of the construction of doubling integrals.

This discussion leads to the following Classification Problem, which illustrates the kind of question that these constructions raise. Let π denote a cuspidal representation of $G(\mathbb{A})$, and σ denote a cuspidal representation of $H(\mathbb{A})$. Find examples of representations Θ defined on a group $L(\mathbb{A})$ as above which satisfy the following two conditions:

1. The dimension equation (22) holds.
2. Suppose that the integral

$$\int_{[G\times H]} \int_{[U]} \varphi_\sigma(h)\varphi_\pi(g)\theta(u(g, h))\, \psi_U(u)\, du\, dg\, dh \tag{27}$$

is not zero for some choice of data. Then the representation π determines the representation σ uniquely.

This classification problem has many solutions as stated. We present an example of how it may be approached in the next Section.

We remark that a similar analysis based on the dimension equation may be applied to the descent integrals of Ginzburg et al. [24].

8 An Example

In this Section we will show how the above considerations can help find possible global integrals of the form given by Eq. (27). To do so, in this Section we will work with the case $G = H = Sp_{2m}$.

The first step is to consider Fourier coefficients whose stabilizer contains the group $Sp_{2m} \times Sp_{2m}$. From the theory of nilpotent orbits, the partition $((2k-1)^{2m}(2r-1)^{2m})$ has this property (see [12]). This leads us to look for a representation Θ defined on the group $Sp_{4m(k+r-1)}(\mathbb{A})$ which is $Sp_{4m(k+r-1)}(F)$ invariant and which satisfies the dimension equation given in (22). Thus, we are seeking representations Θ such that $\mathcal{O}(\Theta) = \mathcal{O}$ with

$$\dim(Sp_{2m}) + \tfrac{1}{2}\dim((2k-1)^{2m}(2r-1)^{2m}) = \tfrac{1}{2}\dim(\mathcal{O}). \tag{28}$$

Here $\mathcal{O}$ corresponds to a partition of the number $4m(k+r-1)$. There are many solutions. For example, if we begin with the orbit $(5^2 3^2)$, that is $m = 1$, $k = 3$ and $r = 2$, then the orbits $\mathcal{O}$ equal to (65^2), $(83^2 2)$, $(6^2 2^2)$, and (8421^2) all satisfy condition (28).

Since this low rank case already offers so many possibilities, this suggests that it is not expeditious to classify all orbits $\mathcal{O}$ which satisfy (28). Experience suggests that a good place to begin a further analysis is to focus on orbits of the form $(n_1^{2l_1} n_2^{2l_2} \ldots n_p^{2l_p})$ such that p is minimal. For example, in the case above, if we begin with the orbit $(5^2 3^2)$, we would seek Θ such that $\mathcal{O}(\Theta) = (6^2 2^2)$. In general we have

Lemma 1. *We have*

$$\dim(Sp_{2m}) + \tfrac{1}{2}\dim((2k-1)^{2m}(2r-1)^{2m}) = \tfrac{1}{2}\dim((2k)^{2m}(2r-2)^{2m}).$$

Proof. Using Eq. (2), one may compute the difference

$$\tfrac{1}{2}\dim((2k)^{2m}(2r-2)^{2m}) - \tfrac{1}{2}\dim((2k-1)^{2m}(2r-1)^{2m})$$

and to show that it is equal to $2m^2 + m = \dim(Sp_{2m})$. We omit the details. □

We observe that the doubling construction of the authors, Cai and Kaplan [11] fits this rubric, indeed, it provides an example of such an integral with $r = 1$. More precisely, the representation Θ used in [11] is an Eisenstein series $E(\cdot, s)$ defined on $Sp_{4mk}(\mathbb{A})$, and it satisfies $\mathcal{O}(E(\cdot, s)) = ((2k)^{2m})$. Then a Fourier coefficient of $E(\cdot, s)$ is taken with respect to a unipotent group U and generic character corresponding to the partition $((2k-1)^{2m}1^{2m})$ of 4 km.

Returning to the general case, the next step is to find an automorphic representation Θ defined on the group $Sp_{4m(k+r-1)}(\mathbb{A})$ which satisfies $\mathcal{O}(\Theta) = ((2k)^{2m}(2r-2)^{2m})$. The main source of examples for such representations are Eisenstein series and their residues. If we have a candidate for Θ which is an Eisenstein series, then by unfolding the Eisenstein series it is often possible to check if the non-vanishing of Eq. (27) implies that π determines σ uniquely. See [11] for an example. However, if the representation Θ is obtained as a *residue* of an Eisenstein series, an unfolding process is not readily available to us unless we attempt to unfold first and then take the residue, a strategy that is often problematic, and in this case, typically only a weaker statement can be checked.

In the rest of this Section we will describe a simple case where the representation Θ is not an Eisenstein series. We will only give the flavor of the construction here; we plan to present the details in a separate paper.

Let τ denote an irreducible cuspidal representation of the group $GL_n(\mathbb{A})$. Suppose that $n > 2$ is even and that the partial L-function $L^S(\tau, \wedge^2, s)$ has a simple pole at $s = 1$. Let $\mathcal{E}_\tau$ denote the generalized Speh representation defined on the group $GL_{3n}(\mathbb{A})$. See [11]. Let $P(GL_{3n})$ be the maximal parabolic subgroup of Sp_{6n} whose Levi part is GL_{3n} (the so-called Siegel parabolic). Let $E_\tau(\cdot, s)$ denote the Eisenstein series defined on the group $Sp_{6n}(\mathbb{A})$ attached to the induced space $Ind_{P(GL_{3n})(\mathbb{A})}^{Sp_{6n}(\mathbb{A})}\mathcal{E}_\tau\delta_P^s$. The poles of this Eisenstein series are determined by the poles of $L^S(\tau, \wedge^2, s)$ and $L^S(\tau, \vee^2, s)$. See [31]. It follows from that reference that if $L^S(\tau, \wedge^2, s)$ has a simple pole at $s = 1$, then the Eisenstein series $E_\tau(\cdot, s)$ has a simple pole at $s_0 = (3n+2)/(6n+2)$. Denote $\Theta_\tau' = \text{Res}_{s=s_0} E_\tau(\cdot, s)$. That is, Θ_τ' is the automorphic representation spanned by the residues of this family of Eisenstein series at the point s_0. Then, it follows from [26] that there is an irreducible constituent of Θ_τ of Θ_τ' such that

$$\mathcal{O}(\Theta_\tau) = ((2n)^2 n^2). \tag{29}$$

Choosing $k = n$, $m = 1$ and $r = (n+2)/2$ in Lemma 1 we conclude that the dimension equation (22) is satisfied, and we may form the global integral (27).

More generally, suppose that the representation Θ in (27) satisfies $\mathcal{O}(\Theta) = ((2n)^2 n^2)$. Suppose moreover that $\Theta = \otimes_\nu' \Theta_\nu$ (restricted tensor product) where for almost all ν the unipotent orbit attached to $(\Theta)_\nu$ is also $((2n)^2 n^2)$. This is true if we take Θ to be Θ_τ. Indeed, in this case the corresponding local statement may be proved by arguing analogously to the global case, i.e. replacing Fourier expansions by the geometric lemma, global root exchange by local root exchange, etc.

Suppose that π is an irreducible cuspidal automorphic representation of the group $SL_2(\mathbb{A})$ whose image under the 'lift' corresponding to Θ contains an irreducible cuspidal representation σ, also on the group $SL_2(\mathbb{A})$. We seek to establish the relation between π and σ assuming that integral (27) is not zero for some choice of data. Since the representation Θ is not an Eisenstein series we cannot simply carry out an unfolding process. However, in this case we can show that the representations π and σ are nearly equivalent. We sketch a local argument for representations in general position, omitting the details.

Suppose that $\pi = \otimes'_\nu \pi_\nu$, $\sigma = \otimes'_\nu \sigma_\nu$. At unramified places suppose that $\pi_\nu = \text{Ind}_B^{SL_2} \chi_\nu \delta_B^{1/2}$ and that $\sigma_\nu = \text{Ind}_B^{SL_2} \mu_\nu \delta_B^{1/2}$, where B is the standard Borel subgroup of SL_2 and χ_ν and μ_ν are unramified characters. Suppose that χ_ν is in general position. If the integral (27) is not zero for some choice of data, then the space

$$\text{Hom}_{SL_2 \times SL_2}(\text{Ind}_B^{SL_2} \chi_\nu \delta_B^{1/2} \times \text{Ind}_B^{SL_2} \mu_\nu \delta_B^{1/2}, J_{U,\psi_U}(\Theta_\nu)) \tag{30}$$

is not zero. Here J_{U,ψ_U} is the local twisted Jacquet module which corresponds to the Fourier coefficient over the group U and the character ψ_U of integral (27). It follows from Frobenius reciprocity that the space (30) is equal to

$$\text{Hom}_{GL_1 \times GL_1}(\chi_\nu \delta_B^{1/2} \mu_\nu \delta_B^{1/2}, J_{U_1,\psi_U}(\Theta_\nu)). \tag{31}$$

Here U_1 is a certain unipotent subgroup which contains U and the character ψ_U is the trivial extension from U to U_1. Then performing root exchanges and using that the unipotent orbit attached to $(\Theta)_\nu$ is $((2n)^2 n^2)$, one can prove that the nonvanishing of the space (31) implies that the space

$$\text{Hom}_{GL_1}(\chi_\nu \delta_B^{1/2} \mu_\nu \delta_B^{1/2}, J_{U_2,\psi_{U_2}}(\Theta_\nu)) \tag{32}$$

is also nonzero. Here GL_1 is embedded in $GL_1 \times GL_1$ diagonally, and $J_{U_2,\psi_{U_2}}$ is the twisted Jacquet module attached to a certain unipotent group U_2 and character ψ_{U_2}. This last space may then be analyzed by using properties of the representation Θ_ν, and to deduce that $\mu_\nu = \chi_\nu^{\pm 1}$. This example illustrates how the dimension equation may be used to suggest new integral kernels.

9 Local Analogues

The unfolding of a Rankin-Selberg integral typically has a local analogue, so when the dimension equation or extended dimension equation is satisfied, it is natural to seek local statements that hold. Once again we view the equation as necessary but not sufficient. In this Section we illustrate the local statements that arise. Let K be a non-archimedean local field whose residue field has cardinality q.

A first example is given by the Rankin-Selberg integrals of Jacquet, Piatetski-Shapiro and Shalika [29]. Let π_1, π_2 be irreducible admissible generic representations of $GL_n(K)$, $GL_k(K)$, resp. We keep the notation of Sect. 3 above. Consider π_2 to be a module for $Y_{n,k}(K)$ via ψ. If $k < n$, then ([29], paragraph (2.11), Proposition) the space of $(GL_k \ltimes Y_{n,k})(K)$-equivariant bilinear forms

$$\mathrm{Bil}_{(GL_k \ltimes Y_{n,k})(K)}(\pi \otimes |\det(\cdot)|^s, \pi')$$

has dimension at most 1, except for finitely many values of q^{-s}. Similarly, if $n = k$, the space

$$\mathrm{Bil}_{GL_n(K)}(\pi \otimes \pi' \otimes |\det(\cdot)|^s, \mathrm{Ind}_{P(K)}^{GL_n(K)} \delta_P^{-1/2})$$

has dimension at most 1, except for finitely many values of q^{-s} ([29], paragraph (2.10), Eqn. (5); see also the Proposition in (2.10) for an equivalent formulation in terms of trilinear forms). We caution the reader that in the literature this is presented using $GL_n(K)$-equivariance rather than $PGL_n(K)$-equivariance, but unless the central characters are chosen compatibly the space is zero. We need to use PGL_n to satisfy the dimension equation. This is comparable to insisting that we choose the group of smallest possible dimension in assigning a dimension to a functional.

As an additional example, in the situation of the generalized doubling integrals of Cai, Kaplan and the authors [11], the study of the global integral (19) leads to the local Hom space

$$\mathrm{Hom}_{G(K)\times G(K)}(J_{U,\psi_U^{-1}}(\mathrm{Ind}_{P(K)}^{H(K)}(W_c(\tau)\delta_P^s), \pi^\vee \otimes \pi)$$

where $J_{U,\psi_U^{-1}}$ denotes a twisted Jacquet module with respect to the group $U(K)$ and character ψ_U^{-1} and the remaining notation is given in [11], see especially (3.2) there. Once again, this space is at most one dimensional except for finitely many values of q^{-s} (see the proof of [11, Theorem 21]). As explained above, a dimension equation holds but only if we use the extended dimension equation and treat $\pi^\vee \otimes \pi$ as having dimension equal to $\dim(G)$.

Finally, when the dimension equation appears in the context of a lifting result, then one may hope to prove the existence of a local correspondence similar to the Howe correspondence for the classical theta representation. The local concerns that arise are illustrated by the treatment of (30) above. In particular, it is natural to seek to extend such a correspondence beyond a matching of the unramified principal series.

Acknowledgments This research was supported by the US-Israel Binational Science foundation, grant number 2016000, and by the NSF, grant number DMS-1801497 (Friedberg).

References

1. D. Barbasch, D.A. Vogan, Unipotent representations of complex semisimple groups. Ann. Math. **121**(1), 41–110 (1985)
2. D. Bump, The Rankin-Selberg method: an introduction and survey, in *Automorphic Representations, L-Functions and Applications: Progress and Prospects*, Ohio State University Mathematics Research Institute Publications, vol. 11 (de Gruyter, Berlin, 2005), pp. 41–73
3. D. Bump, S. Friedberg, *The Exterior Square Automorphic L-Functions on GL(n)*. Festschrift in Honor of I. I. Piatetski-Shapiro on the Occasion of His Sixtieth Birthday, Part II (Ramat Aviv, 1989), Israel Mathematical Conference Proceedings, vol. 3 (Weizmann, Jerusalem, 1990), pp. 47–65
4. D. Bump, S. Friedberg, D. Ginzburg, Whittaker-orthogonal models, functoriality and the Rankin-Selberg method. Invent. Math. **109**, 55–96 (1992)
5. D. Bump, S. Friedberg, D. Ginzburg, Rankin-Selberg integrals in two complex variables. Math. Ann. **313**(4), 731–761 (1999)
6. D. Bump, S. Friedberg, D. Ginzburg, Small representations for odd orthogonal groups. Int. Math. Res. Not. **2003**(25), 1363–1393 (2003)
7. D. Bump, S. Friedberg, D. Ginzburg, Lifting automorphic representations on the double covers of orthogonal groups. Duke Math. J. **131**(2), 363–396 (2006)
8. D. Bump, M. Furusawa, D. Ginzburg, Nonunique models in the Rankin-Selberg method. J. Reine Angew. Math. **468**, 77–111 (1995)
9. D. Bump, D. Ginzburg, Symmetric square L-functions on $GL(r)$. Ann. Math. **136**(1), 137–205 (1992)
10. Y. Cai, Fourier coefficients for degenerate Eisenstein series and the descending decomposition. Manuscripta Math. **156**, 69–501 (2018)
11. Y. Cai, S. Friedberg, D. Ginzburg, E. Kaplan, Doubling constructions and tensor product L-functions: the linear case. Invent. Math. **217**(3), 985–1068 (2019)
12. D.H. Collingwood, W.M. McGovern, *Nilpotent Orbits in Semisimple Lie Algebras*. Van Nostrand Reinhold Mathematics Series (Van Nostrand Reinhold, New York, 1993)
13. Y.Z. Flicker, Twisted tensors and Euler products. Bull. Soc. Math. France **116**(3), 295–313 (1988)
14. W.T. Gan, Doubling zeta integrals and local factors for metaplectic groups. Nagoya Math. J. **208**, 67–95 (2012)
15. W.T. Gan, Theta correspondence: recent progress and applications, in *Proceedings of the International Congress of Mathematicians—Seoul 2014*, vol. II (Kyung Moon Sa, Seoul, 2014), pp. 343–366
16. D. Ginzburg, L-functions for $SO_n \times GL_k$. J. Reine Angew. Math. **405**, 156–180 (1990)
17. D. Ginzburg, Certain conjectures relating unipotent orbits to automorphic representations. Israel J. Math. **151**, 323–355 (2006)
18. D. Ginzburg, Towards a classification of global integral constructions and functorial liftings using the small representations method. Adv. Math. **254**, 157–186 (2014)
19. D. Ginzburg, Classification of some global integrals related to groups of type A_n. J. Number Theory **165**, 169–202 (2016)
20. D. Ginzburg, On the length of global integrals for GL_n. Res. Number Theory **3**, 15pp. (2017)
21. D. Ginzburg, J. Hundley, A doubling integral for G_2. Israel J. Math. **207**(2), 835–879 (2015)
22. D. Ginzburg, D. Jiang, S. Rallis, D. Soudry, L-Functions for symplectic groups using Fourier-Jacobi models, in *Arithmetic Geometry and Automorphic Forms*. Advanced Lectures in Mathematics (ALM), vol. 19 (International Press, Somerville, 2011), pp. 183–207
23. D. Ginzburg, S. Rallis, D. Soudry, Periods, poles of L-functions and symplectic-orthogonal theta lifts. J. Reine Angew. Math. **487**, 85–114 (1997)
24. D. Ginzburg, S. Rallis, D. Soudry, *The Descent Map from Automorphic Representations of GL(n) to Classical Groups* (World Scientific Publishing, Hackensack, 2011)

25. D. Ginzburg, D. Soudry, Integrals derived from the doubling method, Int. Math. Res. Not. **2020**(4), 10553–10596 (2020)
26. D. Ginzburg, D. Soudry, The poles and residues of Eisenstein series induced from Speh representations. arXiv:2012.03717 (2020)
27. R. Godement, H. Jacquet, *Zeta Functions of Simple Algebras*. Lecture Notes in Mathematics, vol. 260 (Springer, Berlin, 1972)
28. N. Gurevich, A. Segal, The Rankin-Selberg integral with a non-unique model for the standard $\mathcal{L}$-function of G_2. J. Inst. Math. Jussieu **14**(1), 149–184 (2015)
29. H. Jacquet, I. Piatetskii-Shapiro, J. Shalika, Rankin-Selberg convolutions. Am. J. Math. **105**(2), 367–464 (1983)
30. H. Jacquet, J. Shalika, On Euler products and the classification of automorphic representations I. Am. J. Math. **103**(3), 499–558 (1981)
31. D. Jiang, B. Liu, L. Zhang, Poles of certain residual Eisenstein series of classical groups. Pac. J. Math. **264**(1), 83–123 (2013)
32. N. Kawanaka, Shintani lifting and Gel'fand-Graev representations, in *The Arcata Conference on Representations of Finite Groups (Arcata, Calif., 1986)*. Proceedings of Symposia in Pure Mathematics, vol. 47, Part 1 (American Mathematical Society, Providence, 1987), pp. 147–163
33. E.M. Lapid, S. Rallis, On the local factors of representations of classical groups, in *Automorphic Representations, L-Functions and Applications: Progress and Prospects*. Ohio State University Mathematics Research Institute Publications, vol. 11 (de Gruyter, Berlin, 2005), pp. 309–359
34. S. Leslie, A generalized theta lifting, CAP representations, and Arthur parameters. Trans. Am. Math. Soc. **372**, 5069–5121 (2019)
35. B. Liu, B. Xu, On top Fourier coefficients of certain automorphic representations of GL_n (preprint 2018). arXiv:1812.03157
36. I. Piatetski-Shapiro, S. Rallis, L -functions for the classical groups, in *Explicit Constructions of Automorphic L-Functions*, ed. by S. Gelbart, I. Piatetski-Shapiro, S. Rallis. Lecture Notes in Mathematics, vol. 1254 (Springer, Berlin, 1987)
37. I. Piatetski-Shapiro, S. Rallis, A new way to get Euler products. J. Reine Angew. Math. **392**, 110–124 (1988)
38. A. Pollack, S. Shah, Multivariate Rankin-Selberg integrals on GL_4 and $GU(2, 2)$. Can. Math. Bull. **61**(4), 822–835 (2018)
39. F. Shahidi, *Eisenstein Series and Automorphic L-Functions*. American Mathematical Society Colloquium Publications, vol. 58 (American Mathematical Society, Providence, 2010)
40. S. Takeda, The twisted symmetric square L-function of $GL(r)$. Duke Math. J. **163**(1), 175–266 (2014)
41. S. Yamana, L-functions and theta correspondence for classical groups. Invent. Math. **196**(3), 651–732 (2014)

Relative Character Identities and Theta Correspondence

Wee Teck Gan and Xiaolei Wan

Abstract We consider the Sakellaridis-Venkatesh program for the rank 1 spherical variety $X = \mathrm{SO}_{n-1}\backslash\mathrm{SO}_n$. Using theta correspondence, we classify the X-distinguished representations in terms of the dual group $X^\vee$ in the L^2-setting, the smooth local setting as well as the global setting. More precisely, we obtain the spectral decomposition of $L^2(X)$, establish local relative character identities and deduce the factorization of global period integrals. In particular, we give an alternative treatment of the theory of transfer of Schwarz functions for rank 1 spherical varieties recently developed by Sakellaridis.

1 Introduction

This paper is inspired by the talk of Yiannis Sakellaridis in the Simons Symposium held at the Schloss Elmau in April 2018. Let us begin by describing the relevant context for his talk.

The study of periods of automorphic forms has been an important theme in the Langlands program, beginning with the early work of Harder-Langlands-Rapoport and Jacquet. In particular, the nonvanishing of certain periods is known to characterize the image of certain Langlands functorial lifting and to be related to the analytic properties of certain automorphic L-functions. An effective approach for proving such results is the technique of relative trace formulae developed by Jacquet. Typically, such an approach involves the comparison of the geometric sides of two relative trace formulae, which results in a global spectral identity and an accompanying family of local relative character identities.

In [SV], Sakellaridis and Venkatesh initiated a general framework for treating such period problems in the context of spherical varieties. In particular, to a spherical variety $X = H\backslash G$ over a local field F or a global field k, they associated

W. T. Gan (✉) · X. Wan
Department of Mathematics, National University of Singapore, Singapore, Singapore
e-mail: matgwt@nus.edu.sg; xiaoleiwan@u.nus.edu

W. Müller et al. (eds.), *Relative Trace Formulas*, Simons Symposia,
https://doi.org/10.1007/978-3-030-68506-5_4

- a Langlands dual group $X^\vee$ (at least when G is split), together with a canonical (up to conjugacy) map

$$\iota : X^\vee \times \mathrm{SL}_2(\mathbb{C}) \longrightarrow G^\vee.$$

- a $\frac{1}{2}\mathbb{Z}$-graded finite-dimensional algebraic representation $V_X = \oplus_d V_X^d$ of $X^\vee$, which gives rise to an L-function

$$L_X(s, \rho) = \prod_d L(s + d, \rho, V_X^d)$$

for each L-parameter ρ valued in $X^\vee$

They then conjectured, among other things, that representations of G (in the automorphic dual) which have nonzero H-periods are those belonging to A-packets whose associated A-parameters factor through ι. This means roughly that the H-distinguished representations of G are those which are Langlands functorial lift via ι from a (split) group G_X whose dual group $G_X^\vee$ is $X^\vee$. Experience shows that it is sometimes more pertinent to regard H-distinguished representations of G as lifted from the *Whittaker variety* $(N_X, \psi)\backslash G_X$, as opposed to the group variety G_X itself.

The conjecture of Sakellaridis–Venkatesh can be made on several fronts. We give a brief description of the various incarnations of their conjecture (at least a first approximation), under some simplifying hypotheses and without using the language of A-parameters. Our description is adapted to the needs of this paper. For the conjecture in its most general form (taking into account Vogan L-packets for example), the reader should consult [SV].

(a) In the context of smooth representation theory of $G(F)$ over a local field F, one is interested in determining $\mathrm{Hom}_H(\pi, \mathbb{C})$ for any $\pi \in \mathrm{Irr}(G(F))$. One expects (in the context of this paper) a map

$$\iota_* : \mathrm{Irr}(G_X(F)) \longrightarrow \mathrm{Irr}(G(F)),$$

such that for any $\pi \in \mathrm{Irr}(G(F))$, there is an isomorphism

$$f : \bigoplus_{\sigma : \iota_*(\sigma) = \pi} \mathrm{Hom}_{N_X}(\sigma, \psi) \cong \mathrm{Hom}_H(\pi, \mathbb{C}).$$

In the smooth setting, the Sakellaridis-Venkatesh conjecture thus gives a precise quantitative formulation of the expectation that H-distinguished representations of G are lifted from G_X.

If further ι_* is injective, there is at most one term on the left hand side, and all these Hom spaces are at most one-dimensional (by the uniqueness of Whittaker models). This will be the favourable situation encountered in this paper. In such instances, if $\ell \in \mathrm{Hom}_{N_X}(\sigma, \psi)$, with corresponding $f(\ell) \in$

$\mathrm{Hom}_H(\iota_*(\sigma), \mathbb{C})$, one can define relative characters $\mathcal{B}_{\sigma,\ell}$ and $\mathcal{B}_{\iota_*(\sigma),f(\ell)}$ which are certain equivariant distributions on $(N_X, \psi)\backslash G_X$ and X respectively. In this case, one might expect a relative character identity relating $\mathcal{B}_{\sigma,\ell}$ and $\mathcal{B}_{\iota_*(\sigma),f(\ell)}$.

(b) In the context of L^2-representation theory, one is interested in obtaining the spectral decomposition of the unitary representation $L^2(X)$ of G (relative to a fixed G-invariant measure on X). By abstract results of functional analysis, one has a direct integral decomposition

$$L^2(X) \cong \int_{\Omega} \pi_{\omega}^{\oplus m(\omega)} \, d\mu_X(\omega)$$

where

- $(\Omega, d\mu_X)$ is some measure space;
- $\pi : \omega \mapsto \pi_\omega$ is a measurable field of irreducible unitary representations of G defined on Ω, giving rise to a measurable map from Ω to the unitary dual $\widehat{G}$ of G;
- $m : \Omega \to \mathbb{N} \cup \{\infty\}$ is a measurable multiplicity function.

There is some fluidity in this direct integral decomposition; for example, given Ω, only the measure class of $d\mu_X$ is well-defined (without explicating the isomorphism).

In this L^2-setting, the crux of the Sakellaridis-Venkatesh conjecture is to provide a canonical candidate for $(\Omega, d\mu_X, \pi)$. Namely, one expects a map

$$\iota_* : \widehat{G}_X \longrightarrow \widehat{G}$$

associated to ι from the unitary dual of G_X to that of G, so that one has a (unitary) isomorphism

$$L^2(X) \cong \int_{\widehat{G}_X} \iota_*(\sigma)^{\oplus m(\sigma)} \, d\mu_{G_X}(\sigma),$$

where $d\mu_{G_X}$ denotes the Plancherel measure of G_X and $m(\sigma)$ is a multiplicity space which is typically isomorphic to the dual space of $\mathrm{Hom}_{N_X}(\sigma, \psi)$. In other words, one may take $(\Omega, d\mu_X, \pi)$ to be $(\widehat{G}_X, d\mu_{G_X}, \iota_*)$. One can think of this as saying that the spectral decomposition of $L^2(X)$ is obtained from the Whittaker-Plancherel theorem

$$L^2(N_X, \psi\backslash G_X) \cong \int_{\widehat{G}_X} \sigma^{\oplus m(\sigma)} \, d\mu_{G_X}(\sigma)$$

by applying ι_*. One consequence of such a spectral decomposition is that it provides a canonical element $\ell_{\iota_*(\sigma)} \in \mathrm{Hom}_H(\iota_*(\sigma), \mathbb{C})$, as we explained in Sect. 2, for $m(-)\cdot d\mu_{G_X}$-almost all σ. Because of the presence of the Plancherel measure $d\mu_{G_X}$ of G_X, only tempered representations σ in $\widehat{G}_X$ need to be considered (though $\iota_*(\sigma)$ may be nontempered).

(c) Globally, when k is a global field with ring of adeles $\mathbb{A}$, one considers the global period integral along H:

$$P_H : \mathcal{A}_{cusp}(G) \longrightarrow \mathbb{C}$$

defined by

$$P_H(\phi) = \int_{H(k)\backslash H(\mathbb{A})} \phi(h)\, dh$$

on the space of cusp forms on G. The restriction of P_H to a cuspidal representation $\Pi = \otimes_v \Pi_v$ of G then defines an element $P_{H,\Pi} \in \mathrm{Hom}_{H(\mathbb{A})}(\Pi, \mathbb{C})$. One is interested in two problems in the global setting:

(i) characterising those Π for which $P_{H,\Pi}$ is nonzero as functorial lifts from G_X via the map ι;
(ii) seeing if $P_{H,\Pi}$ can be decomposed as the tensor product of local functionals.

Such a factorization certainly exists in the instances discussed in this paper since the local Hom spaces $\mathrm{Hom}_{H(F_v)}(\Pi_v, \mathbb{C})$ are at most one-dimensional for all places v. In (a), we have seen that these Hom spaces are nonzero precisely when $\Pi_v = \iota_*(\sigma_v)$ for some $\sigma_v \in \mathrm{Irr}(G_X(k_v))$. Thus, in the context of the first global problem, one would like to show that, if $P_{H,\Pi} \neq 0$, there exists a cuspidal representation Σ of G_X such that $\Sigma_v \cong \sigma_v$ for all v, so that $\Pi \cong \iota_*(\Sigma)$.

On the other hand, in (b), we have remarked that the spectral decomposition of $L^2(X)$ in the local setting gives rise to a canonical basis element $\ell_{\Pi_v} \in \mathrm{Hom}_{H(F_v)}(\Pi_v, \mathbb{C})$. For the second global problem, it is natural to compare the two elements $P_{H,\Pi}$ and $\prod_v^* \ell_{\Pi_v}$. Here the asterisk in the product indicates that there may be a need to normalize the local functionals ℓ_{Π_v} appropriately to ensure that the Euler product $\prod_v \ell_{\Pi_v}(\phi_v)$ converges. More precisely, to see if the Euler product converges, one would need to evaluate $\ell_{\Pi_v}(\phi_v^0)$ where ϕ_v^0 is a spherical unit vector in Π_v for almost all v. This evaluation has been carried by Sakellaridis in [S1, S2] and this is where the L-factor $L_X(s, -)$ associated to the $\frac{1}{2} \cdot \mathbb{Z}$-graded representation V_X enters the picture. Namely, it turns out that with $\Pi_v \cong \iota_*(\Sigma_v)$ for tempered $\Sigma_v \in \widehat{G}_{X,v}$, one has:

$$|\ell_{\Pi_v}(\phi_v^0)|^2 = L^{\#}_{X,v}(\Sigma_v) := \Delta_v(0) \cdot \frac{L_{X,v}(\Sigma_v)}{L(1, \Sigma_v, Ad)} > 0,$$

where

- $\Delta_v(s)$ is itself a product of local L-factors which depends only on X and not on the representation Σ_v;
- $L(s, \Sigma_v, Ad)$ denotes the adjoint L-factor of Σ_v and

- $L_{X,v}(\Sigma_v) := L_{X,v}(0, \Sigma_v) = \prod_d L(d, \Sigma_v, V_X^d)$ (following the convention of the introduction of [S4]).

This necessitates that one defines a normalization of ℓ_{Π_v} by:

$$\ell_{\Pi_v}^{\flat} = \frac{1}{|L_{X,v}^{\#}(\Sigma_v)|^{1/2}} \cdot \ell_{\Pi_v}$$

Then the main issue with the second global problem is to determine the constant $c(\Pi)$ such that

$$|P_{H,\Pi}(\phi)|^2 = c(\Pi) \cdot L_X^{\#}(\Sigma) \cdot \prod_v |\ell_{\Pi_v}^{\flat}(\phi_v)|^2 \qquad \text{for } \phi = \otimes_v \phi_v \in \otimes_v \Pi_v.$$

Here the global L-function $L_X^{\#}(s, \Sigma)$ is defined by the Euler product $\prod_v L_{X,v}^{\#}(s, \Sigma_v)$ for $\mathrm{Re}(s) \gg 0$ and needs to be meromorphically continued so that one can evaluate it at $s = 0$ to give the global L-value $L_X^{\#}(\Sigma)$.

This concludes our brief and simplified description of the Sakellaridis-Venkatesh conjecture. It is instructive to observe the crucial unifying role played by the typically ignored L^2-theory, which supplies the canonical basis elements in the relevant local Hom spaces for use in the factorization of the global periods.

We can now describe the content of Sakellaridis' lecture at the Simons Symposium. In a series of recent papers [S4, S5, S6], Sakellaridis examined aspects of the above program in the context of rank 1 spherical varieties X. There is a classification of such rank 1 X's, but a standard example is $X \cong \mathrm{SO}_{n-1}\backslash\mathrm{SO}_n$, i.e. a hyperboloid (or a sphere) in an n-dimensional quadratic space, and a more exotic example is $\mathrm{Spin}_9\backslash F_4$. In this rank 1 setting, the group G_X is SL_2 or its variants (such as PGL_2 or Mp_2). For example, for $X = \mathrm{SO}_{n-1}\backslash\mathrm{SO}_n$ with n even, $X^\vee \cong \mathrm{PGL}_2(\mathbb{C})$, so that $G_X \cong \mathrm{SL}_2$ and the map ι is given by:

$$\iota : \mathrm{PGL}_2(\mathbb{C}) \times \mathrm{SL}_2(\mathbb{C}) \xrightarrow{\mathrm{Sym}^2 \times \mathrm{Sym}^{n-4}} \mathrm{SO}_3(\mathbb{C}) \times \mathrm{SO}_{n-3}(\mathbb{C}) \longrightarrow \mathrm{SO}_n(\mathbb{C}).$$

On the other hand, if n is odd, then $X^\vee \cong \mathrm{SL}_2(\mathbb{C})$ and we take $G_X \cong \mathrm{Mp}_2$, with the map ι given by

$$\iota : \mathrm{SL}_2(\mathbb{C}) \times \mathrm{SL}_2(\mathbb{C}) \xrightarrow{\mathrm{Sym}^1 \times \mathrm{Sym}^{n-4}} \mathrm{Sp}_2(\mathbb{C}) \times \mathrm{Sp}_{n-3}(\mathbb{C}) \longrightarrow \mathrm{Sp}_{n-1}(\mathbb{C}).$$

In such rank 1 setting, Sakellaridis developed a theory of transfer of test functions from X to $(N, \psi)\backslash G_X$ as a first step towards establishing local relative character identities and effecting a global comparison of the relative trace formula of X and the Kuznetsov trace formula for G_X. The formula for the transfer map he discovered was motivated by considering an analogous transfer for the boundary degenerations of X and $(N_X, \psi)\backslash G_X$. For the hyperboloid $\mathrm{SO}_{n-1}\backslash\mathrm{SO}_n$, the boundary degeneration is simply the cone of nonzero null vectors in the underlying quadratic space. In any case, the transfer map he wrote down differs from the typical transfer map in the theory of endoscopy in two aspects:

- the spaces of test functions may be larger than the space of compactly supported smooth functions;
- the transfer map in endoscopy is carried out via an orbit-by-orbit comparison, whereas the transfer map in this relative setting is more global in nature, involving an integral kernel transformation reminiscent of the Fourier transform.

An ongoing work [JK] of D. Johnstone and R. Krishna establishes the fundamental lemma for the basic functions in the space of test functions; this is necessary for the comparison of relative trace formulae. As an example, in the special case when $n = 4$, one has:

$$X = \mathrm{SO}_3\backslash\mathrm{SO}_4 \cong \mathrm{PGL}_2\backslash(\mathrm{SL}_2 \times \mathrm{SL}_2)/\mu_2^{\Delta}.$$

The relative trace formula for this X is essentially the stable trace formula for SL_2. Thus, the expected comparison of relative trace formulae is between the stable trace formula for SL_2 and the Kuznetsov trace formula for SL_2. The local transfer in this case was first investigated in the thesis work of Z. Rudnick. The discussion of these results was the content of Sakellaridis's lecture in the Simons Symposium.

On the other hand, the spectral analysis of $L^2(X)$ when $X = \mathrm{SO}_{n-1}\backslash\mathrm{SO}_n$ or the analysis of the SO_{n-1}-period for representations of SO_n (both locally and globally) is familiar from the theory of theta correspondence. The L^2-theory was studied in the early work of Strichartz [St] and Howe [H]. In a paper [GG] by the first author and R. Gomez, the L^2-theory was treated using theta correspondence for essentially general rank 1 spherical varieties from the viewpoint of the Sakellaridis-Venkatesh conjecture. For the smooth theory, one can see the recent expository paper [G]. In the case of $X = \mathrm{SO}_{n-1}\backslash\mathrm{SO}_n$, it was known that SO_{n-1}-distinguished representations of SO_n are theta lifts (of ψ-generic representations) from SL_2 or Mp_2 according to whether n is even or odd. Indeed, the theta lifting from SL_2 or Mp_2 to SO_n realises the functorial lifting (at least at the level of unramified representations) predicted by the map $\iota : X^\vee \times \mathrm{SL}_2(\mathbb{C}) \longrightarrow \mathrm{SO}_n(\mathbb{C})$. As such, it is very natural to ask if the results discussed in Sakellaridis' talk can be approached from the viewpoint of the theta correspondence.

This paper is the result of this investigation. In short, its main conclusion is that the theory of transfer developed by Sakellaridis can be very efficiently developed using the theta correspondence. More precisely,

- one can give a conceptual definition of the transfer and the relevant spaces of test functions (Definition 8.1), from which the fundamental lemma (for the basic function and its translate by the spherical Hecke algebra) follows readily (see Lemmas 8.5 and 8.6);
- one can establish the desired relative character identities highlighted in (a) above, without doing a geometric comparison; (see Theorem 9.1)

- one can express this conceptually defined transfer in geometric terms, from which one sees that it agrees with Sakellaridis' formula (see Proposition 11.1);
- one can address the two global problems highlighted in (c) above (see Theorem 12.5).

We leave the precise formulation of the results to the main body of the paper. We would like to remark that, as far as we are aware, the paper [MR] of Mao–Rallis is the first instance where one finds a derivation of relative character identities using the theta correspondence; this approach was followed up by the paper [BLM] of Baruch–Lapid–Mao. The situation treated in this paper is in fact simpler than those in [MR] and [BLM]. In addition, it has been known to practitioners that the theory of theta correspondence is useful for addressing period problems in the smooth local context, the global context, as well as in the local L^2-context [G, GG], with similar computations and parallel treatment in the various settings. One goal of this paper is to demonstrate how the treatment of the 3 different threads can be synthesised into a rather coherent story.

Here is a short summary of the contents of this paper. In Sect. 2, we recall some foundational results of Bernstein [B] on spectral decomposition of $L^2(X)$. These results provide the mechanism for us to navigate between the L^2-setting and the smooth setting. We illustrate Bernstein's general theory in the setting of the Harish-Chandra-Plancherel formula and the Whittaker-Plancherel formula in Sect. 3 and further specialize to the group SL_2 in Sect. 4, where we set up some standard conventions and establish some basic results. In Sect. 5, we recall the setup of theta correspondence, especially a recent result of Sakellaridis [S3] on the spectral decomposition of the Weil representation when restricted to a dual pair. Using the theory of theta correspondence, we address in Sect. 6 the local problems (a) and (b), except for the part involving relative character identities. After recalling the notion of relative characters in Sect. 7, we come to the heart of the paper (Sects. 8–9), where we develop the theory of transfer and establish some of its key properties, culminating in the relative character identity in Sect. 9. In Sect. 10, we place ourselves in the unramified setting and explicitly determine the local L-factor $L_X(s, -)$ using theta correspondence. We verify that our transfer map is the same as that of Sakellaridis' in Sect. 11, where we describe the transfer in geometric terms, as an explicit integral transform. The final Sect. 12 discusses and resolves the global problems.

2 Spectral Decomposition à la Bernstein

Let F be a local field and G a reductive group over F acting transitively on a variety X. We fix a base point $x_0 \in X(F)$, with stabilizer $H \subset G$, so that $g \mapsto g^{-1} \cdot x_0$ gives an identification $H\backslash G \cong X$. For simplicity, we shall write $X = G(F) \cdot x_0 \cong H(F)\backslash G(F)$.

2.1 Direct Integral Decompositions

Suppose that there is a G-invariant measure dx on X, in which case we may consider the unitary representation $L^2(X, dx)$ of G, with G-invariant inner product

$$\langle \phi_1, \phi_2 \rangle_X = \int_X \phi_1(x) \cdot \overline{\phi_2(x)}\, dx.$$

Such a unitary representation admits a direct integral decomposition

$$\iota : L^2(X, dx) \cong \int_\Omega \sigma(\omega)\, d\mu(\omega). \tag{2.1}$$

Here,

- Ω is a measurable space, equipped with a measure $d\mu(\omega)$;
- $\sigma : \omega \mapsto \sigma(\omega)$ is a measurable field of irreducible unitary representations of G over Ω, which we may regard as a measurable map from Ω to the unitary dual $\widehat{G}$ of G (equipped with the Fell topology and the corresponding Borel structure).

In this section, we give an exposition of some results of Bernstein [B] which provide some useful ways of understanding the above direct integral decomposition. This viewpoint of Bernstein underpins the results of this paper.

2.2 Pointwise-Defined and Fine Morphisms

Let $S \subset L^2(X)$ be a dense subspace which is G-stable. Following Bernstein [B, §1.3], one says that the inclusion $S \hookrightarrow L^2(X)$ is *pointwise-defined* (relative to ι) if there exists a family of G-equivariant (continuous) morphisms

$$\alpha_{\sigma(\omega)} : S \longrightarrow \sigma(\omega)$$

for $\omega \in \Omega$ such that for each $\phi \in S$, the element $\iota(\phi)$ in the direct integral decomposition in (2.1) is the measurable section

$$\iota(\phi)(\omega) = \alpha_{\sigma(\omega)}(\phi).$$

In particular, these sections determine the measurable field structure on the right hand side of (2.1). The family $\{\alpha_{\sigma(\omega)} : \omega \in \mathrm{supp}(d\mu)\}$ is essentially unique, in the sense that any two such families differ only on a subset of Ω with measure zero with respect to $d\mu$. Bernstein calls the embedding $S \hookrightarrow L^2(X)$ *fine* if it is pointwise-defined relative to any such isomorphism ι to a direct integral decomposition.

2.3 The Maps $\alpha_{\sigma(\omega)}$ and $\beta_{\sigma(\omega)}$

A basic result of Bernstein [B, Prop. 2.3], obtained as an application of the Gelfand-Kostyuchenko method [B, Thm. 1.5], is that the natural inclusion $C_c^\infty(X) \hookrightarrow L^2(X)$ is fine. For any isomorphism ι as in (2.1), we let

$$\{\alpha_{\sigma(\omega)} : \omega \in \text{supp}(d\mu)\}$$

be the associated family of G-equivariant morphisms as above.

The elements in $C_c^\infty(X)$ are G-smooth vectors and so the image of each $\alpha_{\sigma(\omega)}$ is contained in the space $\sigma(\omega)^\infty$ of smooth vectors in $\sigma(\omega)$. As the map $\alpha_{\sigma(\omega)}$ is nonzero for $d\mu$-almost all ω, its image is dense in $\sigma(\omega)^\infty$, and is in fact equal to $\sigma(\omega)^\infty$ when F is p-adic (where there is no topology considered on $\sigma(\omega)^\infty$). To simplify notation, we shall sometimes write $\sigma(\omega)$ in place of $\sigma(\omega)^\infty$, trusting that the context will make it clear whether one is working with a unitary representation on a Hilbert space or a smooth representation. In particular, $\alpha_{\sigma(\omega)} \in \text{Hom}_G(C_c^\infty(X), \sigma(\omega))$.

If $\alpha_{\sigma(\omega)}$ is nonzero, then by duality, one obtains a G-equivariant embedding

$$\overline{\beta}_{\sigma(\omega)} : \sigma(\omega)^\vee \cong \overline{\sigma(\omega)} \longrightarrow C^\infty(X).$$

Here, the isomorphism $\sigma(\omega)^\vee \cong \overline{\sigma(\omega)}$ is induced by the fixed inner product $\langle -, -\rangle_{\sigma(\omega)}$ and the duality between $C_c^\infty(X)$ and $C^\infty(X)$ is given by the natural pairing induced by integration with respect to the G-invariant measure dx. Taking complex conjugate on $C^\infty(X)$, we obtain a G-equivariant linear map

$$\beta_{\sigma(\omega)} : \sigma(\omega)^\infty \longrightarrow C^\infty(X).$$

The maps $\alpha_{\sigma(\omega)}$ and $\beta_{\sigma(\omega)}$ are thus related by the adjunction formula:

$$\langle \alpha_{\sigma(\omega)}(\phi), v\rangle_{\sigma(\omega)} = \langle \phi, \beta_{\sigma(\omega)}(v)\rangle_X, \quad \text{for } \phi \in C_c^\infty(X) \text{ and } v \in \sigma(\omega)^\infty. \tag{2.2}$$

If we compose $\beta_{\sigma(\omega)}$ with the evaluation-at-x_0 map ev_{x_0}, we obtain

$$\ell_{\sigma(\omega)} := ev_{x_0} \circ \beta_{\sigma(\omega)} \in \text{Hom}_H(\sigma(\omega)^\infty, \mathbb{C}).$$

Thus the direct integral decomposition gives rise to a family of canonical elements $\ell_{\sigma(\omega)} \in \text{Hom}_H(\sigma(\omega)^\infty, \mathbb{C})$ for $\omega \in \text{supp}(d\mu)$. This family depends on the isomorphism ι in (2.1); changing ι will result in another family which differs from the original one by a measurable function $f : \text{supp}(d\mu) \longrightarrow S^1$. Thus, the family

$$\{\alpha_{\sigma(\omega)} \otimes \overline{\alpha_{\sigma(\omega)}} : \omega \in \text{supp}(d\mu)\}$$

is independent of the choice of the isomorphism ι in (2.1). Likewise, the family

$$\{\beta_{\sigma(\omega)} \otimes \overline{\beta_{\sigma(\omega)}} : \omega \in \text{supp}(d\mu)\}$$

is independent of ι.

2.4 Harish-Chandra-Schwartz Space of X

In [B, Pg. 689], Bernstein showed that the space X has a naturally associated Harish-Chandra Schwartz space $\mathcal{C}(X)$ which is G-stable and which contains $C_c^\infty(X)$. Moreover, $\mathcal{C}(X)$ has a natural (complete) topology, such that $C_c^\infty(X)$ is a dense subspace. Indeed, $\mathcal{C}(X)$ is a Frechét space in the Archimedean case and is a strict LF space in the non-Archimedean case. More importantly, he showed in [B, Thm. 3.2] that the inclusion $\mathcal{C}(X) \hookrightarrow L^2(X)$ is fine. Hence, the maps $\alpha_{\sigma(\omega)} : C_c^\infty(X) \to \sigma(\omega)$ defined above extend continuously to the larger space $\mathcal{C}(X)$:

$$\alpha_{\sigma(\omega)} : \mathcal{C}(X) \longrightarrow \sigma(\omega)^\infty.$$

The dual map $\beta_{\sigma(\omega)}$ then takes value in the weak Harish-Chandra Schwartz space $\mathcal{C}^w(X) \subset C^\infty(X)$ (see [B] or [BP2, §2.4] for the group case). The elements $\ell_{\sigma(\omega)} \in \mathrm{Hom}_H(\sigma(\omega), \mathbb{C})$ are called X-tempered forms and the support of $d\mu$ consists precisely of those representations with nonzero X-tempered forms [B, Pg. 689].

2.5 Inner Product

The direct integral decomposition (2.1) leads to a spectral decomposition of the inner product $\langle -, - \rangle_X$ of X:

$$\langle \phi_1, \phi_2 \rangle_X = \int_\Omega J_{\sigma(\omega)}(\phi_1, \phi_2)\, d\mu(\omega), \tag{2.3}$$

where $J_{\sigma(\omega)}$ is a G-invariant positive-semidefinite Hermitian form on $C_c^\infty(X)$ given by:

$$\begin{aligned} J_{\sigma(\omega)}(\phi_1, \phi_2) &= \langle \iota(\phi_1)(\omega), \iota(\phi_2)(\omega) \rangle_{\sigma(\omega)} \\ &= \langle \alpha_{\sigma(\omega)}(\phi_1), \alpha_{\sigma(\omega)}(\phi_2) \rangle_{\sigma(\omega)} \\ &= \langle \beta_{\sigma(\omega)} \alpha_{\sigma(\omega)}(\phi_1), \phi_2 \rangle_X \end{aligned} \tag{2.4}$$

for $d\mu$-almost all ω. In particular, $J_{\sigma(\omega)}$ factors as:

$$J_{\sigma(\omega)} : C_c^\infty(X) \times \overline{C_c^\infty(X)} \xrightarrow{\alpha_{\sigma(\omega)} \otimes \overline{\alpha}_{\sigma(\omega)}} \sigma(\omega) \otimes \overline{\sigma(\omega)} \xrightarrow{\langle -,- \rangle_{\sigma(\omega)}} \mathbb{C}.$$

2.6 *Pointwise Spectral Decomposition*

The fact that the morphism $\mathcal{C}(X) \hookrightarrow L^2(X)$ is fine leads to a pointwise spectral decomposition for elements of $\mathcal{C}(X)$. More precisely, for $\phi \in \mathcal{C}(X)$, one has

$$\phi(x) = \int_{\Omega} \beta_{\sigma(\omega)}\alpha_{\sigma(\omega)}(\phi)(x)\, d\mu(\omega) \tag{2.5}$$

for any $x \in X$. We give a sketch of the derivation of this when F is non-Archimedean. In that case, $\phi \in \mathcal{C}(X)$ is fixed by some open compact subgroup $J \subset G$. The group J also fixes $\beta_{\sigma(\omega)}\alpha_{\sigma(\omega)}(\phi)$ for any σ, since $\alpha_{\sigma(\omega)}$ and $\beta_{\sigma(\omega)}$ are G-equivariant. If 1_{xJ} denotes the characteristic function of the open compact subset $xJ \subset X$, then it follows that

$$\langle \phi, 1_{xJ}\rangle_X = \mathrm{Vol}(xJ; dx) \cdot \phi(x) \quad \text{and}$$
$$\langle \beta_{\sigma(\omega)}\alpha_{\sigma(\omega)}(\phi), 1_{xJ}\rangle_X = \mathrm{Vol}(xJ; dx) \cdot \beta_{\sigma(\omega)}\alpha_{\sigma(\omega)}(\phi)(x)$$

where $\mathrm{Vol}(xJ; dx)$ is the volume of $xJ \subset X$ with respect to the measure dx. Now it follows that

$$\begin{aligned}\phi(x) &= \frac{1}{\mathrm{Vol}(xJ; dx)} \cdot \langle \phi, 1_{xJ}\rangle_X \\ &= \frac{1}{\mathrm{Vol}(xJ; dx)} \int_{\Omega} \langle \beta_{\sigma(\omega)}\alpha_{\sigma(\omega)}(\phi), 1_{xJ}\rangle_X \, d\mu(\omega) \\ &= \int_{\Omega} \beta_{\sigma(\omega)}\alpha_{\sigma(\omega)}(\phi)(x)\, d\mu(\omega),\end{aligned}$$

where the second equality is a consequence of (2.3) and (2.4).

The crux of Bernstein's viewpoint in [B] is that to give the isomorphism ι in the direct integral decomposition (2.1) is equivalent to giving the family $\{\alpha_{\sigma(\omega)} : \omega \in \Omega\}$ (satisfying appropriate properties), together with the measure $d\mu$ on Ω. In the next section, we shall illustrate this in two basic examples.

3 Basic Plancherel Theorems

In this section, we describe two basic Plancherel theorems as an illustration of the abstract theory of Bernstein discussed in the previous section. These are the Harish-Chandra-Plancherel theorem and the Whittaker-Plancherel theorem. The latter will play a crucial role in this paper.

We shall continue to work over a local field F. However, we will implicitly be assuming that F is non-Archimedean. In fact, the results of this paper will hold for Archimedean local fields as well, but greater care is needed in introducing the various objects (such as various spaces of functions and the topologies they carry) and in formulating the results. Thus, there are analytic and topological considerations that need to be addressed in the archimedean case. We refer the reader to the papers [BP1, BP2] and the thesis of the second author [Wan] where these issues are dealt with carefully and content ourselves with treating the nonarchimedean case in the interest of efficiency.

3.1 Harish-Chandra-Plancherel Theorem

The most basic example is the regular representation $L^2(G)$ of a semisimple group $G \times G$ (acting by right and left translation):

$$(g_1, g_2) f(g) = f(g_2^{-1} \cdot g \cdot g_1).$$

Here, we have fixed a Haar measure dg on G which defines the inner product on $L^2(G)$.

Now Harish-Chandra's Plancherel theorem [W, Wal] asserts that there is an explicitly constructed $G \times G$-equivariant isomorphism

$$L^2(G) \cong \int_{\widehat{G}} \overline{\sigma} \widehat{\boxtimes} \sigma \, d\mu_G(\sigma) \tag{3.1}$$

for a specific measure $d\mu_G$ on $\widehat{G}$ known as the Plancherel measure of G (which depends on the Haar measure dg). The support of this measure is precisely the subset $\widehat{G}_{temp}$ of irreducible tempered representations of G. Thus, in this case, one may take the measurable space Ω to be the unitary dual $\widehat{G}$ and the map $\widehat{G} \to \widehat{G \times G}$ is given by $\sigma \mapsto \overline{\sigma} \boxtimes \sigma$. Implicit in the theorem is the data of a measurable field of unitary representations over $\widehat{G}$ whose fiber at $\sigma \in \widehat{G}$ is the representation $\overline{\sigma} \widehat{\boxtimes} \sigma$.

One may describe the above direct integral decomposition (including the isomorphism) from Bernstein's viewpoint. The Hilbert space $\overline{\sigma} \widehat{\boxtimes} \sigma$ is naturally identified with the space $\mathrm{End}_{HS}(\sigma)$ of Hilbert-Schmidt operators on σ, equipped with the Hilbert-Schmidt norm, and its space of $G \times G$-smooth vectors is the space $\overline{\sigma}^\infty \otimes \sigma^\infty = \mathrm{End}_{\mathrm{fin}}(\sigma^\infty)$ of finite rank operators on σ^∞. To describe the direct integral decomposition, one needs to give the family of maps:

$$\alpha_{\overline{\sigma} \boxtimes \sigma} : C_c^\infty(G) \longrightarrow \mathrm{End}_{\mathrm{fin}}(\sigma^\infty),$$

The map $\alpha_{\overline{\sigma} \boxtimes \sigma}$ is given by

$$\alpha_{\overline{\sigma} \boxtimes \sigma}(f) = \sigma(f) := \int_G f(g) \cdot \sigma(g) \, dg$$

and the (conjugate) dual map

$$\beta_{\overline{\sigma}\boxtimes\sigma} : \overline{\sigma}^{\infty} \otimes \sigma^{\infty} \longrightarrow C^{\infty}(G)$$

is given by the formation of matrix coefficients. This data characterizes and explicates the measurable field of unitary representations implicit in the Plancherel theorem: the sections $\sigma \mapsto \sigma(f)$ for $f \in C_c^{\infty}(G)$ generate the family of measurable sections of the Hilbert space bundle $\{\mathrm{End}_{HS}(\sigma) = \overline{\sigma}\widehat{\boxtimes}\sigma\}_{\sigma\in\widehat{G}}$ over $\widehat{G}$.

The associated inner product $J_{\overline{\sigma}\boxtimes\sigma}$ is given by:

$$J_{\overline{\sigma}\boxtimes\sigma}(f_1, f_2) = \mathrm{Tr}(\sigma(f_1)\sigma(f_2^{\vee})),$$

where

$$f_2^{\vee}(g) = \overline{f_2(g^{-1})}.$$

The $\overline{\sigma} \otimes \sigma$-component of $f \in C_c^{\infty}(G)$ in the pointwise spectral decomposition is the function given by

$$(\beta_{\overline{\sigma}\boxtimes\sigma} \circ \alpha_{\overline{\sigma}\boxtimes\sigma})(f)(g) = \mathrm{Tr}(\sigma(g)^{-1} \circ \sigma(f)).$$

In particular,

$$f \mapsto \ell_{\overline{\sigma}\boxtimes\sigma}(\alpha_{\overline{\sigma}\boxtimes\sigma}(f)) = Tr(\sigma(f))$$

is the Harish-Chandra character distribution of σ. For tempered σ, it extends to the (original) Harish-Chandra Schwartz space $\mathcal{C}(G)$.

It is instructive to take note of how the Plancherel measure $d\mu_G$ depends on the Haar measure dg. If we replace dg by $\lambda \cdot dg$ for some $\lambda \in \mathbb{R}_{>0}^{\times}$, then we observe that

$$J_{\overline{\sigma}\boxtimes\sigma} \mapsto \lambda^2 \cdot J_{\overline{\sigma}\boxtimes\sigma}, \quad \alpha_{\overline{\sigma}\boxtimes\sigma} \mapsto \lambda \cdot \alpha_{\overline{\sigma}\boxtimes\sigma} \quad \text{and} \quad \beta_{\overline{\sigma}\boxtimes\sigma} \mapsto \beta_{\overline{\sigma}\boxtimes\sigma}.$$

Hence, we have

$$d\mu_G \mapsto \lambda^{-1} \cdot d\mu_G.$$

We have restricted ourselves to semisimple groups in this subsection for simplicity. When G is a reductive algebraic group and $Z \subset G$ the maximal F-split torus in the center of G, then one may fix a unitary character χ of Z and consider the unitary representation $L^2_{\chi}(G)$ consisting of L^2-functions f which satisfies $f(zg) = \chi(z) \cdot f(g)$ for $z \in Z$ and $g \in G$ and equipped with the unitary structure determined by a Haar measure dg/dz of G/Z. Moreover, one may also consider nonlinear finite central extensions of $G(F)$ by finite cyclic groups. In all these cases, the Harish-Chandra Plancherel theorem continue to hold.

3.2 *Whittaker-Plancherel Theorem*

Our second example is the Whittaker-Plancherel theorem (see [BP2, D, SV, W]), which is a variant of the setting discussed above. Let G be a quasi-split semisimple group with N the unipotent radical of a Borel subgroup. Fix a nondegenerate unitary character ψ_N of N. We consider the *Whittaker variety* $(N, \psi_N)\backslash G$ and its associated unitary representation $L^2(N, \psi_N\backslash G)$ (which depends on fixed Haar measures dg on G and dn on N). This extends the setting we discussed above, as one is considering L^2-sections of a line bundle on the spherical variety $N\backslash G$ instead of L^2-functions, but it is also covered in [B].

It has been shown (see [D, SV, Wa2]) that one has a direct integral decomposition

$$L^2(N, \psi_N\backslash G) \cong \int_{\widehat{G}} \dim \mathrm{Hom}_N(\sigma, \psi_N) \cdot \sigma \; d\mu_G(\sigma), \tag{3.2}$$

where we recall that $d\mu_G$ is the Plancherel measure of G (associated to the fixed Haar measure dg). Thus, in this case, we are taking Ω to be $\widehat{G}$ and the map $\Omega \to \widehat{G}$ is the identity map. The spectral measure $d\mu_{G,\psi_N}$ is equal to $\dim \mathrm{Hom}_N(\sigma, \psi_N) \cdot d\mu_G$, whose support is the subset $\widehat{G}_{temp,\psi_N}$ of ψ_N-generic irreducible tempered representations.

Associated to this direct integral decomposition is the family of morphisms

$$\alpha_\sigma : C_c^\infty(N, \psi_N\backslash G) \longrightarrow \sigma$$

for all $\sigma \in \widehat{G}_{temp,\psi_N}$. Moreover, the map α_σ extends to the Harish-Chandra-Schwarz space $\mathcal{C}(N, \psi_N\backslash G)$. We describe instead the (conjugate) dual map

$$\beta_\sigma \otimes \overline{\beta_\sigma} : \sigma \otimes \overline{\sigma} \longrightarrow C^\infty(N \times N, \psi_N \otimes \overline{\psi_N}\backslash G \times G)$$

as follows. Given $v_1, v_2 \in \sigma$, one has

$$\beta_\sigma \otimes \overline{\beta_\sigma}(v_1 \otimes v_2)(g_1, g_2) = \int_N^* \overline{\psi_N(n)} \cdot \langle \sigma(n \cdot g_1)(v_1), \sigma(g_2)(v_2)\rangle_\sigma \, dn \tag{3.3}$$

where the integral is a regularized one (see [LM, Prop. 2.3], [SV, §6.3] and [BP1]). Here, note that $\beta_\sigma \otimes \overline{\beta_\sigma}$ depends on dn, as it should. The composite of this with the evaluation-at-1 map is thus the Whittaker functional

$$\ell_\sigma \otimes \overline{\ell_\sigma} : v_1 \otimes v_2 \mapsto \int_N^* \overline{\psi_N(n)} \cdot \langle \sigma(n)(v_1), v_2\rangle_\sigma \, dn. \tag{3.4}$$

The associated (positive semidefinite) inner product J_σ on $C_c^\infty(N, \psi_N\backslash G)$ is then given by

$$
\begin{aligned}
&J_\sigma(f_1, f_2)\\
&= \sum_{v\in \mathrm{ONB}(\sigma)} \langle f_1, \beta_\sigma(v)\rangle_{N\backslash G} \cdot \langle \beta_\sigma(v), f_2\rangle_{N\backslash G}\\
&= \sum_{v\in \mathrm{ONB}(\sigma)} \int_{N\times N\backslash G\times G} f_1(g_1)\cdot \overline{f_2(g_2)}\cdot \ell_\sigma(g_2v)\cdot \overline{\ell_\sigma(g_1v)}\, \frac{dg_1}{dn}\,\frac{dg_2}{dn}\\
&= \sum_{v\in \mathrm{ONB}(\sigma)} \int_{N\times N\backslash G\times G} f_1(g_1)\cdot \overline{f_2(g_2)}\cdot \left(\int_N^* \overline{\psi_N(n)}\cdot \langle ng_2v, g_1v\rangle\, dn\right) \frac{dg_1}{dn}\,\frac{dg_2}{dn}.
\end{aligned}
$$

The above maps specify on the right hand side of (3.2) the structure of a measurable field of unitary representations on $\widehat{G}_{temp,\psi_N}$ whose fiber at $\sigma \in \widehat{G}_{temp,\psi_N}$ is the representation σ. We can think of this measurable field of unitary representations as a "tautological" or "universal bundle of unitary representations" over the "moduli space" $\widehat{G}_{temp,\psi_N}$ of irreducible ψ_N-generic tempered representations.

It is again useful to take note of how the various quantities change when one replaces the Haar measure dg of G by $\lambda \cdot dg$ for some $\lambda \in \mathbb{R}^\times_{>0}$. In the Whittaker-Plancherel case, one sees from the above formula that β_σ and ℓ_σ are unchanged whereas

$$J_\sigma \mapsto \lambda^2 \cdot J_\sigma \quad \text{and} \quad \alpha_\sigma \mapsto \lambda \cdot \alpha_\sigma,$$

keeping in mind that the Plancherel measure $d\mu_G$ gets replaced by $\lambda^{-1} \cdot d\mu_G$.

As in the case of the Harish-Chandra Plancherel theorem, we could have worked with a reductive algebraic group (in which case we fix a central character χ as before and consider $L^2_\chi(N, \psi_N\backslash G)$) or a nonlinear finite cover thereof. One has the Whittaker-Plancherel theorem in these settings as well, though we take note that uniqueness of Whittaker models fails for nonlinear covering groups in general.

3.3 Continuity Properties

We now consider the issue of continuity (in σ) for some of the quantities discussed above. We first need to say a few words about the Fell topology on $\widehat{G}_{temp}$.

The unitary dual $\widehat{G}$ is typically non-Hausdorff even though it is still a T1 space for the groups considered here. The tempered dual $\widehat{G}_{temp}$ is still not necessarily Hausdorff, but can often be replaced by a substitute which is Hausdorff. Namely, one can work with the space of equivalence classes of induced representations $\tau = \mathrm{Ind}_P^G \pi$ where P is a parabolic subgroup of G and π a discrete series representation of its Levi factor M. This space was variously denoted by $\mathcal{T}$ in [S3], $\mathrm{Temp}_{\mathrm{Ind}}(G)$ in [BP2] and $\mathcal{X}_{temp}(G)$ in [BP1, X], so we are spoilt for choices! To add to this galore, we shall denote this space by $\widehat{G}^{ind}_{temp}$. Then $\widehat{G}^{ind}_{temp}$ has the structure of an orbifold (given by twisting π by unramified unitary characters of M). There is a natural continuous finite-to-one surjective map

$$\widehat{G}_{temp} \longrightarrow \widehat{G}^{ind}_{temp}$$

sending a tempered irreducible representation σ to the unique induced representation $\mathrm{Ind}_P^G \pi$ containing σ. This map is injective outside a subset of $\widehat{G}_{temp}$ which has measure zero with respect to the Plancherel measure $d\mu_G$.

In the setting of the Harish-Chandra-Plancherel theorem of Sect. 3.1, one could safely replace the integral over $\widehat{G}_{temp}$ in (3.1) by an integral over $\widehat{G}^{ind}_{temp}$. Moreover, we have the Hermitian form $J_{\overline{\sigma}\boxtimes\sigma}(\phi_1, \phi_2)$ for $\phi_i \in C_c^\infty(G)$ (or more generally $\mathcal{C}(G)$) and $\sigma \in \widehat{G}_{temp}$. We can similarly define $J_{\overline{\tau}\boxtimes\tau}(\phi_1, \phi_2)$ for $\tau \in \widehat{G}^{ind}_{temp}$. Then we have [BP2, §2.13]:

Lemma 3.1. *For fixed ϕ_1 and ϕ_2 in $\mathcal{C}(G)$, the maps*

$$\tau \mapsto J_{\overline{\tau}\boxtimes\tau}(\phi_1, \phi_2)$$

is continuous as a $\mathbb{C}$-valued function on $\widehat{G}^{ind}_{temp}$. In particular, the map $\sigma \mapsto J_{\overline{\sigma}\boxtimes\sigma}(\phi_1, \phi_2)$ is continuous on the subset of $\widehat{G}_{temp}$ which maps injectively into $\widehat{G}^{ind}_{temp}$.

In the context of the Whittaker-Plancherel theorem, we are working with the subset $\widehat{G}_{temp,\psi_N}$. When uniqueness of Whitaker models holds (such as for reductive algebraic groups or the metaplectic groups which are two fold covers of symplectic groups), each $\tau = \mathrm{Ind}_P^G \pi$ in $\widehat{G}^{ind}_{temp}$ can have at most one irreducible constituent which is ψ_N-generic. Hence, we see that the composite map

$$\widehat{G}_{temp,\psi_N} \longrightarrow \widehat{G}_{temp} \longrightarrow \widehat{G}^{ind}_{temp}$$

is injective (and continuous). As a consequence of [BP2, §2.14], we have:

Lemma 3.2. *In the context of the Whittaker-Plancherel theorem, for fixed f_1 and f_2 in $\mathcal{C}(N, \psi_N\backslash G)$, the map*

$$\sigma \mapsto J_\sigma(f_1, f_2)$$

is a continuous $\mathbb{C}$-valued function on $\widehat{G}_{temp,\psi_N}$. Likewise, for fixed $f \in \mathcal{C}(N, \psi_N\backslash G)$, the map

$$\sigma \mapsto \beta_\sigma \alpha_\sigma(f)(1)$$

is continuous.

4 GL_2 and SL_2

In this section, we specialize the discussion of the previous section to the case of GL_2 and SL_2. Since the group SL_2 will feature prominently in the rest of the paper, we also take the opportunity to set up some precise conventions which will be used for the rest of the paper.

4.1 *Measures on F and* $F^\times$

Let us first fix a nontrivial additive character

$$\psi : F \longrightarrow S^1.$$

Then ψ determines an additive Haar measure $d_\psi x$ on F, characterized by the requirement that $d_\psi x$ is self-dual with respect to the Fourier transform relative to the pairing $(x, y) \mapsto \psi(xy)$ on F. If F is nonarchimedean with ring of integers $\mathcal{O}_F$ and ψ has conductor $\mathcal{O}_F$, then the Haar measure $d_\psi x$ gives $\mathcal{O}_F$ volume 1. One also obtains a multiplicative Haar measure on $F^\times$ given by

$$d_\psi^\times x = d_\psi x / |x|.$$

More generally, for any algebraic group G over F, a nonzero element ω of $\wedge^{\mathrm{top}}\mathrm{Lie}(G)^*$ and the additive Haar measure $d_\psi x$ together give rise to a right-invariant (or left-invariant) Haar measure $|\omega|_\psi$. Since ψ is fixed throughout, we will often suppress it from the notation and simply write $|\omega|$.

4.2 *The Group* GL_2

We now consider the group GL_2 over F. Let $\tilde{B}$ be the upper triangular Borel subgroup with unipotent radical N (the group of upper triangular unipotent matrices) and consider the diagonal maximal torus. We can write the diagonal maximal torus as $Z \cdot S$ where Z is the center of GL_2 (the scalar matrices) and

$$S = \{s(y) = \begin{pmatrix} y & 0 \\ 0 & 1 \end{pmatrix} : y \in F^\times\}.$$

By Sect. 4.1, we have Haar measures on $N(F) = F$, $Z(F) = F^\times$ and $S(F) = F^\times$ and hence a right-invariant measure on $\tilde{B}$.

Regard the fixed additive character ψ of F as a character of $N(F)$. For a fixed a unitary character χ of Z, the Whittaker-Plancherel theorem for $L^2_\chi(N,\psi\backslash \mathrm{GL}_2)$ gives a family of GL_2-equivariant embeddings

$$\tilde{\beta}_{\tilde{\sigma},\psi} : \tilde{\sigma} \longrightarrow C^\infty_\chi(N,\psi\backslash \mathrm{GL}_2) = \mathrm{Ind}_{Z\cdot N}^{\mathrm{GL}_2} \chi \otimes \psi$$

for irreducible tempered representations $\tilde{\sigma}$ of GL_2 with central character χ. As we have noted, $\tilde{\beta}_{\tilde{\sigma},\psi}$ only depends (up to an element of S^1) on the Haar measure $d_\psi n$ on N (which we have fixed), and does not depend on the choice of the Haar measure on GL_2 which enters into the formulation of the Whittaker-Plancherel theorem.

We may consider the $\tilde{B}$-equivariant map

$$\mathrm{rest} : C^\infty_\chi(N,\psi\backslash \mathrm{GL}_2) \longrightarrow C^\infty_\chi(N,\psi\backslash \tilde{B}) = \mathrm{Ind}_{Z\cdot N}^{\tilde{B}} \chi \otimes \psi$$

given by the restriction of functions. The Haar measures we have fixed endow the latter space with a unitary structure, whose inner product is given by

$$\langle f_1, f_2\rangle_{Z\cdot N\backslash \tilde{B}} = \int_{F^\times} f_1\begin{pmatrix} y & 0 \\ 0 & 1\end{pmatrix} \cdot \overline{f_2\begin{pmatrix} y & 0 \\ 0 & 1\end{pmatrix}}\, d^\times_\psi y.$$

We now note the following basic result, which is a reformulation of [LM, Lemma 4.4] and [BP2, Prop. 2.14.3]. In these references, this result was shown in the setting of GL_n (with appropriate formulation). In any case, this result is the reason why we consider the case of GL_2 in this section.

Proposition 4.1. *The composite map* $\mathrm{rest} \circ \tilde{\beta}_{\tilde{\sigma},\psi}$ *gives an isometric embedding*

$$\mathrm{rest} \circ \tilde{\beta}_{\tilde{\sigma},\psi} : \tilde{\sigma} \longrightarrow L^2_\chi(N,\psi\backslash \tilde{B}).$$

4.3 *The Group* SL_2

Now we turn to the group SL_2. The goal is to deduce from Proposition 4.1 its analog in the setting of SL_2. We first take the opportunity to introduce some conventions for SL_2 which will be enforced throughout the paper.

We first fix the upper triangular Borel subgroup $B = T \cdot N$, where T is the diagonal torus and N the unipotent radical of B, and a maximal compact subgroup K in good relative position with respect to B. For example, when F is non-Archimedean with ring of integers $\mathcal{O}_F$, we can simply take $K = \mathrm{SL}_2(\mathcal{O}_F)$. We have natural identifications $N(F) \cong F$ and $T(F) \cong F^\times$ such that the modulus character of B is given by $\delta_B(t) = |t|^2$. For $a \in F^\times$ and $b \in F$, we write

$$n(b) = \begin{pmatrix} 1 & b \\ 0 & 1 \end{pmatrix} \in N(F) \quad \text{and} \quad t(a) = \begin{pmatrix} a & 0 \\ 0 & a^{-1} \end{pmatrix} \in T(F).$$

Further, the groups $N(F) = F$ and $T(F)^\times = F$ carry the fixed Haar measure $dn = d_\psi x$ and $dt = d_\psi x/|x|$. We also fix a Haar measure dg on $\mathrm{SL}_2(F)$, which in turn determines a Plancherel measure $d\mu_{\mathrm{SL}_2}$ on the unitary dual $\widehat{\mathrm{SL}}_2$.

Now we come to the analog of Proposition 4.1 in the SL_2-setting. The main difference between GL_2 and SL_2 is that, whereas there is a unique equivalence class of Whittaker datum in the case of GL_2, there are $F^\times/F^{\times 2}$-worth of them in the case of SL_2. For any $a \in F^\times$, set $\psi_a(x) = \psi(ax)$ so that ψ_a is a nontrivial additive character of F. Then the two Whittaker data (N, ψ_a) and (N, ψ_b) of SL_2 are equivalent if and only if $a/b \in F^{\times 2}$. In formulating an analog of Proposition 4.1 in the SL_2-setting, it will be necessary to take all the various inequivalent Whittaker data into account.

Henceforth, let us fix a set of representatives $[a]$ for $F^\times/F^{\times 2}$, so that

$$F^\times = \bigsqcup_{[a]\in F^\times/F^{\times 2}} aF^{\times 2}. \tag{4.1}$$

We shall assume $a = 1$ is one of the representatives.

For each $a \in F^\times$, the Whittaker-Plancherel theorem for $L^2(N, \psi_a\backslash\mathrm{SL}_2)$ (relative to the fixed Haar measures dg on SL_2 and $d_\psi x$ on N) then furnishes the maps α_{σ,ψ_a}, β_{σ,ψ_a} and ℓ_{σ,ψ_a} for any irreducible ψ_a-generic tempered representation σ of SL_2. In particular, if σ is ψ_a-generic, then

$$\beta_{\sigma,\psi_a} : \sigma \longrightarrow C^\infty(N, \psi_a\backslash\mathrm{SL}_2)$$

is an SL_2-equivariant embedding. As in the GL_2 case, we may consider the restriction of functions

$$C^\infty(N, \psi_a\backslash\mathrm{SL}_2) \longrightarrow C^\infty(N, \psi_a\backslash B).$$

Let us scale this restriction map a little, by setting

$$\mathrm{rest}_a : C^\infty(N, \psi_a\backslash\mathrm{SL}_2) \longrightarrow C^\infty(N, \psi_a\backslash B)$$

to be

$$\mathrm{rest}_a(f) = |a|^{1/2} \cdot f|_B.$$

Combining all these maps together gives us an B-equivariant map

$$j_\sigma := \bigoplus_{[a]\in F^\times/F^{\times 2}} \mathrm{rest}_a \circ \beta_{\sigma,\psi_a} : \sigma \longrightarrow \bigoplus_{[a]\in F^\times/F^{\times 2}} C^\infty(N, \psi_a\backslash B) \tag{4.2}$$

where here, the map β_{σ,ψ_a} is interpreted to be 0 if σ is not ψ_a-generic. Now we have the following analog of Proposition 4.1, which will play a crucial role later on (in the proof of Proposition 6.4).

Proposition 4.2. *Equip $C^\infty(N,\psi_a\backslash B)\cong C^\infty(T)$ with the unitary structure*

$$\langle f_1, f_2\rangle_{N\backslash B} = \frac{1}{2}\cdot|2|_F\cdot\int_{F^\times} f_1(t(b))\cdot\overline{f_2(t(b))}\,d_\psi^\times b,$$

which is the natural inner product associated to the Haar measures fixed on T scaled by the factor $|2|_F/2$. Then, for any irreducible tempered representation σ of SL_2, *the B-equivariant map defined by (4.2) is an isometry*

$$j_\sigma:\sigma\longrightarrow\bigoplus_{[a]\in F^\times/F^{\times 2}} L^2(N,\psi_a\backslash B).$$

Proof. To deduce this proposition from Proposition 4.1, we naturally regard SL_2 as a subgroup of GL_2. Given an irreducible tempered representation σ of SL_2, we pick a unitary representation $\tilde{\sigma}$ of GL_2 with unitary central character χ such that $\sigma\subset\tilde{\sigma}$. We may assume that the inner product $\langle -,-\rangle_{\tilde{\sigma}}$ restricts to the inner product $\langle -,-\rangle_\sigma$ on σ.

Consider the ψ-Whittaker functional $\tilde{\ell}_{\tilde{\sigma},\psi}$ and the associated map $\tilde{\beta}_{\tilde{\sigma},\psi}$ for $\tilde{\sigma}$. Observe that for any $a\in F^\times$, $\tilde{\ell}_{\tilde{\sigma},\psi}\circ s(a)\in\mathrm{Hom}_N(\tilde{\sigma},\psi_a)$ and the following statements are equivalent:

- the restriction of $\tilde{\ell}_{\tilde{\sigma},\psi}\circ s(a)$ to σ is nonzero;
- σ is ψ_a-generic;

Further, if σ is ψ-generic, then from the formula (3.4), one sees that the restriction of $\tilde{\ell}_{\tilde{\sigma},\psi}$ to σ is equal to $\ell_{\sigma,\psi}$.

Now take $w_1, w_2\in\sigma\subset\tilde{\sigma}$. By Proposition 4.1, one has

$$\langle w_1,w_2\rangle_\sigma=\langle w_1,w_2\rangle_{\tilde{\sigma}}=\int_{F^\times}\tilde{\beta}_{\tilde{\sigma}}(w_1)(s(x))\cdot\overline{\tilde{\beta}_{\tilde{\sigma}}(w_2)(s(x))}\,d_\psi^\times(x),$$

where we have written $\tilde{\beta}_{\tilde{\sigma}}$ in place of $\tilde{\beta}_{\tilde{\sigma},\psi}$ to simplify notation. To evaluate the latter integral, we decompose the domain of integration into square classes as in (4.1) and uniformize each square class $aF^{\times 2}$ by $F^\times$, using the map $b\mapsto ab^2$ (which is a 2-to-1 map). Performing the corresponding change of variables in the integral (i.e. replacing x by ab^2 on $aF^{\times 2}$), we obtain:

$$\begin{aligned}\langle w_1,w_2\rangle_\sigma &= \sum_{[a]\in F^\times/F^{\times 2}}\frac{1}{2}\cdot|2|_F\cdot\int_{F^\times}\tilde{\beta}_{\tilde{\sigma}}(w_1)(s(ab^2))\cdot\overline{\tilde{\beta}_{\tilde{\sigma}}(w_2)(s(ab^2))}\,d_\psi^\times b\\ &= \sum_{[a]\in F^\times/F^{\times 2}}\frac{1}{2}\cdot|2|_F\cdot\int_{F^\times}\tilde{\beta}_{\tilde{\sigma}}(w_1)(s(a)\cdot t(b))\cdot\overline{\tilde{\beta}_{\tilde{\sigma}}(w_2)(s(a)\cdot t(b))}\,d_\psi^\times b\end{aligned}$$

This computation is the source of the factor $|2|_F/2$ appearing in the proposition. Let us define a map

$$j_{\sigma,a} : \sigma \mapsto C^\infty(N, \psi_a \backslash B)$$

by

$$j_{\sigma,a}(w)(t(b)) = \tilde{\beta}_{\tilde{\sigma},\psi}(w)(s(a) \cdot t(b)),$$

noting that $j_{\sigma,a}$ is nonzero if and only if σ is ψ_a-generic. Then we have shown that for $w_1, w_2 \in \sigma$,

$$\langle w_1, w_2 \rangle_\sigma = \sum_{[a] \in F^\times / F^{\times 2}} \langle j_{\sigma,a}(w_1), j_{\sigma,a}(w_2) \rangle_{N \backslash B}.$$

In other words, one has an isometry

$$\bigoplus_{[a] \in F^\times / F^{\times 2}} j_{\sigma,a} : \sigma \longrightarrow \bigoplus_{[a] \in F^\times / F^{\times 2}} L^2(N, \psi_a \backslash B)$$

where the unitary structures of the latter spaces are as given in the proposition.

It remains then to show that

$$j_{\sigma,a} = \mathrm{rest}_a \circ \beta_{\sigma,\psi_a} \quad \text{for each } a \in F^\times,$$

or equivalently

$$\tilde{\ell}_{\tilde{\sigma}} \circ s(a) = |a|^{1/2} \cdot \ell_{\sigma,\psi_a} \quad \text{on } \sigma \subset \tilde{\sigma}.$$

To see this, for $w_1, w_2 \in \sigma \subset \tilde{\sigma}$, we apply (3.4) to get

$$\begin{aligned}
&\tilde{\ell}_{\tilde{\sigma}}(s(a)w_1) \cdot \overline{\tilde{\ell}_{\tilde{\sigma}}(s(a)w_2)} \\
=& \int_F^* \langle n(x) \cdot s(a)w_1, s(a)w_2 \rangle_{\tilde{\sigma}} \cdot \overline{\psi(x)}\, d_\psi x \\
=& \int_N^* \langle s(a) \cdot n(a^{-1}x)w_1, s(a)w_2 \rangle_{\tilde{\sigma}} \cdot \overline{\psi(x)}\, d_\psi x \\
=& \int_N^* \langle n(x)w_1, w_2 \rangle_{\tilde{\sigma}} \cdot \overline{\psi(ax)} \cdot |a|\, d_\psi x \\
=& |a| \cdot \ell_{\sigma,\psi_a}(w_1) \cdot \overline{\ell_{\sigma,\psi_a}(w_2)}.
\end{aligned}$$

Here in the penultimate equality, we have applied a change of variables, replacing x by ax and used the unitarity of $\tilde{\sigma}$. We have thus shown that

$$\tilde{\ell}_{\tilde{\sigma}} \circ s(a) = |a|^{1/2} \cdot \ell_{\sigma,\psi_a}$$

at least up to an element of S^1 (which we may ignore, by absorbing it into ℓ_{σ,ψ_a}). This completes the proof of Proposition 4.2. □

We can describe the image of the isometry j_σ precisely. Observe that the center $Z_{\mathrm{SL}_2} = \mu_2$ of SL_2 induces a decomposition

$$L^2(N, \psi_a, \backslash B) = L^2(N, \psi_a, \backslash B)^+ \oplus L^2(N, \psi_a, \backslash B)^-$$

into two irreducible B-subrepresentations which are the $\pm$-eigenspaces of μ_2. Then one has:

Corollary 4.3. *Let $z_\sigma = \pm$ denote the central character of the irreducible tempered representation σ of SL_2. The map j_σ in (4.2) defines a B-equivariant isometric isomorphism*

$$j_\sigma : \sigma \cong \bigoplus_{[a]\in F^\times/F^{\times 2}} \dim \mathrm{Hom}_N(\sigma, \psi_a) \cdot L^2(N, \psi_a \backslash B)^{z_\sigma}$$

where the unitary structure of the right hand side is as defined in Proposition 4.2.

4.4 Harish-Chandra-Schwartz Space

Finally, we explicate when a function $f \in C^\infty(N, \psi\backslash\mathrm{SL}_2)$ lies in the Harish-Chandra Schwartz space. The measures dg and dn determine an $\mathrm{SL}_2(F)$-invariant measure on $N(F)\backslash\mathrm{SL}_2(F)$, which can be described as follows. An element $f \in C^\infty(N, \psi\backslash\mathrm{SL}_2)$ is determined by its restriction to TK, by the Iwasawa decomposition. Then the integral of f with respect to dg/dn is given by

$$f \mapsto \int_T \int_K f(tk) \cdot \delta_B(t)^{-1} \, dt \, dk$$

for some Haar measure dk of K.

Given a function $f \in C^\infty(N, \psi\backslash\mathrm{SL}_2)$, the smoothness of f implies that the function $t \to f(tk)$ on $T \cong F^\times$ is necessarily rapidly decreasing at $|t| \to \infty$ (indeed, it vanishes on some domain $|t| > C$ in the p-adic case). Thus the analytic properties of f depend on its asymptotics as $|t| \to 0$. We have the following lemma:

Lemma 4.4. *Let $f \in C^\infty(N, \psi\backslash\mathrm{SL}_2)$ and suppose that there exists $C > 0$ and $d > 0$ such that*

$$\sup_{k\in K} |f(tk)| \leq C \cdot |t|^d \quad \textit{as } |t| \to 0.$$

In the Archimedean case, assume the same bound holds for all derivatives of f, i.e. $D \cdot f$ for all D in the (complexified) universal enveloping algebra of SL_2.

(i) *If $d > 1$, then $f \in \mathcal{C}(N, \psi\backslash \mathrm{SL}_2)$.*
(ii) *If $d > 2$, then $f \in L^1(N, \psi\backslash \mathrm{SL}_2)$.*

5 Theta Correspondence

In this section, we recall the setup of the theta correspondence and recall some results of Sakellaridis [S3] on the spectral decomposition of the Weil representation for a dual pair.

5.1 *Weil Representation*

If W is a symplectic vector space and (V, q) a quadratic space over a local field F, then one has a dual reductive pair

$$\mathrm{Sp}(W) \times \mathrm{O}(V) \longrightarrow \mathrm{Sp}(V \otimes W).$$

In this paper, we shall only consider the case where $W = F \cdot e \oplus F \cdot f$ is two-dimensional with $\langle e, f\rangle_W = 1$. With the Witt basis $\{e, f\}$, we may identify $\mathrm{Sp}(W)$ with $\mathrm{SL}_2(F)$, and we let $B = T \cdot N$ be the Borel subgroup which stabilises the line $F \cdot e$ (so that B is upper triangular). In particular, the conventions we have set up in Sect. 4.3 for SL_2 apply to $\mathrm{Sp}(W)$.

Attached to a fixed nontrivial additive character ψ of F and other auxiliary data, this dual pair has a distinguished representation Ω_ψ known as the Weil representation. To be precise, if $\dim V$ is odd, we need to work with the metaplectic double cover $\mathrm{Mp}_2(F)$ of $\mathrm{SL}_2(F)$. To simplify notation, we shall ignore this issue; the reader may assume $\dim V$ is even. We refer the reader to [GQT, GS] for the metaplectic cases.

To describe the Weil representation Ω_ψ, we first need to endow the vector space V with a Haar measure. Let $\langle -, -\rangle$ be the symmetric bilinear form associated to the quadratic form q on V, so that $\langle v_1, v_2\rangle = q(v_1 + v_2) - q(v_1) - q(v_2)$. Then one has an S^1-valued nondegenerate pairing $\psi(\langle -, -\rangle)$ on V. We then equip V with the Haar measure $d_\psi v$ which is self-dual with respect to the Fourier transform defined by this pairing and observe that $d_\psi v$ is $\mathrm{O}(V)$-invariant. The unitary representation Ω_ψ can be realised on $L^2(f \otimes V) = L^2(V)$, where the inner product is defined using the Haar measure $d_\psi v$. The action of various elements of $\mathrm{SL}_2(F) \times \mathrm{O}(V)$ via Ω_ψ is given as follows:

$$\begin{cases} h \cdot \Phi(v) = \Phi(h^{-1} \cdot v), \text{ for } h \in \mathrm{O}(V); \\ n(b) \cdot \Phi(v) = \psi(b \cdot q(v)) \cdot \Phi(v), \text{ for } n(b) = \begin{pmatrix} 1 & b \\ 0 & 1 \end{pmatrix} \in N; \\ t(a) \cdot \Phi(v) = |a|^{\frac{1}{2}\dim V} \chi_{\mathrm{disc}(V)}(a) \cdot \Phi(a^2 v), \text{ for } t(a) = \begin{pmatrix} a & 0 \\ 0 & a^{-1} \end{pmatrix} \in T. \end{cases}$$

Here $\mathrm{disc}(V) \in F^\times/F^{\times 2}$ is the discriminant of (V, q) and $\chi_{\mathrm{disc}(V)}$ is the associated quadratic character of $F^\times$. This describes Ω_ψ as a representation of $B \times \mathrm{O}(V)$. To describe the full action of $\mathrm{SL}_2(F)$, one needs to give the action of a nontrivial Weyl group element

$$w = \begin{pmatrix} 0 & 1 \\ -1 & 0 \end{pmatrix}.$$

Its action is given by a normalized Fourier transform $\mathcal{F}$:

$$w \cdot \Phi(v) = \mathcal{F}(\Phi)(v) := \gamma_{\psi,q} \cdot \int \Phi(v') \cdot \psi(\langle v, v' \rangle)\, d_\psi v' \tag{5.1}$$

where $\gamma_{\psi,q}$ is a root of unity (a Weil index) whose precise value need not concern us here.

One may consider the underlying smooth representation Ω_ψ^∞ which is realized on the subspace $S(V)$ of Schwartz-Bruhat functions on V. Following our convention, we shall use Ω_ψ to denote the Weil representation in both the smooth and L^2-setting when there is no cause for confusion.

5.2 *Smooth Theta Correspondence*

The theory of theta correspondence concerns the understanding of the representation Ω_ψ of $\mathrm{SL}_2(F) \times \mathrm{O}(V)$. One can consider this question on the level of smooth representation theory or L^2-representation theory. In this subsection, we recall the setup of the smooth theory. Henceforth, we shall assume that $\dim V \geq 3$ (and sometimes $\dim V \geq 4$).

For $\sigma \in \mathrm{Irr}(\mathrm{SL}_2)$, the (smooth) big theta lift of σ to $\mathrm{O}(V)$ is:

$$\Theta_\psi(\sigma) := (\Omega_\psi^\infty \otimes \sigma^\vee)_{\mathrm{SL}_2}$$

where we are considering the space of SL_2-coinvariants. With this definition, we have the natural SL_2-invariant and $\mathrm{O}(V)$-equivariant projection map

$$A_\sigma : \Omega_\psi^\infty \otimes \sigma^\vee \longrightarrow \Theta_\psi(\sigma)$$

which gives by duality a canonical $\mathrm{SL}_2 \times \mathrm{O}(V)$-equivariant map

$$\theta_\sigma : \Omega_\psi^\infty \longrightarrow \sigma \boxtimes \Theta_\psi(\sigma).$$

Likewise, for $\pi \in \mathrm{Irr}(\mathrm{O}(V))$, the (smooth) big theta lift of π to SL_2 is:

$$\Theta_\psi(\pi) := (\Omega_\psi^\infty \otimes \pi^\vee)_{\mathrm{O}(V)}$$

where we are considering the space of $\mathrm{O}(V)$-coinvariants.

By the Howe duality principle [GT], one knows that:

- the representations $\Theta_\psi(\sigma)$ and $\Theta_\psi(\pi)$ are finite length representations which (if nonzero) have unique irreducible quotients $\theta_\psi(\sigma)$ and $\theta_\psi(\pi)$ respectively (known as the small theta lifts);
- for any $\sigma_1, \sigma_2 \in \mathrm{Irr}(\mathrm{SL}_2)$,

$$\theta_\psi(\sigma_1) \cong \theta_\psi(\sigma_2) \neq 0 \Longrightarrow \sigma_1 \cong \sigma_2.$$

As a consequence, we see that if $\pi := \theta_\psi(\sigma)$, then $\sigma \cong \theta_\psi(\pi)$.

Composing A_σ and θ_σ with the natural projection $\Theta_\psi(\sigma) \twoheadrightarrow \theta_\psi(\sigma)$, we have canonical equivariant maps (still denoted by the same symbols)

$$A_\sigma : \Omega \otimes \sigma^\vee \longrightarrow \theta_\psi(\sigma) \tag{5.2}$$

and

$$\theta_\sigma : \Omega_\psi^\infty \longrightarrow \sigma \boxtimes \theta_\psi(\sigma). \tag{5.3}$$

The theta correspondence for $\mathrm{SL}_2 \times \mathrm{O}(V)$ (when $\dim V$ is even) and $\mathrm{Mp}_2 \times \mathrm{O}(V)$ (when $\dim V$ is odd) was studied in great detail by Rallis [R3]. His results were supplemented by later results of J.S. Li [Li]. We may summarize their results by:

Proposition 5.1.

(i) Assume that $\dim V \geq 4$ *is even and the Witt index* $\mathrm{Witt}(V)$ *of* V *is* ≥ 2 *(so that one is in the stable range). If* $\sigma \in \mathrm{Irr}(\mathrm{SL}_2)$ *is unitary, then* $\theta(\sigma)$ *is nonzero unitary, so that one has an injective map*

$$\theta_\psi : \widehat{\mathrm{SL}_2} \longrightarrow \widehat{\mathrm{O}(V)}.$$

In general, the theta correspondence gives a map

$$\theta_\psi : \widehat{\mathrm{SL}_{2}}_{temp} \longrightarrow \widehat{\mathrm{O}(V)} \cup \{0\}$$

which is injective on that part of the domain outside the preimage of 0.

(ii) *Assume that* $\dim V \geq 3$ *is odd and* $\mathrm{Witt}(V)$ *is* ≥ 2. *If* $\sigma \in \mathrm{Irr}(\mathrm{Mp}_2)$ *is a unitary genuine representation, then* $\theta_\psi(\sigma)$ *is nonzero unitary, so that one has an injective map*

$$\theta_\psi : \widehat{\mathrm{Mp}_2} \longrightarrow \widehat{\mathrm{O}(V)}$$

where $\widehat{\mathrm{Mp}_2}$ *denotes the* ψ*-generic genuine unitary dual of* $\mathrm{Mp}_2(F)$. *In general, the theta correspondence gives a map*

$$\theta_\psi : \widehat{\mathrm{Mp}_{2}}_{temp} \longrightarrow \widehat{\mathrm{O}(V)} \cup \{0\}$$

which is injective on that part of the domain outside the preimage of 0.

One can in fact describe the map θ_ψ very explicitly but we will not need this description here.

5.3 Doubling Zeta Integral

Using the unitary structures of Ω_ψ and σ (and the Haar measure dg on SL_2), the representation $\theta_\psi(\sigma)$ of $\mathrm{O}(V)$ can be given a unitary structure by the local doubling zeta integral. More precisely, for $\Phi_1, \Phi_2 \in S(V)$ and $v_1, v_2 \in \sigma$, the local doubling zeta integral is given by:

$$Z_\sigma(\Phi_1, \Phi_2, v_1, v_2) = \int_{\mathrm{SL}_2} \langle g \cdot \Phi_1, \Phi_2 \rangle_\Omega \cdot \overline{\langle \sigma(g) \cdot v_1, v_2 \rangle_\sigma}\, dg, \tag{5.4}$$

which converges for tempered σ when $\dim V \geq 3$. It defines a $(\mathrm{SL}_2 \times \mathrm{SL}_2)$-invariant and $\mathrm{O}(V)^\Delta$-invariant map

$$Z_\sigma : \Omega_\psi \otimes \overline{\Omega_\psi} \otimes \overline{\sigma} \otimes \sigma \longrightarrow \mathbb{C}.$$

The inner product $\langle -, - \rangle_\sigma$ on σ gives an isomorphism $\overline{\sigma} \cong \sigma^\vee$. Hence, Z_σ factors through the canonical projection map

$$A_\sigma \otimes \overline{A_\sigma} : \Omega_\psi \otimes \overline{\Omega_\psi} \otimes \sigma^\vee \otimes \overline{\sigma^\vee} \longrightarrow \Theta_\psi(\sigma) \otimes \overline{\Theta_\psi(\sigma)}$$

so that

$$Z_\sigma(\Phi_1, \Phi_2, v_1, v_2) = \langle A_\sigma(\Phi_1, v_1), A_\sigma(\Phi_2, v_2) \rangle_{\theta(\sigma)}. \tag{5.5}$$

for some Hermitian form $\langle -, - \rangle_{\theta(\sigma)}$ on $\Theta_\psi(\sigma)$. We have:

Proposition 5.2. *Suppose that σ is an irreducible tempered representation of* SL_2 *(or* Mp_2*) such that $\theta_\psi(\sigma) \neq 0$. Then the Hermitian form $\langle -, - \rangle_{\theta(\sigma)}$ on $\Theta_\psi(\sigma)$ descends to a nonzero inner product on $\theta_\psi(\sigma)$.*

Proof. We note:

- If the Witt index of V is ≥ 2 (so that one is in the stable range), this is due to [Li].
- In general, it was shown by Rallis [R3, Prop. 6.1] that

$$\theta_\psi(\sigma) \neq 0 \Longleftrightarrow \langle -, - \rangle_{\theta(\sigma)} \neq 0.$$

- In the Archimedean case, it was shown in [He] that $\langle -, - \rangle_{\theta(\sigma)}$ descends to $\theta_\psi(\sigma)$.
- Consider the nonarchimedean case with $\mathrm{Witt}(V) \leq 1$. There are only a few cases of such, all in low rank. In these small number of low rank cases, one can verify that $\Theta_\psi(\sigma)$ is irreducible when σ is tempered.

Taken together, the proposition is proved. □

Henceforth, we shall equip $\theta_\psi(\sigma)$ with this unitary structure; it depends on dg, $\langle -, - \rangle_\sigma$ and $d_\psi v$. By completion, we may regard $\theta_\psi(\sigma)$ as an irreducible unitary representation of $\mathrm{O}(V)$. Observe that the identity (5.5) may be considered as the local analog of the Rallis inner product formula [GQT]. A reformulation, using the map θ_σ instead of A_σ is:

$$\langle \theta_\sigma(\Phi_1), \theta_\sigma(\Phi_2) \rangle_{\sigma \boxtimes \theta(\sigma)} = \sum_{v \in \mathrm{ONB}(\sigma)} Z_\sigma(\Phi_1, \Phi_2, v, v) \tag{5.6}$$

where $\mathrm{ONB}(\sigma)$ denotes an orthonormal basis of σ.

5.4 L^2-Theta Correspondence

Now we consider the theta correspondence in the L^2-setting. Though we are not exactly in the setting discussed in Sect. 2, Bernstein's theory continues to apply here (see [S3]). When $\dim V \geq 3$, it was shown in [GG, S3] that one has a direct integral decomposition of $\mathrm{SL}_2(F) \times \mathrm{O}(V)$-representations:

$$\Omega_\psi = L^2(V) \cong \int_{\widehat{\mathrm{SL}_2}} \sigma \boxtimes \theta_\psi(\sigma) \, d\mu_{\mathrm{SL}_2}(\sigma), \tag{5.7}$$

where $d\mu_{\mathrm{SL}_2}$ is the Plancherel measure of $\mathrm{SL}_2(F)$ (associated to the fixed Haar measure of SL_2). Hence the spectral measure of Ω_ψ as an SL_2-module is absolutely continuous with respect to the Plancherel measure. Indeed, by Propositions 5.1 and 5.2, when $\mathrm{Witt}(V) \geq 2$, the support of Ω_ψ as an SL_2-module is precisely $\widehat{\mathrm{SL}}_{2temp}$.

We need to explicate the spectral decomposition (5.7) here. By the theory of spectral decomposition à la Bernstein, one way to do this is to give a spectral decomposition of the inner product $\langle -, - \rangle_\Omega$. In [S3], Sakellaridis showed that for $\Phi_1, \Phi_2 \in S(V)$,

$$\langle \Phi_1, \Phi_2 \rangle_\Omega = \int_{\widehat{\mathrm{SL}_2}} J_\sigma^\theta(\Phi_1, \Phi_2)\, d\mu_{\mathrm{SL}_2}(\sigma)$$

where

$$J_\sigma^\theta(\Phi_1, \Phi_2) = \langle \theta_\sigma(\Phi_1), \theta_\sigma(\Phi_2) \rangle_{\sigma \boxtimes \theta(\sigma)} = \sum_{v \in \mathrm{ONB}(\sigma)} Z_\sigma(\Phi_1, \Phi_2, v, v) \tag{5.8}$$

where the unitary structure on $\Theta(\sigma)$ is that defined in the previous subsection. What this says is that the family of canonical maps θ_σ defined in (5.3) is precisely the family of maps associated to the direct integral decomposition (5.7) for $\sigma \in \widehat{\mathrm{SL}}_{2temp}$.

Now for $\tau \in \widehat{\mathrm{SL}}_{2temp}^{ind}$, one can define $Z_\tau(\Phi_1, \Phi_2, v_1, v_2)$ by the same formula as in (5.4) and then define J_τ^θ by the formula (5.8). Then it is useful to note [X, Lemma 3.3]:

Lemma 5.3. *For fixed Φ_i, the $\mathbb{C}$-valued function*

$$\tau \mapsto J_\tau^\theta(\Phi_1, \Phi_2)$$

is continuous in $\tau \in \widehat{\mathrm{SL}}_{2temp}^{ind}$.

5.5 The Maps A_σ and $B_{\theta(\sigma)}$

We have seen the canonical maps A_σ and θ_σ in (5.2) and (5.3) which intervene in the spectral decomposition (5.7). Identifying $\sigma^\vee$ with $\overline{\sigma}$ using $\langle -, - \rangle_\sigma$, we may regard A_σ as a map $\Omega_\psi \otimes \overline{\sigma} \longrightarrow \theta_\psi(\sigma)$. Then A_σ and θ_σ are related by:

$$A_\sigma(\Phi, v) = \langle \theta_\sigma(\Phi), v \rangle_\sigma.$$

Likewise, we have a $\mathrm{O}(V)$-invariant and SL_2-equivariant map

$$B_{\theta(\sigma)} : \Omega_\psi \otimes \overline{\theta_\psi(\sigma)} \longrightarrow \sigma$$

characterized by

$$B_{\theta(\sigma)}(\Phi, w) = \langle \theta_\sigma(\Phi), w \rangle_{\theta(\sigma)}.$$

The two maps are related by:

$$\langle A_\sigma(\Phi, v), w \rangle_{\theta(\sigma)} = \langle \theta_\sigma(\Phi), v \otimes w \rangle_{\sigma \otimes \theta(\sigma)} = \langle B_{\theta(\sigma)}(\phi, w), v \rangle_\sigma \tag{5.9}$$

for $\Phi \in \Omega_\psi$, $v \in \sigma$ and $w \in \theta(\sigma)$. Moreover, the inner product J_σ^θ can be expressed in terms of A_σ and $B_{\theta(\sigma)}$ as follows:

$$J_\sigma^\theta(\Phi_1, \Phi_2) = \langle \theta_\sigma(\Phi_1), \theta_\sigma(\Phi_2) \rangle_{\sigma \otimes \theta(\sigma)} = \sum_{v \in \mathrm{ONB}(\sigma)} \langle A_\sigma(\Phi_1, v), A_\sigma(\Phi_2, v) \rangle_{\theta(\sigma)}.$$

and

$$J_\sigma^\theta(\Phi_1, \Phi_2) = \sum_{w \in \mathrm{ONB}(\theta(\sigma))} \langle B_{\theta(\sigma)}(\Phi_1, w), B_{\theta(\sigma)}(\Phi_2, w) \rangle_\sigma.$$

The maps A_σ and $B_{\theta(\sigma)}$ are local versions of global theta lifting considered in Sect. 12.

To summarize, this section discusses the smooth theta correspondence and the L^2-theta correspondence and the relation between them. In particular, through the theory of the doubling zeta integral, we equip $\theta_\psi(\sigma)$ with a unitary structure so that there is a strong synergy between the smooth theory and the L^2-theory.

6 Periods

It is a basic principle that theta correspondence frequently allows one to transfer periods on one member of a dual pair to the other member. For an exposition of this in the setting of smooth theta correspondence, the reader can consult [G]. On the other hand, in the setting of L^2-theta correspondence, this principle has been exploited in [GG] to establish low rank cases of the local conjecture of Sakellaridis-Venkatesh on the unitary spectrum of spherical varieties.

In this section, we shall consider the dual pair $\mathrm{SL}_2 \times \mathrm{O}(V)$ and show how the spectral decomposition à la Bernstein allows one to refine the results of [G] and [GG].

6.1 *Transfer of Periods*

We first consider periods in smooth representation theory. For $a \in F^\times$, fix a vector $v_a \in V$ with $q(v_a) = a$ (if it exists), so that $V = F \cdot v_a \oplus v_a^\perp$. Set

$$X_a = \{v \in V : q(v) = a\} \subset V,$$

which is a Zariski closed subset of V. By Witt's theorem, $\mathrm{O}(V)$ acts transitively on X_a and the stabilizer of v_a in $\mathrm{O}(V)$ is $\mathrm{O}(v_a^\perp)$. Hence

$$X_a \cong \mathrm{O}(v_a^\perp)\backslash \mathrm{O}(V)$$

via $h \mapsto h^{-1} \cdot v_a$. If v_a does not exist, we understand X_a to be empty (i.e. the algebraic variety has no F-points). To fix ideas, we shall assume that v_1 exists; this is not a serious hypothesis. We also set $\psi_a(x) = \psi(ax)$, so that ψ_a is a nontrivial additive character of F, and write $S(X_a)$ for the space of Schwartz-Bruhat functions on X_a, so that $S(X_a) = C_c^\infty(X_a)$ if F is nonarchimedean.

The following proposition essentially resolves the local problem (a) in the smooth setting for the Sakellaridis-Venkatesh conjecture highlighted in the introduction, except for the part about relative character identities. It is essentially a folklore result and a proof has been written down in [G] in a more general setting. We recount the proof here to explicate the isomorphism f_a in the proposition.

Proposition 6.1. *Let π be an irreducible smooth representation of $\mathrm{O}(V, q)$ and let $\Theta_\psi(\pi) = (\Omega_\psi \otimes \pi^\vee)_{\mathrm{SL}_2}$ be its big theta lift to SL_2 (or Mp_2 if $\dim V$ is odd). For $a \in F^\times$, there is an explicit isomorphism (to be described in the proof)*

$$f_a : \mathrm{Hom}(\Theta_\psi(\pi)_{N,\psi_a}, \mathbb{C}) \cong \mathrm{Hom}_{\mathrm{O}(V)}(S(X_a), \pi) \cong \mathrm{Hom}_{\mathrm{O}(v_a^\perp)}(\pi^\vee, \mathbb{C}),$$

where the second isomorphism is by Frobenius reciprocity. Here, the right hand side is understood to be 0 if X_a is empty. In particular, when π is such that $\sigma := \Theta_\psi(\pi)$ is irreducible, we see that σ is ψ_a-generic if and only if π is $\mathrm{O}(v_a^\perp)$-distinguished, in which case

$$\dim \mathrm{Hom}_{\mathrm{O}(v_a^\perp)}(\pi^\vee, \mathbb{C}) = 1.$$

Proof. We describe the proof when F is nonarchimedean. The Archimedean case is based on the same ideas, and the reader can consult [GZ, Z] for a careful treatment.

We prove the proposition by computing the space

$$\mathrm{Hom}_{N\times \mathrm{O}(V)}(\Omega_\psi, \psi_a \boxtimes \pi)$$

in two different ways.

On one hand, let us fix any equivariant surjective map

$$\theta : \Omega_\psi \longrightarrow \Theta_\psi(\pi) \boxtimes \pi.$$

Then θ induces an isomorphism

$$\theta^* : \mathrm{Hom}_N(\Theta_\psi(\pi), \psi_a) \cong \mathrm{Hom}_{N\times \mathrm{O}(V)}(\Omega_\psi, \psi_a \otimes \pi).$$

On the other hand, for $a \in F^{\times}$, consider the surjective restriction map

$$\mathrm{rest} : \Omega_{\psi} = S(V) \longrightarrow S(X_a).$$

This map induces an equivariant isomorphism

$$\mathrm{rest} : \Omega_{N,\psi_a} \cong S(X_a).$$

Hence, we have an induced isomorphism

$$\mathrm{rest}^* : \mathrm{Hom}_{\mathrm{O}(V)}\, S(X_a), \pi) \cong \mathrm{Hom}_{\mathrm{O}(V)}(\Omega_{N,\psi_a}, \pi) \cong \mathrm{Hom}_{N\times \mathrm{O}(V)}(\Omega_{\psi}, \psi_a \otimes \pi).$$

Since

$$S(X_a) \cong \mathrm{ind}_{\mathrm{O}(v_a^{\perp})}^{\mathrm{O}(V)} \mathbb{C},$$

it follows by Frobenius reciprocity that one has the desired isomorphism:

$$f_a^{-1} : \mathrm{Hom}_{\mathrm{O}(v_a^{\perp})}(\pi^{\vee}, \mathbb{C}) \xrightarrow{\mathrm{Frob}} \mathrm{Hom}_{\mathrm{O}(V)}(S(X_a), \pi) \xrightarrow{(\theta^*)^{-1} \circ \mathrm{rest}^*} \mathrm{Hom}_N(\Theta_{\psi}(\pi), \psi_a).$$

This proves the proposition. □

The purpose of recounting the proof of the proposition is to bring forth the point that the isomorphism

$$f_a : \mathrm{Hom}(\Theta_{\psi}(\pi)_{N,\psi_a}, \mathbb{C}) \cong \mathrm{Hom}_{\mathrm{O}(v_a^{\perp})}(\pi^{\vee}, \mathbb{C})$$

essentially depends only on the choice of the equivariant projection map

$$\theta : \Omega_{\psi} \longrightarrow \Theta_{\psi}(\pi) \boxtimes \pi.$$

On the other hand, when σ is an irreducible tempered representation of SL_2 with $\theta_{\psi}(\sigma) \neq 0$, we have seen in (5.7) that there is a canonical map

$$\theta_{\sigma} : \Omega_{\psi} \longrightarrow \sigma \boxtimes \theta_{\psi}(\sigma).$$

Repeating the proof of the proposition using this map θ_{σ}, we obtain an injective map

$$f_a : Hom_N(\sigma, \psi_a) \longrightarrow \mathrm{Hom}_{\mathrm{O}(V)}(S(X_a), \theta_{\psi}(\sigma)). \tag{6.1}$$

It is an isomorphism if $\Theta_{\psi}(\theta_{\psi}(\sigma)) \cong \sigma$ (by Proposition 6.1) or if σ is ψ_a-generic (since the target space has dimension at most 1 by [AGRS])

6.2 Unitary Structure on $L^2(X_a)$

We may begin our investigation of the local problem (b) in the Sakellaridis-Venkatesh conjecture, i.e. in the L^2-setting. With $a \in F^\times$, we have seen that

$$X_a = \{v \in V : q(v) = a\} \cong \mathrm{O}(v_a^\perp)\backslash\mathrm{O}(V),$$

under $h^{-1}v_a \longleftrightarrow h$ (assuming $X_a(F)$ has a point v_a). We may equip X_a with an $\mathrm{O}(V)$-invariant measure and consider the space $L^2(X_a)$; of course the space $L^2(X_a)$ does not depend on the choice of the $\mathrm{O}(V)$-invariant measure but the unitary structure does. In [GG], using the spectral decomposition (5.7) of Ω_ψ, one obtains a spectral decomposition of $L^2(X_a)$. However, we wish to refine the results of [GG] by being more precise about the unitary structures and invariant measures used here.

The hyperboloids X_a are precisely the fibers of the $\mathrm{O}(V)$-invariant map given by the quadratic form $q : V \longrightarrow F$. This map is submersive at all points of V outside the zero vector. In particular, if we ignore the null cone X_0 and consider the map q over the Zariski open subset $F^\times \subset F$, the Haar measures $d_\psi v$ and $d_\psi x$ we have already fixed for V and F induces an $\mathrm{O}(V)$-invariant measure $|\omega_a|$ for each fiber X_a (with $a \in F^\times$), characterized by: for any compactly-supported smooth functions f on $V \setminus X_0$,

$$\int_V f(v)\, d_\psi(v) = \int_{F^\times} \left(\int_{X_a} f \cdot |\omega_a| \right) d_\psi a,$$

where the function of $a \in F^\times$ defined by the inner integral on the right-hand-side is smooth and compactly supported. It is these measures $|\omega_a|$ that we shall use on X_a. Hence, we shall be considering $L^2(X_a, |\omega_a|)$.

The map $q : V \longrightarrow F$ is $F^\times$-equivariant where $t \in F^\times$ acts by scaling on V and via $x \mapsto t^2x$ on F. The measure $d_\psi v$ on V is $\mathrm{O}(V)$-invariant and satisfies: for $b \in F^\times$,

$$d_\psi(bv) = |b|^{dim V} \cdot d_\psi v.$$

This homogeneity property implies the following property of the family of measures $|\omega_a|$. For $b \in F^\times$, scalar multiplication-by-b gives an isomorphism of varieties $\lambda_b : X_a \longrightarrow X_{ab^2}$ and one may consider the pushforward measure $(\lambda_b)_*(|\omega_a|)$ on X_{ab^2}.

Lemma 6.2. *In the above context, one has:*

$$(\lambda_b)_*(|\omega_a|) = |b|^{2-\dim V} \cdot |\omega_{ab^2}|$$

for any $a, b \in F^\times$.

Proof. For $f \in C_c^\infty(V \setminus X_0)$ and fixed $b \in F^\times$, we have

$$\int_V f(bv)\,dv = |b|^{-\dim V} \cdot \int_V f(v)\,dv. \tag{6.2}$$

The left-hand-side of (6.2) is given by

$$\begin{aligned}\int_V f(bv)\,dv &= \int_{F^\times}\int_{X_a} f(bx)\cdot|\omega_a(x)|\,d_\psi a\\ &= \int_{a\in F^\times}\int_{X_a} \lambda_b^*(f)(x)\cdot\cdot|\omega_a(x)|\,d_\psi a\\ &= \int_{a\in F^\times}\int_{X_{ab^2}} f(x)\cdot|(\lambda_b)_*(\omega_a|)(x)|\,d_\psi a\\ &= \int_{c\in F^\times}\int_{X_c} f(x)\cdot|(\lambda_b)_*(\omega_{cb^{-2}})|\cdot|b|^{-2}\,d_\psi c\end{aligned}$$

where in the last equality, we have made a change of variables by setting $c = ab^2$, so that $d_\psi c = |b|^2 \cdot d_\psi a$.

On the other hand, the right hand side of (6.2) is given by

$$|b|^{-\dim V} \cdot \int_V f(v)\,dv = |b|^{-\dim V} \cdot \int_{c\in F^\times}\int_{X_c} f(x)\cdot|\omega_c(x)|\,d_\psi c.$$

Comparing the two sides, one obtains:

$$|(\lambda_b)_*(\omega_{cb^{-2}})| = |b|^{2-\dim V}\cdot|\omega_c|$$

which is the desired assertion. □

Now observe that (on the level of F-valued points),

$$V\setminus X_0 = \bigcup_{[a]\in F^{\times 2}\setminus F^\times} F^\times \cdot X_a \subset V$$

is open dense with complement of measure 0. Hence the measure $d_\psi v$ induces $\mathrm{O}(V)$-invariant measures on each of the open sets $Y_a := F^\times \cdot X_a$ and we have

$$L^2(V) = \bigoplus_{[a]\in F^{\times 2}\setminus F^\times} L^2(Y_a)$$

We would like a more direct description of the unitary structures on the Hilbert spaces $L^2(Y_a)$ in terms of appropriate invariant measures on $F^\times \times X_a$.

Consider the natural surjective map

$$m : F^{\times} \times X_a \longrightarrow Y_a = F^{\times} \cdot X_a,$$

defined by $m(t, x). \mapsto t \cdot x$. This map m induces an isomorphism $\mu_2 \backslash (F^{\times} \times X_a) \cong Y_a$ where $\mu_2 = \{\pm 1\}$ acts diagonally on $F^{\times} \times X_a$ by scaling on each factor. In terms of the identification $X_a = \mathrm{O}(v_a^{\perp}) \backslash \mathrm{O}(V)$, this action of μ_2 on X_a is the left-translation action of $\mathrm{O}(v_a) = \mu_2$ (where $\mathrm{O}(v_a)$ is the orthogonal group of the one-dimensional quadratic space $F \cdot v_a$) which commutes with the right-translation action of $\mathrm{O}(V)$. In any case, via $f \mapsto m^*(f)$, we may identify functions on Y_a with functions on $F^{\times} \times X_a$ invariant under the action of μ_2. Now we note:

Lemma 6.3. *For a smooth compactly supported function f on Y_a, one has:*

$$\int_{Y_a} f(v)\, d_{\psi} v = \frac{1}{2} \cdot |2a|_F \cdot \int_{b \in F^{\times}} \left(\int_{x \in X_a} f(b \cdot x) \cdot |\omega_a(x)| \right) \cdot |b|^{\dim V} \frac{d_{\psi} b}{|b|}.$$

Hence, if we define

$$\phi_f : F^{\times} \times X_a \longrightarrow \mathbb{C}$$

by

$$\phi_f(b, x) = \sqrt{|a|} \cdot |b|^{\frac{\dim V}{2}} \cdot f(b \cdot x),$$

then one has

$$\langle f, f \rangle_{\Omega} = \frac{1}{2} \cdot |2|_F \cdot \int_{F^{\times}} \int_{X_a} |\phi_f(b, x)|^2\, |\omega_a(x)|\, d_{\psi}^{\times} b.$$

In particular, the map $f \mapsto \phi_f$ defines an isometric isomorphism

$$L^2(Y_a, d_{\psi} v) \longrightarrow L^2(F^{\times} \times X_a)^{\Delta \mu_2}$$

where the unitary structure on the left is defined by $d_{\psi} v$ and that on the right is defined by the $\mathrm{O}(V)$-invariant measure $|\omega_a|$ on X_a and the measure $|2|_F/2 \cdot d_{\psi}^{\times} t$ of $F^{\times}$ (defined in Sect. 4.3). Further, for $\epsilon = \pm$, if we let $L^2(Y_a)^{\epsilon}$ and $L^2(F^{\times})^{\epsilon}$ denote the ϵ-eigenspace of the μ_2-action, then

$$L^2(Y_a) \cong \bigoplus_{\epsilon = \pm} L^2(F^{\times})^{\epsilon} \widehat{\otimes} L^2(X_a)^{\epsilon}.$$

Proof. We consider $f \in C_c^{\infty}(Y_a)$. Then

$$\int_{Y_a} f(y)\, dy = \int_{t \in aF^{\times 2}} \int_{X_t} f(x) \cdot |\omega_t(x)|\, d_{\psi} t$$

$$= \frac{1}{2} \int_{b \in F^\times} \int_{X_{ab^2}} f(x) \cdot |\omega_{ab^2}(x)| \cdot |2ab|_F \, d_\psi b$$

$$= \frac{1}{2} \cdot |2a|_F \cdot \int_{b \in F^\times} \int_{X_{ab^2}} f(x) \cdot |(\lambda_b)_*(|\omega_a|) \cdot |b|^{\dim V} \frac{d_\psi b}{|b|}$$

$$= \frac{1}{2} \cdot |2a|_F \cdot \int_{b \in F^\times} \int_{X_a} \lambda_b^*(f)(x) \cdot |\omega_a(x)| \cdot |b|^{\dim V} \, d_\psi^\times b$$

$$= \frac{1}{2} \cdot |2a|_F \cdot \int_{b \in F^\times} \int_{X_a} f(b \cdot x) \cdot |\omega_a(x)| \cdot |b|^{\dim V} \, d_\psi^\times b. \tag{6.3}$$

Here, in the second equality, we have made a change of variables, replacing t by ab^2, so that $d_\psi t = |2ab|_F \cdot d_\psi b$, whereas in the third equality, we have applied Lemma 6.2. This establishes the lemma. □

6.3 Spectral Decomposition of $L^2(X_a)$

We are now ready to show the direct integral decomposition of the unitary representation $L^2(X_a, |\omega_a|)$ of $\mathrm{O}(V)$, where the unitary structure is determined by the $\mathrm{O}(V)$-invariant measures $|\omega_a|$. Observe that $L^2(X_a, |\omega_a|)$ is in fact a representation of $\mathrm{O}(v_a) \times \mathrm{O}(V)$, where $\mathrm{O}(v_a) \cong \mu_2$ acts by left translation on $\mathrm{O}(v_a^\perp)\backslash\mathrm{O}(V)$ (this is the action of scaling by -1 on X_a). This gives a decomposition

$$L^2(X_a) = L^2(X_a)^+ \oplus L^2(X_a)^-$$

into $\mathrm{O}(V)$-submodules which are the $\pm$-eigenspaces of the $\mathrm{O}(v_a)$-action.

As mentioned before, the spectral decomposition of $L^2(X_a)$ as an $\mathrm{O}(V)$-module has been obtained in [GG]. The following proposition is a special case of the results in [GG]; we recount the proof here to explicate certain isomorphisms used in the course of the proof.

Proposition 6.4. *We have an explicit isomorphism (to be described in the proof):*

$$L^2(X_a, |\omega_a|) \cong \int_{\widehat{\mathrm{SL}_2}} \dim \mathrm{Hom}_N(\sigma, \psi_a) \cdot \theta_\psi(\sigma) \, d\mu_{\mathrm{SL}_2}(\sigma).$$

Proof. We shall exploit the spectral decomposition of the unitary Weil representation Ω_ψ of $\mathrm{SL}_2(F) \times \mathrm{O}(V)$ on $L^2(V)$ given in (5.7). More precisely, we shall consider its restriction to $B \times \mathrm{O}(V)$. We have seen that

$$\bigcup_{[a] \in F^{\times 2}\backslash F^\times} Y_a = \bigcup_{[a] \in F^{\times 2}\backslash F^\times} F^\times \cdot X_a \subset V$$

is open dense (with complement of measure 0), so that

$$L^2(V) = \bigoplus_{a\in F^{\times 2}\backslash F^\times} L^2(Y_a)$$

From the formulae for the action of $B \times \mathrm{O}(V)$ on Ω_ψ, one sees that the subspace $L^2(Y_a)$ is stable under the action of $B \times \mathrm{O}(V)$ and thus this is a decomposition of $B \times \mathrm{O}(V)$-modules. Moreover, with $T \cong F^\times$ acting on V by scaling, we see that $T \times \mathrm{O}(V)$ acts transitively on Y_a and the stabilizer of $v_a \in X_a$ is the subgroup

$$\mu_2^\Delta \times \mathrm{O}(v_a^\perp) \subset Z \times \mathrm{O}(v_a) \times \mathrm{O}(v_a^\perp),$$

where $Z = \mu_2$ is the center of $\mathrm{SL}_2(F)$ and $\mathrm{O}(v_a) = \mu_2$. Thus, the isometry $f \mapsto \phi_f$ described in Lemma 6.3 gives an $B \times \mathrm{O}(V)$-equivariant isometric map

$$\begin{aligned} L^2(Y_a, d_\psi v) &\cong \bigoplus_{\epsilon=\pm} \left(\mathrm{ind}_{ZN}^B \epsilon \otimes \psi_a\right) \widehat{\otimes} L^2(X_a)^\epsilon \\ &= \bigoplus_{\epsilon=\pm} L^2(N, \psi_a\backslash B)^\epsilon \widehat{\otimes} L^2(X_a)^\epsilon = L^2(F^\times \times X_a)^{\mu_2}, \end{aligned}$$

where the unitary structure on $L^2(N, \psi_a\backslash B)^\epsilon$ is given by that defined in Lemma 6.3 or equivalently in Proposition 4.2. Hence, we conclude that

$$\Omega_\psi \cong \bigoplus_{a\in F^{\times 2}\backslash F^\times} L^2(F^\times \times X_a)^{\Delta\mu_2} \cong \bigoplus_{a\in F^{\times 2}\backslash F^\times} \bigoplus_{\epsilon=\pm} L^2(N, \psi_a\backslash B)^\epsilon \widehat{\otimes} L^2(X_a)^\epsilon \tag{6.4}$$

via

$$f \mapsto \left(\phi_{f|Y_a}\right)_{a\in F^{\times 2}\backslash F^\times}.$$

On the other hand, by (5.7), one has:

$$\iota : L^2(V) \cong \int_{\widehat{\mathrm{SL}_2}} \sigma \otimes \theta_\psi(\sigma)\, d\mu_{\mathrm{SL}_2}(\sigma)$$

as $\mathrm{SL}_2 \times \mathrm{O}(V)$-modules. Restricting from SL_2 to B, Corollary 4.3 gives an isometric B-equivariant isomorphism

$$j_\sigma = \bigoplus_{a\in F^\times\backslash F^{\times 2}} j_{\sigma,a} : \sigma|_B \cong \bigoplus_{[a]\in F^{\times 2}\backslash F^\times} \dim \sigma_{N,\psi_a} \cdot L^2(N, \psi_a\backslash B)^{z_\sigma}$$

where $z_\sigma = \pm$ denotes the central character of σ and the unitary structure on the right-hand-side is as in Lemma 6.3. Hence, via $j_\sigma \circ \iota$ for each σ, one has a unitary isomorphism:

$$\Omega_\psi \cong \bigoplus_{a\in F^{\times 2}\backslash F^\times} \bigoplus_{\epsilon=\pm} L^2(N,\psi_a\backslash B)^\epsilon \widehat{\otimes} \int_{\widehat{\mathrm{SL}_2}} \dim \sigma_{N,\psi_a} \cdot \theta_\psi(\sigma)\cdot 1(z_\sigma=\epsilon)\, d\mu_{\mathrm{SL}_2}(\sigma). \tag{6.5}$$

Comparing the two descriptions of $\Omega = L^2(V)$ as a $B \times \mathrm{O}(V)$-module given in (6.4) and (6.5), one obtains an isomorphism

$$L^2(X_a)^\epsilon \cong \int_{\widehat{\mathrm{SL}_2}} \dim \sigma_{N,\psi_a} \cdot \theta_\psi(\sigma)\cdot 1(\omega_\sigma=\epsilon)\, d\mu_{\mathrm{SL}_2}(\sigma).$$

for $\epsilon = \pm$. Summing over ϵ, we obtain the desired isomorphism in the proposition. □

6.4 A Commutative Diagram

Examining the proof of Proposition 6.4, the unitary isomorphism there can be explicated as follows. Given $f \in S(X_a)$, we first chooce $\Phi \in S(V)$ such that

$$|a|^{1/2} \cdot \mathrm{rest}(\Phi) = f.$$

Then the image of f under the isomorphism of Proposition 6.4 is represented by the measurable section of the direct integral decomposition given by

$$\sigma \mapsto |a|^{1/2} \cdot \ell_{\sigma,\psi_a}(\theta_\sigma(\Phi)),$$

where θ_σ is as given in (5.3) and ℓ_{σ,ψ_a} is the ψ_a-Whittaker functional arising from the Whittaker-Plancherel theorem for $L^2(N,\psi_a\backslash \mathrm{SL}_2)$. In other words, we have:

Proposition 6.5. *For each $\sigma \in \widehat{\mathrm{SL}}_{2temp,\psi_a}$, there is a commutative diagram:*

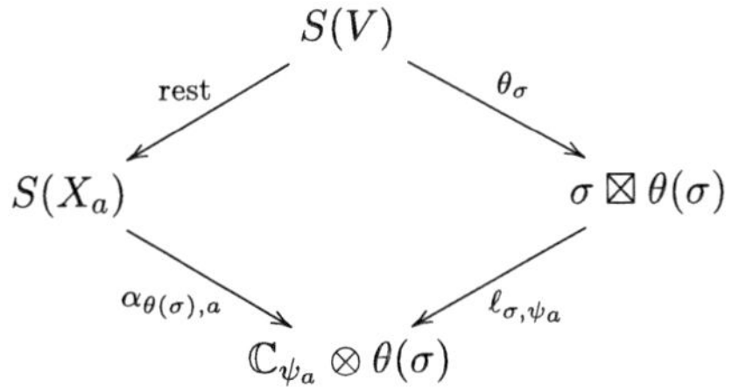

where $\alpha_{\theta(\sigma),a}$ is the morphism associated to the direct integral decomposition of Proposition 6.4.

This proposition gives a precise relation between the transfer of periods in the smooth setting and the spectral decomposition of $L^2(N,\psi_a\backslash \mathrm{SL}_2)$, $L^2(X_a, |\omega_a|)$ and Ω_ψ in the L^2-theory. Indeed, it is fairly clear that one has a commutative diagram as in the proposition up to scalars. The point of the proposition is to explicate the scalar. More precisely, specializing to the case $a = 1$, one has:

Corollary 6.6. *Under the isomorphism*

$$f_\sigma : \mathrm{Hom}_N(\sigma, \psi) \cong \mathrm{Hom}_{O(V)}(C_c^\infty(X_1), \theta_\psi(\sigma))$$

given in (6.1) for $\sigma \in \widehat{\mathrm{SL}_{2}}_{temp,\psi}$ *(which is induced by the map* θ_σ *of (5.3)), one has:*

$$f_\sigma(\ell_\sigma) = \alpha_{\theta(\sigma)} \tag{6.6}$$

where the Whittaker functional ℓ_σ *is the one in (3.4) which intervenes in the Whittaker-Plancherel theorem for* $(N, \psi)\backslash \mathrm{SL}_2$ *and the morphism* $\alpha_{\theta(\sigma)}$ *is the one which intervenes in the spectral decomposition of* $L^2(X_1, |\omega_1|)$ *obtained in Proposition 6.4.*

Another way to interpret Corollary 6.6 is that if one defines the elements $\alpha_{\theta(\sigma)}$ by (6.6) or equivalently by requiring that the diagram in Proposition 6.5 be commutative, then the family $\{\alpha_{\theta(\sigma)} : \sigma \in \widehat{\mathrm{SL}_{2}}_{temp,\psi}\}$ induces the spectral decomposition of $L^2(X_1, |\omega_1|)$ in Proposition 6.4.

We conclude this section with a few formal consequences of Proposition 6.5. The commutative diagram in Proposition 6.5 gives an identity in $\theta(\sigma)$. If we pair both sides of the identity with a vector in $\theta_\psi(\sigma)$, using the inner product on $\theta_\psi(\sigma)$, we obtain:

Corollary 6.7. *For any* $\Phi \in S(V) = \Omega_\psi$ *and* $w \in \theta(\sigma)$*, one has*

$$\ell_\sigma(B_{\theta(\sigma)}(\Phi, w)) = \langle \Phi|_X, \beta_{\theta(\sigma)}(w)\rangle_X,$$

where $B_{\theta(\sigma)}$ *was defined in Sect. 5.5.*

Proof. We have

$$\begin{aligned}\langle \Phi|_X, \beta_{\theta(\sigma)}(w)\rangle_X &= \langle \alpha_{\theta(\sigma)}(\Phi|_X), w\rangle_{\theta(\sigma)} = \langle \ell_\sigma(\theta_\sigma(\Phi)), w\rangle_{\theta(\sigma)} \\ &= \ell_\sigma\left(\langle \theta_\sigma(\Phi)), w\rangle_{\theta(\sigma)}\right) = \ell_\sigma(B_{\theta(\sigma)}(\Phi, w)).\end{aligned}$$

□

We may also "double-up" the commutative diagram in Proposition 6.5 and contract the resulting doubled identity using the inner product on $\theta_\psi(\sigma)$. This gives:

Corollary 6.8. *For* $\Phi_1, \Phi_2 \in S(V) = \Omega_\psi$*, one has:*

$$J_{\theta(\sigma)}(\Phi_1|_X, \Phi_2|_X) = \int_N^* \overline{\psi(n)} \cdot J_\sigma^\theta(n \cdot \Phi_1, \Phi_2)\, dn.$$

For fixed $\Phi_1|_X$ *and* $\Phi_2|_X$ *in* $S(X)$*, the* $\mathbb{C}$*-valued function* $\sigma \mapsto J_{\theta(\sigma)}(\Phi_1|_X, \Phi_2|_X)$ *is continuous in* $\sigma \in \widehat{\mathrm{SL}_{2}}_{temp,\psi}$.

Proof. We have

$$\begin{aligned}
&J_{\theta(\sigma)}(\Phi_1|_X, \Phi_2|_X) \\
=&\langle \alpha_{\theta(\sigma)}(\Phi_1|_X), \alpha_{\theta(\sigma)}(\Phi_2|_X)\rangle_{\theta(\sigma)} && \text{(definition of } J_{\theta(\sigma)}) \\
=&\langle \ell_\sigma(\theta_\sigma(\Phi_1)), \ell_\sigma(\theta_\sigma(\Phi_2))\rangle_{\theta(\sigma)} && \text{(by Proposition 6.5)} \\
=&\ell_\sigma \otimes \overline{\ell_\sigma}\left(\langle \theta_\sigma(\Phi_1), \theta_\sigma(\Phi_2)\rangle_{\theta(\sigma)}\right) && \text{(clear)} \\
=&\int_N^* \overline{\psi(n)} \cdot \langle n \cdot \theta_\sigma(\Phi_1), \theta_\sigma(\Phi_2)\rangle_{\sigma\otimes\theta(\sigma)}\, dn && \text{(formula for } \ell_\sigma \otimes \overline{\ell_\sigma}) \\
=&\int_N^* \overline{\psi(n)} \cdot J_\sigma^\theta(n \cdot \Phi_1, \Phi_2)\, dn && \text{(definition of } J_\sigma^\theta)
\end{aligned}$$

The continuity of $\sigma \mapsto J_{\theta(\sigma)}(\Phi_1|_X, \Phi_2|_X)$ follows from the above formula, together with Lemma 3.2 and Lemma 5.3. □

The last two corollaries thus give different variants of the identity in Proposition 6.5.

7 Relative Characters

In this section, we briefly recall the notion of the relative character associated to a period in its various incarnations.

7.1 Relative Characters

Suppose that, for $i = 1$ or 2, $H_i \subset G$ is a subgroup of G and $\chi_i : H_i(F) \to S^1$ a unitary character of $H_i(F)$. We fix also Haar measures dg on G and dh_i on H_i. For any $\pi \in \widehat{G}$ and $L_i \in \mathrm{Hom}_{H_i}(\pi, \chi_i)$, one can associate a distribution on G as follows. Given $(f_1, f_2) \in C_c^\infty(G) \times C_c^\infty(G)$, one sets:

$$\mathcal{B}_{\pi,L_1,L_2}(f_1, f_2) = \sum_{v\in \mathrm{ONB}(\pi)} \overline{L_1(\pi(\overline{f_1})(v))} \cdot L_2(\pi(\overline{f_2})(v)) \tag{7.1}$$

where the sum runs over an orthonormal basis of π. The sum defining $\mathcal{B}_{\pi,L_1,L_2}(f_1, f_2)$ is independent of the choice of the orthonormal basis. It gives a linear map

$$\mathcal{B}_{\pi,L_1,L_2} : C_c^\infty(G) \otimes \overline{C_c^\infty(G)} \longrightarrow \mathbb{C}$$

which is $G(F)^\Delta$-invariant (and which depends on the Haar measure dg).

The distribution $\mathcal{B}_{\pi,L_1,L_2}$ is called the relative character of π with respect to (L_1, L_2). Note that, in the literature, it is frequent to find a different convention in the definition of the relative character, using instead the sum

$$\sum_{v\in\mathrm{ONB}(\pi)} L_1(\pi(f_1)(v)) \cdot \overline{L_2(\pi(f_2)(v))}.$$

The difference between the two conventions is merely one of form rather than substance, and it is easy to convert from one convention to the other using complex conjugation. We choose the normalisation given above so as to avoid the appearance of multiple complex conjugations in later formulae.

Now a short computation gives

$$\overline{L_i(\pi(\overline{f_i})v)} = \int_{H_i\backslash G} \overline{L_i(\pi(g_i)(v))} \cdot (f_i)_{H_i,\chi_i}(g)\, \frac{dg}{dh_i}$$

with

$$(f_i)_{H_i,\chi_i}(g) = \int_{H_i} f(h_i g) \cdot \overline{\chi_i(h_i)}\, dh_i.$$

Hence one deduces that the linear form $\mathcal{B}_{\pi,L_1,L_2}$ factors as

$$C_c^\infty(G) \otimes \overline{C_c^\infty(G)} \twoheadrightarrow C_c^\infty(H_1, \chi_1\backslash G) \otimes \overline{C_c^\infty(H_2, \chi_2\backslash G)} \longrightarrow \mathbb{C}.$$

We may think of $C_c^\infty(H_i, \chi_i\backslash G(F))$ as the space of compactly supported smooth sections of the line bundle on $X_i = H_i\backslash G$ determined by χ_i and denote this space by the alternative notation $C_c^\infty(X_i, \chi_i)$. Then we shall think of $\mathcal{B}_{\pi,L_1,L_2}$ as a G^Δ-invariant linear form on $C_c^\infty(X_1, \chi_1) \otimes \overline{C_c^\infty(X_2, \chi_2)}$; this now depends on the Haar measures dg and dh_i, or rather on the G-invariant measure dg/dh_i on $H_i\backslash G$.

Let us write:

$$f_{L,v}(g) = L(\pi(g)v)$$

for the matrix coefficient associated to $L \in \pi^*$ and $v \in \pi$. Then the distribution $\mathcal{B}_{\pi,L_1,L_2}$ is given by the formula

$$\mathcal{B}_{\pi,L_1,L_2}(\phi_1, \phi_2) = \sum_{v\in\mathrm{ONB}(\pi)} \langle \phi_1, f_{L_1,v}\rangle_{X_1} \cdot \langle f_{L_2,v}, \phi_2\rangle_{X_2}, \tag{7.2}$$

where $\langle -, -\rangle_{X_i}$ is the inner product defined by the measure dg/dh_i.

7.2 Alternative Incarnation

We can also give an alternative formulation of the notion of relative characters. Continuing with the context of Sect. 7.1, it is not difficult to verify that

$$\mathcal{B}_{\pi,L_1,L_2}(f_1, f_2) = \sum_{v \in \mathrm{ONB}(\pi)} \overline{L_1(\pi(\overline{f_1 * f_2^\vee})(v))} \cdot L_2(v),$$

where

$$f_2^\vee(g) = \overline{f_2(g^{-1})},$$

and

$$(f_1 * f_2)(g) = \int_G f_1(gx^{-1}) \cdot f_2(x)\, dx$$

is the convolution of f_1 and f_2.

Thus, we may alternatively define $\mathcal{B}_{\pi,L_1,L_2}$ as a linear form

$$\mathcal{B}_{\pi,L_1,L_2} : C_c^\infty(G) \longrightarrow \mathbb{C}$$

given by the formula

$$\mathcal{B}_{\pi,L_1,L_2}(f) = \sum_{v \in \mathrm{ONB}(\pi)} \overline{L_1(\pi(\overline{f})(v)} \cdot L_2(v).$$

As in Sect. 7.1, this linear form factors as:

$$\mathcal{B}_{\pi,L_1,L_2} : C_c^\infty(G) \twoheadrightarrow C_c^\infty(X_1, \chi_1) \longrightarrow \mathbb{C},$$

so that we may regard it as a linear form on $C_c^\infty(X_1, \chi_1)$, given by the formula

$$\mathcal{B}_{\pi,L_1,L_2}(\phi) = \sum_{v \in \mathrm{ONB}(\pi)} \langle \phi, f_{L_1,v} \rangle_{X_1} \cdot f_{L_2,v}(1). \tag{7.3}$$

In fact, it further factors as:

$$\mathcal{B}_{\pi,L_1,L_2} : C_c^\infty(X_1, \chi_1) \twoheadrightarrow C_c^\infty(X_1, \chi_1)_{H_2,\chi_2} \longrightarrow \mathbb{C}.$$

From this alternative description of the relative character, we can recover the previous version discussed in the previous subsection by using the fact that any $f \in C_c^\infty(G)$ can be expressed as a finite linear combination of $f_1 * f_2$. This is clear in the nonarchimedean case and is a result of Dixmier-Malliavin in the Archimedean case.

7.3 J_σ as a Relative Character

We shall now relate the notion of relative character with the theory of direct integral decomposition.

We shall focus on the case when $H_1 = H_2 = H$ and $\chi_1 = \chi_2 = \chi$ and such that $\dim \mathrm{Hom}_H(\pi, \chi) \leq 1$. With $X = H\backslash G$, equipped with a G-invariant measure dx, suppose one has a direct integral decomposition:

$$L^2(X, \chi, dx) := L^2(H, \chi\backslash G) = \int_\Omega \sigma(\omega)\, d\mu(\omega)$$

with associated families of maps $\{\alpha_{\sigma(\omega)}\}$ and $\{\beta_{\sigma(\omega)}\}$ (see Sect. 2.3) and associated decomposition of inner product as in (2.3):

$$\langle -, -\rangle_X = \int_\Omega J_{\sigma(\omega)}(-, -)\, d\mu(\omega).$$

Observe that the positive semidefinite Hermitian form $J_{\sigma(\omega)}$ is a G^Δ-invariant linear form

$$J_{\sigma(\omega)} : C_c^\infty(X, \chi) \otimes \overline{C_c^\infty(X, \chi)} \longrightarrow \mathbb{C}.$$

This suggests that $J_{\sigma(\omega)}$ may be regarded as a relative character according to our definition in Sect. 7.1. Indeed, one has:

Lemma 7.1. *One has* $J_{\sigma(\omega)} = \mathcal{B}_{\sigma(\omega), \ell_{\sigma(\omega)}, \ell_{\sigma(\omega)}}$, *where* $\ell_{\sigma(\omega)} = ev_1 \circ \beta_{\sigma(\omega)} \in \mathrm{Hom}_H(\sigma(\omega), \chi)$.

Proof. Since ω is fixed in the proposition, we shall write $\sigma = \sigma(\omega)$ for simplicity. Now we have:

$$\begin{aligned} J_\sigma(\phi_1, \phi_2) = \langle \alpha_\sigma(\phi_1), \alpha_\sigma(\phi_2)\rangle_\sigma &= \sum_{v \in \mathrm{ONB}(\sigma)} \langle \alpha_\sigma(\phi_1), v\rangle_\sigma \cdot \langle v, \alpha_\sigma(\phi_2)\rangle_\sigma \\ &= \sum_{v \in \mathrm{ONB}(\sigma)} \langle \phi_1, \beta_\sigma(v)\rangle_X \cdot \langle \beta_\sigma(v), \phi_2\rangle_X. \end{aligned}$$

Noting that

$$\beta_\sigma(v)(g) = ev_1 \circ \beta_\sigma(\sigma(g)(v)) = \ell_\sigma(\sigma(g)(v)) = f_{\ell_\sigma, v}(g),$$

we see that the lemma follows by Eq. (7.2). □

We can also work with the alternative context of Sect. 7.2. In this incarnation, one has:

Lemma 7.2. *As a linear form on $C_c^\infty(X, \chi)$, one has*

$$\mathcal{B}_{\sigma(\omega),\ell_{\sigma(\omega)},\ell_{\sigma(\omega)}}(\phi) = \ell_{\sigma(\omega)}(\alpha_{\sigma(\omega)}(\phi)) = \beta_{\sigma(\omega)}\alpha_{\sigma(\omega)}(\phi)(1).$$

Proof. We write $\sigma = \sigma(\omega)$ for simplicity. Then we have

$$\alpha_\sigma(\phi) = \sum_{v \in \mathrm{ONB}(\sigma)} \langle \alpha_\sigma(\phi), v\rangle_\sigma \cdot v = \sum_v \langle \phi, f_{\ell_\sigma, v}\rangle_X \cdot v.$$

Hence

$$\ell_\sigma(\alpha_\sigma(\phi)) = \sum_v \langle \phi, f_{\ell_\sigma, v}\rangle_X \cdot \ell_\sigma(v),$$

so that the lemma follows by Eq. (7.3). □

Corollary 7.3. *Let $\mathcal{C}(X, \chi)$ be the Harish-Chandra-Schwarz space of $X = (H, \chi)\backslash G$. Then the relative character $\mathcal{B}_{\sigma,\ell_\sigma,\ell_\sigma}$ extends to $\mathcal{C}(X, \chi)$.*

7.4 Space of Orbital Integrals

Set

$$\mathcal{I}(X, \chi) := C_c^\infty(X, \chi)_{H,\chi}.$$

We think of this as "the space of orbital integrals" on X. Indeed, given a H-orbit on X, the associated orbital integral factors to $\mathcal{I}(X, \chi)$. As noted above, the relative character $\mathcal{B}_{\sigma,\ell_\sigma,\ell_\sigma}$ factors to give a linear form on $\mathcal{I}(X, \chi)$. Henceforth, we will write $\mathcal{B}_{\sigma,\ell_\sigma}$ in place of $\mathcal{B}_{\sigma,\ell_\sigma,\ell_\sigma}$ to simplify notation.

8 Transfer of Test Functions

If two periods on the two members of a dual pair are related by theta correspondence as in Proposition 6.1, then one might ask if the associated relative characters are related in a precise way. Such a relation is called a relative character identity. To compare the two relative characters in question, which are distributions on different spaces, we first need to define a correspondence of the relevant spaces of test functions.

8.1 A Correspondence of Test Functions

The considerations of the previous sections suggest that one considers the following maps. Set

$$p : \mathcal{S}(V) = \Omega^{\infty} \longrightarrow C^{\infty}(N, \psi\backslash \mathrm{SL}_2)$$

given by

$$p(\Phi)(g) := (g \cdot \Phi)(v_1).$$

This map is $\mathrm{O}(v_1^{\perp})$-invariant and SL_2-equivariant. Let us set

$$\mathcal{S}(N, \psi\backslash \mathrm{SL}_2) := \text{image of } p, \tag{8.1}$$

noting that it is a SL_2-submodule. Likewise, consider the $\mathrm{O}(V)\times(N, \psi)$-equivariant *surjective* restriction map

$$q = \text{rest} : \mathcal{S}(V) \longrightarrow \mathcal{S}(X_1)$$

so that

$$q(\Phi)(h) = (h \cdot \Phi)(v_1) = \Phi(h^{-1} \cdot v_1) \quad \text{for } h \in \mathrm{O}(v_1^{\perp})\backslash \mathrm{O}(V) \cong X_1.$$

We have already seen and used the map q in the setting of smooth theta correspondence, seeing that it induces an $\mathrm{O}(V)$-equivariant isomorphism

$$q : \mathcal{S}(V)_{N,\psi} \cong \mathcal{S}(X_1)$$

Hence we have the diagram:

$$\begin{array}{ccccc} & & \mathcal{S}(V) & & \\ & \swarrow^{p} & & \searrow^{q} & \\ \mathcal{S}(N, \psi\backslash \mathrm{SL}_2) & & & & \mathcal{S}(X_1) \end{array} \tag{8.2}$$

We now make a definition:

Definition 8.1. Say that $f \in \mathcal{S}(N, \psi\backslash \mathrm{SL}_2)$ and $\phi \in \mathcal{S}(X_1)$ are in correspondence (or are transfers of each other) if there exists $\Phi \in \mathcal{S}(V)$ such that $p(\Phi) = f$ and $q(\Phi) = \phi$.

Our goal in this section is to establish some basic properties of the spaces of test functions and the transfer defined above. We start with the following simple observation.

Proposition 8.2. *Every $f \in \mathcal{S}(N, \psi\backslash \mathrm{SL}_2)$ has a transfer $\phi \in \mathcal{S}(X_1)$ and vice versa.*

Proof. This is simply because the maps p and q above are surjective. □

We also note:

Lemma 8.3. *The space $\mathcal{S}(N, \psi\backslash \mathrm{SL}_2)$ is contained in the Harish-Chandra-Schwarz space of the Whittaker variety $(N, \psi)\backslash \mathrm{SL}_2$. In particular, for any $\sigma \in \widehat{\mathrm{SL}_2}_{temp,\psi}$, the associated relative character $\mathcal{B}_{\sigma,\ell_\sigma}$ extends to a linear form on $\mathcal{S}(N, \psi\backslash \mathrm{SL}_2)$.*

Proof. From the formula defining the Weil representation, we see that for $f = p(\Phi)$,

$$|f(t(a)k)| = |a|^{\dim V/2} \cdot |(k \cdot \Phi)(a \cdot v_1)|.$$

It follows by Lemma 4.4(i) that $f \in \mathcal{C}(N, \psi\backslash \mathrm{SL}_2)$ if $\dim V \geq 3$. □

8.2 Basic Function and Fundamental Lemmas

We shall now place ourselves in the unramified situation. Namely, let us assume that:

- F is a nonarchimedean local field of residual characteristic different from 2;
- the conductor of the additive character $\psi : F \to S^1$ is the ring of integers $\mathcal{O}_F$ of F;
- the quadratic space V contains a self-dual lattice L and $v_1 \in L$.

Under these hypotheses, we have:

- the measure $d_\psi x$ of F is such that the volume of $\mathcal{O}_F$ is 1;
- the measure $d_\psi v$ on V is such that the volume of L is 1;
- the stabilizer $K' = K'_L$ of L in $\mathrm{O}(V)$ is a hyperspecial maximal compact subgroup.

We let $\Phi_0 \in S(V)$ be the characteristic function of L, so that Φ_0 is a unit vector in Ω_ψ. Here is a basic definition:

Definition 8.4. Set

$$f_0 = p(\Phi_0) \qquad \text{and} \qquad \phi_0 = q(\Phi_0).$$

We call these the basic functions in the relevant space of test functions.

Observe that $\phi_0 = q(\Phi_0)$ is the characteristic function of $X_1(F) \cap L$. On the other hand, f_0 is not compactly supported. Indeed: f_0 is determined by its value on T and we have

$$f_0(t(a)) = (t(a)\cdot\Phi)(v_1) = \begin{cases} |a|^{\dim V/2}, \text{ if } |a| \le 1; \\ 0, \text{ if } |a| > 1. \end{cases}$$

It is also immediate from definition that one has the following "fundamental lemma":

Lemma 8.5. *The basic functions f_0 and ϕ_0 correspond.*

Let $K = \mathrm{SL}_2(\mathcal{O}_F) \subset \mathrm{SL}_2(F)$ and $K' = K'_L \subset \mathrm{O}(V)$, so that they are hyperspecial maximal compact subgroups which fix the unramified vector Φ_0. Endow SL_2 and $\mathrm{O}(V)$ with Haar measures such that the volumes of K and K' are 1. Then we have the corresponding spherical Hecke algebras $\mathcal{H}(\mathrm{SL}_2, K)$ and $\mathcal{H}(G, K')$. These are commutative unital algebras whose units are the characteristic functions 1_K and $1_{K'}$ respectively.

It was shown by Howe (see [MVW, Chap. 5, Thm. I.4, Pg 103] and [MVW, Pg 107]) that

$$\Omega_\psi^K = C_c^\infty(\mathrm{O}(V))\cdot\Phi_0 \quad \text{and} \quad \Omega_\psi^{K'} = C_c^\infty(\mathrm{SL}_2)\cdot\Phi_0. \tag{8.3}$$

By applying $1_{K'}$ and 1_K respectively to these two equations, it follows that

$$(\Omega^\infty)^{K\times K'} = \mathcal{H}(\mathrm{SL}_2, K)\cdot\Phi_0 = \mathcal{H}(\mathrm{O}(V), K')\cdot\Phi_0.$$

It also follows from (8.3) that if one has a nonzero equivariant map

$$\Omega^\infty \twoheadrightarrow \sigma\otimes\pi \in \mathrm{Irr}(\mathrm{SL}_2\times\mathrm{O}(V)),$$

then σ is K-unramified if and only if $\pi = \theta_\psi(\sigma)$ is K'-unramified. Indeed, it was shown by Rallis [R, §6] that there is an algebra morphism

$$c : \mathcal{H}(G, K') \longrightarrow \mathcal{H}(\mathrm{SL}_2, K)$$

such that for any $f \in \mathcal{H}(G, K')$, one has

$$f\cdot\Phi_0 = c(f)\cdot\Phi_0.$$

From this, one easily deduces the following "fundamental lemma for spherical Hecke algebras":

Lemma 8.6. *For any $f \in \mathcal{H}(G, K')$, the element $f\cdot\phi_0 \in C_c^\infty(X_1)$ corresponds to the element $c(f)\cdot f_0 \in \mathcal{S}(N, \psi\backslash\mathrm{SL}_2)$.*

8.3 Relation with Adjoint L-Factors

We shall see that the space $\mathcal{S}(N, \psi\backslash \mathrm{SL}_2)$ is intimately related to the standard (degree 3) L-factor of irreducible representations of SL_2. Let us recall a certain Rankin-Selberg local zeta integral for this particular L-factor, due to Gelbart-Jacquet [GJ]. It requires the following 3 pieces of data:

- a ψ-generic $\sigma \in \mathrm{Irr}(\mathrm{SL}_2)$ with ψ-Whittaker model $\mathcal{W}_\sigma$,
- the Weil representation ω_ψ of Mp_2 acting on the space $C_c^\infty(F)$ (regarding F as a one-dimensional quadratic space equipped with the quadratic form $x \mapsto x^2$);
- a principal series representation $I_\psi(\chi, s)$ of Mp_2, consisting of genuine functions $\phi_s : N\backslash \mathrm{Mp}_2 \to \mathbb{C}$ such that $\phi_s(t(a)g) = \chi_\psi(a) \cdot \chi(a) \cdot |a|^{1+s} \cdot \phi_s(g)$ (where χ_ψ is a genuine character of the diagonal torus of Mp_2 defined using the Weil index).

Then for $f \in \mathcal{W}_\sigma$, $\varphi \in C_c^\infty(F)$ and a section $\phi_s \in I(s)$, one can consider the local zeta integral

$$Z(f, \varphi, \phi_s) = \int_{N\backslash \mathrm{SL}_2} \overline{f(g)} \cdot (g \cdot \varphi)(1) \cdot \phi_s(g) \, \frac{dg}{dn}.$$

This converges when $Re(s) \gg 0$, and when σ is tempered, it converges for $Re(s) > 0$. Moreover, the GCD of this family of local zeta integrals is used to define the local twisted adjoint L-factor $L(s + \frac{1}{2}, \sigma, Ad \times \chi) = L(s + \frac{1}{2}, \sigma, std \times \chi)$.

Hence, the (twisted) adjoint L-value $L(s + \frac{1}{2}, \sigma, Ad \times \chi)$ is obtained by considering the integrals of $f \in \mathcal{W}_\sigma$ against a space of functions $\mathcal{S}_s(N, \psi\backslash \mathrm{SL}_2)$ of the form

$$g \mapsto \phi_s(g) \cdot (g \cdot \varphi)(1).$$

Moreover, as an SL_2-module, $\mathcal{S}_s(N, \psi\backslash \mathrm{SL}_2)$ is a quotient of $\omega_\psi \otimes I_\psi(\chi, s)$.

Now let us return to our space of test functions $\mathcal{S}(N, \psi\backslash \mathrm{SL}_2)$. Let us write

$$V = \langle v_1 \rangle \oplus U, \qquad \text{with } U = v_1^\perp.$$

Then $C_c^\infty(V) = C_c^\infty(Fv_1) \otimes C_c^\infty(U)$. Here, $C_c^\infty(Fv_1)$ affords the Weil representation ω_ψ of $\mathrm{Mp}_2 \times \mathrm{O}(v_1)$ whereas $C_c^\infty(U)$ afford a Weil representation of $\mathrm{Mp}_2 \times \mathrm{O}(U)$. If $\Phi \in C_c^\infty(V)$ is of the form $\Phi_1 \otimes \Phi'$, then

$$p(\Phi)(g) = (g \cdot \Phi_1)(v_1) \cdot (g \cdot \Phi')(0).$$

The function

$$g \mapsto \phi_{\Phi'}(g) = (g \cdot \Phi')(0)$$

belongs to the principal series $I_\psi(\chi_{\mathrm{disc}(U)}, \frac{1}{2}(\dim V - 3))$. Indeed, by a result of Rallis [R2, Thm. II.1.1], the map

$$\Phi' \mapsto \phi_{\Phi'}$$

gives an $\mathrm{O}(U)$-invariant, Mp_2-equivariant injective map

$$0 \neq C_c^\infty(U)_{\mathrm{O}(U)} \hookrightarrow I_\psi(\chi_{\mathrm{disc}(U)}, \frac{1}{2}(\dim V - 3)).$$

This result of Rallis underlies the theory of the doubling seesaw and the Siegel-Weil formula. When $\dim V \geq 4$, this injective map is surjective as well. Indeed, when $\dim V > 4$, the relevant principal series $I_\psi(\chi_{\mathrm{disc}(U)}, \frac{1}{2}(\dim V - 3))$ is irreducible. If $\dim V = 4$, the principal series $I_\psi(\chi_{\mathrm{disc}(U)}, 1/2)$ has length 2 with unique irreducible quotient the even Weil representation $\omega^+_{\psi,\chi_{\mathrm{disc}(U)}}$ and unique submodule a (twisted) Steinberg representation. The above map is nonetheless surjective, as the small theta lift of the trivial representation of $\mathrm{O}(U)$ is equal to $\omega^+_{\psi,\chi_{\mathrm{disc}(U)}}$.

To summarise, we have more or less shown:

Proposition 8.7. *When* $\dim V \geq 4$*, the map* p *factors as*

$$C_c^\infty(V) \twoheadrightarrow C_c^\infty(V)_{\mathrm{O}(U)} \cong \omega_\psi \otimes I_\psi(\chi_{\mathrm{disc}(U)}, \frac{1}{2}(\dim V - 3)) \twoheadrightarrow \mathcal{S}(N, \psi\backslash \mathrm{SL}_2),$$

so that

$$\mathcal{S}(N, \psi\backslash \mathrm{SL}_2) = \mathcal{S}_{\frac{1}{2}(\dim V - 3)}(N, \psi\backslash \mathrm{SL}_2).$$

Proof. One has an $\mathrm{Mp}_2 \times \mathrm{Mp}_2$-equivariant isomorphism

$$S(V)_{\mathrm{O}(U)} = S(Fv_1)\widehat{\otimes} S(U)_{\mathrm{O}(U)} \cong \omega_\psi \widehat{\otimes} I_\psi(\chi_{\mathrm{disc}(U)}, \frac{1}{2}(\dim V - 3)).$$

The rest of the proposition follows from our preceding discussion. □

Question. Is the surjective map in Proposition 8.7 in fact an isomorphism

$$C_c^\infty(V)_{\mathrm{O}(U)} \cong \mathcal{S}(N, \psi\backslash \mathrm{SL}_2)?$$

8.4 Orbital Integrals

Let us set

$$\mathcal{I}(N, \psi\backslash \mathrm{SL}_2) := \mathcal{S}(N, \psi\backslash \mathrm{SL}_2)_{N,\psi}$$

so that by definition, a linear functional on $\mathcal{I}(N, \psi\backslash \mathrm{SL}_2)$ is a (N, ψ)-equivariant linear form on $\mathcal{S}(N, \psi\backslash \mathrm{SL}_2)$. One may think of $\mathcal{I}(N, \psi\backslash \mathrm{SL}_2)$ as the space of orbital integrals (with respect to the (N, ψ)-period) and write $\mathcal{I}(f)$ for the image of f in $\mathcal{I}(N, \psi\backslash \mathrm{SL}_2)$. Likewise, we set

$$\mathcal{I}(X_1) = S(X_1)_{\mathrm{O}(U)}.$$

This may be regarded as the space of orbital integrals with respect to the $\mathrm{O}(U)$-period and we write $\mathcal{I}(\phi)$ for the image of ϕ in $\mathcal{I}(X_1)$.

The following proposition summaries the properties of the transfer of test functions:

Proposition 8.8. *The composite map*

$$S(V) \xrightarrow{\ p\ } \mathcal{S}(N, \psi\backslash \mathrm{SL}_2) \longrightarrow \mathcal{I}(N, \psi\backslash \mathrm{SL}_2)$$

factors through q, i.e.. it induces a linear map

$$S(X_1) \longrightarrow \mathcal{I}(N, \psi\backslash \mathrm{SL}_2).$$

Hence the transfer correspondence descends to a linear map when one passes to the space of orbital integrals in the target. Indeed, it further descends to give a surjective linear map

$$t_\psi : \mathcal{I}(X_1) \longrightarrow \mathcal{I}(N, \psi\backslash \mathrm{SL}_2).$$

Proof. The composite map in question is (N, ψ)-invariant, and hence factors through $S(V)_{N,\psi} \cong S(X_1)$. But it is also $\mathrm{O}(U)$-invariant and so further factors through

$$(S(V)_{N,\psi})_{\mathrm{O}(U)} = \mathcal{I}(X_1)$$

as desired. □

Likewise, one may consider the composite

$$S(V) \xrightarrow{\ p\ } \mathcal{S}(N, \psi\backslash \mathrm{SL}_2) \longrightarrow \mathcal{I}(N, \psi\backslash \mathrm{SL}_2)$$

which as above factors through $S(V)_{\mathrm{O}(U)}$. But now we do not know if $S(V)_{\mathrm{O}(U)} \cong \mathcal{S}(N, \psi\backslash \mathrm{SL}_2)$; see the Question at the end of the previous subsection. If the answer to that question is Yes, then we will likewise conclude that the above composite map induces a linear map

$$\mathcal{S}(N, \psi\backslash \mathrm{SL}_2) \longrightarrow \mathcal{I}(X_1),$$

which descends further to

$$\mathcal{I}(N,\psi\backslash \mathrm{SL}_2) \longrightarrow \mathcal{I}(X_1),$$

In that case, this linear map will be inverse to the one in the proposition, and hence we will have an isomorphism of vector spaces:

$$t_\psi : \mathcal{I}(N,\psi\backslash \mathrm{SL}_2) \cong \mathcal{I}(X_1).$$

In other words, the transfer correspondence would give an isomorphism of the space of orbital integrals (for the relevant spaces of test functions). As it stands, we only have the surjective transfer map

$$t_\psi : \mathcal{I}(X_1) \twoheadrightarrow \mathcal{I}(N,\psi\backslash \mathrm{SL}_2)$$

given in the above proposition.

9 Relative Character Identities

Finally, we are ready to establish the following relative character identity, which is the main local result of this paper.

Theorem 9.1. *Suppose that*

- $f \in \mathcal{S}(N,\psi\backslash \mathrm{SL}_2)$ *and* $\phi \in \mathcal{S}(X_1)$ *are in correspondence;*
- $\sigma \in \widehat{\mathrm{SL}_2}_{temp,\psi}$ *with (nonzero) theta lift* $\theta_\psi(\sigma) \in \widehat{\mathrm{O}(V)}$*;*
- $\ell_\sigma \in \mathrm{Hom}_N(\sigma,\psi)$ *is the canonical element determined in (3.4) by the Whittaker-Plancherel theorem;*
- $\ell_{\theta(\sigma)} = f_\sigma(\ell_\sigma) \in \mathrm{Hom}_{\mathrm{O}(v_1^\perp)}(\theta(\sigma),\mathbb{C})$ *is the canonical element determined by the spectral decomposition in Proposition 6.4 (which is in turn determined by* ℓ_σ *and* θ_σ*).*

Then one has the character identity:

$$\mathcal{B}_{\sigma,\ell_\sigma}(f) = \mathcal{B}_{\theta(\sigma),\ell_{\theta(\sigma)}}(\phi).$$

More succintly, one has the identity

$$\mathcal{B}_{\sigma,\ell_\sigma} \circ t_\psi = \mathcal{B}_{\theta(\sigma),\ell_{\theta(\sigma)}}$$

of linear forms on $\mathcal{I}(X_1)$ *or equivalently the identity*

$$\mathcal{B}_{\sigma,\ell_\sigma} \circ p = \mathcal{B}_{\theta(\sigma),\ell_{\theta(\sigma)}} \circ q$$

of linear forms on $\mathcal{S}(V)$.

See Sect. 7, especially Sect. 7.4, for the definition of $\mathcal{B}_{\sigma,\ell_\sigma}$ and $\mathcal{B}_{\theta(\sigma),\ell_{\theta(\sigma)}}$.

9.1 Proof of Theorem 9.1

This subsection is devoted to the proof of the theorem. With f and ϕ as given in the theorem, choose $\Phi \in C_c^\infty(V)$ such that $f = p(\Phi)$ and $\phi = q(\Phi)$. We shall now find two different expressions for $\Phi(v_1)$.

On one hand, by the direct integral decomposition given in Proposition 6.4, (2.5) gives

$$\Phi(v_1) = \phi(v_1) = \int_{\widehat{\mathrm{SL}_2}} \beta_{\theta(\sigma)}\alpha_{\theta(\sigma)}(\phi)(v_1)\, d\mu_{\mathrm{SL}_2,\psi}(\sigma).$$

On the other hand, by the Whittaker-Plancherel theorem for $(N, \psi)\backslash\mathrm{SL}_2$, (2.5) gives

$$\Phi(v_1) = f(1) = \int_{\widehat{\mathrm{SL}_2}} \beta_\sigma\alpha_\sigma(f)(1)\, d\mu_{\mathrm{SL}_2,\psi}(\sigma).$$

Comparing the two expressions, we deduce that

$$\int_{\widehat{\mathrm{SL}_2}} \mathcal{B}_{\theta(\sigma),\ell_{\theta(\sigma)}}(\phi)\, d\mu_{\mathrm{SL}_2,\psi}(\sigma) = \int_{\widehat{\mathrm{SL}_2}} \mathcal{B}_{\sigma,\ell_\sigma}(f)\, d\mu_{\mathrm{SL}_2,\psi}(\sigma). \tag{9.1}$$

We would like to remove the integral sign in the above identity. For this, we will apply a Bernstein center argument.

Given an arbitrary element z in the Bernstein center (or the ring of Arthur multipliers in the Archimedean case) of $\mathrm{SL}_2 \times \mathrm{O}(V)$, the element z acts on the irreducible representation $\sigma \boxtimes \theta_\psi(\sigma)$ by a scalar $z(\sigma \boxtimes \theta_\psi(\sigma))$. This implies that one has a commutative diagram

$$\begin{array}{ccc} \Omega_\psi & \xrightarrow{\ z\ } & \Omega_\psi \\ {\scriptstyle \theta_\sigma}\big\downarrow & & \big\downarrow{\scriptstyle \theta_\sigma} \\ \sigma \boxtimes \theta_\psi(\sigma) & \xrightarrow{z(\sigma\boxtimes\theta_\psi(\sigma))} & \sigma \boxtimes \theta_\psi(\sigma) \\ {\scriptstyle \lambda}\big\downarrow & & \big\downarrow{\scriptstyle \lambda} \\ \sigma & \xrightarrow{z(\sigma\boxtimes\theta_\psi(\sigma))} & \sigma \end{array} \tag{9.2}$$

for any linear form λ on $\theta_\psi(\sigma)$. One has an analogous commutative diagram where one takes λ to be any linear form on σ (so the last row of the commutative diagram has $\theta_\psi(\sigma)$ in place of σ).

Now what we would like to show is that there are commutative diagrams

$$\begin{array}{ccc} \Omega_\psi & \xrightarrow{z} & \Omega_\psi \\ {\scriptstyle p}\downarrow & & \downarrow{\scriptstyle p} \\ \mathcal{S}(N,\psi\backslash \mathrm{SL}_2) & & \mathcal{S}(N,\psi\backslash \mathrm{SL}_2) \\ {\scriptstyle \alpha_\sigma}\downarrow & & \downarrow{\scriptstyle \alpha_\sigma} \\ \sigma & \xrightarrow{z(\sigma\boxtimes\theta_\psi(\sigma))} & \sigma \end{array} \quad \text{and} \quad \begin{array}{ccc} \Omega_\psi & \xrightarrow{z} & \Omega_\psi \\ {\scriptstyle q}\downarrow & & \downarrow{\scriptstyle q} \\ \mathcal{S}(X_1) & & \mathcal{S}(X_1) \\ {\scriptstyle \alpha_\sigma}\downarrow & & \downarrow{\scriptstyle \alpha_\sigma} \\ \theta_\psi(\sigma) & \xrightarrow{z(\sigma\boxtimes\theta_\psi(\sigma))} & \theta_\psi(\sigma) \end{array} \qquad (9.3)$$

We shall explain how the commutativity of the diagram on the left follows from the commutativity of the diagram in (9.2); a similar argument works for the diagram on the right.

Since the map $\alpha_\sigma \circ p$ is SL_2-equivariant, it factors through $\theta_\sigma : \Omega_\psi \longrightarrow \sigma \boxtimes \theta_\psi(\sigma)$, i.e. there is a $\lambda : \theta_\psi(\sigma) \to \mathbb{C}$ such that

$$\alpha_\sigma \circ p = \lambda \circ \theta_\sigma .$$

Using this, we see that the desired commutativity of the left diagram in (9.3) is reduced to the commutativity of the diagram in (9.2).

Now we shall apply the identity (9.1) to the pair of test functions arising from $z \cdot \Phi$. Note that

$$\mathcal{B}_{\sigma,\ell_\sigma}(p(z\cdot\Phi)) = \beta_\sigma \alpha_\sigma(p(z\Phi))(1) = z(\sigma \boxtimes \theta_\psi(\sigma)) \cdot \mathcal{B}_{\sigma,\ell_\sigma}(f)$$

and

$$\mathcal{B}_{\theta(\sigma),\ell_{\theta(\sigma)}} = \beta_{\theta(\sigma)}\alpha_{\theta(\sigma)}(q(z\cdot\Phi))(x_1) = z(\sigma\boxtimes\theta_\psi(\sigma)) \cdot \mathcal{B}_{\theta(\sigma),\ell_{\theta(\sigma)}}(\phi).$$

Hence the identity (9.1), when applied to $z \cdot \Phi$, reads:

$$\int_{\widehat{SL}_2} z(\sigma\boxtimes\theta_\psi(\sigma)) \cdot \big(\mathcal{B}_{\sigma,\ell_\sigma}(f) - \mathcal{B}_{\theta(\sigma),\ell_{\theta(\sigma)}}(\phi)\big)\, d\mu_{\mathrm{SL}_2,\psi}(\sigma) = 0. \qquad (9.4)$$

Now note that there is a natural homomorphism from the Bernstein center of SL_2 to the Bernstein center for $\mathrm{SL}_2 \times \mathrm{O}(V)$. Hence we may take z to be an element in the (tempered) Bernstein center of SL_2. Then $z(\sigma \boxtimes \theta_\psi(\sigma)) = z(\sigma)$. When regarded as $\mathbb{C}$-valued functions on $\widehat{\mathrm{SL}}_{2temp,\psi}$, the elements z of the (tempered) Bernstein center of SL_2, are dense in the space of all Schwarz functions on $\widehat{\mathrm{SL}}_{2temp,\psi}$. Hence, (9.4) implies that for $d\mu_{\mathrm{SL}_2,\psi}$-almost all σ, one has

$$\mathcal{B}_{\sigma,\ell_\sigma}(f) = \mathcal{B}_{\theta(\sigma),\ell_{\theta(\sigma)}}(\phi).$$

To obtain the equality for all $\sigma \in \widehat{\mathrm{SL}}_{2temp,\psi}$, we note that both sides of the identity are continuous as functions of $\sigma \in \widehat{\mathrm{SL}}_{2temp,\psi}$ by Lemma 3.2 and Corollary 6.8. This completes the proof of Theorem 9.1.

9.2 Some Consequences

We shall now give some consequences of the relative character identity shown in Theorem 9.1. Let us consider the following diagram:

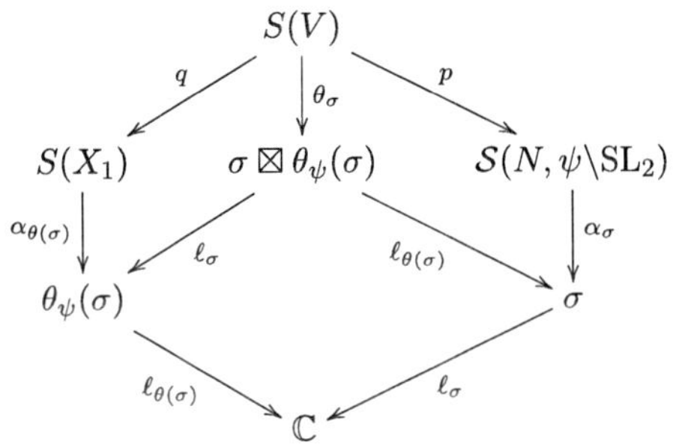

In this diagram, the rhombus at the bottom is clearly commutative. Now the parallelogram at the upper left side is precisely the commutative diagram in Proposition 6.5. On the other hand, the parallelogram at the upper right side is commutative up to a scalar since

$$\dim \mathrm{Hom}_{\mathrm{O}(v_1^\perp)\times \mathrm{SL}_2}(\Omega_\psi, \mathbb{C} \boxtimes \sigma) = 1 \quad \text{for } \sigma \in \widehat{\mathrm{SL}_2}_{temp,\psi}$$

and both $\alpha_\sigma \circ p$ and $\ell_{\theta(\sigma)} \circ \theta_\sigma$ are nonzero elements of this space. We would like to show that it is in fact commutative.

To deduce this, we observe that the composite of the three maps along the left boundary of the hexagon is simply the relative character $\mathcal{B}_{\theta(\sigma)} \circ q$, whereas the composite of the three maps along the right boundary of the hexagon is the relative character $\mathcal{B}_\sigma \circ p$. The relative character identity of Theorem 9.1 says that the boundary of the diagram is commutative! From this, we deduce the following counterpart of Proposition 6.5:

Proposition 9.2. *The following diagram is commutative:*

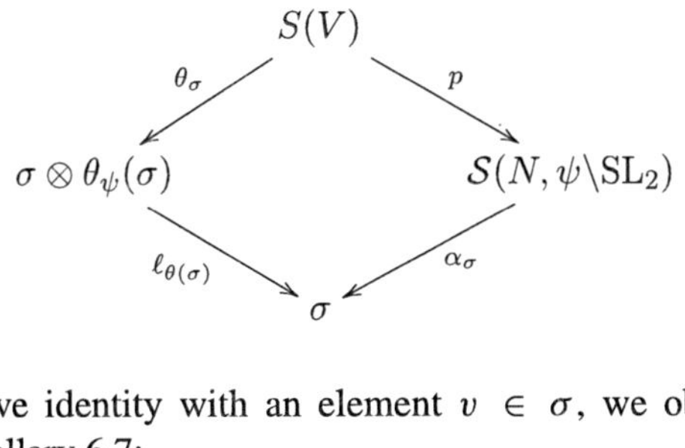

Pairing the above identity with an element $v \in \sigma$, we obtain the following counterpart of Corollary 6.7:

Corollary 9.3. *For any $\Phi \in \Omega_\psi$ and $v \in \sigma$, one has:*

$$\ell_{\theta(\sigma)}(A_\sigma(\Phi, v)) = \langle p(\Phi), \beta_\sigma(v)\rangle_{N\backslash \mathrm{SL}_2}$$

where A_σ is the map defined in Sect. 5.5.

10 Local L-Factor

In this section, we are going to examine the local L-factor $L_X(s, -)$ associated to a spherical variety X. As we explained in the introduction, this local L-factor is associated to a $\frac{1}{2}\mathbb{Z}$-graded representation $V_X = \oplus_d V_X^d$ of the dual group $X^\vee$ and its value has been computed by Sakellaridis [S1, S2] in great generality. However, we shall show that for the particular case treated in this paper, the results developed thus far through theta correspondence can be used to compute $L_X(s, -)$ in terms of the analogous local L-factor for the Whitaker variety $(N, \psi)\backslash \mathrm{SL}_2$.

10.1 Unramified Setting

We place ourselves in the unramified setting of Sect. 8.2, so that

- F is a nonarchimedean local field of residual characteristic not 2;
- ψ has conductor $\mathcal{O}_F$, so that the associated measure $d_\psi x$ of F gives $\mathcal{O}_F$ volume 1;
- $L \subset V$ is a self-dual lattice with stabilizer $K' = K'_L$, so that the measure $d_\psi v$ on V gives L volume 1.
- the vector v_1 lies in the lattice L.

Hence, $v_1 \in X(\mathcal{O}_F) = X(F) \cap L$ and we have an orthogonal decomposition

$$L = \mathcal{O}_F v_1 \oplus (v_1^\perp \cap L).$$

The lattices L and $L \cap v_1^\perp$ (which are both self-dual) endow $\mathrm{O}(V)$ and $\mathrm{O}(v_1^\perp)$ with $\mathcal{O}_F$-structures so that they become smooth group schemes over $\mathcal{O}_F$; in particular, $K' = \mathrm{O}(V)(\mathcal{O}_F)$. Then the map $h \mapsto h^{-1} \cdot v_1$ defines an $\mathrm{O}(V)$-equivariant isomorphism

$$X \longrightarrow \mathrm{O}(v_1^\perp)\backslash \mathrm{O}(V)$$

of smooth schemes over $\mathcal{O}_F$. Moreover, as a consequence of Hensel's lemma, K' acts transitively on $X(\mathcal{O}_F)$, so that

$$X(\mathcal{O}_F) \cong \mathrm{O}(v_1^\perp)(\mathcal{O}_F)\backslash \mathrm{O}(V)(\mathcal{O}_F).$$

Moreover, the Haar measure $|\omega|$ on X is associated to an $\mathrm{O}(V)$-invariant differential of top degree which has nonzero reduction on the special fiber. Further, if we equip the smooth group schemes $\mathrm{O}(V)$ and $\mathrm{O}(v_1^\perp)$ over $\mathcal{O}_F$ with invariant differentials $\omega_{\mathrm{O}(V)}$ and $\omega_{\mathrm{O}(v_1^\perp)}$ of top degree with nonzero reduction on the special fibers, then

$$|\omega| = \frac{|\omega_{\mathrm{O}(V)}|}{|\omega_{\mathrm{O}(v_1^\perp)}|}.$$

This means that

$$\mathrm{Vol}(X(\mathcal{O}_F); |\omega|) := \int_{X(\mathcal{O})} |\omega| = q^{-\dim X} \cdot \#X(\kappa_F) = q^{-\dim X} \cdot \frac{\#\mathrm{O}(V)(\kappa_F)}{\#\mathrm{O}(v_1^\perp)(\kappa_F)} \tag{10.1}$$

where κ_F is the residue field of F and $q = \#\kappa_F$.

10.2 The L-Factor $L_X^\#$

From the spectral decomposition of $L^2(X, |\omega|)$ obtained in Proposition 6.4, we have a family of $\mathrm{O}(v_1^\perp)$-invariant linear functionals $\ell_{\theta(\sigma)}$ on $\theta_\psi(\sigma)$ for $\sigma \in \widehat{\mathrm{SL}}_{2temp,\psi}$. We remind the reader that even in this unramifed setting that we have placed ourselves, the linear functional $\ell_{\theta(\sigma)}$ depends on the Haar measure dg on SL_2. We have also specified an inner product $\langle -, - \rangle_{\theta(\sigma)}$ in Proposition 5.2 (using the doubling zeta integral) and this depends on the Haar measure dg on SL_2 as well.

Let us assume that $\sigma \in \widehat{\mathrm{SL}}_{2temp,\psi}$ is K-unramified, where $K = \mathrm{SL}_2(\mathcal{O}_F)$. Fix $v_0 \in \sigma$ such that

$$\langle v_0, v_0 \rangle_\sigma = 1 \quad \text{and} \quad K \cdot v_0 = v_0.$$

Then $\theta_\psi(\sigma)$ is K'-unramified and we fix $w_0 \in \theta_\psi(\sigma)$ with

$$\langle w_0, w_0 \rangle_{\theta(\sigma)} = 1 \quad \text{and} \quad K' \cdot w_0 = w_0.$$

We then set

$$L_X^\#(\sigma) := |\ell_{\theta(\sigma)}(w_0)|^2 \in \mathbb{R}_{\geq 0}.$$

Our goal is to determine this non-negative valued function defined on the K-unramified part of $\widehat{\mathrm{SL}}_{2temp,\psi}$, where $K = \mathrm{SL}_2(\mathcal{O}_F)$.

According to the conjecture of Sakellaridis and Venkatesh, one should have

$$L_X^{\#}(\sigma) = \Delta(0) \cdot \frac{L_X(0,\sigma)}{L(1,\sigma,Ad)}$$

where $\Delta(s)$ is a product of local zeta factors (which is independent of the representation σ), $L(s,\sigma,Ad)$ is the adjoint L-factor of σ and

$$L_X(s,\sigma) = \prod_d L(s+d,\sigma,V_X^d)$$

is the L-factor of σ associated to a $1/2\cdot\mathbb{Z}$-graded representation $V_X = \oplus_d V_X^d$ of $X^\vee$. This is the essential part of $L_X^{\#}(\sigma)$. The computation of $L_X(0,\sigma)$ is thus equivalent to the precise determination of $L_X^{\#}(\sigma)$.

10.3 Some Constants

We shall determine $L_X^{\#}(\sigma)$ in terms of the analogous quantity for the Whittaker variety $(N,\psi)\backslash \mathrm{SL}_2$. To this end, let $\Phi_0 = 1_L \in S(V)$ be the characteristic function of L which is a unit vector in Ω_ψ and is $K \times K'$-invariant. Then under the canonical map

$$\theta_\sigma : \Omega_\psi \longrightarrow \sigma \otimes \theta_\psi(\sigma),$$

we have

$$\theta_\sigma(\Phi_0) = c_\sigma \cdot v_0 \otimes w_0 \in \sigma \otimes \theta_\psi(\sigma). \tag{10.2}$$

for some nonzero constant c_σ.

On the other hand, recall that we have the basic functions

$$p(\Phi_0) = f_0 \in S(N,\psi\backslash \mathrm{SL}_2) \quad \text{and} \quad q(\Phi_0) = \phi_0 \in S(X).$$

We have observed that

$$\phi_0 = 1_{X(\mathcal{O}_F)}.$$

Now we define a constant $\lambda_{\theta(\sigma)}$ by:

$$\alpha_{\theta(\sigma)}(\phi_0) = \lambda_{\theta(\sigma)} \cdot w_0. \tag{10.3}$$

10.4 Key Computations

We now perform the following computations:

(a) Taking inner product of both sides of (10.3) with w_0 gives:

$$\lambda_{\theta(\sigma)} = \langle \alpha_{\theta(\sigma)}(\phi_0), w_0 \rangle_{\theta(\sigma)}.$$

Now the right hand side of this identity is equal to

$$\langle \phi_0, \beta_{\theta(\sigma)}(w_0) \rangle_X = \langle 1_{X(\mathcal{O}_F)}, \beta_{\theta(\sigma)}(w_0) \rangle_X = \beta_{\theta(\sigma)}(w_0)(v_1) \cdot \mathrm{Vol}(X(\mathcal{O}_F), |\omega|).$$

Hence we have

$$\lambda_{\theta(\sigma)} = \ell_{\theta(\sigma)}(w_0) \cdot \mathrm{Vol}(X(\mathcal{O}_F), |\omega|).$$

(b) On the other hand, applying the commutative diagram in Proposition 6.5 to the left hand side of (10.3) and using (10.2) gives:

$$c_\sigma \cdot \ell_\sigma(v_0) \cdot w_0 = \ell_\sigma(\theta_\sigma(\Phi_0)) = \lambda_{\theta(\sigma)} \cdot w_0,$$

so that

$$\lambda_{\theta(\sigma)} = c_\sigma \cdot \ell_\sigma(v_0)$$

(c) Finally, taking inner product of both sides of (10.2) with v_0 gives:

$$A_\sigma(\Phi_0, v_0) = \langle \theta_\sigma(\Phi_0), v_0 \rangle_\sigma = c_\sigma \cdot w_0$$

Computing inner product of both sides gives:

$$|c_\sigma|^2 = \langle A_\sigma(\Phi_0, v_0), A_\sigma(\Phi_0, v_0) \rangle_{\theta(\sigma)} = Z_\sigma(\Phi_0, \Phi_0, v_0, v_0),$$

where the last equality is (5.5) and $Z_\sigma(-)$ is the doubling zeta integral.

Combining the last identities resulting from (a), (b) and (c) above, we obtain:

$$|\ell_{\theta(\sigma)}(w_0)| = |\ell_\sigma(v_0)| \cdot |Z(\Phi_0, \Phi_0, v_0, v_0)|^{1/2} \cdot \mathrm{Vol}(X(\mathcal{O}_F), |\omega|)^{-1}.$$

Hence, it remains to determine the 3 quantities on the right hand side. We have already determined the volume of $X(\mathcal{O}_F)$ in (10.1). On the other hand, we have:

Lemma 10.1.

(i) Suppose $\dim V = 2n \geq 4$ *is even. Then*

$$|\ell_\sigma(v_0)|^2 = \frac{\zeta_F(2)}{L(1, \sigma, Ad)}$$

and

$$Z(\Phi_0, \Phi_0, v_0, v_0) = \mathrm{Vol}(K; dg) \cdot \frac{L(n-1, \sigma \times \chi_{\mathrm{disc}V}, std)}{\zeta_F(2n-2) \cdot L(n, \chi_{\mathrm{disc}V})}$$

where Ad = std *is the adjoint = standard L-factor for* SL_2.

(ii) Suppose $\dim V = 2n+1 \geq 3$ *is odd, so that one is working with* Mp_2 *instead of* SL_2. *Then*

$$|\ell_\sigma(v_0)|^2 = \frac{\zeta_F(2) \cdot L_\psi(1/2, \sigma, std)}{L_\psi(1, \sigma, Ad)}$$

and

$$Z(\Phi_0, \Phi_0, v_0, v_0) = \mathrm{Vol}(K; dg) \cdot \frac{L_\psi(n-\frac{1}{2}, \sigma \times \chi_{\mathrm{disc}V}, std)}{\zeta_F(2n)}$$

Proof. The determination of $|\ell_\sigma(v_0)|$ was carried out in [LM, Prop. 2.14 and §2.6] whereas that of the doubling zeta factor can be found in [LR, Prop. 3] and [G2, Prop. 6.1]. □

From this lemma, one sees that dependence of $\ell_{\theta(\sigma)}$ on the Haar measure dg. In the unramified setting, it is customary to take the Haar measure dg such that $\mathrm{Vol}(K; dg) = 1$. However, in view of the global applications later on, we prefer to take the Haar measure associated to an invariant differential of top degree on SL_2 over $\mathbb{Z}$. In that case, we have

$$\mathrm{Vol}(K, dg) = q^{-3} \cdot \#\mathrm{SL}_2(\kappa_F) = \zeta_F(2)^{-1}.$$

Putting everything together, we have shown:

Proposition 10.2.

(i) When $\dim V = 2n$ *is even, one has*

$$L_X^\#(\sigma) = |\ell_{\theta(\sigma)}(w_0)|^2 = \frac{L(n-1, \sigma \times \chi_{\mathrm{disc}V}, std)}{L(1, \sigma, Ad)} \cdot \frac{L(n, \chi_{\mathrm{disc}V})}{\zeta_F(2n-2)},$$

taking note that Ad = std *for* SL_2.

(ii) When $\dim V = 2n+1$ *is odd, so that one is working with* Mp_2,

$$L_X^\#(\sigma) = |\ell_{\theta(\sigma)}(w_0)|^2$$
$$= \frac{L_\psi(n-\frac{1}{2}, \sigma \times \chi_{\mathrm{disc}V}, std) \cdot L_\psi(1/2, \sigma, std)}{L_\psi(1, \sigma, Ad)} \cdot \frac{\zeta_F(2n)}{L(n, \chi_{\mathrm{disc}V})^2}.$$

As an illustration, when $\dim V = 3$ (so $n = 1$) and $\mathrm{disc}(V) = 1$, one gets

$$L_X^{\#}(\sigma) = \frac{L_\psi(1/2, \sigma, std)^2}{L_\psi(1, \sigma, Ad)} \cdot \frac{\zeta_F(2)}{\zeta_F(1)^2}$$

whereas when $\dim V = 4$ (so $n = 2$) and $\mathrm{disc}(V) = 1$, one has

$$L_X^{\#}(\sigma) = 1.$$

The reader should compare these values with those given the table (3) in the introduction of [S4].

10.5 General Case

So far, we have placed ourselves in the unramified setting. We now return to the general setting and define

$$L_X^{\#} : \widehat{\mathrm{SL}_2} \longrightarrow \mathbb{C} \quad \text{or} \quad \widehat{\mathrm{Mp}_2} \longrightarrow \mathbb{C}$$

by using the formulae given in Proposition 10.2, depending on whether $\dim V$ is even or odd.

11 Transfer in Geometric Terms

We have defined the transfer of test functions and established a relative character identity without making any geometric comparison. This is not so surprising, as the theta correspondence is a means of transferring spectral data from one group to another. Nonetheless, one can ask for an explicit formula for the transfer map

$$t_\psi : \mathcal{I}(X_1) \longrightarrow \mathcal{I}(N, \psi\backslash \mathrm{SL}_2).$$

For example, we may wonder if one could describe t_ψ as an integral transform. We shall derive such a formula in this section, assuming that F is nonarchimedean (with ring of integers $\mathcal{O}_F$ and uniformizer ϖ). We also assume for simplicity that the conductor of the additive character ψ is $\mathcal{O}_F$ and the discriminant of V is 1. In particular, the measure $d_\psi x = dx$ on F gives $\mathcal{O}_F$ volume 1.

Recall that we have called the domain and target of t_ψ the spaces of orbital integrals. To describe t_ψ geometrically, we shall appeal to incarnations of these spaces as concrete spaces of functions. Consider for example the case of $\mathcal{I}(N, \psi\backslash \mathrm{SL}_2) = \mathcal{S}(N, \psi\backslash \mathrm{SL}_2)_{N,\psi}$. Given a function $f \in \mathcal{S}(N, \psi\backslash \mathrm{SL}_2)$, we may consider its literal (N, ψ)-orbital integral:

$$\mathcal{I}(f)(a) = \int_F f(wt(a)n(b)) \cdot \overline{\psi(n)}\, db.$$

Assuming this converges, it defines a smooth function on the open Bruhat cell NwB which is (N, ψ)-invariant on both sides. Hence it is determined by its value on wT and we may regard it as a function on $F^\times$. The map $\mathcal{I}$ factors as:

$$\mathcal{S}(N, \psi\backslash \mathrm{SL}_2) \longrightarrow \mathcal{I}(N, \psi\backslash \mathrm{SL}_2) \longrightarrow C^\infty(F^\times),$$

and we view it as giving an incarnation of the elements of $\mathcal{I}(N, \psi\backslash \mathrm{SL}_2)$ as functions on $F^\times$. Likewise, we shall later see an incarnation of the elements of $\mathcal{I}(X_1)$, as functions on a set of generic $\mathrm{O}(v_1^\perp)$-orbits on X_1.

Given $\Phi \in \mathcal{S}(V)$, we would thus like to compute the (N, ψ)-orbital integral of $p(\Phi) = f \in \mathcal{S}(N, \psi\backslash \mathrm{SL}_2)$:

$$\mathcal{I}(f)(a) = \int_F f(wt(a)n(b)) \cdot \overline{\psi(n)}\, db = \int_F (wt(a)n(b) \cdot \Phi)(v_1) \cdot \overline{\psi(b)}\, db.$$

We should perhaps say a few words about the convergence of this integral. Let us identify $N\backslash \mathrm{SL}_2$ with $W^* = F^2 \setminus O$ (where O is the origin of F^2) via $g \mapsto (0, 1) \cdot g$. Then $|f|$ is a function on W^* which vanishes on a neighbourhood of O. Now the element $Nwt(a)n(b) \in N\backslash \mathrm{SL}_2$ corresponds to the element $(-a, -ab) \in W^*$. For fixed $a \in F^\times$, the function

$$b \mapsto f(wt(a)n(b))$$

is thus not necessarily compactly supported on F. However, if we had assumed that $f \in C_c^\infty(N, \psi\backslash \mathrm{SL}_2)$ (which is a dense subspace of $\mathcal{C}(N, \psi\backslash \mathrm{SL}_2)$), then $|f|$ would in addition vanish outside a compact set of W, so that the above function of b is compactly supported on F and the integral defining $\mathcal{I}(f)(a)$ would have been convergent. This suggests that if we let $U_n = \varpi^{-n}\mathcal{O}_F$ and set

$$\mathcal{I}_n(f)(a) = \int_{U_n} (wt(a)n(b) \cdot \Phi)(v_1) \cdot \overline{\psi(b)}\, db,$$

then the value $\mathcal{I}_n(f)(a)$ should stabilize for sufficiently large n (and this does happen for $f \in C_c^\infty(N, \psi\backslash \mathrm{SL}_2)$). With this motivation, we shall define

$$\mathcal{I}(f)(a) := \lim_{n\to\infty} \mathcal{I}_n(f)(a)$$

and shall show below that the right hand side indeed stabilizes.

For this, we will perform an explicit computation:

$$\begin{aligned}
&\mathcal{I}_n(f)(a)\\
&= \int_{U_n} (wt(a)n(b)\cdot\Phi)(v_1)\cdot\overline{\psi(b)}\,db\\
&= \int_{U_n} \mathcal{F}(t(a)n(b)\cdot\Phi)(v_1)\cdot\overline{\psi(b)}\,db\\
&= \gamma_{\psi,q}\cdot\int_{U_n}\int_V (t(a)n(b)\cdot\Phi)(y)\cdot\psi(\langle v_1,y\rangle)\cdot\overline{\psi(b)}\,dy\,db\\
&= \gamma_{\psi,q}\cdot\int_{U_n}\int_V |a|^{\frac{1}{2}\dim(V)}\cdot(n(b)\cdot\Phi)(ay)\cdot\psi(\langle v_1,y\rangle)\cdot\overline{\psi(b)}\,dy\,db\\
&= \gamma_{\psi,q}\cdot\int_{U_n}\int_{a^{-1}\omega} |a|^{\frac{1}{2}\dim(V)}\cdot\Phi(ay)\cdot\psi(a^2bq(y))\cdot\psi(\langle v_1,y\rangle)\cdot\overline{\psi(b)}\,dy\,db\\
&= \gamma_{\psi,q}\cdot\int_{\omega} |a|^{-\frac{1}{2}\dim(V)}\cdot\Phi(x)\cdot\psi(\langle v_1,a^{-1}x\rangle)\cdot\left(\int_{U_n}\psi(b(q(x)-1))\,db\right)dx
\end{aligned}$$

where $\omega = \mathrm{supp}(\Phi)$ is compact and we have made the substitution $x = ay$ in the last step. Recall also that $\mathcal{F}$ is a normalized Fourier transform giving the action of the standard Weyl group element w on the Weil representation and $\gamma_{\psi,q}$ is a root of unity (a Weil index).

Now let us consider the inner integral

$$\int_{U_n}\psi(b(q(x)-1))\,db$$

If $q(x)-1\notin\varpi^n\mathcal{O}_F$, then the integrand is a nontrivial character of U_n and hence the integral is 0. On the other hand, if $q(x)-1\in\varpi^n\mathcal{O}_F$, the integral gives the volume of $U_n=\varpi^{-n}\mathcal{O}_F$. Since ψ is assumed to have conductor $\mathcal{O}_F$, the volume of U_n with respect to the measure db is q^n (where q is the size of the residue field of F). Hence

$$\int_{U_n}\psi(b(q(x)-1))\,db = q^n\cdot 1(q(x)\in 1+\varpi^n\mathcal{O}_F)$$

and so

$$\begin{aligned}
&\gamma_{\psi,q}^{-1}\cdot\mathcal{I}_n(f)(a)\\
&= q^n\cdot|a|^{-\frac{1}{2}\dim(V)}\cdot\int_V\Phi(x)\cdot\psi(\langle v_1,a^{-1}x\rangle)\cdot 1(q(x)-1\in\varpi^n\mathcal{O}_F)\,dx
\end{aligned}$$

$$= q^n \cdot |a|^{-\frac{1}{2}\dim(V)} \cdot \int_{q^{-1}(1+\varpi^n \mathcal{O}_F)} \Phi(x) \cdot \psi(\langle v_1, a^{-1}x\rangle)\, dx.$$

Now this last expression is a quantity which appears in the theory of local densities in the theory of quadratic forms over local fields. Indeed, consider the map

$$q : q^{-1}(1+\varpi^n \mathcal{O}_F) \longrightarrow 1+\varpi^n \mathcal{O}_F$$

of p-adic manifolds. Since every point in the base is a regular value of the map q, or equivalently q is submersive at every point of the domain, the integral of the compactly supported and locally constant integrand over $q^{-1}(1+\varpi^n \mathcal{O}_F)$ can be performed by first integrating over the fibers of q followed by integration over the base. Indeed, this was how we had defined the measures $|\omega_a|$ on each fiber X_a (for $a \in F^{\times}$). In other words for any locally constant compactly supported φ,

$$\int_{q^{-1}(1+\varpi^n \mathcal{O}_F)} \varphi(x)\, dx = \int_{1+\varpi^n \mathcal{O}_F} q_*(\varphi)(z)\, dz$$

where

$$q_*(\varphi)(z) = \int_{q^{-1}(z)} \varphi(x) \cdot |\omega_z(x)|$$

But $q_*(\varphi)$ is a locally constant function on the base. Hence for n sufficiently large, the above integral is simply equal to

$$\mathrm{Vol}(\varpi^n \mathcal{O}_F) \cdot q_*(\varphi)(1) = q^{-n} \cdot \int_{X_1} \varphi(x) \cdot |\omega_1(x)|.$$

Applying this to the integral of interest, we thus deduce that the sequence $\mathcal{I}_n(f)(a)$ stabilizes for large n and

$$\mathcal{I}(f)(a) = \gamma_{\psi,q} \cdot |a|^{-\frac{1}{2}\dim(V)} \cdot \int_{X_1} \Phi(x) \cdot \psi(\langle v_1, a^{-1}x\rangle) \cdot |\omega_1(x)|$$

Now observe that the map

$$\gamma : X_1 = \mathrm{O}(U)\backslash \mathrm{O}(V) \longrightarrow F$$

given by

$$x = h^{-1}v_1 \mapsto \langle v_1, x\rangle = \langle v_1, h^{-1}v_1\rangle$$

is $O(U)$-invariant (on the right). Moreover, for $\xi \in F$, the preimage of ξ is equal to

$$\gamma^{-1}(\xi) = \{x = \xi \cdot v_1 + v : v \in U \quad \text{and} \quad q(v) = 1 - \xi^2\}.$$

Outside $\xi^2 = 1$, the map γ is submersive at all points and it follows by Witt's theorem that the fiber $\gamma^{-1}(\xi)$ is a homogeneous space under $O(U)$. For $x_\xi \in \gamma^{-1}(\xi)$ (with $\xi^2 \neq 1$), its stabilizer in $O(U)$ is $O(U \cap x_\xi^\perp)$. Thus, if $F^\heartsuit = F \smallsetminus \{\pm 1\}$ and $X_1^\heartsuit = \gamma^{-1}(F^\heartsuit)$, then

$$X_1^\heartsuit / O(U) \cong F^\heartsuit.$$

Moreover, the measures $|\omega_1|$ on $X_1^\heartsuit$ and $d\xi$ on $F^\heartsuit$ that we have fixed give rise to measures $|\nu_\xi|$ on the fibers $\gamma^{-1}(\xi)$. The $O(U)$-orbital integrals of functions on $S(X_1)$ are thus obtained via integration on the fibres of γ and are smooth functions on $F^\heartsuit$, These functions give an incarnation of the elements of $\mathcal{I}(X_1) = S(X_1)_{O(U)}$, so that we have a map

$$\mathcal{I} : S(X_1) \longrightarrow \mathcal{I}(X_1) \longrightarrow C^\infty(F^\heartsuit).$$

Hence, continuing with our computation, we have:

$$\begin{aligned}\mathcal{I}(f)(a) &= \gamma_{\psi,q} \cdot |a|^{-\frac{1}{2}\dim(V)} \cdot \int_{X_1^\heartsuit} \Phi(x) \cdot \psi(a^{-1}\gamma(x)) \cdot |\omega_1(x)| \\ &= \gamma_{\psi,q} \cdot |a|^{-\frac{1}{2}\dim(V)} \cdot \int_{F^\heartsuit} \int_{\gamma^{-1}(x_\xi)} \Phi(y) \cdot \psi(a^{-1}\xi) \cdot |\nu_\xi(y)| \, d\xi \\ &= \gamma_{\psi,q} \cdot |a|^{-\frac{1}{2}\dim(V)} \cdot \int_{F^\heartsuit} \mathcal{I}(\phi)(\xi) \cdot \psi(a^{-1}\xi) \, d\xi\end{aligned}$$

where $\phi = q(\Phi) \in S(X_1)$ (so that ϕ and f are transfers of each other) and $\mathcal{I}(\phi)$ is the orbital integral of ϕ defined by the inner integral over the fibers of γ over $F^\heartsuit$. We have shown:

Proposition 11.1. *The transfer map $t_\psi : \mathcal{I}(X_1) \longrightarrow \mathcal{I}(N, \psi\backslash SL_2)$ is given by the integral transform:*

$$t_\psi(\phi)(a) = \gamma_{\psi,q} \cdot |a|^{-\frac{1}{2}\dim(V)} \cdot \int_{F^\heartsuit} \phi(\xi) \cdot \psi(a^{-1}\xi) \, d\xi.$$

where we have regarded $\mathcal{I}(X_1)$ and $\mathcal{I}(N, \psi\backslash SL_2)$ as spaces of functions on $F^\heartsuit$ and $F^\times$ respectively.

Comparing with the formula for the transfer defined in [S4], we see that our transfer map t_ψ essentially agrees with that of [S4]. In particular, our approach gives an alternative proof of the transfer theorem of [S4] in the setting of hyperboloids.

We close this section with another remark. As mentioned in the introduction, the transfer map in [S4] was first defined and studied on the level of the boundary degenerations of the rank 1 spherical varieties and then one uses essentially the same formula in the setting of the spherical varieties themselves. For the case treated in this paper, the boundary degeneration of X_1 is simply the nullcone (with vertex removed)

$$X_0 = \{0 \neq x \in V : q(x) = 0\}$$

of nonzero isotropic vectors. This is a homogeneous $\mathrm{O}(V)$-variety and one can carry out essentially all the analysis of the earlier sections with X_0 in place of X_a with $a \neq 0$. One would then be describing the spectrum of X_0 in terms of the spectrum of the basic affine space $N\backslash\mathrm{SL}_2$. Indeed, since the map $q : V \setminus \{0\} \longrightarrow F$ is submersive at all points, the derivation of the formula for the transfer map given in this section can be carried out essentially uniformly for X_a with any $a \in F$. In other words, the Weil representation allows one to construct a coherent family of transfer maps relating $\mathcal{S}(N, \psi_a\backslash\mathrm{SL}_2)$ and $\mathcal{S}(X_a)$ varying smoothly with $a \in F$ (though $\mathcal{S}(X_0) \neq C_c^\infty(X_0)$ in the nonarchimedean case), which explains in some sense why "the same formula works". We leave the analysis of the transfer map for the boundary degeneration X_0 as an exercise for the interested reader.

12 Factorization of Global Periods

In this final section, we turn to the global setting, where we examine the question of factorisation of global period integrals, in the context of the periods considered in the earlier sections. We first need to introduce the global analogs of various constructions encountered in the local setting.

12.1 *Tamagawa Measures*

Let k be a number field with ring of adèles $\mathbb{A}$. We fix a nontrivial unitary character

$$\psi : F\backslash\mathbb{A} \longrightarrow S^1.$$

This has a factorization $\psi = \prod_v \psi_v$ where ψ_v is a nontrivial character of the local field k_v for each place v of k. Then ψ_v determines a self-dual Haar measure $d_{\psi_v}x$ on k_v such that for almost all v, the volume of the ring of integers $\mathcal{O}_{k_v}$ relative to $d_{\psi_v}x$ is 1. The product measure $dx := \prod_v d_{\psi_v}x$ then gives a measure on $\mathbb{A}$. This is the Tamagawa measure of $\mathbb{A}$: it is independent of ψ (by the Artin-Whaples product formula) and satisfies

$$\int_{F\backslash\mathbb{A}} dx = 1.$$

If G is a (smooth) algebraic group over k, we may consider the adelic group $G(\mathbb{A})$. It is a restricted direct product $\prod'_v G(k_v)$, taken with respect to a sequence $\{K_v = G(\mathcal{O}_{k_v})\}$ of open compact subgroups determined by any $\mathcal{O}_k$-structure on G. For almost all v, K_v is a hyperspecial maximal compact subgroup of $G(k_v)$.

Now suppose ω_G is a nonzero invariant differential of top degree on G over k. Then for each place v of k, the pair (ω_G, ψ_v) determines a Haar measure $|\omega_{G,\psi_v}|_v$ of $G(k_v)$. We would like to consider the product measure $\prod_v |\omega_{G,\psi_v}|_v$ on $G(\mathbb{A})$. For this, one needs to assume that $\prod_v \mathrm{Vol}(K_v; |\omega_{G,\psi_v}|_v)$ is finite. This is the case for unipotent groups or semisimple groups. If the infinite product is not convergent (e.g. if $G = \mathbb{G}_m$), one can still deal with this by introducing "normalization factors"; we will not go into this well-documented story here. In any case, this product measure on $G(\mathbb{A})$ is independent of ψ and ω_G. It is the so-called Tamagawa measure of G. When $G = \mathbb{G}_a$ is the additive group, the Tamagawa measure on $G(\mathbb{A}) = \mathbb{A}$ is precisely the measure $dx = \prod_v d_{\psi_v}x$ defined above (so that the terminology is used consistently).

More generally, let $X = H\backslash G$ be a homogeneous G-variety over k (with G acting on the right). Assume for simplicity that $X(k_v) = H(k_v)\backslash G(k_v)$ for each place v of k and $X(\mathbb{A}) = H(\mathbb{A})\backslash G(\mathbb{A})$. Suppose further that ω_X is a nonzero G-invariant differential form of top degree on X over k. Then for each v, one has a $G(k_v)$-invariant measure $|\omega_{X,\psi_v}|_v$ on $X(k_v)$. We shall call the product measure $\prod_v |\omega_{X,\psi_v}|_v$ (when it makes sense) the Tamagawa measure of $X(\mathbb{A})$. It is independent of ψ and ω_X. Moreover, it is simply the quotient of the Tamagawa measures of $G(\mathbb{A})$ and $H(\mathbb{A})$. Indeed, one can construct an invariant differential ω_X of top degree as a quotient of (right-)invariant differentials ω_G and ω_H of top degree on G and H.

In short, when working with adelic groups or the adelic points of homogeneous G-varieties, we shall always use such Tamagawa measures.

12.2 Automorphic Forms

For a reductive group G defined over k, we shall write $[G]$ for the quotient $G(k)\backslash G(\mathbb{A})$ and equip it with its Tamagawa measure dg (divided by the counting measure on the discrete subgroup $G(k)$).

Let $C^\infty_{mod}([G])$ denote the space of smooth functions on $[G]$ which are of (uniform) moderate growth. It is a representation of $G(\mathbb{A})$ containing the $G(\mathbb{A})$-submodule $\mathcal{A}(G)$ of (smooth) automorphic forms on G, which in turn contains the submodule $\mathcal{A}_{cusp}(G)$ of cusp forms:

$$\mathcal{A}_{cusp}(G) \subset \mathcal{A}(G) \subset C^\infty_{mod}([G]).$$

When the group G has a nontrivial split torus in its center, we shall fix a unitary automorphic central character χ and consider the space $\mathcal{A}_\chi(G)$ of automorphic forms with central character χ; we shall suppress this technical issue in the following discussion.

On $\mathcal{A}_{cusp}(G)$, we have the Petersson inner product $\langle -, - \rangle_G$ (defined using the Tamagawa measure dg). Indeed, the Petersson inner product defines a pairing between $\mathcal{A}_{cusp}(G)$ and the larger space $C^\infty_{mod}([G])$. Hence, we have a canonical projection map

$$C^\infty_{mod}([G]) \longrightarrow \mathcal{A}_{cusp}(G).$$

In particular, for an irreducible cuspidal representation $\Sigma \subset \mathcal{A}_{cusp}(G)$, we have a projection

$$\mathrm{pr}_\Sigma : C^\infty_{mod}([G]) \longrightarrow \Sigma.$$

We denote the restriction of the Petersson inner product on Σ by $\langle -, - \rangle_\Sigma$.

12.3 Global Periods

Let $H \subset G$ be a subgroup so that $X = H\backslash G$ is quasi-affine. Fix a unitary Hecke character χ of H. Then we may consider the global (H, χ)-period:

$$P_{H,\chi} : \mathcal{A}_{cusp}(G) \longrightarrow \mathbb{C}$$

defined by

$$P_{H,\chi}(\phi) = \int_{[H]} \overline{\chi(h)} \cdot \phi(h)\, dh$$

where dh is the Tamagawa measure of $H(\mathbb{A})$. For a cuspidal representation $\Sigma \subset \mathcal{A}_{cusp}(G)$, we may thus consider the restriction of $P_{H,\chi}$ to Σ, denoting it by $P_{H,\chi,\Sigma}$.

12.4 The Maps α_{Aut} and β_{Aut}

We shall now introduce the global analog of the maps α_σ and β_σ introduced in Sect. 2.3 in the local setting. Set $X_\mathbb{A} = H(\mathbb{A})\backslash G(\mathbb{A})$, equipped with its Tamagawa measure (which is the quotient of the Tamagawa measures of $G(\mathbb{A})$ and $H(\mathbb{A})$). We have a $G(\mathbb{A})$-equivariant map

$$\theta : C^\infty_c(X_\mathbb{A}, \chi) \longrightarrow C^\infty_{mod}([G])$$

defined by

$$\theta(f)(g) = \sum_{x \in X_k} f(x \cdot g).$$

The map θ is called the formation of theta series. Hence, we may define a composite map

$$\alpha_{\Sigma}^{\mathrm{Aut}} : C_c^{\infty}(X_{\mathbb{A}}, \chi) \xrightarrow{\ \theta\ } C_{mod}^{\infty}([G]) \xrightarrow{\ \mathrm{pr}_{\Sigma}\ } \Sigma.$$

Concretely, we have:

$$\alpha_{\Sigma}^{\mathrm{Aut}}(f) = \sum_{\phi \in \mathrm{ONB}(\Sigma)} \langle \theta(f), \phi \rangle_G \cdot \phi.$$

On the other hand, we have the $G(\mathbb{A})$-equivariant map

$$\beta_{\Sigma}^{\mathrm{Aut}} : \Sigma \longrightarrow C^{\infty}(X_{\mathbb{A}})$$

defined by

$$\beta_{\Sigma}^{\mathrm{Aut}}(\phi)(g) = P_{H,\chi}(g \cdot \phi).$$

One has the following adjunction formula, which is the global analog of (2.2):

Lemma 12.1. *For $f \in C_c^{\infty}(X_{\mathbb{A}}, \chi)$ and $\phi \in \Sigma$, one has*

$$\langle \alpha_{\Sigma}^{\mathrm{Aut}}(f), \phi \rangle_G = \langle f, \beta_{\Sigma}^{\mathrm{Aut}}(\phi) \rangle_X$$

Proof. We have:

$$\begin{aligned}\langle \alpha_{\Sigma}^{\mathrm{Aut}}(f), \phi \rangle_G &= \int_{[G]} \theta(f)(g) \cdot \overline{\phi(g)}\, dg = \int_{[G]} \sum_{\gamma \in H(k)\backslash G(k)} f(\gamma g) \cdot \overline{\phi(g)}\, dg \\ &= \int_{H(k)\backslash G(\mathbb{A})} f(g) \cdot \overline{\phi(g)}\, dg = \int_{X_{\mathbb{A}}} \int_{[H]} f(hg) \cdot \overline{\phi(hg)}\, dh\, \frac{dg}{dh} \\ &= \int_{X_{\mathbb{A}}} f(x) \cdot \overline{P_{H,\chi}(\phi)(x)}\, dx = \langle f, \beta_{\Sigma}^{\mathrm{Aut}}(\phi) \rangle_X,\end{aligned}$$

as desired. □

12.5 Global Relative Characters

We may also introduce the global analog of the inner product J_σ:

$$J_\Sigma^{\mathrm{Aut}}(f_1, f_2) := \langle \alpha_\Sigma^{\mathrm{Aut}}(f_1), \alpha_\Sigma^{\mathrm{Aut}}(f_2)\rangle_\Sigma = \langle \beta_\Sigma^{\mathrm{Aut}}\alpha_\Sigma^{\mathrm{Aut}}(f_1), f_2\rangle_X.$$

Then

$$\begin{aligned} J_\Sigma^{\mathrm{Aut}}(f_1, f_2) &= \sum_{\phi\in \mathrm{ONB}(\Sigma)} \langle \alpha_\Sigma^{\mathrm{Aut}}(f_1), \phi\rangle_\Sigma \cdot \langle \phi, \alpha_\Sigma^{\mathrm{Aut}}(f_2)\rangle_\Sigma \\ &= \sum_{\phi\in \mathrm{ONB}(\Sigma)} \langle f_1, \beta_\Sigma^{\mathrm{Aut}}(\phi)\rangle_X \cdot \langle \beta_\Sigma^{\mathrm{Aut}}(\phi), f_2\rangle_X. \end{aligned}$$

By analog with the local case, we may introduce the global relative character $\mathcal{B}_\Sigma^{\mathrm{Aut}}$ as an equivariant distribution on $C_c^\infty(X_{\mathbb{A}}, \chi)$, defined by

$$\mathcal{B}_\Sigma^{\mathrm{Aut}}(f) = \beta_\Sigma^{\mathrm{Aut}}\left(\alpha_\Sigma^{\mathrm{Aut}}(f)\right)(1) = \sum_{\phi\in \mathrm{ONB}(\Sigma)} \langle f, \beta_\Sigma^{\mathrm{Aut}}(\phi)\rangle_X \cdot P_{H,\chi}(\phi)$$

for $f \in C_c^\infty(X_{\mathbb{A}}, \chi)$. When pulled back to give a distribution on $G(\mathbb{A})$, one has

$$\mathcal{B}_\Sigma^{\mathrm{Aut}}(f) = \sum_{\phi\in \mathrm{ONB}(\Sigma)} \overline{P_{H,\chi}(\Sigma(\overline{f})\phi)} \cdot P_{H,\chi}(\phi),$$

for $f \in C_c^\infty(G(\mathbb{A}))$.

12.6 Quadratic Spaces and Hyperboloids

Suppose now that (V, q) is a quadratic space over k. Then as an additive group scheme over k, $V \cong \mathbb{G}_a^k$ and so $V(\mathbb{A})$ has its canonical Tamagawa measure. We would like to compare this Tamagawa measure with the measures we considered in the local setting.

If $\langle -, -\rangle$ is the symmetric bilinear form associated to q and $\psi = \prod_v \psi_v$ is our fixed additive character of $F\backslash\mathbb{A}$, then the pair $(\langle -, -\rangle, \psi_v)$ determines a Haar measure $d_{\psi_v} v$ on $V_v = V \otimes_k k_v$ (the self-dual measure with respect to the pairing $\psi(\langle -, -\rangle)$). This is the measure on V_v that we have been using in the local setting. If $L \subset V$ is any $\mathcal{O}_k$-lattice, which endows V with an $\mathcal{O}_k$-structure, then for almost all places v, the volume of $L_v = L \otimes_{\mathcal{O}_k} \mathcal{O}_{k_v}$ with respect to $d_{\psi_v} v$ is 1. We may thus consider the product measure

$$dv_{\mathbb{A}} = \prod_v d_{\psi_v} v \quad \text{on } V_{\mathbb{A}}.$$

As the notation suggests, it is independent of the choice of ψ. Moreover, $dv_{\mathbb{A}}$ is equal to the Tamagawa measure on $V_{\mathbb{A}}$.

Suppose that the quadratic space (V, q) contains a vector $v_1 \in V_k$ with $q(v_1) = 1$. By changing L if necessary, we may assume that v_1 lies in the lattice L. Let

$$X_1 = \{x \in V : q(x) = 1\}$$

be a hyperboloid. Then the map $h \mapsto h^{-1} \cdot v_1$ gives an isomorphism

$$\mathrm{O}(v_1^{\perp})\backslash \mathrm{O}(V) \cong X_1$$

of $\mathrm{O}(V)$-varieties over k. Moreover, in this case, one has (by Witt's theorem):

$$\mathrm{O}(v_1^{\perp})(\mathbb{A})\backslash \mathrm{O}(V)(\mathbb{A}) \cong X_1(\mathbb{A}).$$

Recall that we have equipped both sides with their Tamagawa measures which are respected by this isomorphism. Now we would like to relate the Tamagawa measure on $X_1(\mathbb{A})$ with the measures we have been using in the local case.

As noted, the additive character $\psi = \prod_v \psi_v$ gives us decompositions of Tamagawa measures

$$dv_{\mathbb{A}} = \prod_v d_{\psi_v} v \quad \text{and} \quad dx_{\mathbb{A}} = \prod_v d_{\psi_v} x$$

on $V_{\mathbb{A}}$ and $\mathbb{A}$ respectively. Using the submersive map

$$q : V \setminus X_0 \longrightarrow F \setminus \{0\},$$

the local measures $d_{\psi_v} v$ and $d_{\psi_v} x$ determine an $\mathrm{O}(V_v)$-invariant measure $|\omega_{1,v}|$ on $X_1(k_v)$: this is the measure on $X_1(k_v)$ that we have been using in the local setting. We observe that the product measure

$$|\omega_{1,\mathbb{A}}| := \prod_v |\omega_{1,v}|$$

is equal to the Tamagawa measure of $X_1(\mathbb{A}) = \prod'_v X_1(k_v)$ where the restricted direct product is taken with respect to the family $\{X_1(\mathcal{O}_{k_v}) = X_1(k_v) \cap L_v\}$ for almost all v.

12.7 *Global Weil Representation*

We now consider the dual pair $\mathrm{SL}_2 \times \mathrm{O}(V)$ and recall its global Weil representation. We have fixed an $\mathcal{O}_k$-lattice $L \subset V$. For almost all v, $L_v = L \otimes_{\mathcal{O}_k} \mathcal{O}_{k_v}$ is a self dual lattice of volume 1 with respect to $d_{\psi_v} v$. Let

$$K'_v = \text{stabilizer of } L \text{ in } \mathrm{O}(V_v) \quad \text{and} \quad \Phi_{0,v} = 1_{L_v} \in S(V_v).$$

For each v, we have the (smooth) Weil representation Ω_{ψ_v} of $\mathrm{SL}_2(k_v) \times \mathrm{O}(V_v)$ realized on the space $S(V_v)$ of Schwarz-Bruhat functions on $V_v = V \otimes_k k_v$. For almost all v, $\Phi_{0,v}$ is a unit vector which is fixed by $K_v \times K'_v = \mathrm{SL}_2(\mathcal{O}_{k_v}) \times K'_v$. The restricted tensor product

$$\Omega_\psi = \otimes'_v \Omega_{\psi_v}$$

with respect to the family of vectors $\Phi_{0,v}$ is the Weil representation of the adelic dual pair $\mathrm{SL}_2(\mathbb{A}) \times \mathrm{O}(V)(\mathbb{A})$. It is realised on the space

$$S(V_\mathbb{A}) = S(V_\infty) \otimes \left(\otimes'_{v<\infty} S(V_v)\right)$$

of Schwarz-Bruhat functions on $V_\mathbb{A}$ (where $V_\infty = V \otimes_\mathbb{Q} \mathbb{R}$).

The Weil representation Ω_ψ has a canonical automorphic realization

$$\theta : \Omega_\psi = S(V_\mathbb{A}) \longrightarrow C^\infty_{mod}([\mathrm{SL}_2 \times \mathrm{O}(V)])$$

defined by

$$\theta(\Phi)(g, h) = \sum_{v \in V_k} (\Omega_\psi(g, h)\Phi)(v).$$

12.8 Global Theta Lifting

For an irreducible cuspidal representation $\Sigma \subset \mathcal{A}_{cusp}(\mathrm{SL}_2)$ of SL_2, we may consider its global theta lift to $\mathrm{O}(V)$. More precisely, given $\Phi \in \Omega_\psi$ and $f \in \Sigma$, one defines the $\mathrm{SL}_2(\mathbb{A})$-invariant and $\mathrm{O}(V)(\mathbb{A})$-equivariant map

$$A^{\mathrm{Aut}}_\Sigma : \Omega_\psi \otimes \overline{\Sigma} \longrightarrow \mathcal{A}(\mathrm{O}(V))$$

by

$$A^{\mathrm{Aut}}_\Sigma(\Phi, f)(h) = \int_{[\mathrm{SL}_2]} \theta(\Phi)(gh) \cdot \overline{f(g)}\, dg.$$

The image of A^{Aut}_Σ is the global theta lift of Σ, which we denote by

$$\Pi = \Theta^{\mathrm{Aut}}(\Sigma) \subset \mathcal{A}(\mathrm{O}(V)).$$

If Π is cuspidal and nonzero, then it follows by the Howe duality conjecture that Π is an irreducible cuspidal representation.

Conversely, assume that $\Pi \subset \mathcal{A}_{cusp}(\mathrm{O}(V))$ is irreducible cuspidal. Then we may consider the global theta lift of Π to SL_2. More precisely, given $\Phi \in \Omega_\psi$ and $\phi \in \Pi$, one defines the $\mathrm{O}(V)(\mathbb{A})$-invariant and $\mathrm{SL}_2(\mathbb{A})$-equivariant map

$$B_\Pi^{\mathrm{Aut}} : \Omega_\psi \otimes \overline{\Pi} \longrightarrow \Sigma \subset \mathcal{A}(\mathrm{SL}_2)$$

by

$$B_\Pi^{\mathrm{Aut}}(\Phi, \phi)(g) = \int_{[\mathrm{O}(V)]} \theta(\Phi)(gh) \cdot \overline{\phi(h)}\, dh.$$

By computing constant term, one can show that the image of B_Π^{Aut} is necessarily cuspidal.

12.9 The Maps A_Σ^{Aut} and $B_{\Theta(\Sigma)}^{\mathrm{Aut}}$

We continue with the setting of the previous subsection. If $\Pi = \Theta^{\mathrm{Aut}}(\Sigma)$ is nonzero cuspidal, then the image of $B_{\Theta(\Sigma)}^{\mathrm{Aut}}$ is Σ (because the cuspidal spectrum of SL_2 is multiplicity-free [Rama]). In this case, the maps A_Σ^{Aut} and $B_{\Theta(\Sigma)}^{\mathrm{Aut}}$ are global analogs of the maps A_σ and $B_{\Theta(\sigma)}$ introduced in Sect. 5.5. By an exchange of the order of integration, we have the following global analog of (5.9):

$$\langle A_\Sigma^{\mathrm{Aut}}(\Phi, f), \phi \rangle_{\Theta(\Sigma)} = \langle B_{\Theta(\Sigma)}^{\mathrm{Aut}}(\Phi, \phi), f \rangle_\Sigma \tag{12.1}$$

for $\Phi \in \Omega_\psi$, $f \in \Sigma$ and $\phi \in \Theta^{\mathrm{Aut}}(\Sigma)$.

12.10 Global Transfer of Periods

For $\Phi \in \Omega_\psi$ and $\phi \in \Pi$, we may compute the ψ-Whittaker coefficient of $B_\Pi^{\mathrm{Aut}}(\Phi, \phi)$. One has the following global analog of Corollary 6.7:

Proposition 12.2. *For $\Phi \in \Omega_\psi$ and $\phi \in \Pi$,*

$$P_{N,\psi}(B_\Pi^{\mathrm{Aut}}(\Phi, \phi)) = \int_{H(\mathbb{A})\backslash G(\mathbb{A})} \Phi(g^{-1}v_1)\cdot P_{\mathrm{O}(v_1^\perp)}(\phi)(g)\, \frac{dg}{dh} = \langle \Phi|_X, \beta_\Pi^{\mathrm{Aut}}(\phi) \rangle_{X_\mathbb{A}}.$$

In particular, a cuspidal representation Π of $\mathrm{O}(V)$ has nonzero $\mathrm{O}(v_1^\perp)$-period if and only if its global theta lift to SL_2 is globally ψ-generic.

We omit the proof as it is based on a standard computation.

12.11 Decompositions of Unitary Representations

Suppose now that $\Sigma \subset \mathcal{A}_{cusp}(\mathrm{SL}_2)$ is an irreducible *tempered* ψ-generic cuspidal representation of SL_2 and $\Theta^{\mathrm{Aut}}(\Sigma) \subset \mathcal{A}_{cusp}(\mathrm{O}(V))$ is a nonzero (irreducible) cuspidal representation of $\mathrm{O}(V)$. In this case, $\Theta({}^{\mathrm{Aut}}\Sigma)$ is globally $\mathrm{O}(v_1^{\perp})$-distinguished. We would like to factor the global $\mathrm{O}(v_1^{\perp})$-period of $\Theta^{\mathrm{Aut}}(\Sigma)$ as a product of the local functionals constructed in the earlier part of the paper. To carry this out, we need to set things up precisely and systematically.

On the side of SL_2, we begin by fixing a decomposition of the Tamagawa measures

$$dg = \prod_v dg_v$$

To be concrete, we take an invariant differential ω_{SL_2} of top degree on SL_2 over $\mathbb{Z}$ (which is well-defined up to ± 1), which together with the $d_{\psi_v}x$ on k_v gives a measure

$$dg_v := |\omega_{\mathrm{SL}_2,\psi_v}|_v \quad \text{on } \mathrm{SL}_2(k_v),$$

for which

$$\int_{\mathrm{SL}_2(\mathcal{O}_{k_v})} dg_v = \zeta_{k_v}(2)^{-1}.$$

This is the measure we used on $\mathrm{SL}_2(k_v)$ in Sect. 10 when we computed the local L-factor $L_X^{\#}$. The product of these measures is then equal to the Tamagawa measure dg.

We also fix an isomorphism

$$\Sigma \cong \otimes_v' \Sigma_v \tag{12.2}$$

and a decomposition of the Petersson inner product

$$\langle -,-\rangle_{\Sigma} = \prod_v \langle -,-\rangle_{\Sigma_v}. \tag{12.3}$$

This equips Σ_v with a unitary structure $\langle -,-\rangle_{\Sigma_v}$. Alternatively, we could work with the tautological measurable field of unitary representations provided by the local Whittaker-Plancherel theorem (so that each Σ_v comes equipped with a unitary structure already), in which case one would require the isomorphism in (12.2) to be an isometry, so that one has (12.3) as a consequence. These are two viewpoints with no difference in content. The restricted tensor product in (12.2) is with respect to a family of unit vectors $v_{0,v}$ which is fixed by $K_v = \mathrm{SL}_2(\mathcal{O}_{k_v})$.

For the Weil representation, we have already seen in Sect. 12.7 the decompositions:

$$S(V_{\mathbb{A}}) = S(V_{\infty}) \otimes \left(\otimes'_{v<\infty} S(V_v)\right) \tag{12.4}$$

and

$$dv_{\mathbb{A}} = \prod_v d_{\psi_v} v \quad \text{of Haar measures on } V_{\mathbb{A}},$$

inducing compatible unitary structures on the global and local Weil representations. In particular, the restricted direct product in (12.4) is with respect to the family of unit vectors $\Phi_{0,v} = 1_{L_v}$ for almost all v and

$$\langle -, - \rangle_{\Omega_\psi} = \prod_v \langle -, - \rangle_{\Omega_{\psi_v}}.$$

Once these decompositions are fixed as above, we see that for each place v, we have the local big theta lift

$$\Theta_{\psi_v}(\Sigma_v) := (\Omega_{\psi_v} \otimes \Sigma_v^\vee)_{\mathrm{SL}_2(k_v)}$$

and its unique irreducible quotient

$$\theta_{\psi_v}(\Sigma_v) \in \mathrm{Irr}(\mathrm{O}(V_v))$$

Moreover, $\theta_{\psi_v}(\Sigma_v)$ inherits an inner product $\langle -, - \rangle_{\theta(\Sigma_v)}$ defined via the local doubling zeta integral. We let $w_{0,v} \in \theta_{\psi_v}(\Sigma_v)$ be a unit vector which is fixed by K'_v for almost all v. Then we may form the abstract global theta lift

$$\Theta^{\mathbb{A}}(\Sigma) = \otimes'_v \theta_{\psi_v}(\Sigma_v)$$

where the restricted tensor product is with respect to the family of unit vectors $w_{0,v}$. The abstract global theta lift $\Theta^{\mathbb{A}}(\Sigma)$ inherits a unitary structure

$$\langle -, - \rangle_{\Theta^{\mathbb{A}}(\Sigma)} = \prod_v \langle -, - \rangle_{\theta(\Sigma_v)}$$

from that of its local components. Hence we may fix an isometric isomorphism

$$\Theta^{\mathrm{Aut}}(\Sigma) \cong \Theta^{\mathbb{A}}(\Sigma)$$

so that

$$\langle -, - \rangle_{\Theta^{\mathrm{Aut}}(\Sigma)} = \prod_v \langle -, - \rangle_{\theta(\Sigma_v)}$$

where the inner product on the left is that defined by the Petersson inner product.

12.12 Adelic Periods

Having fixed the various decompositions in the previous subsection, we can now introduce the adelic versions of various period maps. We have introduced various global or automorphic quantities associated to Σ and $\Theta^{\mathrm{Aut}}(\Sigma)$, namely

- the maps $\alpha_{\Sigma}^{\mathrm{Aut}}$, $\beta_{\Sigma}^{\mathrm{Aut}}$, $P_{N,\psi,\Sigma}$ and $J_{\Sigma}^{\mathrm{Aut}}$ related to the global Whittaker period with respect to (N, ψ);
- the maps $\alpha_{\Theta(\Sigma)}^{\mathrm{Aut}}$, $\beta_{\Theta(\Sigma)}^{\mathrm{Aut}}$, $P_{\mathrm{O}(v_1^{\perp}),\Theta(\Sigma)}$ and $J_{\Theta(\Sigma)}^{\mathrm{Aut}}$ related to the $\mathrm{O}(v_1^{\perp})$-period;
- the maps $A_{\Sigma}^{\mathrm{Aut}}$, $B_{\Theta(\Sigma)}^{\mathrm{Aut}}$ related to global theta lifting.

All the above global objects have local counterparts, relative to the decompositions fixed in the previous subsection. Namely, for each place v of k, we have:

- the maps α_{Σ_v}, β_{Σ_v}, ℓ_{Σ_v} and J_{Σ_v} given by the Whittaker-Plancherel theorem;
- the maps $\alpha_{\Theta(\Sigma_v)}$, $\beta_{\Theta(\sigma_v)}$, $\ell_{\Theta(\Sigma_v)}$ and $J_{\Theta(\Sigma_v)}$ given by the spectral decomposition of $L^2(X_v, |\omega_v|)$;
- the maps A_{Σ_v} and $B_{\Theta(\Sigma_v)}$ given by the spectral decomposition of the Weil representation Ω_{ψ_v}.

We may take the Euler product of the above local quantities. As an example, we set:

$$\beta_{\Sigma}^{\mathbb{A}} := \prod_v^{*} \beta_{\Sigma_v} \quad \text{and} \quad \ell_{\Sigma}^{\mathbb{A}} := \prod_v^{*} \ell_{\Sigma_v}.$$

Here, the Euler product has to be understood as a regularized product, via meromorphic continuation if necessary, as discussed in the introduction. Let us illustrate this in three instances, assuming that $\dim(V) = 2n$ and $\mathrm{disc}(V) = 1 \in k^{\times}/k^{\times 2}$ (for simplicity):

- Suppose we want to define $\ell_{\Sigma}^{\mathbb{A}}$. Given a decomposable vector $u = \otimes_v u_v \in \Sigma$, with $u_v = v_{0,v}$ for almost all v, we need to examine the convergence of the product

$$\prod_v \ell_{\Sigma_v}(u_v).$$

This is determined by the value $\ell_{\Sigma_v}(v_{0,v})$ for almost all v. In Lemma 10.1(i), we have noted that

$$|\ell_{\Sigma_v}(v_{0,v})|^2 = \frac{\zeta_{k_v}(2)}{L(1, \Sigma_v, Ad)}.$$

Thus the Euler product may not converge because of the denominator. However, we may normalize ℓ_{Σ_v} by setting

$$\ell_{\Sigma_v}^{\flat} = \lambda_v^{-1} \cdot \ell_{\Sigma_v} \quad \text{for each place } v$$

with

$$|\lambda_v|^2 = \frac{\zeta_{k_v}(2)}{L(1, \Sigma_v, Ad)}$$

and

$$\ell^{\flat}_{\Sigma_v}(v_{0,v}) = 1 \quad \text{for almost all } v.$$

Then the product $\prod_v \ell^{\flat}_{\Sigma_v}(u_v)$ certainly converges and we set

$$\ell^{\mathbb{A}}_{\Sigma} := \prod_v^{*} \ell_{\Sigma_v} := \left(\frac{|\zeta_k(2)|}{|L(1, \Sigma, Ad)|}\right)^{1/2} \cdot \prod_v \ell^{\flat}_{\Sigma_v}$$

where the global L-value is defined by meromorphic continuation of the global L-function.

- Suppose we want to define

$$A^{\mathbb{A}}_{\Sigma} : \Omega_{\psi} \otimes \overline{\Sigma} \longrightarrow \Theta^{\mathbb{A}}(\Sigma).$$

Hence we need to consider the infinite product

$$\prod_{v \notin S} A_{\Sigma_v}(\Phi_{0,v}, v_{0,v})$$

for some finite set S of places of k. By Lemma 10.1(i), we see that for almost all v,

$$\begin{aligned} &\langle A_{\Sigma_v}(\Phi_{0,v}, v_{0,v}), A_{\Sigma_v}(\Phi_{0,v}, v_{0,v})\rangle_{\theta(\Sigma_v)} \\ &= \frac{L(n-1, \Sigma_v, std)}{\zeta_{k_v}(2) \cdot \zeta_{k_v}(n) \cdot \zeta_{k_v}(2n-2)} \cdot \langle w_{0,v}, w_{0,v}\rangle_{\theta(\Sigma_v)} \end{aligned}$$

where $\dim V = 2n$ and $w_{0,v}$ is a unit vector. Since Σ is tempered, the relevant Euler product actually converges absolutely when $n > 2$, in which case we can simply define

$$A^{\mathbb{A}}_{\Sigma} = \prod_v A_{\Sigma_v}.$$

In general, we would set

$$A^{\flat}_{\Sigma_v} = \lambda_v^{-1} \cdot A_{\Sigma_v}$$

with

$$|\lambda_v|^2 = \frac{|L(n-1, \Sigma_v, std)|}{|\zeta_{k_v}(2)| \cdot |\zeta_{k_v}(n)| \cdot |\zeta_{k_v}(2n-2)|} \quad \text{for all } v,$$

and

$$A^{\flat}_{\Sigma_v}(\Phi_{0,v}, v_{0,v}) = w_{0,v} \quad \text{for almost all } v.$$

Then we set

$$A^{\mathbb{A}}_{\Sigma} = \left(\frac{|L(n-1, \Sigma, std)|}{|\zeta_k(2)| \cdot |\zeta_k(n)| \cdot |\zeta_k(2n-2)|} \right)^{1/2} \cdot \prod_v A^{\flat}_{\Sigma_v}.$$

- Suppose we want to define $\ell^{\mathbb{A}}_{\Theta(\Sigma)} : \Theta^{\mathbb{A}}(\Sigma) \longrightarrow \mathbb{C}$. Then we need to consider the Euler product

$$\prod_{v \notin S} \ell_{\theta(\Sigma_v)}(w_{0,v}) \quad \text{for some finite set } S.$$

In Proposition 10.2, we have seen that

$$|\ell_{\theta(\Sigma_v)}(w_{0,v})|^2 = \frac{L(n-1, \Sigma_v, std)}{L(1, \sigma, Ad)} \cdot \frac{\zeta_{k_v}(n)}{\zeta_{k_v}(2n-2)}$$

for almost all v (with $\dim V = 2n$). If $n = 2$, the right hand side is equal to 1, so the relevant Euler product converges. For $n > 2$, we set

$$\ell^{\flat}_{\theta(\Sigma_v)} = \lambda_v^{-1} \cdot \ell_{\theta(\Sigma_v)}$$

with

$$|\lambda_v|^2 = \frac{L(n-1, \Sigma_v, std)}{L(1, \Sigma_v, Ad)} \cdot \frac{\zeta_{k_v}(n)}{\zeta_{k_v}(2n-2)} \quad \text{for all } v$$

and

$$\ell^{\flat}_{\theta(\Sigma_v)}(w_{0,v}) = 1 \quad \text{for almost all } v.$$

Then we set

$$\ell^{\mathbb{A}}_{\Theta^{\mathbb{A}}(\Sigma)} = \left(\frac{|L(n-1, \Sigma, std)|}{|L(1, \Sigma, Ad)|} \cdot \frac{|\zeta_k(n)|}{|\zeta_k(2n-2)|} \right)^{1/2} \cdot \prod_v \ell^{\flat}_{\theta(\Sigma_v)}.$$

After these three illustrative examples, we leave the precise definition of other adelic period maps to the reader.

12.13 Comparison of Automorphic and Adelic Periods

We can now compare the various adelic period maps with their automoprhic counterparts, using the decompositions

$$\Omega_{\psi} \cong \otimes_v' \Omega_{\psi}, \quad \Sigma \cong \otimes_v' \Sigma_v \quad \text{and} \quad \Theta^{\mathrm{Aut}}(\Sigma) \cong \Theta^{\mathbb{A}}(\Sigma) := \otimes_v' \theta_{\psi_v}(\Sigma_v)$$

fixed in Sect. 12.11.

Since both $\ell_{\Sigma}^{\mathbb{A}}$ and $P_{\Sigma,N,\psi}$ are nonzero elements of the one-dimensional space $\mathrm{Hom}_{N(\mathbb{A})}(\Sigma, \psi)$, there is a constant $c(\Sigma) \in \mathbb{C}^{\times}$ such that

$$P_{\Sigma,N,\psi} = c(\Sigma) \cdot \ell_{\Sigma}^{\mathbb{A}},$$

so that

$$\beta_{\Sigma}^{\mathrm{Aut}} = c(\Sigma) \cdot \beta_{\Sigma}^{\mathbb{A}}$$

Likewise, we have $c(\Theta(\Sigma)) \in \mathbb{C}^{\times}$ such that

$$P_{\Theta(\Sigma),\mathrm{O}(v_1^{\perp})} = c(\Theta(\Sigma)) \cdot \ell_{\Theta(\Sigma)}^{\mathbb{A}}$$

so that

$$\beta_{\Theta(\Sigma)}^{\mathrm{Aut}} = c(\Theta(\Sigma)) \cdot \beta_{\Theta(\Sigma)}^{\mathbb{A}},$$

Similarly, we have $a(\Sigma)$ and $b(\Sigma) \in \mathbb{C}^{\times}$ such that

$$A_{\Sigma}^{\mathrm{Aut}} = a(\Sigma) \cdot A_{\Sigma}^{\mathbb{A}} \quad \text{and} \quad B_{\Theta(\Sigma)}^{\mathrm{Aut}} = b(\Sigma) \cdot B_{\Theta(\Sigma)}^{\mathbb{A}}$$

12.14 Global Result

The main global problem is to determine the constant $|c(\Theta(\Sigma))|^2$. We shall resolve this by relating $c(\Theta(\Sigma))$ to the other constants $c(\Sigma)$, $a(\Sigma)$, and $b(\Sigma)$.

Proposition 12.3. *We have:*

$$\overline{c(\Theta(\Sigma))} = c(\Sigma) \cdot b(\Sigma).$$

Proof. This follows by combining the global Proposition 12.2 and the local Corollary 6.7. □

It remains then to compute $c(\Sigma)$ and $b(\Sigma)$.

Proposition 12.4. *We have:*

$$a(\Sigma) = b(\Sigma) \quad \textit{and} \quad |a(\Sigma)| = 1.$$

Moreover,

$$c(\Sigma) = \begin{cases} 1, & \textit{if } \Sigma \textit{ is non-endoscopic;} \\ 1/2, & \textit{if } \Sigma \textit{ is an endoscopic lift from } \mathrm{O}_2 \times \mathrm{O}_1; \\ 1/4, & \textit{if } \Sigma \textit{ is an endoscopic lift from } \mathrm{O}_1 \times \mathrm{O}_1 \times \mathrm{O}_1. \end{cases}$$

Proof. The equality of $a(\Sigma)$ and $b(\Sigma)$ follows by the global equation (12.1) and the local equation (5.9). On the other hand, the Rallis inner product formula [GQT] gives:

$$\langle A_\Sigma^{\mathrm{Aut}}(\Phi, f), A_\Sigma^{\mathrm{Aut}}(\Phi, f)\rangle_{\Theta(\Sigma)} = Z_\Sigma(\Phi, \Phi, f, f) = \langle A_\Sigma^{\mathbb{A}}(\Phi, f), A_\Sigma^{\mathbb{A}}(\Phi, f)\rangle$$

where $Z_\Sigma(-)$ is the global doubling zeta integral (evaluated at the point $s = (\dim V - 3)/2$, where it is holomorphic). Combining this with the local equation (5.5), we deduce that $|a(\Sigma)| = 1$.

Finally, the value of $c(\Sigma)$ (for the group SL_2) was determined in [LM, Cor. 4.3 and §6.1]. □

As a consequence, we have:

Theorem 12.5. *Let Σ be a globally ψ-generic cuspidal representation of SL_2 such that $\Pi = \Theta(\Sigma) \subset \mathcal{A}_{cusp}(\mathrm{O}(V))$. Then $|c(\Theta(\Sigma))| = |c(\Sigma)|$, so that*

$$|P_{\Pi,\mathrm{O}(v_1^\perp)}(\phi)|^2 = |c(\Sigma)|^2 \cdot |\ell_\Pi^{\mathbb{A}}(\phi)|^2$$

for all $\phi \in \Pi$.

12.15 Avoiding Rallis Inner Product

In proving Theorem 12.5, we have pulled the Rallis inner product formula out of the hat to deduce that $|a(\Sigma)| = 1$ in Proposition 12.4. In fact, it is possible to avoid the Rallis inner product formula, as we briefly sketch in this subsection.

Just as Proposition 12.2 is the global analog of the local Corollary 6.7, one can establish a global analog of Corollary 9.3, namely:

Proposition 12.6. *For $\Phi \in \Omega_\psi$ and $f \in \Sigma$, one has*

$$P_{\mathrm{O}(v_1^\perp)}(A_\Sigma^{\mathrm{Aut}}(\Phi, f)) = \langle p(\Phi), \beta_\Sigma^{\mathrm{Aut}}(f)\rangle_{N\backslash \mathrm{SL}_2}.$$

Proof. The proof relies on the see-saw diagram

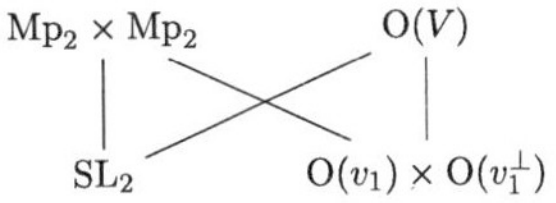

which gives rise to a global see-saw identity. More precisely, if we take

$$\Phi = \Phi_1 \otimes \Phi' \in \mathcal{S}(\mathbb{A}v_1) \otimes \mathcal{S}((v_1^{\perp})_{\mathbb{A}}),$$

then the see-saw identity reads:

$$P_{\Pi,\mathrm{O}(v_1^{\perp})}(A_{\Sigma}^{\mathrm{Aut}}(\Phi, f)) = \int_{[\mathrm{SL}_2]} \overline{f(g)} \cdot \theta(\Phi_1)(g) \cdot I(\Phi')(g)\, dg$$

where

- $\theta(\Phi_1)$ is the theta function associated to $\Phi_1 \in \mathcal{S}(\mathbb{A}v_1)$ which affords the Weil representation (associated to ψ) for $\mathrm{Mp}_2 \times \mathrm{O}(v_1)$;
- $I(\Phi')$ is the theta integral

$$I(\Phi')(g) := \int_{[\mathrm{O}(v_1^{\perp})]} \theta(\Phi')(gh)\, dh$$

  which belongs to the global theta lift of the trivial representation of $\mathrm{O}(v_1^{\perp})$ to Mp_2.

The theta integral converges absolutely when $\dim V > 4$ (it is in the so-called Weil's convergence range) or when $\mathrm{O}(v_1^{\perp})$ is anisotropic. When $\dim V = 4$ and $\mathrm{O}(v_1^{\perp})$ is split, one needs to regularise the theta integral following Kudla-Rallis (see [GQT, §3]). Since our intention here is to indicate an alternative approach to a result which we have shown, we will ignore this analytic complication in the following exposition.

In Sect. 8.3, we have seen that the map $\phi' \mapsto \phi_{\Phi'}$, where

$$\phi_{\Phi'}(g) = (g \cdot \Phi')(0),$$

gives an isomorphism

$$\mathcal{S}((v_1^{\perp})_{\mathbb{A}})_{\mathrm{O}(v_1^{\perp})} \longrightarrow I_{\psi}(\chi_{\mathrm{disc}(v_1^{\perp})}, (\dim V - 3)/2)$$

of the $\mathrm{O}(v_1^{\perp})$-coinvariant of the Weil representation of $\mathrm{O}(v_1^{\perp}) \times \mathrm{Mp}_2$ to a principal series representation of Mp_2. Now the Siegel-Weil formula shows that

$$I(\Phi') = E(\phi_{\Phi'})$$

where $E(\phi_{\Phi'})$ is the Eisenstein series associated to $\phi_{\Phi'}$. Again, when $\dim V > 4$, the sum defining the Eisenstein series is convergent, but when $\dim V = 4$, it is defined by meromorphic continuation. Further, if $\mathrm{O}(v_1^\perp)$ is also split, then the Eisenstein series does have a pole at the point of interest, and we need to invoke the second term identity of the Siegel-Weil formula [GQT]. As mentioned before, we omit these extra (though interesting) details in this proof. The reader may assume that $\dim V > 4$ in the rest of the proof.

Hence, we have the following identity:

$$P_{\Pi,\mathrm{O}(v_1^\perp)}(A_\Sigma^{\mathrm{Aut}}(\Phi, f)) = \int_{[\mathrm{SL}_2]} \overline{f(g)} \cdot \theta(\Phi_1)(g) \cdot E(\phi_{\Phi'})\, dg.$$

Now the right hand side is the value at $s = (\dim V - 3)/2$ of the global zeta integral

$$Z(f, \Phi_1, \phi_s) = \int_{[\mathrm{SL}_2]} \overline{f(g)} \cdot \theta(\Phi_1)(g) \cdot E(\phi_s)\, dg$$

for $\phi_s \in I_\psi(\chi_{\mathrm{disc}(v_1^\perp)}, s)$. This is the global analog of the local zeta integrals we discussed in Sect. 8.3 and represents the (twisted) adjoint L-function of Σ. The unfolding of this global zeta integral, for $\mathrm{Re}(s)$ sufficiently large, gives:

$$Z(s, f, \Phi_1, \phi) = \int_{N(\mathbb{A})\backslash \mathrm{SL}_2(\mathbb{A})} \overline{P_{N,\psi}(g \cdot f)} \cdot (g \cdot \Phi_1)(v_1) \cdot \phi_s(g)\, dg.$$

Specializing to $s = (\dim V - 3)/2$ gives the Proposition. □

By combining Proposition 12.6 and Corollary 9.3, we deduce:

Corollary 12.7. *One has:*

$$\overline{c(\Sigma)} = c(\Theta(\Sigma)) \cdot a(\Sigma).$$

Combining Corollary 12.7 with Propositions 12.3 and 12.4, we see that

$$\overline{c(\Sigma)} = c(\Theta(\Sigma)) \cdot a(\Sigma) = \overline{c(\Sigma)} \cdot \overline{b(\Sigma)} \cdot a(\Sigma) = \overline{c(\Sigma)} \cdot |a(\Sigma)|^2$$

from which we deduce that

$$|a(\Sigma)| = 1.$$

There is a good reason for avoiding the use of the Rallis inner product formula in the treatment of the global problem. Indeed, the viewpoint and techniques developed in this paper should carry over to essentially all the low rank spherical varieties treated in [GG]. Many of these (such as $\mathrm{Spin}_9\backslash F_4$, $G_2\backslash\mathrm{Spin}_8$ or $F_4\backslash E_6$ to name a few) would involve the exceptional theta correspondence. Unfortunately, in the setting of the exceptional theta correspondence, an analog of the Rallis inner product formula is not known. The argument in this subsection, however, shows that this lack need not be an obstruction in the exceptional setting.

12.16 Global Relative Character Identity

We can also establish the global analog of the relative character identity. One has the diagram

$$\begin{array}{ccc} & \mathcal{S}(V_{\mathbb{A}}) & \\ {}^{p}\swarrow & & \searrow^{q} \\ \mathcal{S}(N(\mathbb{A}),\psi\backslash \mathrm{SL}_2(\mathbb{A})) & & \mathcal{S}(X_{\mathbb{A}}) \end{array}$$

which is the adelic analog of the diagram (8.2). The space $\mathcal{S}(N(\mathbb{A}), \psi\backslash\mathrm{SL}_2(\mathbb{A}))$ (which is the image of p) is the restricted tensor product of the local spaces $\mathcal{S}(N(k_v), \psi_v\backslash\mathrm{SL}_2(k_v))$ of test functions defined in (8.1), where the restricted tensor product is with respect to the family of basic functions $\{f_{0,v}\}$ given in Definition 8.4. Likewise, the space $S(X_{\mathbb{A}})$ is the restricted tensor product of $S(X_{k_v})$ (which is $C_c^\infty(X_{k_v})$ at finite places) with respect to the family of basic functions $\{\phi_{0,v}\}$ in Definition 8.4. As in the local case, one says that $f \in \mathcal{S}(N(\mathbb{A}), \psi\backslash\mathrm{SL}_2(\mathbb{A}))$ and $\phi \in S(X_{\mathbb{A}})$ are in correspondence or are transfers of each other if there exists $\Phi \in S(V_{\mathbb{A}})$ such that $f = p(\Phi)$ and $\phi = q(\Phi)$. The fundamental Lemma 8.5 ensures that every f has a transfer ϕ and vice versa.

Before formulating the global relative character identity, we need to address an additional subtle point here. In our general discussion in Sect. 12.4, we have considered the maps

$$\alpha_{\Sigma}^{\mathrm{Aut}} : C_c^\infty(N(\mathbb{A}), \psi\backslash\mathrm{SL}_2(\mathbb{A})) \longrightarrow \Sigma \subset \mathcal{A}_{cusp}(\mathrm{SL}_2)$$

and

$$\alpha_{\Theta^{\mathrm{Aut}}(\Sigma)}^{\mathrm{Aut}} : C_c^\infty(X(\mathbb{A})) \longrightarrow \Theta^{\mathrm{Aut}}(\Sigma) \subset \mathcal{A}_{cusp}(\mathrm{O}(V)).$$

The point here is that their domains consist of smooth compactly supported functions on the adelic points of the relevant spherical varieties. Likewise, in Sect. 12.5, the global relative character is given as a distribution on the space of smooth compactly supported functions. Now in the case of the hyperboloid X, this is fine since the basic function $\phi_{0,v}$ belongs to $C_c^\infty(X(k_v))$. However, this is not sufficient for the case of the Whittaker variety $(N, \psi)\backslash\mathrm{SL}_2$ since the basic function $f_{0,v}$ is not compactly supported. In particular, while it is true that

$$C_c^\infty(N(k_v), \psi_v\backslash\mathrm{SL}_2(k_v)) \subset \mathcal{S}(N(k_v), \psi_v\backslash\mathrm{SL}_2(k_v)) \quad \text{for each place } v,$$

one has in the adelic setting:

$$C_c^\infty(N(\mathbb{A}), \psi\backslash \mathrm{SL}_2(\mathbb{A})) \not\subseteq \mathcal{S}(N(\mathbb{A}), \psi\backslash \mathrm{SL}_2(\mathbb{A})).$$

Indeed, these adelic spaces have nothing much to do with each other.

To define global relative characters for the Whittaker variety, we thus need to ensure that the map $\alpha_\Sigma^{\mathrm{Aut}} = \mathrm{pr}_\Sigma \circ \theta$ can be defined on $\mathcal{S}(N(\mathbb{A}), \psi\backslash \mathrm{SL}_2(\mathbb{A}))$.

The main issue is to ensure that the formation of theta series θ can be applied to $f = \otimes'_v f_v \in \mathcal{S}(N(\mathbb{A}), \psi\backslash \mathrm{SL}_2)$. Recall that

$$\theta(f)(g) = \sum_{\gamma \in N(k)\backslash \mathrm{SL}_2(k)} f(\gamma g) \quad \text{for } g \in \mathrm{SL}_2(\mathbb{A}). \tag{12.5}$$

Hence, if $f = p(\Phi)$ with $\Phi \in \mathcal{S}(V_\mathbb{A})$, we would like to show the convergence of the sum

$$\sum_{\gamma \in B(k)\backslash \mathrm{SL}_2(k)} \sum_{\lambda \in k^\times} |f(t(\lambda)\gamma g)| = \sum_{\gamma \in B(k)\backslash \mathrm{SL}_2(k)} \sum_{\lambda \in k^\times} |(\gamma g \cdot \Phi)(\lambda \cdot v_1)|.$$

Let us denote the inner sum by

$$F_\Phi(g) := \sum_{\lambda \in k^\times} |(g \cdot \Phi)(\lambda \cdot v_1)|.$$

which is certainly convergent since Φ is a Schwatz function on $V_\mathbb{A}$. Then we need to show the convergence of

$$\sum_{\gamma \in B(k)\backslash \mathrm{SL}_2(k)} F_\Phi(\gamma g). \tag{12.6}$$

This looks very much like the sum defining an Eisenstein series on SL_2. Indeed, observe that F_Φ is a function on $T(k) \cdot N(\mathbb{A})\backslash \mathrm{SL}_2(\mathbb{A})$ and for $a \in \mathbb{A}^\times$ and $k \in K = \prod_v K_v$, we have:

$$F_\Phi(t(a))k) = |a|_\mathbb{A}^{\frac{\dim V}{2}} \cdot \sum_{\lambda \in k^\times} |(k \cdot \Phi)(a\lambda \cdot v_1)|.$$

To understand the asymptotic of F_Φ as $|a|_\mathbb{A}$ tends to 0 or ∞, we note:

Lemma 12.8. *Let $\phi \in S(\mathbb{A})$ and define a function Ψ on $\mathbb{A}^\times$ by*

$$\Psi(a) = \sum_{\lambda \in k^\times} \phi(a\lambda).$$

Then Ψ is rapidly decreasing as $|a|_{\mathbb{A}} \to \infty$. Moreover, as $|a|_{\mathbb{A}} \to 0$,

$$|\Psi(a)| \leq C \cdot |a|_{\mathbb{A}}^{-1} \quad \textit{for some constant } C.$$

Proof. The rapid decrease of $\Psi(a)$ as $|a|_{\mathbb{A}} \to \infty$ follows from the fact that $\phi \in S(V_{\mathbb{A}})$. For the asymptotic as $|a|_{\mathbb{A}} \to 0$, we need to apply the Poisson summation formula to the sum defining Ψ. Writing $\widehat{\phi}$ for the Fourier transform of ϕ (with respect to $d_{\psi}x$), we have:

$$\begin{aligned}\Psi(a) &= -\phi(0) + \sum_{\lambda \in k} \phi(a\lambda) \\ &= -\phi(0) + |a|_{\mathbb{A}}^{-1} \cdot \sum_{\lambda \in k} \widehat{\phi}(a^{-1} \cdot \lambda) \\ &= -\phi(0) + |a|_{\mathbb{A}}^{-1} \cdot \widehat{\phi}(0) + |a|_{\mathbb{A}}^{-1} \cdot \sum_{\lambda \in k^{\times}} \widehat{\phi}(a^{-1} \cdot \lambda)\end{aligned}$$

As $|a|_{\mathbb{A}}^{-1} \to \infty$, the last sum tends to 0 rapidly, so the third term is bounded. Hence we see that the asymptotic of $\Psi(a)$ as $|a|_{\mathbb{A}} \to 0$ is governed by the second term in the last expression. □

Applying the lemma to $\phi(a) = (k \cdot \Phi)(a \cdot v_1)$, we deduce that $F_{\Phi}(t(a)k)$ is rapidly decreasing as $|a|_{\mathbb{A}} \to \infty$ and

$$|F_{\Phi}(t(a)k)| \leq C \cdot |a|_{\mathbb{A}}^{\frac{\dim V}{2}-1} \qquad \text{as } |a|_{\mathbb{A}} \to 0$$

for some C which can be taken to be independent of $k \in K$. In other words, the sum in (12.6) is dominated by the sum defining a spherical Eisenstein series associated to the principal series representation $I(\frac{\dim V}{2} - 2)$ of SL_2. Hence, when $\dim V > 6$, the above sum does converge to give a smooth function on $[SL_2]$ of moderate growth (c.f. [Bor, Thm. 11.2]), so that (12.5) defines a $SL_2(\mathbb{A})$-equivariant map

$$\theta : \mathcal{S}(N(\mathbb{A}), \psi \backslash SL_2) \longrightarrow C^{\infty}_{mod}([SL_2]).$$

One then has the map $\alpha_{\Sigma}^{\mathrm{Aut}} = \mathrm{pr}_{\Sigma} \circ \theta$ such that one still has the adjunction formula

$$\langle \alpha_{\Sigma}^{\mathrm{Aut}}(f), \phi \rangle_{\Sigma} = \langle f, \beta_{\Sigma}^{\mathrm{Aut}}(\phi) \rangle_{N(\mathbb{A})\backslash SL_2(\mathbb{A})},$$

for $f \in \mathcal{S}(N(\mathbb{A}), \psi \backslash SL_2(\mathbb{A}))$ and $\phi \in \Sigma$.

Presumably, one can show that θ and $\alpha_{\Sigma}^{\mathrm{Aut}}$ are also defined when $3 \leq \dim V \leq 6$ by a more careful analysis, involving the meromorphic continuation of pseudo-Eisenstein series, but we have not pursued this further. One now has the following global relative character identity.

Theorem 12.9. *Assume that* $\dim V > 6$. *If* $f \in \mathcal{S}(N(\mathbb{A}), \psi\backslash \mathrm{SL}_2(\mathbb{A}))$ *and* $\phi \in S(X_{\mathbb{A}})$ *are transfer of each other, then for a cuspidal representation* Σ *of* SL_2 *with cuspidal theta lift* $\Theta^{\mathrm{Aut}}(\Sigma)$ *on* $\mathrm{O}(V)$, *one has:*

$$\mathcal{B}_{\Sigma}^{\mathrm{Aut}}(f) = \mathcal{B}_{\Theta(\Sigma)}^{\mathrm{Aut}}(\phi).$$

Proof. We have defined $c(\Sigma)$ by

$$\beta_{\Sigma}^{\mathrm{Aut}} = c(\Sigma) \cdot \beta_{\Sigma}^{\mathbb{A}}.$$

By the adjunction formulae for the pairs $(\alpha_{\Sigma}^{\mathrm{Aut}}, \beta_{\Sigma}^{\mathrm{Aut}})$ and $(\alpha_{\Sigma}^{\mathbb{A}}, \beta_{\Sigma}^{\mathbb{A}})$, we deduce that

$$\alpha_{\Sigma}^{\mathrm{Aut}} = \overline{c(\Sigma)} \cdot \alpha_{\Sigma}^{\mathbb{A}}.$$

Hence, one has

$$\mathcal{B}_{\Sigma}^{\mathrm{Aut}}(f) = (\beta_{\Sigma}^{\mathrm{Aut}}\alpha_{\Sigma}^{\mathrm{Aut}}(f))(1) = |c(\Sigma)|^2 \cdot (\beta_{\Sigma}^{\mathbb{A}}\alpha_{\Sigma}^{\mathbb{A}}(f))(1) = |c(\Sigma)|^2 \cdot \mathcal{B}_{\Sigma}^{\mathbb{A}}(f).$$

Likewise, we have

$$\mathcal{B}_{\Theta(\Sigma)}^{\mathrm{Aut}}(\phi) = |c(\Theta(\Sigma))|^2 \cdot \mathcal{B}_{\Theta(\Sigma)}^{\mathbb{A}}(\phi).$$

Since $|c(\Sigma)| = |c(\Theta(\Sigma))|$, the desired result follows from the local relative character identity of Theorem 9.1. □

12.17 End Remarks

We end this paper with some comparisons with the relative trace formula approach. The spectral side of a relative trace formula is essentially a sum of the relevant global relative characters over all cuspidal representations. One then hopes to separate the different spectral contributions by using the action of the spherical Hecke algebra at almost all places. The main global output of a comparison of (the geometric side of) two such relative trace formulae is typically a global relative character identity as in Theorem 12.9, as a consequence of which one deduces Proposition 12.2 and the local relative character identities in Theorem 9.1, which in turn implies Proposition 6.1. It is interesting to compare this with the approach via theta correspondence which we have pursued in this paper.

Acknowledgments The first author thanks Sug Woo Shin, Nicholas Templier and Werner Mueller for their kind invitation to participate in the Simons Symposium and the Simons Foundation for providing travel support. He also thanks Yiannis Sakellaridis for helpful conversations on the various topics discussed in this paper. We thank the referees for their careful work and helpful suggestions, especially the suggestion to be absolutely precise about normalization of measures. The first author is partially supported by a Singapore government MOE Tier 2 grant R146-000-233-112, whereas the second author is supported by an MOE Graduate Research Scholarship.

References

[AGRS] A. Aizenbud, D. Gourevitch, S. Rallis, O. Schiffmann, Multiplicity one theorems. Ann. Math. (2) **172**(2), 1407–1434 (2010)

[BLM] M. Baruch, E. Lapid, Z.Y. Mao, A Bessel identity for the theta correspondence. Israel J. Math. **194**(1), 225–257 (2013)

[B] J.N. Bernstein, On the support of Plancherel measure. J. Geom. Phys. 5 **1988**(4), 663–710 (1989)

[BP1] R. Beuzart-Plessis, A local trace formula for the Gan-Gross-Prasad conjecture for unitary groups: the Archimedean case. Asterisque **418** (2020)

[BP2] R. Beuzart-Plessis, Plancherel formula for $\mathrm{GL}_n(F)\backslash\mathrm{GL}_n(E)$ and applications to the Ichino-Ikeda and formal degree conjectures for unitary groups. arXiv:1812.00047 (2020)

[Bor] A. Borel, Introduction to automorphic forms, in *Algebraic Groups and Discontinuous Subgroups*. Proceedings of Symposia in Pure Mathematics, Boulder (American Mathematical Society, Providence, 1965), pp. 199–210

[D] P. Delorme, Formule de Plancherel pour les fonctions de Whittaker sur un groupe réductif p-adique. Ann. Inst. Fourier (Grenoble) **63**(1), 155–217 (2013)

[G2] W.T. Gan, Doubling zeta integrals and local factors for metaplectic groups. Nagoya Math. J. **208**, 67–95 (2012)

[G] W.T. Gan, Periods and theta correspondence, in *Proceedings of Symposia in Pure Mathematics*, vol. 101 (2019), pp. 113–132

[GG] W.T. Gan, R. Gomez, A conjecture of Sakellaridis-Venkatesh on the unitary spectrum of spherical varieties, in *Symmetry: Representation Theory and Its Applications*, pp. 185–226. Progress in Mathematics, vol. 257 (Birkhauser/Springer, New York/Berlin, 2014)

[GT] W.T. Gan, S. Takeda, A proof of the Howe duality conjecture. J. Amer. Math. Soc. **29**(2), 473–493 (2016)

[GS] W.T. Gan, G. Savin, Representations of metaplectic groups I: epsilon dichotomy and local Langlands correspondence. Compositio Math. **148**(6), 1655–1694 (2012)

[GQT] W.T. Gan, Y.N. Qiu, S. Takeda, The regularized Siegel-Weil formula (the second term identity) and the Rallis inner product formula. Invent. Math. **198**(3), 67–95 (2014)

[GJ] S. Gelbart, H. Jacquet, A relation between automorphic representations of GL(2) and GL(3). Ann. Sci. École Norm. Sup. (4) **11**(4), 471–542 (1978)

[GZ] R. Gomez, C.B. Zhu, Local theta lifting of generalized Whittaker models associated to nilpotent orbits. Geom. Funct. Anal. **24**, 796–853 (2014)

[He] H.Y. He, Theta correspondence. I. Semistable range: construction and irreducibility. Commun. Contemp. Math. **2**(2), 255–283 (2000)

[H] R. Howe, On some results of Strichartz and Rallis and Schiffman. J. Funct. Anal. **32**(3), 297–303 (1979)

[JK] D. Johnstone, R. Krishna, Beyond endoscopy for spherical varieties of rank one: the fundamental lemma. In preparation (2021)

[LM] E. Lapid, Z.Y. Mao, A conjecture on Whittaker-Fourier coefficients of cusp forms. J. Number Theory **146**, 448–505 (2015)

[LR] E. Lapid, S. Rallis, On the local factors of representations of classical groups, in *Automorphic Representations, L-Functions and Applications: Progress and Prospects*. Ohio State University Mathematics Research Institute Publications, vol. 11 (de Gruyter, Berlin, 2005), pp. 309–359

[Li] J.-S. Li, Singular unitary representations of classical groups. Invent. Math. **97**(2), 237–255 (1989)

[MR] Z.Y. Mao, S. Rallis, Jacquet modules of the Weil representations and families of relative trace identities. Compos. Math. **140**(4), 855–886 (2004)

[MVW] C. Moeglin, M.-F. Vigneras, J.-L. Waldspurger, *Correspondances de Howe sur un corps p-adique*. Lecture Notes in Mathematics, vol. 1291 (Springer, Berlin, 1987), viii+163 pp.

[R] S. Rallis, Langlands' functoriality and the Weil representation. Am. J. Math. **104**(3), 469–515 (1982)

[R2] S. Rallis, On the Howe duality conjecture. Compos. Math. **51**(3), 333–399 (1984)

[R3] S. Rallis, *L-Functions and the Oscillator Representation*. Lecture Notes in Mathematics, vol. 1245 (Springer, Berlin, 1987), 239 pp.

[Rama] D. Ramakrishnan, Modularity of the Rankin-Selberg L-series, and multiplicity one for SL(2). Ann. Math. (2) **152**(1), 45–111 (2000)

[S1] Y. Sakellaridis, On the unramified spectrum of spherical varieties over p-adic fields. Compos. Math. **144**(4), 978–1016 (2008)

[S2] Y. Sakellaridis, Spherical functions on spherical varieties. Amer. J. Math. **135**(5), 1291–1381 (2013)

[S3] Y. Sakellaridis, Plancherel decomposition of Howe duality and Euler factorization of automorphic functionals, in *Representation Theory, Number Theory, and Invariant Theory*. Progr. Math., vol. 323 (Birkhauser/Springer, Cham, 2017), pp. 545–585

[S4] Y. Sakellaridis, Functorial transfer between relative trace formulas in rank one (2018). arXiv 1808.09358

[S5] Y. Sakellaridis, Transfer operators and Hankel transforms between relative trace formulas, I: character theory (2018). arXiv 1804.02383

[S6] Y. Sakellaridis, Transfer operators and Hankel transforms between relative trace formulas, II: Rankin–Selberg theory (2018). arXiv 1805.04640

[St] R.S. Strichartz, Harmonic analysis on hyperboloids. J. Funct. Anal. **12**, 341–383 (1973)

[SV] Y. Sakellaridis, A. Venkatesh, Periods and harmonic analysis on spherical varieties. Astérisque No. 396 (2017), viii+360 pp.

[W] J.-L. Waldspurger, La formule de Plancherel pour les groupes *p-adiques (d'après Harish-Chandra)*. J. Inst. Math. Jussieu **2**(2), 235–333 (2003)

[Wa1] N. Wallach, *Real reductive groups. I*. Pure and Applied Mathematics, vol. 132 (Academic Press Inc., Boston, 1988), xx+412 pp.

[Wa2] N. Wallach, *Real reductive groups. II*. Pure and Applied Mathematics, vol. 132-II (Academic Press Inc., Boston, 1992), xiv+454 pp.

[Wan] X.L. Wan, The Sakellaridis-Venkatesg conjecture for $U_2\backslash SO_5$, PhD Thesis (2020). National University of Singapore

[X] H. Xue, Arithmetic theta lifts and the arithmetic Gan-Gross-Prasad conjecture for unitary groups. Duke Math. J. **168**(1), 127–185 (2019)

[Z] C.B. Zhu, Local theta correspondence and nilpotent invariants, in *Representations of Reductive Groups*. Proceedings of Symposia in Pure Mathematics, vol. 101 (American Mathematical Society, Providence, 2019), pp. 427–450

Incoherent Definite Spaces and Shimura Varieties

Benedict H. Gross

Abstract In this paper, we define incoherent definite quadratic spaces over totally real number fields and incoherent definite Hermitian spaces over CM fields. We use the neighbors of these spaces to study the local points of orthogonal and unitary Shimura varieties.

1 Introduction

In [G04] we considered the infinite set of quaternion algebras $B(v)$ over a totally real field k whose ramification locus has distance one from a finite set Σ of places of k which has odd cardinality. These algebras are indexed by the places v of k; the algebra $B(v)$ is ramified at the set $\Sigma \cup \{v\}$ when v is not contained in Σ, and is ramified at the set $\Sigma - \{v\}$ if v is contained in Σ. When the set Σ contains all the real places of k, each neighboring algebra $B(v)$ gives local information on a single Shimura curve S which is defined over k. For example, if v is a real place, the algebra $B(v)$ is split at v and ramified at all other real places, and Σ-arithmetic subgroups of the multiplicative group of $B(v)$ give an analytic description of the points of S over the quadratic extension $K_v = \mathbb{C}$ of the completion $k_v = \mathbb{R}$ (See [Sh61, Sh67] for the construction of a canonical model for S over $k \hookrightarrow \mathbb{C}$ and [DN67] for the analytic description of S at other real places). If $v = \mathfrak{p}$ is a finite place where the curve has good reduction, the algebra $B(\mathfrak{p})$ is ramified at all real places and Σ-arithmetic subgroups of its multiplicative group give an analytic description of the points of S which have "supersingular" reduction modulo $\mathfrak{p}$ (cf. [C86, §11]).

In this note, we generalize the notion of an odd set Σ of places of k, containing all the real places (which can be viewed as an incoherent definite quaternion data over k) to incoherent definite orthogonal and Hermitian data over k. The former is a collection of local orthogonal spaces V_v over the completions k_v of fixed rank $n \geq 3$

B. H. Gross (✉)
Department of Mathematics, Harvard University, Cambridge, MA, USA
e-mail: gross@math.harvard.edu

W. Müller et al. (eds.), *Relative Trace Formulas*, Simons Symposia,
https://doi.org/10.1007/978-3-030-68506-5_5

and determinant d in k^*/k^{*2}. We insist that the spaces V_v are positive definite for all real places of k (so the determinant d is totally positive in k^*), that the Hasse-Witt invariants of V_v are equal to +1 for almost all places v, and that the product of the Hasse-Witt invariants is equal to -1. There is no global orthogonal space over k with these completions at all places, for then the product of the Hasse-Witt invariants would be equal to +1 by Hilbert's reciprocity law. However, for each place v of k there is a global orthogonal space $V(v)$ of rank n and discriminant d which is locally isomorphic to V_u at all places $u \neq v$, and **not** locally isomorphic to V_v at v. The last condition determines the local orthogonal space when v is finite; for a real place we also insist that the signature is equal to $(n-2, 2)$. The global space $V(v)$ is determined up to isomorphism by its localizations, by the Hasse-Minkowski theorem. We call the global orthogonal spaces $V(v)$, indexed by the places v of k, the neighbors of the incoherent definite data $\{V_v\}$.

An odd set Σ of places of k which contains all the real places gives incoherent definite orthogonal data of dimension $n = 3$ and determinant $d \equiv 1$. Indeed, for each place v, let B_v be the local quaternion algebra over k_v which is split if v is not in Σ and is ramified if v is in Σ. Let V_v be the local orthogonal space given by the norm form on elements of trace zero in B_v. The Hasse-Witt invariant of V_v is the product of the Hasse invariant of the quaternion algebra B_v with the class $(-1, -1)_v$ in the local Brauer group, so almost all of the Hasse-Witt invariants are equal to $+1$ and the product of Hasse-Witt invariants is equal to -1. The elements of trace zero in the nearby quaternion algebras $B(v)$ defined above give the neighboring global orthogonal spaces $V(v)$.

There is a similar definition of an incoherent definite Hermitian data and its neighbors, where we fix the rank $n \geq 1$ of the space and a totally complex quadratic extension field K of k. In this case, we only define a neighboring Hermitian space $V(v)$ at the places v of k which are not split in K. After working through the definitions, we show how incoherent definite orthogonal data of dimension $n \geq 3$ determines a Shimura variety S of orthogonal type and dimension $n-2$, with field of definition k. We show how incoherent definite Hermitian data of dimension $n \geq 1$ determines a Shimura variety S of unitary type and dimension $n-1$, with field of definition K. The special orthogonal groups of the neighbors of orthogonal data at real places v can be used to describe S over the complex quadratic extension K_v of the completion k_v. Similarly, the unitary groups of the neighbors of Hermitian data at real places v can be used to describe S over the complex completions K_v.

At primes $\mathfrak{p}$ of k where the orthogonal Shimura variety S has good reduction, we propose to use the neighboring orthogonal space $V(\mathfrak{p})$ to describe the points over the unramified quadratic extension $K_\mathfrak{p}$ of the completion $k_\mathfrak{p}$ which reduce to a finite set of special points over the residue field $\mathbb{F}_{\mathfrak{p}^2}$. When $k = \mathbb{Q}$, these special points should form the 0-dimensional stratum of the supersingular locus. They should correspond to the moduli of orthogonal motives of rank n and weight 2 over $\mathbb{F}_{p^2}$ whose crystalline cohomology is a non-degenerate lattice of rank n over $\mathbb{Z}_{p^2}$ with a semi-linear endomorphism ϕ. The $\mathbb{Z}_p$ sublattice where $\phi = p$ should have rank n and discriminant lattice isomorphic to $\mathbb{F}_{p^2}$ with its norm form.

When the incoherent orthogonal data has dimension $n = 21$ over $\mathbb{Q}$ and is split at all finite primes, the Shimura variety S parametrizes polarized $K3$ surfaces, and the stratification of the supersingular locus modulo $\mathfrak{p}$ was first studied by Michael Artin [A74]. In that case, the 0-dimensional stratum of special points consists of the moduli of supersingular $K3$ surfaces X with Artin invariant $\sigma(X) = 1$. When $k \neq \mathbb{Q}$ we do not have a good definition of the special locus modulo $\mathfrak{p}$, as there is no modular description of the points of S modulo $\mathfrak{p}$. But our special locus agrees with the superspecial locus, defined group-theoretically by Viehmann and Wedhorn [ViW13].

Similarly at the primes $\mathfrak{p}$ of k which are inert in K and where the unitary Shimura variety S has good reduction, we propose to use the neighboring Hermitian spaces $V(\mathfrak{p})$ to describe the points over the completion $K_{\mathfrak{p}}$ which reduce to a finite set of points in the special locus modulo $\mathfrak{p}$. In the Hermitian case, the points of a Shimura variety closely related to S have a modular interpretation, classifying abelian varieties with extra structure [RSZ19], and the special points form the 0-dimensional stratum of the supersingular locus [VW11]. A supersingular abelian variety A is isogenous to a product of supersingular elliptic curves, and its modulus lies the the special locus precisely when $\dim \mathrm{Hom}(\alpha_p, A[p]) = \dim A$. This stratification of the supersingular locus for the moduli space of polarized abelian varieties was first studied by Franz Oort (cf. [O75] and the historical remarks in [LO98]).

This paper represents my attempt to approach some of the work of Kudla and Rapoport on incoherent Eisenstein series and the arithmetic intersection of cycles on Shimura varieties [KR99, KR00]. It is an expanded version of a letter that I wrote to Deligne in 2009. Defining these Shimura varieties via incoherent quadratic and Hermitian spaces, rather than their neighboring special orthogonal and unitary groups at a real place v of k, was suggested by the arithmetic conjectures in [GGP12].

2 Local and Global Orthogonal Spaces

In this section, we review the classification of orthogonal spaces over number fields and their completions. The main results are due to Minkowski, Witt, and Hasse. Some good references are [L80, MH73] and [S73, Ch IV].

We begin with a review of some invariants of an orthogonal space over a general field k, of characteristic not equal to 2. An orthogonal space (V, q) is a finite dimensional vector space V over k together with a quadratic form $q : V \to k$. The bilinear form

$$\langle v, w\rangle = q(v + w) - q(v) - q(w)$$

is symmetric, and we will always assume that it is non-degenerate (so gives an isomorphism from V to its dual). The first invariant of the space (V, q) is the rank, defined as the dimension of V.

We can choose a basis $\{e_1.e_2, \ldots e_n\}$ of V whose elements are pairwise orthogonal so that

$$q(\sum x_i e_i) = \sum a_i x_i^2.$$

The coefficients a_i are all non-zero elements in k and their product

$$d = a_1 a_2 \cdots a_n$$

is well-defined in k^*/k^{*2}, independent of the orthogonal basis chosen. The class of d in k^*/k^{*2} is called the determinant of the quadratic space (V, q), and is the second invariant we will use. If the rank is equal to 1, the determinant d determines the orthogonal space up to isomorphism—the space is isometric to k with the quadratic form $q(x) = dx^2$.

For two classes a_i and a_j in $H^1(k, \mu_2) = k^*/k^{*2}$, we let (a_i, a_j) denote their cup product in a $H^2(k, (\mu_2)^{\otimes 2}) = H^2(k, \mu_2) = Br_2(k)$. Here $Br(k) = H^2(k, \mathbb{G}_m)$ is the Brauer group of the field k. The Witt invariant of the quadratic space (V, q) is defined by choosing an orthogonal basis and taking the sum in the Brauer group

$$w = \sum_{i<j} (a_i, a_j).$$

An important theorem, due to Witt, states that this class is again independent of the orthogonal basis chosen. This is the third invariant of (V, q) we will need.

If the rank of (V, q) is equal to 2, the determinant d and the Witt invariant determine the orthogonal space up to isomorphism. In this case, the two dimensional k vector space V has the structure of a free module of rank one over the étale algebra $K = k[t]/(t^2 + d)$ (in two different ways). Indeed, any element in the special orthogonal group of V which is not a scalar satisfies a quadratic polynomial whose roots generate the quadratic extension K. Picking one of the roots gives the action of K on V (the other root gives the conjugate action). Then V is a free module of rank 1 over K with basis e, $q(e) = a$ is non-zero in k, and we have $q(\alpha.e) = a.N(\alpha)$ for all α in K^*, where $N : K^* \to k^*$ is the norm. The Witt invariant is determined by the class of a in $Br_2(K/k) = k^*/N(K^*)$. Using the basis 1 and t for K over k, and writing $\alpha = x + yt$ we see that the quadratic form on V is given by $ax^2 + ady^2$ and that V has Witt invariant $w = (a, ad) = (a, -d)$.

We call the choice of a homomorphism $i : K \to \mathrm{End}(V)$ of k-algebras such that $q(i(\alpha).v) = N(\alpha)q(v)$ for all $\alpha \in K^*$ and $v \in V$ an **orientation** of the two dimensional space V. The orientations on V form a principal homogenous space for the group $O(V)/\mathrm{SO}(V) = \mu_2$, and the group preserving the quadratic form on V and a fixed orientation is $\mathrm{SO}(V)$. When K is a field, we can also give an orientation of V by choosing one of the two isomorphisms $(\mathrm{Res}_{K/k}\,\mathbb{G}_m)_{N=1} \cong \mathrm{SO}(V)$ of non-split one dimensional tori over k. Indeed, the K^*-action on V is determined by the action of the elements of norm 1, so by the elements in $\mathrm{SO}(V)$. Finally, the choice of an orientation determines, and is determined by the choice of an isotropic line in the split orthogonal space $V \otimes K$ (where there are two isotropic lines). From an orientation and an element α in $\mathrm{SO}(V) \cong K^*_{N=1}$, we obtain an isotropic vector $e \otimes \alpha + \alpha(e) \otimes 1$ from any non-zero vector $e \in V$. Conversely, the choice of an isotropic line in $V \otimes K$ gives an isometric action of the group K^* on this space, where $\alpha \in K^*$ acts by multiplication by α on the chosen isotropic line and by multiplication by α^{-1} on the other isotropic line. When α has norm 1, this action preserves the space V and gives an isomorphism $(\mathrm{Res}_{K/k}\,\mathbb{G}_m)_{N=1} \cong \mathrm{SO}(V)$.

As a final example in in low rank, let $B = k + ki + kj + k(ij)$ be a quaternion algebra over k, with products $i^2 = a$, $j^2 = b$, $ij = -ji$. Let V be the orthogonal space of dimension three over k, given by the norm form on elements of trace zero in B. Then

$$q(xi + yj + z(ij)) = -ax^2 - by^2 + abz^2.$$

The determinant of the $d(V, q)$ is $\equiv 1$ in k^*/k^{*2}, and the Witt invariant of (V, q) is given by the sum $(a, b) + (-1, -1)$ in $Br_2(k)$.

When $k = k_v$ is a local field which is not isomorphic to $\mathbb{C}$, the group $Br_2(k)$ is isomorphic to $\langle \pm 1 \rangle$ and the cup-product of two classes in k^*/k^{*2} is given by the Hilbert symbol $(a_i, a_j)_v = \pm 1$. In this case, the Witt invariant is often referred to as the Hasse-Witt invariant, and is written as a product of signs:

$$\epsilon_v(V, q) = \prod_{i<j} (a_i, a_j)_v = \pm 1.$$

When $k_v = \mathbb{C}$ we put $\epsilon_v = +1$.

When $k_v = \mathbb{R}$, there is a further invariant of (V, q)—the signature (r, s). Here r is the dimension of a maximal positive definite subspace of V and s is the dimension of a maximal negative definite subspace of V. We have $r + s = \dim(V)$ and there is an orthogonal basis with

$$q(\sum x_i e_i) = x_1^2 + x_2^2 + \ldots + x_r^2 - x_{r+1}^2 - x_{r+2}^2 + \ldots - x_n^2.$$

We have

$$d_v \equiv (-1)^s \qquad \epsilon_v = (-1)^{s(s-1)/2}.$$

Over the complex numbers, the dimension is a complete invariant of (V, q) and over the real numbers, the signature is a complete isomorphism invariant of (V, q). Over a non-Archimedean field, the dimension, determinant, and Hasse-Witt invariant are complete isomorphism invariants of (V, q). The latter two invariants may be chosen arbitrarily in k^*/k^{*2} and $\langle \pm 1 \rangle$, provided that $\dim(V) \geq 3$. In particular, for a fixed dimension $n \geq 3$ and determinant d there are precisely two isomorphism classes of orthogonal spaces over a non-Archimedean field, which are distinguished by the value of their Hasse-Witt invariant.

With these local invariants in hand, we can now describe the isomorphism classes of global orthogonal spaces (V, q) over a number field k, with fixed dimension $n \geq 3$ and determinant d in k^*/k^{*2}. At each real place v choose a signature (r_v, s_v) with $d_v \equiv (-1)^s$, and put $\epsilon_v = (-1)^{s(s-1)/2}$. At each finite place v, choose ϵ_v in $\langle \pm 1 \rangle$, with $\epsilon_v = +1$ for almost all v. The choice of local invariants gives us, for each place v, a orthogonal space over k_v which is unique up to isomorphism. A necessary and sufficient condition for the existence of a global orthogonal space (V, q) over k of dimension of n and determinant d, with these localizations is given by Hilbert's reciprocity law: $\prod_v \epsilon_v = +1$. In this case, the theorem of Hasse and Minkowski states that the global space (V, q) is determined up to isomorphism over k by its localizations.

We may summarize these results as follows.

Theorem 2.1. *Let k be a number field, let $n \geq 3$ be an integer and let d be a class in k^*/k^{*2}. For each place v of k, let (V_v, q_v) be a local orthogonal space over k_v of dimension n and determinant d. Assume that the local Hasse-Witt invariants $\epsilon_v = \epsilon_v(V_v, q_v)$ are equal to +1 for almost all places v.*

Then a necessary and sufficient condition for the existence of a global orthogonal space (V, q) over k with these localizations is

$$\prod_v \epsilon_v = +1.$$

If a global orthogonal space (V, q) exists with these localizations, it is unique up to isomorphism.

3 Local and Global Hermitian Spaces

Let k be a number field, and let K be a quadratic field extension of k. In this section, we review the classification of Hermitian spaces over k and over the completions k_v. These results are due to Landherr, and a reference is [MH73]. We begin with the theory over a general field k and a separable quadratic field extension K. Let τ be the non-trivial involution of K over k.

A Hermitian space (V, ϕ) is a finite dimensional vector space over K together with a Hermitian symmetric form $\phi : V \times V \to K$. We assume that ϕ is K-linear in the first variable and satisfies $\phi(w, v) = \phi(v, w)^\tau$. Then ϕ is K-anti-linear in the second variable and the quadratic form $q(v) = \phi(v, v)$ takes values in k. We also assume that ϕ is non-degenerate, so gives an isomorphism from V to its dual. If $A = (\phi(e_i, e_j))$ is the matrix of the values of ϕ on a basis for V, then A is Hermitian symmetric and $d = \det(A)$ is an element of k^*. Let $N(K^*)$ be the subgroup of k^* consisting of those elements which are norms from K^*. Then the value of d in the quotient group $k^*/N(K^*)$ is independent of the basis chosen, and is called the Hermitian determinant of (V, ϕ). A one dimensional Hermitian space is determined up to isomorphism by its Hermitian determinant—the space is isomorphic to K with form $\phi(z, w) = d.zw^\tau$.

One can also define Hermitian spaces when K is the étale quadratic algebra $k \times k$ with involution $(a, b)^\tau = (b, a)$. In this case, for each dimension $n \geq 1$ there is a unique non-degenerate space up to isomorphism. The free K-module V of rank n has the form to $V = W + W'$, where he W is a vector space of dimension n over k and W' is its dual space. The pairing ϕ is defined by

$$\phi(w + w', u + u') = (\langle w, u'\rangle, \langle u, w'\rangle)$$

In this case, all elements of k^* are norms and $d \equiv 1$.

If $k = k_v$ is a non-Archimedean local field, the dimension and the Hermitian determinant are complete isomorphism invariants of (V, ϕ). When $K = K_v$ is a field, the group $k^*/N(K^*)$ has order two and there are precisely two Hermitian spaces of each dimension $n \geq 1$. Moreover, when K_v is the unramified quadratic field extension of k_v, the subgroup of norms are just the elements of k^* with even valuation. If K_v is not a field, there is a unique Hermitian space of each dimension. If $k_v = \mathbb{R}$ and $K_v = \mathbb{C}$, there is a further invariant of a Hermitian space, its signature (r, s). Here r is the dimension of a maximal positive definite subspace and s is the dimension of a maximal negative definite subspace. We have $r + s = \dim(V)$ and $d \equiv (-1)^s$. The signature is a complete isomorphism invariant in the real case.

Finally, let k be a number field and let K be a quadratic field extension of k. We want to describe all Hermitian spaces of a fixed dimension n over K, up to isomorphism. At each real place v of k which is complex in K, choose a signature (r_v, s_v) with $r_v + s_v = n$ and let $\epsilon_v = (-1)^{s_v}$. At each finite place v of k which is not split in K, choose a class ϵ_v in $k_v^*/N(K_v^*) = \langle \pm 1\rangle$ with $\epsilon_v = +1$ for almost all v. This choice gives us local Hermitian spaces of dimension n for all places v, and the condition that these are the localizations of a fixed global space over k is given by: $\prod_v \epsilon_v = +1$. When a global space exists, it is uniquely determined up to isomorphism by its local invariants. We summarize this result as follows.

Theorem 3.1. *Let k be a number field and let K be a quadratic field extension of k. Fix a dimension $n \geq 1$. For each place v of k let (V_v, ϕ_v) be a local Hermitian space over K_v of dimension n. Assume that the Hermitian determinant d_v is a norm from K_v^* for almost all places v, or equivalently that the Hilbert symbol $(d_v, \mathrm{disc}_{K/k})_v = +1$ for almost all v.*

Then a necessary and sufficient condition for the existence of a global Hermitian space (V, ϕ) over K with these localizations is given by:

$$\prod_v (d_v, \mathrm{disc}_{K/k})_v = +1$$

If a global Hermitian space (V, ϕ) exists with these localizations, it is unique up to isomorphism.

4 Incoherent Definite Orthogonal and Hermitian Data

We henceforth assume that k is a totally real number field, and in the Hermitian case, that K is a totally complex quadratic field extension of k. Let k_+^* denote the elements in k^* which are positive at all real places.

Incoherent definite orthogonal data consists of a dimension $n \geq 3$, a determinant d in k_+^*/k^{*2}, and for each place v of k a local orthogonal space V_v of dimension n and determinant d over k_v. We assume that V_v is positive definite for all real places v, that the Hasse-Witt invariant $\epsilon_v = \epsilon_v(V_v)$ is equal to $+1$ for almost all v, and that $\prod_v \epsilon_v = -1$.

Incoherent definite Hermitian data consists of a dimension $n \geq 1$ and for each place v of k a local Hermitian space V_v over $K_v = K \otimes k_v$ of dimension n. We assume that V_v is positive definite for all real places v, that the Hermitian determinant d_v of V_v is a norm from K_v^* for almost all v, and that the number of places where d_v is not a norm is odd. Equivalently, we assume that the Hilbert symbol $(d_v, \mathrm{disc}_{K/k})_v$ is equal to $+1$ for almost all v and that $\prod_v (d_v, \mathrm{disc}_{K/k})_v = -1$.

We will use the notation $\{V_v\}$ for incoherent definite data, either orthogonal or Hermitian.

Given incoherent definite orthogonal data over k, we construct for each place v of k a global orthogonal space $V(v)$ of dimension n and determinant d over k which we call the neighbor at v. This space has localizations V_u for all places $u \neq v$. At the place v we insist that

$$\epsilon_v(V(v)) = -\epsilon_v(V_v)$$

This determines the localization of $V(v)$ at a finite place v. At a real place, we insist further that the signature of $V(v)$ at v is equal to $(n-2, 2)$. Since the product of local epsilon factors is now equal to $+1$, there is a global space $V(v)$ with these localizations, which is unique up to isomorphism. Finally, for each place v of k we let $G(v) = \mathrm{SO}(V(v))$ be the special orthogonal group of the neighboring orthogonal space.

Given incoherent definite Hermitian data over k relative to the quadratic extension K, we construct for each place v of k which is not split in K a global Hermitian space $V(v)$ of dimension n over K which we call the neighbor at v. This space has localizations V_u for all places $u \neq v$. At the place v we insist that

$$(d(V(v)), \mathrm{disc}_{K/k})_v = -(d(V_v), \mathrm{disc}_{K/k})_v$$

This determines the localization of $V(v)$ at a finite place v. At a real place, we insist further that the signature of $V(v)$ at the place v is equal to $(n-1,1)$. Since the product of local Hilbert symbols is now equal to $+1$, there is a global Hermitian space $V(v)$ with these localizations, which is unique up to isomorphism. Finally, for each place v of k which is not split in K, we let $G(v) = U(V(v))$ be the unitary group of the neighboring Hermitian space.

5 Incoherent Data in Codimension One

Let k be a totally real field and let $\{V_v\}$ be incoherent definite orthogonal data of dimension $n \geq 4$ and determinant d, and let a be any class in k_+^*/k^{*2}. Then a is represented by all of the local quadratic spaces V_v. Let U_v denote the orthogonal complement of the corresponding one dimensional subspace of V_v, so that

$$V_v = U_v \oplus \langle a \rangle.$$

Then $\{U_v\}$ is incoherent definite orthogonal data of dimension $n-1$ and determinant ad. Clearly the subspace U_v is positive definite at all real places v. But we also have the formula for the Hasse-Witt invariants

$$\epsilon_v(U_v) = \epsilon_v(V_v) \cdot (da, a)_v.$$

with $(da, a)_v = +1$ for almost all v and $\prod_v (da, a)v = +1$. Hence $\epsilon_v(U_v) = +1$ for almost all v and $\prod_v \epsilon_v(U_v) = -1$. It follows from a calculation of Hasse-Witt invariants that the neighboring spaces $U(v)$ all have the property that

$$U(v) \oplus \langle a \rangle \simeq V(v).$$

Conversely, if we are given incoherent definite data $\{U_v \subset V_v\}$ of codimension 1, with determinants $d(U)$ and $d(V)$, let $a = d(V)/d(U)$ in k_+^*/k^{*2}. Then $U_v^\perp = \langle a \rangle$ is coherent definite orthogonal data of dimension 1. Hence we have shown that

Theorem 5.1. *For incoherent definite orthogonal data $V = \{V_v\}$ of dimension $n \geq 4$, the classes of incoherent definite orthogonal data $U = \{U_v\}$ of codimension one in V correspond bijectively to classes a in k_+^*/k^{*2}.*

There is a similar result when the dimension of V is 3, but one has to restrict to classes a in k_+^*/k^{*2} which are locally representable at the finite set of places v where the space V_v is anisotropic. Also, incoherent definite orthogonal data $\{U_v\}$ of dimension 2 and determinant d is better viewed as incoherent definite Hermitian data of dimension 1 over the imaginary quadratic extension $K = k(\sqrt{-d})$ of k. In the Hermitian case, we have the following.

Theorem 5.2. *Let K be an imaginary quadratic extension of the totally real field k. For incoherent definite Hermitian data $\{V_v\}$ of dimension $n \geq 2$ over K, the classes of incoherent definite Hermitian data $\{U_v\}$ of codimension one in V correspond bijectively to classes a in $k_+^*/N(K^*)$.*

Indeed, the subspaces $U_v^\perp = \langle a_v \rangle$ give coherent definite Hermitian data of dimension 1. Since $\prod_v (a_v, \operatorname{disc} K/k) = +1$, the class of (a_v) is trivial in the quotient group $\mathbb{A}_k^*/N(\mathbb{A}_K^*)k^*$, which has order 2. Hence the local classes a_v can be written as a product of a global class a and a norm from K_v^*, and a is well-defined in the quotient group $k_+^*/N(K^*)$.

6 Orthogonal Lattices

In this section, k is a local non-Archimedian field, with ring of integers A, uniformizing parameter π, and finite residue field $\mathbb{F} = A/\pi A$ of order q. We will assume further that the characteristic of $\mathbb{F}$ is odd.

Let $V = (V, q)$ be a non-degenerate orthogonal space over k of dimension n, with associated bilinear form $\langle v, w \rangle = q(v + w) - q(v) - q(w)$. A lattice $L \subset V$ is by definition a free A-module of rank n which spans V over k, such that the bilinear form $\langle , \rangle$ takes integral values on L. The dual lattice $L^\vee$ is the free A-module of vectors in V with integral pairing against all elements of L. Then L is a submodule of $L^\vee$ and the quotient $L^\vee/L$ has finite length, by the non-vanishing of the determinant d of V.

We say that the lattice L is **non-degenerate** (some say self-dual) if $L = L^\vee$. A necessary condition for the existence of a non-degenerate lattice in V is that the determinant of V has even valuation, so d lies in the subgroup A^*k^{*2}/k^{*2} of k^*/k^{*2}. In this case, we say that the determinant of V is equivalent to a unit. This condition is clearly sufficient for the existence of a non-degenerate lattice in V when the dimension $n = 1$, but is not sufficient when $n \geq 2$.

To see this in the two dimensional case, let D be a unit in A^* which is not a square, so $K = k(\sqrt{D})$ is the unramified quadratic extension of k. Let $V = K$ with quadratic form $q(v) = N(v)$. Then $q(x, y) = x^2 - Dy^2$, the determinant of V is equivalent to the unit $-D$ in k^*/k^{*2} and $\epsilon(V) = +1$. In this case V contains the non-degenerate lattice $A_K = A + \sqrt{D}A$ given by the ring of integers in K. On the other hand, if $V' = K$ with quadratic form $q(v) = \pi.N(v)$, then the determinant of V' is still equivalent to the unit $-D$ but now $\epsilon(V') = -1$. In this case, there are no non-degenerate lattices in V', but $L = A_K$ is a lattice with dual lattice $L^\vee = (\pi)^{-1}A_K$. The quadratic space $L^\vee/L$ has rank 2 over the residue field $\mathbb{F}$ and is isomorphic to the quadratic extension field $A_K/\pi A_K$ of order q^2 with its norm form. More generally, we have the following.

Theorem 6.1. *Let V be an orthogonal space over k of dimension n whose determinant d is equivalent to a unit.*

If $\epsilon(V) = +1$ then V contains a non-degenerate lattice L. All such lattices are conjugate under the action of the group $\mathrm{SO}(V)$.

If $\epsilon(V) = -1$ then there are no non-degenerate lattices in V. However, V contains lattices L with the property that $L^\vee/L$ is a two dimensional orthogonal space over the residue field $\mathbb{F}$, isomorphic to the quadratic extension field of $\mathbb{F}$ with its norm form. All such lattices are conjugate under the action of the group $\mathrm{SO}(V)$.

This is well-known—the lattices L in the theorem are called maximal lattices by Eichler, as they are maximal with the property that the bilinear form takes values in A. We will sketch a proof which involves embedding two dimensional spaces W in V, which will be useful later in this paper. The theorem is easily proved (using the construction above) when the dimension n of V is ≤ 2, so we will assume that $n \geq 3$.

Theorem 6.2. *Assume that the dimension of V satisfies $n \geq 3$ and that the determinant d of V is equivalent to a unit. Let D be a unit which is not a square, let K be the unramified quadratic extension $k(\sqrt{D})$ and let W be an orthogonal space of dimension 2 and determinant $-D$, which satisfies*

$$\epsilon(W) = \epsilon(V).$$

Then W embeds isometrically as a subspace of V, and all embeddings are conjugate under $\mathrm{SO}(V)$. *The orthogonal complement U of W in V has invariants*

$$\dim(U) = n - 2 \quad d(U) \equiv -dD \quad \epsilon(U) = +1$$

Note that the isomorphism class of the orthogonal complement U depends only on the dimension and discriminant of V. This is similar to what happens when $k = \mathbb{R}$, for the signatures we specified. The space V has signature $(n, 0)$ or $(n-2, 2)$, so has $d \equiv 1$. If we take $K = \mathbb{C}$, so $D \equiv -1$, and insist that the two dimensional space W has determinant $-D$ and the same Hasse-Witt invariant as V, then W has signature $(2, 0)$ in the first case and signature $(0, 2)$ in the second case. Hence W embeds in V, and in both cases the orthogonal complement U has signature $(n - 2, 0)$.

To prove Theorem 6.2, we will use some standard results [S73, Ch IV] on which elements of k^* are represented by quadratic forms. We begin with the case when the dimension of V is equal to 3. If a two dimensional space W of determinant $-D$ embeds into V, its orthogonal complement U would be one dimensional, of determinant $-dD$. But the quadratic form q on V represents the unit $-dD$, as it represents all elements of k^* except perhaps $-d$, and the discriminant D of K is not equivalent to 1. Taking the orthogonal complement of this line spanned by a vector v with $q(v) = -dD$ gives a two dimensional subspace W of V which has determinant $-D$. The quadratic form on W is given by $ax^2 - aDy^2$ and on V is given by $ax^2 - aDy^2 - Ddz^2$. Since $-D$ and $-dD$ are both units and the residue characteristic is odd, $(-D, -dD) = +1$ and the Hasse-Witt invariants of both W and V are both equal to $(a, -D)$, This proves our claim on the embedding when the dimension of V is three. The orthogonal group of V acts transitively on these two dimensional subspaces, by Witt's extension theorem. Since the stabilizer contains a reflection, the special orthogonal group also acts transitively.

One can also prove the special case when the dimension of V is 3 and the determinant $d \equiv 1$ using the theory of local quaternion algebras. Indeed, in this case the space V is given by the elements of trace zero in a quaternion algebra B over k, and the quadratic form is the norm form. The unramified quadratic extension K of k embeds as a subfield of B, and its orthogonal complement W is the desired two dimensional subspace of V. We obtain an orientation of the plane W in V by choosing one of the two vectors on the line $W^{\perp}$ with $q(v) = -D$, which also gives W a $K = k.1 + k.v$ linear structure.

Next consider the case when the dimension of V is equal to four. The quadratic space V represents every non-zero class a and the three dimensional orthogonal complement of the line spanned by a vector v with $q(v) = a$ has determinant ad. To obtain an embedding of W, we need to check that this three dimensional space represents $-aD$. But any three dimensional quadratic space of determinant ad represents all classes not equivalent to $-ad$, Hence we have an embedding of both orthogonal spaces of dimension 2 and determinant $-D$, except in the case when the determinant of V is equivalent to D. In that case, we can construct a four dimensional space V of determinant D from a two dimensional space W of determinant $-D$ by taking the direct sum of W with a hyperbolic plane. The quadratic form on V has the form $ax^2 - aDy^2 + z^2 - t^2$, so the Hasse-Witt invariant of both W and V are both equal to $(a, -D)$.

When the dimension of V is at least five, then the quadratic form on V represents any non-zero class a and the orthogonal complement of the line spanned by a vector v with $q(v) = a$ represents $-aD$. Hence we may embed both quadratic spaces W of dimension 2 and determinant $-D$ in V. The special orthogonal group of V acts transitively on the subspaces W with $\epsilon(W) = \epsilon(V)$, by Witt's extension theorem. Hence the orthogonal complement U of W is well-defined up to isomorphism over k. Since $V = W \oplus U$ we find

$$d(U) \equiv -dD \quad \epsilon(U) = +1$$

as claimed.

We can now prove Theorem 6.1 by an induction on the dimension of V. If $\epsilon(V) = +1$, we embed the two dimensional subspace W with determinant $-D$ and $\epsilon(W) = +1$. We have seen that there is a non-degenerate lattice in W, and by induction there is a non-degenerate lattice in its orthogonal complement U. The direct sum of these two lattices gives a non-degenerate lattice in V. If $\epsilon(V) = -1$ we embed the two dimensional subspace W with determinant $-D$ and $\epsilon(W) = -1$. By induction there is a non-degenerate lattice in the orthogonal complement U, and we have seen that there is a lattice L in W with $L^{\vee}/L$ isomorphic to the quadratic extension of $\mathbb{F}$ with its norm form. The direct sum of these lattices gives the desired lattice in V.

The fact that the group SO(V) acts transitively on the lattices described in Theorem 5.1 is proved in [Sh61] and [GHY01] They are the maximal lattices in V, where the bilinear form is integral. In the case when $\epsilon = +1$ and the lattice L is non-degenerate, the stabilizer of L is a hyperspecial maximal compact subgroup of $SO(V)$. When $n = 2m + 1$ is odd, its reductive quotient is the split group SO_{2m+1} over $\mathbb{F}$. When $n = 2m$ is even and $(-1)^m d \equiv 1$ its reductive quotient is the split group SO_{2m} over $\mathbb{F}$. When $n = 2m$ is even and $(-1)^m d \equiv D$ its reductive quotient is the quasi-split group SO^{ϵ}_{2m} over $\mathbb{F}$.

In the case when $\epsilon = -1$, let $T = SO^{\epsilon}_2$ be the one dimensional torus over $\mathbb{F}$ which is split by the quadratic extension of $\mathbb{F}$. Then the stabilizer of L is a parahoric subgroup of $SO(V)$ whose reduction is as follows. When $n = 2m + 1$ is odd, the reductive quotient is the group $T \times SO_{2m-1}$ over $\mathbb{F}$. When $n = 2m$ is even and $(-1)^m d \equiv 1$, the reductive quotient is the group $T \times SO^{\epsilon}_{2m-2}$ over $\mathbb{F}$. When $n = 2m$ is even and $(-1)^m d \equiv D$, the reductive quotient is the group $T \times SO_{2m-2}$ over $\mathbb{F}$. When $n \geq 3$ the parahoric subgroup $SO(L)$ is not hyperspecial. It is maximal except when $2m = 4$ and $d \equiv D$, when it is an Iwahori subgoup of $SO(V) \cong (\mathrm{Res}_{K/k} SL_2)/\mu_2$. See [GHY01] for details.

7 Hermitian Lattices

In this section, we again assume that k is a local non-Archimedian field, with ring of integers A, uniformizing parameter π, and finite residue field $\mathbb{F} = A/\pi A$ of order q. Let $K = k(\sqrt{D})$ be the unramified quadratic extension of k, where D is a unit in A^* which is not a square, and let τ be the non-trivial involution of K which fixes k. We make no assumption on the characteristic of $\mathbb{F}$.

Let V be a non-degenerate Hermitian space over K of dimension n, with associated Hermitian symmetric form $\phi(v, w)$. A lattice $L \subset V$ is by definition a free A_K-module of rank n which spans V over K, such that the Hermitian form ϕ takes integral values on L. The dual lattice $L^\vee$ is the free A_K-module of vectors in V with integral pairing against all elements of L. Then L is an A_K-submodule of $L^\vee$ and the quotient $L^\vee/L$ has finite length, by the non-vanishing of the Hermitian determinant d of V. We say that the lattice L is **non-degenerate** (some say self-dual) if $L = L^\vee$. A necessary condition for the existence of a non-degenerate lattice in V is that the determinant has even valuation, so d lies in the subgroup $N(K^*)$ of index 2 of k^*. In this case, we say that the determinant of V is a norm.

When the dimension of V over K is equal to 1 and the determinant is not a norm, then the Hermitian space is isomorphic to K with form $\phi(x, y) = \pi x y^\tau$. In this case, there are no non-degenerate lattices in V. However the lattice $L = A_K$ has dual lattice $L^\vee = \pi^{-1} A_K$, and the quotient module $L^\vee/L$ is isomorphic to the quadratic extension field of $\mathbb{F}$ with its standard Hermitian form.

Theorem 7.1. *Let V be a non-degenerate Hermitian space of dimension $n \geq 1$ over the unramified quadratic extension $K = k(\sqrt{D})$ of k.*

If the Hermitian determinant d of V is a norm, then V contains a non-degenerate lattice L. All such lattices are conjugate under the action of the unitary group $U(V)$.

If the Hermitian determinant d of V is not a norm, then there are no non-degenerate lattices in V. However, V contains lattices L with the property that $L^\vee/L$ is a one dimensional Hermitian space over the quadratic extension of the residue field $\mathbb{F}$. All such lattices are conjugate under the action of the unitary group $U(V)$.

The proof is similar to the orthogonal case, but simpler as it involves the embedding of a suitable one dimensional subspace W in V. Let W be a Hermitian space of dimension 1 with the property that

$$d(W) \equiv d(V)$$

in $k^*/N(K^*)$. It is then a simple matter to show that W embeds isometrically as a subspace of V, and that all such subspaces are conjugate under the action of $U(V)$. The orthogonal complement of W in V has Hermitian determinant which is a norm, so by induction contains non-degenerate lattice. The direct sum of that lattice with the lattice constructed above (in dimension 1) gives the desired lattice in V.

The fact that the unitary group of V acts transitively on lattices of this type follows from the theory of Iwahori and Bruhat-Tits on parahoric subgroups. Indeed, when the determinant of V is a norm, the stabilizer of a non-degenerate lattice L is a hyperspecial maximal compact subgroup of the locally compact group $U(V)$ whose reductive quotient is the group U_n over $\mathbb{F}$. When the determinant of V is not a norm, the stabilizer is again a maximal compact subgroup, but is not hyperspecial once $n \geq 2$. Its reductive quotient is the group $U_1 \times U_{n-1} = T \times U_{n-1}$ over $\mathbb{F}$.

8 The Homogeneous Spaces X, Y, and Z

We continue with the notation of the previous two sections: k is a non-Archimedian local field with ring of integers A and finite residue field $\mathbb{F} = A/\pi A$. In the orthogonal case, we assume that the characteristic of $\mathbb{F}$ is odd. Let $K = k(\sqrt{D})$ be the unramified quadratic extension of k, and let A_K be the ring of integers of K.

Our aim is to generalize the following construction, where $k = \mathbb{R}$ and $K = \mathbb{C}$. If V is a real orthogonal space of dimension $n \geq 3$ and signature $(n-2, 2)$, the set $X = \mathscr{D}^\pm$ of oriented negative definite two planes W in V is a complex manifold of dimension $n-2$ which admits a transitive action of the group $G = \mathrm{SO}(V)$. Indeed, an oriented negative definite two plane W determines a homomorphism $h : T \to \mathrm{SO}(V)$ over $\mathbb{R}$, where T is the one dimensional torus $(\mathrm{Res}_{\mathbb{C}/\mathbb{R}}\, \mathbb{G}_m)_{N=1}$, and the centralizer of h is the compact subgroup

$$H = \mathrm{SO}(W) \times \mathrm{SO}(W^\perp).$$

The conjugacy class of h is isomorphic to the quotient space $X = G/H$. This quotient is isomorphic to two copies of the symmetric space of the connected real group $\mathrm{SO}(n-2,2)^+$, as the subgroup H has index two in its normalizer $S(O(W)\times O(W^\perp))$ (which is a maximal compact subgroup of G). Hence every coset for the normalizer breaks into two cosets for the subgroup H. The space X has a Hermitian symmetric structure, as the tangent space to the identity coset of G/H is the orthogonal representation $\mathrm{Hom}(W,W^\perp)\cong W\otimes W^\perp$ of H, which has a complex Hermitian structure from the chosen orientation of W. The involution of X which reverses the orientation of W switches the two components and is anti-holomorphic. Finally, the complex manifold X embeds as an open G-orbit (with two components) in the complex quadric $Q(\mathbb{C})$ defined by the vanishing of the quadratic form on the projective space of $V\otimes\mathbb{C}$.

In the non-Archimedean case, we start with an orthogonal space V over k of dimension $n\geq 3$, with unit determinant d and Hasse-Witt invariant $\epsilon=-1$. Let X be the set of oriented two planes W in V with determinant $-D$ and satisfying $\epsilon(W)=-1$. We have seen that such oriented two planes exist, and are permuted transitively by $G=\mathrm{SO}(V)$. An oriented two plane W determines a homomorphism $h:T\to\mathrm{SO}(V)$ over k, where T is the one dimensional torus $(\mathrm{Res}_{K/k}\mathbb{G}_m)_{N=1}$, and the centralizer of h is the closed subgroup

$$H=\mathrm{SO}(W)\times\mathrm{SO}(W^\perp).$$

The conjugacy class X of such homomorphisms h, which can be identified with the coset space G/H once an oriented plane W has been chosen, has the structure of a homogeneous analytic manifold over k. Indeed, if G is a k-analytic group and H is a k-analytic closed subgroup, the coset space G/H has the structure of a k-analytic manifold, satisfying the expected universal properties, with tangent space $\mathrm{Lie}(G)/\mathrm{Lie}(H)$ at the identity [S06, Ch III,§12]. In this case, the tangent space of X at the oriented plane W is the k-vector space $\mathrm{Hom}(W,W^\perp)=W^\vee\otimes W^\perp\cong W\otimes W^\perp$, which affords an orthogonal representation of the stabilizer. Since the subspace W has K-linear structure given by the orientation, the tangent plane at each point is a K-vector space. Hence X is a K-analytic manifold of dimension $n-2$.

As in the real case, there is a fixed point free involution of X, which acts K-anti-linearly on the tangent spaces and commutes with the action of $\mathrm{SO}(V)$. This is given by the action of normalizer of h in G, which contains the centralizer with index 2 and maps each homomorphism $h:T\to\mathrm{SO}(V)$ to its composition with the inverse homomorphism on T. Finally, the analytic manifold X embeds as an open G-orbit in the K-points of the smooth quadric Q of dimension $n-2$ defined by the vanishing of the quadratic form q on the projective space of $V\otimes K$. The embedding sends the oriented two plane $W=K.e$ in V spanned by the orthogonal vectors e and $f=\sqrt{D}.e$ to the line in $V\otimes K$ spanned by the vector $e\otimes 1+f\otimes(\sqrt{D})^{-1}$. Since $q(f)=N(\sqrt{D})q(e)=-Dq(e)$ this line is isotropic, and defines a point on $Q(K)$.

Unlike the real case, the subgroup H need not be compact. It is compact only when the factor $\mathrm{SO}(W^\perp)$ is compact, which can only occur when the dimension of V is three or four. To obtain a compact subgroup of H, we need to choose a non-degenerate lattice L in the subspace $W^{\perp}$ (which has unit determinant and $\epsilon(W^{\perp}) = +1$). Writing $W = K.e$ with $q(e) = \pi$ we obtain a canonical lattice $A_K.e$ in W and the product

$$J = \mathrm{SO}(A_K.e) \times \mathrm{SO}(L)$$

is a hyperspecial maximal compact subgroup of H. The choice of L also gives us a maximal lattice $\Lambda = A_K.e + L$ in V, and the orientation of the two plane W gives an orientation of the two dimensional orthogonal space $\Lambda^{\vee}/\Lambda$ over the residue field $\mathbb{F} = A/\pi A$.

We define Y as the set of pairs (W, L), where W is an oriented two plane with determinant $-D$ and Hasse-Witt invariant $\epsilon(W) = -1$ and L is a non-degenerate lattice in $W^{\perp}$. The group $G = \mathrm{SO}(V)$ acts transitively on the set Y, which can be identified with the coset space G/J, once a fixed pair (W, L) with stabilizer J has been chosen. There is s a fixed point free involution of Y given by the normalizer of J in G, which contains J with index 2. This changes the orientation of the plane W and preserves the lattice L.

We define Z as the set of maximal lattices Λ in V, together with an orientation of the orthogonal space $\Lambda^{\vee}/\Lambda$ over $A/\pi A$. The group $G = \mathrm{SO}(V)$ acts transitively on the set Z, and the stabilizer of a fixed maximal lattice with orientation is a parahoric subgroup N (which is maximal if the dimension of V is at least 5). The reductive quotient of N is a group of the type $\mathrm{SO}(2) \times \mathrm{SO}(n-2)$ [GHY01], and is isomorphic to the reductive quotient of the hyperspecial maximal compact subgroup J of H. We can identify the set Z with the homogeneous space G/N once a fixed oriented maximal lattice Λ with stabilizer N has been chosen. There is a fixed point free involution of Z given by the normalizer of N in G, which contains N with index 2 and is a maximal compact subgroup once the dimension of V is at least 5. This preserves the lattice Λ and changes the orientation of $\Lambda^{\vee}/\Lambda$.

The map that takes the pair (W, L) to the oriented two plane W is a surjective G-equivariant map

$$Y \to X \hookrightarrow Q(K)$$

which is usually of infinite degree. The map that takes the pair (W, L) to the oriented maximal lattice $\Lambda = A_K.e + L$ is a G-equivariant map

$$f : Y \to Z.$$

The map f is a morphism in the category of G sets which commutes with the involutions we have defined on Y and Z. Since G acts transitively on Z, it is surjective. We will see in the next section that the fibers of f, which are all isomorphic to the quotient space N/J, have a simple A_K analytic structure.

The parahoric subgroups of the simply-connected cover $\mathrm{Spin}(V)$ of $\mathrm{SO}(V)$ can be read off the affine diagram, together with its Frobenius automorphism. In the case when the dimension $n = 2m + 1 \geq 7$ is odd, the determinant is a unit, and $\epsilon(V) = -1$, we have the diagram

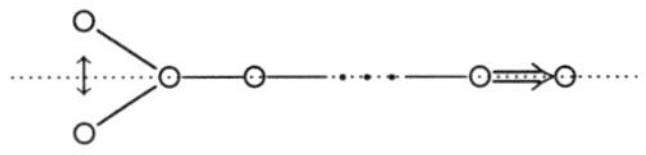

The parahoric N corresponds to the two vertices interchanged by Frobenius at the left end of the diagram. In the case when the dimension $n = 2m \geq 8$ is even, the determinant is $\equiv (-1)^m$ and $\epsilon(V) = -1$ we have the diagram

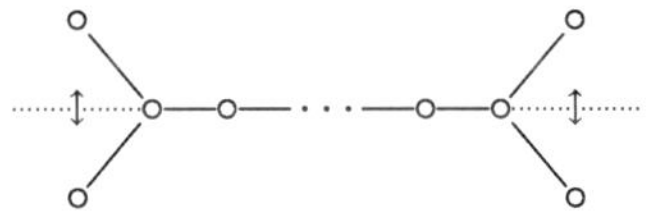

and the parahoric N corresponds to two vertices interchanged by Frobenius at the left end of the diagram. In the case when the dimension $n = 2m$ is even and the determinant is $\equiv (-1)^m D$ we have the diagram

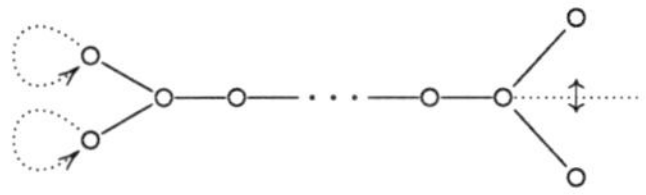

When $\epsilon(V) = -1$ the parahoric N corresponds to the two vertices interchanged by Frobenius at the right end of the diagram.

In all cases, when the dimension of V is at least 5, the reductive quotient of N has center isomorphic to the non-split torus T of dimension 1 over $\mathbb{F}$. The finite group T lifts to N, this lifting is well-defined up to conjugacy, and the centralizer of any lifting is conjugate to the subgroup J.

A similar theory works in the Hermitian case. First consider the situation over the real numbers. If V is a complex Hermitian space of dimension n and signature $(n - 1, 1)$, then the set X of negative lines W in V is a complex manifold of dimension $n - 1$ which admits a transitive action of the group $G = U(V)$. Indeed, such a line determines a homomorphism $h : T \to U(V)$ over $\mathbb{R}$, and the centralizer of h is the compact subgroup

$$H = U(W) \times U(W^{\perp}).$$

The quotient $X \cong G/H$ is isomorphic to the symmetric space of $U(n - 1, 1)$. It has a Hermitian symmetric structure, as the tangent space to the identity is the representation $W \otimes W^{\perp}$ of H, of dimension $n - 1$ over $\mathbb{C}$. Finally, the complex manifold embeds as an open G orbit in the projective space of V over $\mathbb{C}$.

In the non-Archimedian case, we start with a Hermitian space V over K of dimension $n \geq 1$ whose Hermitian determinant is not a norm. Let X be the set of lines W in V with Hermitian determinant equal to the Hermitian determinant of V. We have seen that such lines exist and are permuted transitively by $G = U(V)$. Such a line determines a homomorphism $h : T \to U(V)$, where T is the one dimensional torus $(\mathrm{Res}_{K/k}\, \mathbb{G}_m)_{N=1}$, and the centralizer of h is the closed subgroup

$$H = U(W) \times U(W^{\perp}).$$

The conjugacy class X of such homomorphisms h is isomorphic to the quotient space G/H and X embeds as an open G orbit in the K valued points of the projective space of V.

The subgroup H can only be compact when the dimension of V is one or two. To obtain a suitable compact subgroup of H, we need to choose a non-degenerate lattice L in the subspace $W^{\perp}$. Writing $W = K.e$ with $\phi(e, e) = \pi$, we obtain a canonical lattice $A_K.e$ in $W = K.e$. The product

$$J = U(A_K.e) \times U(L)$$

is a hyperspecial maximal compact subgroup of H and the homogeneous space $Y = G/J$ indexes the lines W in V with Hermitian determinant equal to the determinant of V together with a non-degenerate lattice L in $W^{\perp}$. The map

$$Y \to X \hookrightarrow \mathbb{P}(V \otimes K)$$

which assigns to a pair (W, L) the line W is usually of infinite degree.

Let Z be the set of maximal lattices Λ in V. The map

$$f : Y \to Z$$

associates to the line W with the correct Hermitian determinant and the non-degenerate lattice L in $W^{\perp}$ the maximal lattice $\Lambda = A_K.e + L$. The group $G = U(V)$ acts transitively on the space Z of maximal lattices and the map f is G-equivariant, hence surjective. Let $N = U(\Lambda)$ be the stabilizer of the maximal lattice Λ. This is a parahoric subgroup of G whose reductive quotient is isomorphic to $U(1) \times U(n-1)$ over the residue field [GHY01] and is isomorphic to the reductive quotient of the compact subgroup J. We can identify the space Z with the quotient space G/N, once a maximal lattice has been chosen. In the next section we will see that the fibers of the map f, which are all isomorphic to the quotient space N/J, have a simple A_K analytic structure.

The parahoric subgroups of $U(V)$ can be read off the affine diagram of $SU(V)$ with its automorphism given by Frobenius. When the dimension $n = 2m \geq 4$ of V is even and the determinant is not a norm, we have the diagram

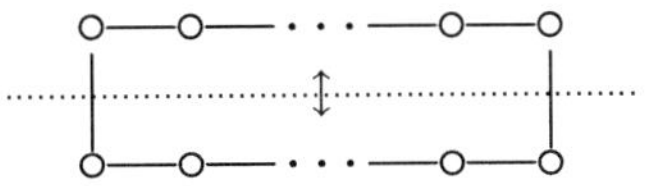

and the parahoric N corresponds to the two vertices interchanged by Frobenius at either end of the diagram. When the dimension $n = 2m + 1 \geq 5$ of V is odd, we have the diagram

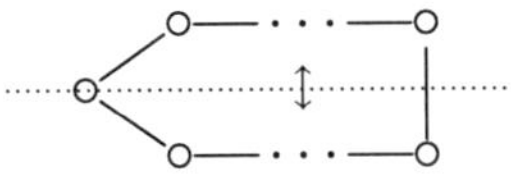

When the determinant of V is not a norm, the parahoric N corresponds to the two vertices interchanged by Frobenius at the right of the diagram.

9 The Fibers of the Map $f : Y \to Z$

In this section, we continue to assume that k is a non-Archimedian local field. If V is an orthogonal space over k of dimension $n \geq 3$, with unit determinant and Hasse-Witt invariant $\epsilon = -1$, we have defined a parahoric subgroup N up to conjugacy in $G = \mathrm{SO}(V)$, as well as an unramified compact subgroup J up to conjugacy in N. Similarly to a Hermitian space V over the unramified quadratic extension k of K, whose Hermitian determinant is not a norm, we have defined a parahoric subgroup N up to conjugacy in $G = U(V)$, as well as an unramified compact subgroup J up to conjugacy in N. The groups J and N have the same reductive quotients. We let $Y = G/J$ and $Z = G/N$ be the associated homogeneous spaces and let $f : Y \to Z$ be the covering map. Fix a point $z \in Z$. In this section we will describe an A_K-analytic structure on the inverse image $f^{-1}(z) \cong N/J$ in Y.

The fact that there is a natural analytic structure on the fibre, whereas Y and Z only have the structure of sets, follows from the fact that the compact groups N and J are both the A valued points of smooth parahoric group schemes $\mathbf{N}$ and $\mathbf{J}$ over A. Since A is a complete discrete valuation ring, the quotient $\mathbf{N}/\mathbf{J}$ has the structure of a formal scheme over A. Since $\mathbf{J}$ is a reductive group scheme over A, $H^1(A, \mathbf{J}) = 0$. Hence $(\mathbf{N}/\mathbf{J})(A) = \mathbf{N}(A)/\mathbf{J}(A) = N/J$. We will determine the formal scheme and analytic space structure of the fiber explicity, and see that it has an A_K structure.

In the orthogonal case, fix a point $y \in Y$ with $f(y) = z$. This expresses the lattice Λ as a direct sum $\Lambda = A_K.e + L$. To parametrize the other points in $f^{-1}(z)$, we need to determine all other such decompositions of the lattice Λ, or equivalently the rank two A-submodules $M = Av + Aw$ of Λ which are isometric to $A_K.e$. The key observation is that $\pi\Lambda^\vee/\pi\Lambda$ is the radical of the quadratic space $\Lambda/\pi\Lambda$ over $\mathbb{F}$, and that a rank 2 submodule M isometric to $A_K.e$ must have this reduction modulo π. Consequently M has a basis of the form

$$v = \alpha.e + \lambda$$

$$w = \beta.e + \mu$$

where α and β are units of A_K which are independent in the $A/\pi A$ vector space $A_K/\pi A_K$ and λ and μ are elements of the sublattice πL. We then find inner products

$$\langle v, v\rangle=2N(\alpha)\pi+O(\pi^2) \quad \langle w, w\rangle=2N(\beta)\pi+O(\pi^2) \quad \langle v, w\rangle=Tr(\alpha.\overline{\beta})\pi+O(\pi^2)$$

and from this it follows that $M = Av + Aw$ is a lattice (oriented, and isometric to $A_K.e$) in a two dimensional orthogonal space over k with determinant $-D$ and Hasse-Witt invariant -1. Taking the orthogonal complement $M^\perp$ of M in Λ, we obtain another point y^* in $f^{-1}(z)$. We still need to check that the lattice $M^\perp$ is non-degenerate. This argument is a bit simpler in the Hermitian case, and we postpone it for the moment.

We can parametrize the data determining the decomposition $\Lambda = M + M^\perp$ by the element

$$\alpha^{-1} \otimes \lambda + \beta^{-1} \otimes \mu \in A_K \otimes \pi L$$

Picking a basis of L over A, this is just the set of points $(\pi A_K)^{n-2}$ in the formal scheme $Spf\, A_K[[z_1, z_2, \dots, z_{n-2}]]$. So we have established the following.

Theorem 9.1. *Assume that V is an orthogonal space over k of dimension $n \geq 3$ over k with unit determinant and Hasse-Witt invariant -1. Let Y be the homogeneous space parametrizing pairs (W, L) consisting of an oriented two plane W in V with determinant $-D$ and Hasse-Witt invariant -1, together with a non-degenerate lattice L in $W^\perp$. Let Z be the homogeneous space of maximal lattices Λ in V together with an orientation of the lattice $\Lambda^\vee/\Lambda$ over the residue field of k.*

Let $f : Y \to Z$ be the $\mathrm{SO}(V)$-equivariant map taking the pair $y = (W, L)$ to the non-degenerate lattice $z = \Lambda = A_K.e + L$ with orientation coming from the orientation of W. Then the inverse image $f^{-1}(z)$ in Y can be parametrized by the points of $A_K \otimes \pi L \cong (\pi A_K)^{n-2}$.

We have a filtration of the parahoric N by normal subgroups N_m of finite index which act trivially on the quotient lattice $\Lambda/\pi^m\Lambda$. This gives a filtration of the quotient set

$$N/J > N_1J/J > N_2J/J > \dots$$

If we identify $f^{-1}(z) = N/J$ with the points of $(\pi_K)^{n-2}$, then the subset $N_m J/J$ corresponds to the points in $(\pi_K^{m+1})^{n-2}$. Indeed, the quotient groups $N_m J/N_{m+1} J$ are isomorphic to $(A_K/\pi A_K)^{n-2}$ for all m.

In the simplest case, when $n = 3$ and $d \equiv 1$ the space V is given by the elements of trace zero in the quaternion division algebra B over k. Let R be the unique maximal order in B. Then A_K embeds as a subring of R, and we have the orthogonal decomposition

$$R = A_K + A_K.\Pi$$

where Π is a uniformizing parameter of R, which normalizes A_K and whose square is a uniformizing element of A. The special orthogonal group $\mathrm{SO}(V) = B^*/k^*$ is compact, and contains the parahoric subgroup $N = R^*/A^*$ with index 2. Hence the space Z consists of two points. The subgroup $H = K^*/k^* = A_K^*/A^*$ is compact in this special case, so $H = J$ and $X = Y = B^*/A_K^*$ is isomorphic to two copies of the maximal ideal πA_K.

In the Hermitian case, a point $y \in Y$ corresponds to a line W in the Hermitian space V and a non-degenerate lattice L in the orthogonal complement. The direct sum $\Lambda = A_K.e + L$ is a maximal Hermitian lattice in V with $\Lambda^\vee/\Lambda = \pi^{-1}A_K.e/A_K.e$, and the point $z = f(y)$ in Z remembers only the lattice Λ, not the decomposition. To find all points in $f^{-1}(z)$, we need o determine the rank 1 sublattices $M = A_K.v$ in Λ which are isometric to $A_K.e$. Since $M/\pi M = A_K.e/\pi A_K.e$ is the radical of $\Lambda/\pi\Lambda$, the vector v of Λ must have the form

$$v = \alpha.e + \lambda$$

where α is a unit in A_K and λ lies the the sublattice $\pi.L$. Let $M = A_K.v$ be the one dimensional lattice spanned by v, and let $M^\perp$ denote its orthogonal complement in the lattice Λ. We need to check that the lattice $M^\perp$ of rank $n-1$ is non-degenerate.

Since α is a unit, there is no loss of generality in assuming that the basis for M has the form $v = e + \mu$ with $\mu = \pi\nu \in \pi.L$. Let $\{f_1, f_2, \ldots, f_{n-1}\}$ be an orthonormal basis for L over A_K, and write $\nu = \sum \beta_i.f_i$. Then the vectors $-\overline{\beta_i}.e + f_i$ in Λ give a basis for $M^\perp$ over A_K, and the Gram matrix of their inner products has determinant $\equiv 1$ modulo π. This establishes the non-degeneracy of the lattice $M^\perp$ (a slightly more complicated version of this argument also works in the orthogonal case).

We can parametrize the decomposition $\Lambda = M + M^\perp$ by the element

$$\alpha^{-1} \otimes \lambda \in \pi L.$$

Picking a basis for L over A_K, this is the set $(\pi_K)^{n-1}$. In summary, we have shown the following.

Theorem 9.2. *Assume that V is a Hermitian space of dimension $n \geq 2$ over K whose Hermitian determinant is not a norm. Let Y be the homogeneous space parametrizing pairs (W, L) consisting of a line W in V whose determinant is not* *a norm together with a non-degenerate lattice L in $W^{\perp}$. Let Z be the homogenous space parametrizing the non-degenerate lattices Λ in V.*

Let $f : Y \to Z$ be the $U(V)$-equivariant map which takes the pair $y = (W, L)$ to the lattice $z = \Lambda = A_K.e + L$. Then the inverse image $f^{-1}(z)$ in Y can be parametrized by the points of $\pi L \cong (\pi A_K)^{n-1}$.

Again, we have a filtration of the parahoric N by normal subgroups N_m of finite index which act trivially on the quotient lattice $\Lambda/\pi^m\Lambda$. This gives a filtration of the quotient set

$$N/J > N_1J/J > N_2J/J > \dots$$

If we identify $f^{-1}(z) = N/J$ with the points of $(\pi_K)^{n-1}$, then the subset N_mJ/J corresponds to the points in $(\pi_K^{m+1})^{n-1}$. Indeed, the quotient groups $N_mJ/N_{m+1}J$ are isomorphic to $(A_K/\pi A_K)^{n-1}$ for all m.

10 Shimura Varieties Associated to Incoherent Definite Data

Let k be a totally real number field with ring of integers A, and fix incoherent definite orthogonal data $\{V_v\}$ over k of dimension $n \geq 3$ and determinant d. For each place v of k let $V(v)$ be the neighboring global orthogonal space, and let $G(v)$ be the special orthogonal group of $V(v)$ over k. Let $\mathbb{A}$ be the ring of adeles of k let $\mathbb{A}^f$ be the subring of finite adeles, and for a finite place $\mathfrak{p}$ of k let $\mathbb{A}^{f,\mathfrak{p}}$ denote the subring of finite adeles away from $\mathfrak{p}$.

To define the orthogonal Shimura variety S associated to this data, we first choose a real place v of k, and begin by defining S over the complex quadratic extension $K_v \cong \mathbb{C}$ of the completion $k_v = \mathbb{R}$. Let L be an integral A lattice in the neighboring orthogonal space $V(v)$, which has signature in $(n-2, 2)$ at v and signature $(n, 0)$ at all other real places w of k. We note that $L \otimes A_{\mathfrak{p}}$ is non-degenerate for almost all primes $\mathfrak{p}$. Then $M = \prod \mathrm{SO}(L_{\mathfrak{p}})$ is an open compact subgroup of $G(v)(\mathbb{A}^f)$, and $M_{\mathfrak{p}} = \mathrm{SO}(L_{\mathfrak{p}})$ is a hyperspecial maximal compact subgroup of $\mathrm{SO}(V_{\mathfrak{p}})$ for almost all primes p.

Let $T = \mathrm{Res}_{K_v/k_v}\mathbb{G}_m/\mathbb{G}_m$ be the one dimensional non-split torus over $k_v = \mathbb{R}$, and let $h : T \to G(v)_{\mathbb{R}}$ be the homomorphism described in the previous section, associated to negative oriented two dimensional subspaces of $V(v)$ over k_v. Then the conjugacy class X_v of h has the structure of two conjugate copies of the Hermitian symmetric space of $\mathrm{SO}(n-2, 2)$. We define [D71]

$$S_M(K_v) = G(v)(k)\backslash X_v \times G(v)(\mathbb{A}^f)/M$$

This double coset space is the disjoint union of a finite number of connected components each of the form $\Gamma\backslash X_v$, where Γ is an arithmetic subgroup of the algebraic group $G(v)(k)$. As such, it is a complex analytic orbifold, which has an algebraic structure (This algebraic structure is unique if M is small enough, so that the subgroups Γ have no non-trivial torsion). The complex algebraic varieties S_M form a projective system, for $M' \subset M$, and one defines S as the projective limit. This complex pro-scheme has a right action of the group $G(v)(\mathbb{A}^f)$, and S_M is the quotient of S by the open compact subgroup M.

The theory of canonical models [D79] shows that S descends canonically to k, viewed as a subfield of $k_v = \mathbb{R}$ in $K_v = \mathbb{C}$. We note that the descent to $k_v = \mathbb{R}$ is given by the anti-holomorphic involution of complex conjugation on X_v, and the action of the group $G(v)(\mathbb{A}^f)$ on S is defined over k on the canonical model. This defines the Shimura variety S associated to the incoherent definite orthogonal data $\{V_v\}$. The irreducible components of S are rational over the maximal abelian extension E of k with exponent 2, and are permuted simply-transitively by the quotient $\mathbb{A}^*/k^*\mathbb{A}^{*2}$ of $G(v)(\mathbb{A}^f)$, which is isomorphic to the Galois group of E over k by the reciprocity homomorphism of global class field theory. To see that this Galois group is a quotient of $G(v)(\mathbb{A}^f)$, note that the spinor norm maps $G(v)(\mathbb{A}^f)$ onto the group $(\mathbb{A}^f)^*/(\mathbb{A}^f)^{*2}$. The quotient of this group by the group k_+^* of totally positive elements in k^* is the group $\mathbb{A}^*/k^*\mathbb{A}^{*2}$.

The reason that we use incoherent data to define S, rather than just the orthogonal group $G(v)$ with its real conjugacy class X_v, is that the latter depends on the choice of a real place of k, and only defines S over a subfield of $\mathbb{C}$. However, for any other real place w of k, we have an isomorphism of adelic groups $G(w)(\mathbb{A}^f) \cong G(v)(\mathbb{A}^f)$ and an isomorphism complex varieties

$$S_M(K_w) \cong G(w)(k)\backslash X_w \times G(w)(\mathbb{A}^f)/M$$

When $n = 3$, so S_M is a Shimura curve, this isomorphism is due to Doi and Naganuma [DN67]. In the general case, it follows from results of Delgine and Borovoi on the conjugation of Shimura varieties. Therefore it is more symmetrical to use the incoherent definite orthogonal data to define S. We will speculate on the use of the neighboring orthogonal spaces $V(\mathfrak{p})$ for finite places of k in the next section.

A similar definition works for the unitary Shimura varieties associated to incoherent definite Hermitian data $\{V_v\}$ of dimension $n \geq 1$ over the imaginary quadratic extension K of the totally real field k. We choose a real place v of k and let $G(v)$ be the unitary group of the neighboring Hermitian space $V(v)$ over k. We begin by defining the Shimura variety S over the complex quadratic extension $K_v = \mathbb{C}$ of $k_v = \mathbb{R}$. Let L be a Hermitian A_K lattice in the neighboring space $V(v)$, which has signature $(n-1, 1)$ at v and signature $(n, 0)$ at all the other infinite places w. Then $M = \prod U(L_\mathfrak{p})$ is an open compact subgroup of $G(v)(\mathbb{A}^f)$ and $M_\mathfrak{p} = U(L_\mathfrak{p})$ is a hyperspecial maximal compact subgroup of $U(V_\mathfrak{p})$ for almost all primes $\mathfrak{p}$.

Let $T = \mathrm{Res}_{K_v/k_v}\mathbb{G}_m/\mathbb{G}_m$ be the one dimensional non-split torus over $k_v = \mathbb{R}$, and let $h : T \to G(v)_\mathbb{R}$ be the homomorphism associated to a negative one dimensional subspace of $V(v)$ over K_v. Then the conjugacy class X_v of h is isomorphic to the Hermitian symmetric space of $U(n-1, 1)$. We define

$$S_M(K_v) = G(v)(k)\backslash X_v \times G(v)(\mathbb{A}^f)/M$$

Again, this is a complex orbifold which has an algebraic structure, and the associated complex varieties S_M form a projective system with limit S. The limit has a right action of the group $G(v)(\mathbb{A}^f)$, and S_M is the quotient of S by the open compact subgroup M.

The theory of canonical models shows that S descends canonically to a pro-scheme over K, viewed as a subfield of $K_v = \mathbb{C}$, and the action of the group $G(v)(\mathbb{A}^f)$ on S is defined over K on the canonical model. This is the Shimura variety S associated to the incoherent definite Hermitian data $\{V_v\}$. The irreducible components of S are rational over the maximal abelian extension E of K which is "dihedral" over k. They are permuted simply-transitively by the quotient $\mathbb{A}_K^*/(\mathbb{A}^*.K^*.\prod_{v|\infty} K_v^*)$ of $G(v)(\mathbb{A}^f)$, which is isomorphic the Galois group of E over K by the reciprocity homomorphism of global class field theory.

The reason that we use incoherent data to define S, rather than just the unitary group $G(v)$ with its real conjugacy class X_v, is that the latter depends on the choice of a real place of k, and only defines S over a subfield of $\mathbb{C}$. However, for any other real place w of k, we have an isomorphism of adelic groups $G(w)(\mathbb{A}^f) \cong G(v)(\mathbb{A}^f)$ and an isomorphism complex varieties

$$S_M(K_w) \cong G(w)(k)\backslash X_w \times G(w)(\mathbb{A}^f)/M$$

This follows from general results of Deligne and Borovoi on the conjugation of Shimura varieties. Therefore it is more symmetric to use the incoherent definite Hermitian data to define S.

11 The Special Locus Modulo $\mathfrak{p}$

Let k be a totally real number field, with ring of integers A. Let $\{V_v\}$ be definite incoherent orthogonal data of dimension n and determinant d for k. At each real place v we have used the neighboring orthogonal space $V(v)$, its special orthogonal group $G(v) = \mathrm{SO}(V(v))$ over k, and the complex analytic space $X_v = G(v)(k_v)/H(v)(k_v)$ to describe the points of the Shimura variety S_M over the quadratic extension $K_v \cong \mathbb{C}$ of the corresponding completion $k_v = \mathbb{R}$. It is reasonable to ask if something similar occurs at the finite places $\mathfrak{p}$ of k. Namely, can we use the special orthogonal group $G(\mathfrak{p}) = \mathrm{SO}(V(\mathfrak{p}))$ of the neighboring space at $\mathfrak{p}$ to say something about the $K_\mathfrak{p}$ points of S_M, where $K_\mathfrak{p}$ is the unramified quadratic extension of the completion $k_\mathfrak{p}$?

In this section, we will attempt to do so, under the assumptions that the residual characteristic is not equal to 2 and that the A-lattice L which we have used to define the open compact subgroup $M = M_\mathfrak{p} \times M^\mathfrak{p}$ of $G(v)(\mathbb{A}^f)$ is non-degenerate at the prime $\mathfrak{p}$. By this we mean that $L \otimes A_\mathfrak{p}$ is a non-degenerate lattice in the orthogonal space $V_{\mathfrak{p}}$. This assumption, which is true for almost all primes $\mathfrak{p}$, implies that d is a unit at $\mathfrak{p}$, that $\epsilon(V_{\mathfrak{p}}) = +1$, and that $M_{\mathfrak{p}}$ is a hyperspecial maximal compact subgroup of $G(v)(k_{\mathfrak{p}})$. In this case, it is known that the Shimura variety S_M, associated to the incoherent data and the choice of M, has a model over $A_{\mathfrak{p}}$ with good reduction modulo $\mathfrak{p}$ [HPR19]. The neighboring orthogonal space $V(\mathfrak{p})$ is positive definite at all real places of k. At $\mathfrak{p}$ its determinant is a unit and $\epsilon_{\mathfrak{p}}(V(\mathfrak{p})) = -1$.

Let $K_{\mathfrak{p}} = k_{\mathfrak{p}}(\sqrt{D})$ be the unramified quadratic extension of $k_{\mathfrak{p}}$, let $W_{\mathfrak{p}}$ be an two dimensional orthogonal space over $k_{\mathfrak{p}}$ of determinant $-D$ with $\epsilon(W_{\mathfrak{p}}) = -1$ and let T be the one dimensional torus $(\mathrm{Res}_{K_{\mathfrak{p}}/k_{\mathfrak{p}}} \mathbb{G}_m)_{N=1}$. Recall that we have proved the existence and conjugacy of isometric embeddings $W_{\mathfrak{p}} \to V(\mathfrak{p})$ over $k_{\mathfrak{p}}$. Choosing an orientation of $W_{\mathfrak{p}}$ gives a homomorphism $h : T \to G(\mathfrak{p})$ over $k_{\mathfrak{p}}$ as in the real case. Writing $V(\mathfrak{p}) = W_{\mathfrak{p}} \oplus W_{\mathfrak{p}}^{\perp}$ as an orthogonal direct sum, and choosing a non-degenerate lattice $L_{\mathfrak{p}}$ in $W_{\mathfrak{p}}^{\perp}$, we obtain a hyperspecial maximal compact subgroup $J_{\mathfrak{p}} = \mathrm{SO}(W_{\mathfrak{p}}) \times \mathrm{SO}(L_{\mathfrak{p}})$ of the centralizer of h. We then defined the set $Y_{\mathfrak{p}} \cong G(\mathfrak{p})(k_{\mathfrak{p}})/J_{\mathfrak{p}}$, which parametrizes the pairs $(W_{\mathfrak{p}}, L_{\mathfrak{p}})$ of oriented planes of this type with the choice of a non-degenerate lattice in the orthogonal complement.

We also defined a map $f : Y_{\mathfrak{p}} \to Z_{\mathfrak{p}}$ to the space of maximal lattices $\Lambda_{\mathfrak{p}}$ in $V(\mathfrak{p})$, with an orientation. Fix a vector e in $W_{\mathfrak{p}}$ with $q(e) = \pi$ a uniformizing element of $k_{\mathfrak{p}}$ and let $A_{K_{\mathfrak{p}}}.e$ be the corresponding lattice in $W_{\mathfrak{p}}$. Then $\Lambda_{\mathfrak{p}} = A_{K_{\mathfrak{p}}}.e + L_{\mathfrak{p}}$ is an oriented lattice in $V(\mathfrak{p})$, whose stabilizer is the parahoric subgroup $N_{\mathfrak{p}}$ with the same reductive quotient as $J_{\mathfrak{p}}$.

We would like an interpretation of the double coset spaces

$$C = G(\mathfrak{p})(k)\backslash(Z_{\mathfrak{p}} \times G(\mathfrak{p})(\mathbb{A}^{f,\mathfrak{p}})/M^{\mathfrak{p}}) \cong G(\mathfrak{p})(k)\backslash G(\mathfrak{p})(\mathbb{A}^f)/(N_{\mathfrak{p}} \times M^{\mathfrak{p}}).$$

$$D = G(\mathfrak{p})(k)\backslash(Y_{\mathfrak{p}} \times G(\mathfrak{p})(\mathbb{A}^{f,\mathfrak{p}})/M^{\mathfrak{p}}) \cong G(\mathfrak{p})(k)\backslash G(\mathfrak{p})(\mathbb{A}^f)/(J_{\mathfrak{p}} \times M^{\mathfrak{p}}).$$

Since the group $G(\mathfrak{p})(k_v)$ is compact at all the real places v of k, and the product $(N_{\mathfrak{p}} \times M^{\mathfrak{p}})$ is open in $G(\mathfrak{p})(\mathbb{A}^f)$, the double coset space C is finite. At each point c of C, we have a finite arithmetic subgroup Γ_c of $G(\mathfrak{p})(k)$, which is defined by the intersection

$$\Gamma_c = G(\mathfrak{p})(k) \cap c.(N_{\mathfrak{p}} \times M^{\mathfrak{p}}).c^{-1}$$

The intersection takes place inside the group $G(\mathfrak{p})(\mathbb{A}^f)$.

The double coset space D has the structure of a $A_{K_{\mathfrak{p}}}$-analytic orbifold. Via the map $F : Y_{\mathfrak{p}} \to Z_{\mathfrak{p}}$,we obtain a map $F_D : D \to C$. The fiber of the map $F : Y_{\mathfrak{p}} \to Z_{\mathfrak{p}}$ over a point $z \in Z_{\mathfrak{p}}$ is isomorphic to the polydisc $(\pi A_{K_{\mathfrak{p}}})^{n-2} \cong N_{\mathfrak{p}}/J_{\mathfrak{p}}$. Hence the fiber of the map F_D over the point $c \in C$ is isomorphic to the quotient of the polydisc of dimension $n-2$ over $A_{K_{\mathfrak{p}}}$ by the finite group Γ_c. There is an involution of the finite set C given by the action of the normalizer of $N_{\mathfrak{p}}$, and this is compatible with the involution of D given by the normalizer of $J_{\mathfrak{p}}$.

Conjecture 1. *The finite set C parametrizes the set of special points on the orthogonal Shimura variety S_M modulo $\mathfrak{p}$. These points are all rational over the quadratic extension $\mathbb{F}_{\mathfrak{p}^2}$ of the residue field $\mathbb{F}_{\mathfrak{p}}$ of $k_{\mathfrak{p}}$, and the action of the Galois group is given by the normalizer of $N_{\mathfrak{p}}$.*

The $A_{K_{\mathfrak{p}}}$-analytic orbifold D parametrizes the set of points of the orthogonal Shimura variety S_M over $A_{K_{\mathfrak{p}}}$ which have special reduction modulo $\mathfrak{p}$. The reduction map is given by F_D and the action of the Galois group of $A_{K_{\mathfrak{p}}}/A_{\mathfrak{p}}$ is given by the normalizer of $J_{\mathfrak{p}}$.

When $n = 3$, so S_M is a Shimura curve over k, the special points correspond to supersingular data, and the conjecture is true by results of Carayol [C86].

In the Hermitian case, we expect a similar conjecture to hold, mutatis mutandis. Here we assume that k is totally real, that K is a totally imaginary quadratic extension of k, and that $\mathfrak{p}$ is a finite prime of k which remains inert in K. We start with incoherent Hermitian data, choose a real place v, and assume that the A_K-lattice in $V(v)$ that we have used to define the open compact subgroup $M = M_{\mathfrak{p}} \times M^{\mathfrak{p}}$ is non-degenerate. This implies that the Hermitian determinant d is a norm locally at the prime $\mathfrak{p}$ and that $M_{\mathfrak{p}}$ is a hyperspecial compact subgroup of $U(v)(k_{\mathfrak{p}})$. In this case, the unitary Shimura variety S_M, associated to the incoherent data and the choice of M, is defined over K and has a model over $A_{K_{\mathfrak{p}}}$ with good reduction modulo p [RSZ19]. The neighboring Hermitian space $V(\mathfrak{p})$ is positive definite at all complex places of K. At the prime $\mathfrak{p}$ its Hermitian determinant is not a norm. We would like to use the corresponding global unitary group $G(\mathfrak{p}) = U(V(\mathfrak{p}))$ and the two local homogeneous spaces $Z_{\mathfrak{p}}$ and $Y_{\mathfrak{p}}$ for $G(\mathfrak{p})(k_{\mathfrak{p}})$ to parametrize the finite set of special points over $\mathbb{F}_{\mathfrak{p}^2}$ as well as the points in $A_{K_{\mathfrak{p}}}$ which reduce to special points modulo $\mathfrak{p}$.

As in the orthogonal case, we consider the double coset spaces

$$C = G(\mathfrak{p})(k)\backslash(Z_{\mathfrak{p}} \times G(\mathfrak{p})(\mathbb{A}^{f,\mathfrak{p}})/M^{\mathfrak{p}}) \cong G(p)(k)\backslash G(p)(\mathbb{A}^f)/(N_p \times M^p).$$

$$D = G(\mathfrak{p})(k)\backslash(Y_{\mathfrak{p}} \times G(\mathfrak{p})(\mathbb{A}^{f,\mathfrak{p}})/M^{\mathfrak{p}}) \cong G(\mathfrak{p})(k)\backslash G(\mathfrak{p})(\mathbb{A}^f)/(J_{\mathfrak{p}} \times M^{\mathfrak{p}}).$$

The double coset space C is finite, and at each point c of C, we have a finite arithmetic subgroup Γ_c of $G(\mathfrak{p})(k)$, which is defined by the intersection

$$\Gamma_c = G(\mathfrak{p})(k) \cap c.(N_{\mathfrak{p}} \times M^{\mathfrak{p}}).c^{-1}$$

The intersection takes place inside the group $G(\mathfrak{p})(\mathbb{A}^f)$.

This double coset space D has the structure of a $A_{K_{\mathfrak{p}}}$-analytic orbifold. Via the map $F : Y_{\mathfrak{p}} \to Z_{\mathfrak{p}}$, we obtain a map $F_D : D \to C$. Since the fiber of the map $F : Y_{\mathfrak{p}} \to Z_{\mathfrak{p}}$ over a point $z \in Z_{\mathfrak{p}}$ is isomorphic to the polydisc $(\pi A_{K_{\mathfrak{p}}})^{n-1} \cong N_{\mathfrak{p}}/J_{\mathfrak{p}}$, the fiber of the map F_D over the point $c \in C$ is isomorphic to the quotient of the polydisc of dimension $n - 1$ over $A_{K_{\mathfrak{p}}}$ by the finite group Γ_c.

Conjecture 2. *The finite set C parametrizes the set of special points on the unitary Shimura variety S_M modulo $\mathfrak{p}$. These points are all rational over the residue field $\mathbb{F}_{\mathfrak{p}^2}$ of $\mathfrak{p}$ in A_K.*

The $A_{K_\mathfrak{p}}$-analytic orbifold D parametrizes the set of points of the unitary Shimura variety S_M over $A_{K_\mathfrak{p}}$ which have special reduction modulo $\mathfrak{p}$. The reduction map is given by F_D.

Finally, note that when $M_\mathfrak{p}$ is hyperspecial, the components of the orthogonal Shimura variety S_M are all rational over the quadratic extension $K_\mathfrak{p}$ of the localization $k_\mathfrak{p}$. This is certainly necessary for the existence of any $K_\mathfrak{p}$ rational points, let alone points reducing to the special locus! Indeed in the orthogonal case, the components of S_M are rational over a finite abelian extension E of k of exponent 2. This extension depends on the image of the spinor norm from the open compact subgroup M of $G(v)(\mathbb{A}^f)$. When $M_\mathfrak{p}$ is hyperspecial, the image of the spinor norm contains the unit classes in $k_\mathfrak{p}^*/k_\mathfrak{p}^{*2}$ and the extension E/k is unramified at the place $\mathfrak{p}$. Hence the decomposition group of $\mathfrak{p}$ in the Galois group $\mathrm{Gal}(E/k)$ is cyclic, so has order 1 or 2, and the components of S_M are all rational over the unramified quadratic extension $K_\mathfrak{p}$ of $k_\mathfrak{p}$. A similar result holds in the Hermitian case, as a prime $\mathfrak{p}$ which is inert in the quadratic extension K where $M_\mathfrak{p}$ is hyperspecial splits completely in the ring class extension E of K where the components of S_M are rational.

12 A Mass Formula

The sum over the double coset space C (which conjecturally parametrizes the special points of S_M modulo $\mathfrak{p}$) of the reciprocals of the order of the corresponding finite groups Γ_c

$$\mathrm{Mass}(C) = \sum_{c \in C} 1/\#\Gamma_c$$

is given by a "mass formula", using the L-function of the motive and the Tamagawa number of $G(\mathfrak{p})$ (cf. [Gr97, GHY01]).

Since the motive and Tamagawa number of $G(\mathfrak{p})$ are equal to the motive and Tamagawa number of its inner form $G(v)$ over k, there is a simple relation between this mass and the virtual Euler characteristic χ of the complex orbifold $S_M(K_v)$, as defined by Serre [S06]

$$\#(W_G/W_H) \times \mathrm{Mass}(C) = \#G_0(\mathbb{F}_\mathfrak{p})/\#N_0(\mathbb{F}_\mathfrak{p}).(-q)^{\dim S} \times \chi(S_M(K_v)).$$

where W_G denotes the Weyl group of $G = G(v)$, W_H denotes the Weyl group of the centralizer of h, and $q = \#F_\mathfrak{p}$.

For example, when $\{V_v\}$ is incoherent definite orthogonal data of odd dimension $2n+1$, we find the formula

$$2n \times \mathrm{Mass}(C) = (1 - q + q^2 - q^3 + \ldots + (-q)^{2n-1}) \times \chi(S_M(K_v)).$$

In the case when $n = 1$, S_M is a Shimura curve. If we assume that M is small enough so that all the finite groups Γ_c are trivial, the mass is just the number of supersingular points on the curve S_M modulo $\mathfrak{p}$. Since these points are all rational over the quadratic extension of $\mathbb{F}_{\mathfrak{p}}$, which is the field of q^2 elements, the formula implies

$$2 \times \#S_M(q^2) \geq (1-q) \times (2-2g)$$

where g is the genus of the complex curve S_M. Hence the tower of curves S_M realize the asymptoptic Drinfeld-Vladut bound [DV83] of $q-1$, for the ratio of the number of points for a curve over a field with q^2 elements to its genus g.

References

[A74] M. Artin, Supersingular K3 surfaces. Ann. Sci. École Norm. Sup. **7**, 543–567 (1974)

[C86] H. Carayol, Sur la mauvais réduction des courbes de Shimura. Compos. Math. **59**, 151–230 (1986)

[D71] P. Deligne, *Travaux de Shimura*. Lecture Notes in Mathematics, vol. 244 (Springer, Berlin, 1971)

[D79] P. Deligne, Variétés de Shimura. Proc. Symp. Pure Math. **33**, 247–290 (1979)

[DN67] K. Doi, H. Naganuma, On the algebraic curves uniformized by arithmetical automorphic functions. Ann. Math. **86**, 449–460 (1967)

[DV83] V.G. Drinfeld, S.G. Vladut, Number of points of an algebraic curve. Funct. Anal. Appl. **17**, 53–54 (1983)

[GHY01] W.-T. Gan, J. Hanke, J.-K. Yu, On an exact mass formula of Shimura. Duke Math. J. **107**, 103–133 (2001)

[GGP12] W.-T. Gan, B. Gross, D. Prasad, Symplectic local root numbers, central critical L-values, and restriction problems in the representation theory of classical groups. Sur les conjectures de Gross et Prasad Astérisque **346**, 1–109 (2012)

[Gr97] B. Gross, On the motive of a reductive group. Invent. Math. **130**, 287–313 (1997)

[G04] B. Gross, *Heegner Points and Representation Theory*. Heegner Points and Rankin L-Series, vol. 49 (MSRI Publication, Berkeley, 2004), pp. 37–65

[HPR19] X. He, G. Pappas, M. Rapoport, Good and semi-stable reduction of Shimura varieties (2018). ArXiv: 1804.09615

[KR99] S. Kudla, M. Rapoport, Arithmetic Hirzebruch-Zagier cycles. J. Reine Angew. Math. **515**, 155–244 (1999)

[KR00] S. Kudla, M. Rapoport, Cycles on Siegel threefolds and derivatives of Eisenstein series. Ann. Sci. École Norm. Sup. **33**, 695–756 (2000)

[L80] T. Lam, *Algebraic Theory of Quadratic Forms* (Addison-Wesley, Boston, 1980)

[LO98] K-Z. Li, F. Oort, Moduli of Supersingular Abelian Varieties. Springer Lecture Notes in Mathematics, vol. 1680 (Springer, Berlin, 1998)

[MH73] J. Milnor, D. Husemoller, *Symmetric Bilinear Forms*. Springer Ergebnisse, vol. 73 (Springer, Berlin, 1973)

[O75] F. Oort, Which abelian surfaces are products of elliptic curves? Math. Ann. **214**, 35–47 (1975)
[RSZ19] M. Rapoport, B. Smithling, W. Zhang, On Shimura varieties for unitary groups. arXiv: 1906.12346, to appear in PAMQ
[S73] J.-P. Serre, *A Course in Arithmetic*. Springer GTM, vol. 7 (Springer, Berlin, 1973)
[S06] J.-P. Serre, *Lie Algebras and Lie Groups*. Springer Lecture Notes in Mathematics, vol. 1500 (Springer, Berlin, 2006)
[Sh61] G. Shimura, On the zeta functions of the algebraic curves uniformized by certain automorphic functions. J. Math. Soc. Japan **13**, 275–331 (1961)
[Sh67] G. Shimura, Construction of class fields and zeta functions of algebraic curves. Ann. Math. **85**, 58–159 (1967)
[ViW13] E. Viehmann, T. Wedhorn, Ekedahl-Oort and Newton strata for Shimura varieties of PEL type. Math. Ann. **356**, 1493–1550 (2013)
[VW11] I. Vollaard, T. Wedhorn, The supersingular locus of the Shimura variety of $GU(n-1,1)$. Invent. Math. **184**, 591–627 (2011)

Shimura Varieties for Unitary Groups and the Doubling Method

Michael Harris

Abstract The theory of Galois representations attached to automorphic representations of $GL(n)$ is largely based on the study of the cohomology of Shimura varieties of PEL type attached to unitary similitude groups. The need to keep track of the similitude factor complicates notation while making no difference to the final result. It is more natural to work with Shimura varieties attached to the unitary groups themselves, which do not introduce these unnecessary complications; however, these are of abelian type, not of PEL type, and the Galois representations on their cohomology differ slightly from those obtained from the more familiar Shimura varieties.

Results on the critical values of the L-functions of these Galois representations have been established by studying the PEL type Shimura varieties. It is not immediately obvious that the automorphic periods for these varieties are the same as for those attached to unitary groups, which appear more naturally in applications of relative trace formulas, such as the refined Gan-Gross-Prasad conjecture (conjecture of Ichino-Ikeda and N. Harris). The present article reconsiders these critical values, using the Shimura varieties attached to unitary groups, and obtains results that can be used more simply in applications.

1 Introduction and Overview

The critical values of L-functions of motives over number fields are the subject of many conjectures and a small number of theorems. In [D79], Deligne formulated a conjecture relating these critical values to periods relating rational structures on de Rham and topological (Betti) cohomology. For the motives conjecturally attached to cohomological automorphic representations of $GL(n)$ over CM fields—imaginary quadratic extensions of totally real fields—the L-functions can be realized explicitly

M. Harris (✉)
Department of Mathematics, Columbia University, New York, NY, USA
e-mail: harris@math.columbia.edu

W. Müller et al. (eds.), *Relative Trace Formulas*, Simons Symposia,
https://doi.org/10.1007/978-3-030-68506-5_6

by the doubling method of Garrett and Piatetski-Shapiro-Rallis [Ga84, PSR], provided the automorphic representations descend to automorphic representations of unitary groups. In this setting, versions of this conjecture have been established in a series of papers, including [H97, Gu16, GL].[1]

Applications of the doubling method to special values of L-functions are based on interpreting special values of the integral representation as cup products in coherent cohomology of Shimura varieties. In the work cited above, the Shimura varieties are of PEL type: they parametrize families of polarized abelian varieties with endomorphisms by (orders in) a CM field F and level structure, and the various structures satisfy natural compatibilities. Combining standard techniques of complex geometry with the theory of relative Lie algebra cohomology, the coherent cohomology classes on these varieties can be identified with automorphic forms on the reductive algebraic groups attached to the Shimura varieties—in the event, they are essentially the groups of similitudes of a hermitian vector space over F, with similitude factor assumed to be rational. The doubling method, however, was originally devised to apply to the *symmetry groups* of bilinear or hermitian forms—in this case, to the unitary groups themselves, rather than to groups of unitary similitudes. While the methods of [Ga84, PSR] can easily be adapted to the similitude groups, this introduces (rather mild) technical complications as well as additional notation that has no clear connection to the original question. The unwelcome similitude factors pop up elsewhere in the theory, notably in the construction of Galois representations attached to automorphic forms on $GL(n)$ (e.g. in [HT01, M10, Sh11, CHLN, HLTT, Sch, B]) as a kind of fee charged for the right to use the theory of Shimura varieties in order to draw arithmetic conclusions.

It has been known for some time, however—though this author only learned about it a few years ago—that Shimura varieties can be attached to the unitary groups themselves, with no need to introduce the parasitic similitude factor. The Shimura data are described in §27 of [GGP] and have since been used in work on generalizations of the Gross-Zagier formula. These are of abelian type but not of PEL type, and the theory of their L-functions has only been established recently [KSZ], in the setting of the Langlands-Kottwitz method. The purpose of the present paper is primarily to work out the analogue of the results of [H97, Gu16, GL] using these Shimura varieties; secondarily, as a form of penance for the author's failure to do so earlier.

There is nothing really new in this paper, but it is hoped that it will serve as a reference for future work on special values, and specifically on p-adic L-functions. The results of [EHLS], for example, are proved using the PEL Shimura varieties attached to unitary similitude groups, but they can just as well be proved in the

---

[1]Using very similar methods, Shimura obtained versions of these results in [Sh97] and in subsequent papers; his results are limited to scalar valued automorphic forms, but are more precise in a number of respects. There has also been important work by various authors relating the critical values of these L-functions to automorphic periods that have no direct motivic interpretation; the present paper has nothing to say about this.

current framework. Such applications will require at least that lip service be paid to the moduli theory underlying their canonical models.

The paper follows the pattern established in [H97], but no attempt has been made to relate the results obtained here to the motivic periods that arise in Deligne's conjecture. Section 2 introduces the Shimura varieties that are used to relate special values of zeta integrals to motivic periods. Section 3 describes the parameters for coherent cohomology—in fact, holomorphic automorphic forms—of these Shimura varieties, with coefficients in automorphic vector bundles. Section 4 defines the Eisenstein series used in the doubling method, and interprets appropriately normalized holomorphic (and nearly holomorphic) Eisenstein series as coherent cohomology classes. Section 5 reinterprets the differential operators studied in [H97, Gu16] (and elsewhere) in terms of the Shimura data without the similitude factor. Section 6 recalls the theory of the doubling integral of [Ga84, PSR]. As in the earlier papers, the main result on critical values is proved in Sect. 7 by composing a paragraph that mentions all the objects introduced in the earlier sections.

2 Shimura Varieties for Unitary Groups

2.1 Notation and Conventions

We let F be any CM-field of degree $2d = \dim_{\mathbb{Q}} F$ and set of real places $S_\infty = S(F)_\infty$. Each place $v \in S_\infty$ refers to an ordered pair of conjugate complex embeddings $(\iota_v, \bar{\iota}_v)$ of F, where we will drop the subscript "v" if it is clear from the context. This fixes a choice of a CM-type $\Sigma = \{\iota_\sigma : \sigma \in S_\infty\}$. When there is no danger of ambiguity, we write $\sigma \in \Sigma$ instead of ι_σ. The maximal totally real subfield of F is denoted F^+, and we let ε_{F/F^+} denote the quadratic character of the adeles of F^+ attached to this extension. The set of real places of F^+ is identified with S_∞, identifying a place σ with its first component embedding $\iota_\sigma \in \Sigma$ and we let $\mathrm{Gal}(F/F^+) = \{1, c\}$. The ring of adeles over F (resp. over F) is denoted $\mathbf{A}_F$ (resp. $\mathbf{A}_{F^+}$) and $\mathbf{A}_{\mathbb{Q}}$ for $\mathbb{Q}$. We write $\mathcal{O}_F$ (resp. $\mathcal{O}_{F^+}$) for the respective rings of integers and drop the subscript, if no confusion is possible.

Let $(V, \langle \cdot, \cdot \rangle)$ be an n-dimensional non-degenerate c-hermitian space over F, $n \geq 2$. If V is understood, we denote the corresponding unitary group by $G = G_V := U(V)$ over F^+.

We define the *rational similitude group* $\tilde{G} := \tilde{G} := GU$ over $\mathbb{Q}$ as follows: If $GU_{F^+}(V_n)$ is the subgroup of $GL(V)$ that preserves the hermitian form up to a scalar multiple $\nu(g) \in \mathbb{G}_{m,F^+}$,

$$GU_{F^+}(V) := \{g \in GL(V) | \langle gv, gw \rangle = \nu(g) \cdot \langle v, w \rangle\}$$

we let $GU(V)$ denote the fiber product

$$GU(V) := GU_{F^+}(V) \times_{R_{F^+/\mathbb{Q}}\mathbb{G}_{m,F^+}} \mathbb{G}_{m,\mathbb{Q}}$$

where the map $GU_{F^+}(V) \to R_{F^+/\mathbb{Q}}\mathbb{G}_{m,F^+}$ is the similitude map ν and $\mathbb{G}_{m,\mathbb{Q}} \to R_{F^+/\mathbb{Q}}\mathbb{G}_{m,F^+}$ is the natural inclusion.

When working with several hermitian spaces we sometimes index V by its dimension n; thus we write $V = V_n$. If V_k is some non-degenerate subspace of V_n, we view $U(V_k)$ (resp. $GU(V_k)$) as a natural F^+-subgroup of $U(V_n)$ (resp. $\mathbb{Q}$-subgroup of $GU(V_n)$). If $n = 1$ we write $U(1) = U(V_1)$, $GU(1) = GU(V_1)$. As an algebraic group $U(1)$ is isomorphic to the kernel of the norm map $N_{F/F^+} : R_{F/F^+}(\mathbb{G}_m)_F \to (\mathbb{G}_m)_{F^+}$, and is thus independent of V_1.

Although $U(V)$ is viewed as an F^+-group, we will occasionally abuse notation and identify $U(V)$ with $R_{F^+/\mathbb{Q}}U(V)$, and do the same with related groups over F^+. Thus we can write $U(V)(\mathbb{R})$ for $R_{F^+/\mathbb{Q}}U(V)(\mathbb{R})$, for example.

The theory of *rationality* for automorphic vector bundles, understood as in [H86] and [H90], will be used without comment. Conventions for holomorphic and antiholomorphic modules, or highest K-type modules, are as in [EHLS, §4.4.1]. Thus holomorphic (resp. anti-holomorphic) modules are the Archimedean local components of automorphic representations corresponding to sections of automorphic vector bundles (resp. to coherent cohomology of automorphic vector bundles in the top degree.) Some of the author's earlier papers use the opposite convention. The dual of an automorphic vector bundle $\mathcal{E}$ is denoted $\mathcal{E}^\vee$, and the notation $^\vee$ is used more generally for duality.

2.2 Shimura Data

For each $\sigma \in \Sigma$ we let (r_σ, s_σ) denote the signature of the hermitian form induced by $\langle \cdot, \cdot \rangle$ on the complex vector space

$$V_\sigma := V \otimes_{F,\sigma} \mathbb{C}.$$

Thus $r_\sigma + s_\sigma = n$ for all σ. Let $\mathbb{S} = R_{\mathbb{C}/\mathbb{R}}\mathbb{G}_{m,\mathbb{C}}$, so that $\mathbb{S}(\mathbb{R}) = \mathbb{C}^\times$, canonically. Define a map

$$h_V = (h_{V,\sigma}, \sigma \in \Sigma) : \mathbb{S} \to GU(V)(\mathbb{R}) \subset \prod_{\sigma \in \Sigma} GU(V_\sigma)$$

componentwise, so that

$$h_{V,\sigma}(z) = \begin{pmatrix} zI_{r_\sigma} & 0 \\ 0 & \bar{z}I_{s_\sigma} \end{pmatrix}; \tag{2.1}$$

clearly this map to $\prod_{\sigma \in \Sigma} GU(V_\sigma)$ has image in the subgroup $GU(V)(\mathbb{R})$. We write h_V although the definition clearly depends on the CM type Σ as well.

We assume $(V_1)_\sigma$ has signature $(0, 1)$ for all $\sigma \in \Sigma$, and define $h_{V_1} = h_\Sigma : \mathbb{S} \to GU(V_1)(\mathbb{R})$ by analogy with (2.1). Let $G''_V \subset GU(1) \times GU(V)$ be the subgroup of (t, g) with $\nu(t) = \nu(g)$. Then the map

$$h''_V : \mathbb{S} \to [GU(V_1) \times GU(V)](\mathbb{R}); h''_V(z) = (h_{V_1}(z), h_V(z))$$

has image contained in $G''_V(\mathbb{R})$; we use the notation h''_V to designate the map $\mathbb{S} \to G''_V(\mathbb{R})$.

Let X''_V (resp. X_V, resp X_1) denote the $G''_V(\mathbb{R})$-conjugacy class (resp. $GU(V)(\mathbb{R})$-conjugacy class, resp. $GU(1)(\mathbb{R})$-conjugacy class) of homomorphisms $h : \mathbb{S} \to G''_V(\mathbb{R})$ containing h''_V (resp. h_V, resp. h_{V_1}). Then (G''_V, X''_V), $(GU(V), X_V)$, and $(GU(1), X_1)$ are all Shimura data, and the inclusion map $G''_V \hookrightarrow GU(V) \times GU(1)$ induces a morphism of Shimura data

$$(G''_V, X''_V) \to (GU(V) \times GU(1), X_V \times X_1).$$

There is a natural map

$$u : G''(V) \to U; u(t, g) = t^{-1}g, \forall (t, g) \in G''(V) \subset GU(1) \times GU(V) \quad (2.2)$$

The map taking $h \in X''_V$ to $u \circ h : \mathbb{S} \to U(V)(\mathbb{R})$ then defines a map of Shimura data

$$(G''_V, X''_V) \to (U(V), Y_V)$$

where Y_V is the $U(V)(\mathbb{R})$-conjugacy class defined by this map. Explicitly, the base point $y_V = u \circ h''_V \in Y_V$ is given by

$$y_{V,\sigma}(z) = \begin{pmatrix} (z/\bar{z})I_{r_\sigma} & 0 \\ 0 & I_{s_\sigma} \end{pmatrix} \quad (2.3)$$

When $\dim V = 1$ we write Y_1 or $Y_{\Sigma(V)}$ for Y_V, where $\Sigma(V)$ is the set of σ such that V_σ has signature $(0, 1)$.

The following is then obvious; we record it here in order to define parameters for automorphic vector bundles in the next section.

Lemma 2.3. *Let $y \in Y_V$. The stabilizer $K_y \in U(V)(\mathbb{R})$ is isomorphic to $\prod_{\sigma\in\Sigma} U(r_\sigma) \times U(s_\sigma)$.*

Here, for any m, we denote by $U(m)$ the compact real form of $GL(m)$. Later we will fix a base point $y \in Y_V$ and let $U_\sigma = K_y \cap U(V \otimes_{F,\sigma} \mathbb{C}) \xrightarrow{\sim} U(r_\sigma) \times U(s_\sigma)$ with respect to this base point.

The Shimura varieties attached to $(GU(V), X_V)$ and $(GU(1), X_1)$ belong to one of the families of PEL type Shimura varieties originally studied (in the form now called *connected Shimura varieties*) by Shimura [Sh64]. The Shimura variety attached to $(U(V), Y_V)$, which we denote $Sh(V, \Sigma)$, parametrizes Hodge structures of weight 0—the homomorphisms $y \in Y_V$ are trivial on the subgroup $\mathbb{R}^\times \subset \mathbb{C}^\times$— and are thus of abelian type but not of Hodge type.

The reflex field $E(V, \Sigma) := E(U(V), Y_V)$ is the subfield of the Galois closure of F over $\mathbb{Q}$ determined as the stabilizer of the cocharacter κ_V with σ-component $\kappa_{V,\sigma}(z) = \begin{pmatrix} zI_{r_\sigma} & 0 \\ 0 & I_{s_\sigma} \end{pmatrix}$. In particular, suppose there is $\sigma_0 \in \Sigma$ such that $r_{\sigma_0} > 0$ but $r_\sigma = 0$ for $\sigma \in \Sigma \setminus \sigma_0$. Then $E(V, \Sigma)$ is the subfield $\sigma_0(F) \subset \mathbb{C}$.

Remark 2.4. Starting in Sect. 3.4.1, the period invariants for the Shimura varieties attached to $U(V)$ will be distinguished from those for $GU(V)$ by the subscript U.

2.5 *Measures and Discriminants*

The group G is viewed as an algebraic group over F^+, and by restriction of scalars as an algebraic group over $\mathbb{Q}$. We choose Haar measures on the local and adelic groups $G(F_v^+)$ and $G(\mathbf{A})$ as in the introduction to [H97]. Thus if v is a finite place of F^+, we choose a local Haar measure dg_v that is rational in the sense that any open compact subgroup of $G(F_v^+)$ has rational measure; if G is unramified at v we also assume that a hyperspecial maximal compact subgroup has measure 1.

We choose a maximal compact subgroup

$$K_\infty = \prod K_\sigma \subset G_\infty = \prod_{\sigma \in S_\infty} G(F_\sigma^+)$$

that stabilizes a CM point x in the locally symmetric space X_V introduced in Sect. 2.2 above. We can thus write

$$X_V = \prod_{\sigma \in S_\infty} G(F_\sigma^+)/K_\sigma := \prod_{\sigma \in S_\infty} X_\sigma.$$

If σ is a real place of F^+, then we write $dg_\sigma = dk_\sigma \cdot dx_\sigma$ with respect to this factorization. Although it is not necessary, we may assume that x is the image of a map of Shimura data

$$\prod_i (U(1), Y_{\Sigma_i})^n \to (G_V, Y_V),$$

where we have written $V = \oplus V_i$, $\dim V_i = 1$, $\Sigma_i = \Sigma_{V_i}$ in the above notation. We let $E_i = E(U(1), Y_{\Sigma_i})$, so that x is defined over the compositum $E_x = E(\prod_i (U(1), Y_{\Sigma_i}))$ of the E_i. Then at each $\sigma \in S_\infty$ (which we identify with the CM type Σ) the Harish-Chandra decomposition

$$\mathfrak{g}_\sigma := Lie(G(F_\sigma^+)) = \mathfrak{p}_\sigma^+ \oplus \mathfrak{p}_\sigma^- \oplus Lie(K_\sigma)$$

is rational over the field $F_{\sigma,x} = E_x \cdot \sigma(F) \subset \mathbb{C}$; $F_{\sigma,x}$ is CM and quadratic over its maximal totally real subfield $F_{\sigma,x}^+$. We choose an element $\daleth \in F$ with $Tr_{F/F^+}(\daleth) = 0$, let $\sigma : F \to \mathbb{C}$ denote the element of Σ associated to σ, and define a top differential

$$\omega_\sigma = \sigma(\daleth)^{-\dim X_\sigma} \bigwedge_{i=1}^{\dim X_\sigma} dz_{\sigma,i} \wedge d\bar{z}_{\sigma,i} \in \wedge^{\dim X_\sigma} \mathfrak{p}_\sigma^+ \otimes \wedge^{\dim X_\sigma} \mathfrak{p}_\sigma^-, \tag{2.4}$$

where $dz_{\sigma,i}$ is an $F_{\sigma,x}$-basis of $\mathfrak{p}_\sigma^+$ and $d\bar{z}_{\sigma,i}$ is the complex conjugate basis of $\mathfrak{p}_\sigma^-$. Similarly, for dk_σ we take a Haar measure defined by an $F_{\sigma,x}^+$-basis of $\wedge^{\dim K_\sigma} Lie(K_\sigma)$. Finally, we let

$$dx_\sigma = (2\pi)^{-\dim X_\sigma} \omega_\sigma. \tag{2.5}$$

The factorization of the adelic Tamagawa measure dg as a product of local measures introduces an additional normalizing constant that was not present in [H97]. We can write $dg = dg_\infty \cdot dg_f$, with $dg_f = \prod_\sigma dg_\sigma$ (product over finite places). However, dg_∞ is given by a $\mathbb{Q}$-basis of $\wedge^{\dim R_{F^+/\mathbb{Q}} G} Lie(G)$, and is thus a $\mathbb{Q}$-rational differential form, whereas $\prod_{\sigma \in S_\infty} dg_\sigma$ is not $\mathbb{Q}$-rational. Instead, we have

$$dg_V := (2\pi)^{-\dim X_V} \cdot dg_\infty = \sqrt{D_{F^+}}^{-n^2} \prod_\sigma dg_\sigma, \tag{2.6}$$

up to an E_x-rational factor, where now the product is taken over all places of F^+. The E_x-rational factor is inevitable because the reflex field $E(G_V, Y_V)$, which is contained in E_x, is precisely the stabilizer of the set of signatures together with the CM type Σ, whereas we have constructed the ω_σ as forms over E_x. However, letting $Gal(\overline{\mathbb{Q}}/\mathbb{Q})$ act on the set of Shimura varieties conjugate to $Sh(V, \Sigma)$—in other words, on the signatures and CM types—the collection (dg_V) can be chosen consistently with the $Gal(\overline{\mathbb{Q}}/\mathbb{Q})$-action.

3 Coherent Cohomology

3.1 Automorphic Vector Bundles

This paper is primarily devoted to the critical values of standard L-functions of automorphic representations of $U(V)$ attached to holomorphic modular forms. These modular forms are viewed as sections of automorphic vector bundles; as in

[Gu16] (see also [H97]), the relevant automorphic vector bundles on $Sh(V, \Sigma)$ are parametrized by irreducible representations of the stabilizer of a chosen base point in Y_V. By Lemma 2.3, we thus parametrize automorphic vector bundles by highest weights of irreducible representations of $\prod_{\sigma\in\Sigma} U(r_\sigma) \times U(s_\sigma)$. In §3.3 of [Gu16] the parametrization of automorphic vector bundles on the Shimura variety attached to $(GU(V), X_V)$ is given by

$$((a_{\sigma,1}, \ldots, a_{\sigma,n})_{\sigma\in\Sigma}; a_0) \text{ such that } \forall\sigma\ a_{\sigma,1} \geq \cdots \geq a_{\sigma,r_\sigma};\quad a_{\sigma,r_\sigma+1} \geq \cdots \geq a_{\sigma,n}. \tag{3.1}$$

The parameters for automorphic vector bundles on $Sh(V, \Sigma)$ are the same as those for (3.1) except that the parameter a_0 is absent:

$$(\kappa_{\sigma,1}, \ldots, \kappa_{\sigma,n})_{\sigma\in\Sigma} \text{ such that } \forall\sigma\ \kappa_{\sigma,1} \geq \cdots \geq \kappa_{\sigma,r_\sigma};\quad \kappa_{\sigma,r_\sigma+1} \geq \cdots \geq \kappa_{\sigma,n}. \tag{3.2}$$

Let κ denote a parameter in (3.2), and let $\mathcal{E}_\kappa$ denote the corresponding automorphic vector bundle on $Sh(V, \Sigma)$. Over $\mathbb{C}$, we have

$$\mathcal{E}_\kappa = \varprojlim_{K_f}[U(V)(F^+)\backslash U(V)(\mathbf{A}_{F^+}) \times W_\kappa/(K_{y_V} \times K_f)]$$

where the terms on the right hand side are vector bundles over the finite-level Shimura variety

$${}_{K_f}Sh(V, \Sigma)(\mathbb{C}) = U(V)(F^+)\backslash U(V)(\mathbf{A}_{F^+})/(K_{y_V} \times K_f).$$

Here $K_f \subset U(V)(\mathbf{A}_f)$ is a compact open subgroup (that is small enough to guarantee that $\mathcal{E}_\kappa$ is in fact a vector bundle) and $W_\kappa = \otimes_{\sigma\in\Sigma} W_{\kappa,\sigma}$ is the representation of $\prod_\sigma U(r_\sigma) \times U(s_\sigma)$ with highest weight $(\kappa_{\sigma,1} \geq \cdots \geq \kappa_{\sigma,r_\sigma}; \kappa_{\sigma,r_\sigma+1} \geq \cdots \geq \kappa_{\sigma,n})$, and the group K_{y_V} acts diagonally on $U(V)(\mathbb{R}) \times W_\kappa$.

3.2 Coherent Cohomology and Period Invariants

We fix a level subgroup K_f as above, and let ${}_{K_f}Sh(V, \Sigma) \hookrightarrow {}_{K_f}\ Sh(V, \Sigma)^{tor}$ denote a toroidal compactification. We may and do assume the compactification is smooth and projective, and the boundary divisor D has normal crossings. The automorphic vector bundle $\mathcal{E}_\kappa$ has two natural extensions to vector bundles over ${}_{K_f}Sh(V, \Sigma)^{tor}$: the canonical extension $\mathcal{E}_\kappa^{can}$ and the subcanonical extension $\mathcal{E}_\kappa^{sub}$, defined as in [H90]. We write

$$H_!^*(Sh(V, \sigma), \mathcal{E}_\kappa) :=$$

$$\varinjlim_{K_f} Im[H^*({}_{K_f}Sh(V, \Sigma)^{tor}, \mathcal{E}_\kappa^{sub}) \rightarrow H^*({}_{K_f}Sh(V, \Sigma)^{tor}, \mathcal{E}_\kappa^{can})], \tag{3.3}$$

where for each K_f the toroidal compactification is chosen to be adapted to the level. The space $H_!^*(Sh(V,\Sigma),\mathcal{E}_\kappa)$ defines an admissible and semisimple representation of $U(V)(\mathbf{A}_f)$.

Let π_f be an irreducible representation of $U(V)(\mathbf{A}_f)$ that can be completed to a cuspidal automorphic representation π of $U(V)(\mathbf{A})$ whose base change to $GL(n)_F$ is cuspidal and cohomological. It thus follows in particular (from a long series of partial results, culminating in Theorem 1.2 of [Ca]) that π is everywhere tempered. Fix a degree q and let $H_!^q(Sh(V,\Sigma),\mathcal{E}_\kappa)[\pi]$ denote the π_f-isotypic component of $H_!^q(Sh(V,\Sigma),\mathcal{E}_\kappa)$. If this π_f-isotypic component is non-trivial, it follows from [H90] (in particular Theorem 4.6.2) and [KMSW] that

Theorem 3.4.

(a) *The representation π_f occurs with multiplicity one in $H_!^q(Sh(V,\Sigma),\mathcal{E}_\kappa)$, and π is uniquely determined by π_f and the bundle $\mathcal{E}_\kappa$.*
(b) *In particular, $H_!^*(Sh(V,\Sigma),\mathcal{E}_\kappa)[\pi]$ determines a rational structure on π_f over some number field $E(\pi)$.*

Remark 3.3. We always choose $E(\pi)$ to contain an extension $E(\kappa)$ of the reflex field $E(V,\Sigma)$ over which the bundle $\mathcal{E}_\kappa$ has a rational model. As in §1.1 of the Erratum to [H13], $E(\kappa)$ in general strictly contains the fixed field E_κ of the stabilizer in $Gal(\overline{\mathbb{Q}}/\mathbb{Q})$ of the isomorphism class of $\mathcal{E}_\kappa$, and moreover is not uniquely determined; it is chosen to eliminate a Brauer obstruction to realizing the bundle over E_κ.

The main Theorem 7.1 relating periods to critical values of L-functions takes the form

$$\text{Critical value of } L(s,\pi,\alpha,St) \sim_{E(\pi,\alpha)} \text{Normalized period invariant.} \tag{3.4}$$

In Deligne's conjecture the place of $E(\pi,\alpha)$ is taken by the field of coefficients $E(M(\pi,\alpha))$ of the motive whose L-function is $L(s,\pi,\alpha,St)$. The previous paragraph indicates that $E(\pi,\alpha)$ is in general a non-trivial extension of the hypothetical $E(M(\pi,\alpha))$. However, the relation (3.4) is equivariant with respect to $Gal(E(\kappa)/E_\kappa)$ (this can be chosen to be an abelian extension) so the extension of the coefficient field is harmless.

Remark 3.4. The results of [KMSW] are conditional on results that have been claimed, not only by the authors, but not published. Careful readers may therefore prefer to view the results that make use of Theorem 3.4 as conditional. In [H97] and [Gu16] the period invariants introduced in the following paragraph are denoted $Q(\pi,\beta)$, where β is a marker that accounts for possible multiplicity greater than one. It can then be proved, using the main identity relating critical values of L-functions to periods, that the period is independent of β, up to appropriate algebraic factors, assuming $L(s,\pi,\alpha)$ has non-vanishing critical values. Thus the multiplicity one hypothesis is only used for convenience.

Returning to Theorem 3.4, the L_2-inner product on the space of cusp forms on $U(V)(\mathbf{A})$ restricts to a non-degenerate hermitian form on π_f. As in [H13], §3.4.2, one obtains a period invariant, denoted $Q(\pi)$ in *loc. cit.*, in $(E(\pi) \otimes_{\mathbb{Q}} \mathbb{C})^\times / E(\pi)^\times$. Each embedding $\tau : E(\pi) \rightarrow \mathbb{C}$ then determines a period invariant $Q(\pi, \tau) \in \mathbb{C}^\times / \tau(E(\pi))^\times$. The period invariant is characterized by Proposition 3.19 of [H13]: suppose $v_1, v_2 \in \pi$, identified with classes in $H_!^q(Sh(V, \Sigma), \mathcal{E}_\kappa)[\pi]$ by the trivialization as discussed in §3.4 of [H13]. Define

$$\langle v_1, v_2 \rangle = (\langle v_1, v_2 \rangle_\tau)_{\tau : E(\pi) \rightarrow \mathbb{C}} = \int_{[U(V)]} v_1(g) \bar{v}_2(g) dg \in E(\pi) \otimes_{\mathbb{Q}} \mathbb{C} \qquad (3.5)$$

as a vector in $E(\pi) \otimes_{\mathbb{Q}} \mathbb{C} = \prod_{\tau : E(\pi) \rightarrow \mathbb{C}} \mathbb{C}$, as in [H13, §3.4.2]. Then v_1 corresponds to an $E(\pi)$-rational class if and only if, for all $E(\pi)$-rational v_2, we have

$$Q(\pi, \tau)^{-1} \cdot \langle v_1, v_2 \rangle_\tau \in \tau(E(\pi)) \qquad (3.6)$$

Here rationality of coherent cohomology is understood as in Theorem 3.4 (b) and [H90, H13].

For the purposes of applications in [GHL], we assume $E(\pi)$ is given as a subfield $\tau(E(\pi))$ of $\mathbb{C}$ and write $P(\pi, V, \Sigma)$ for $Q(\pi, \tau)$. In this paper we will only consider π contributing to $H_!^0(Sh(V, \Sigma), \mathcal{E}_\kappa)$ or $H_!^{\dim Y_V}(Sh(V, \Sigma), \mathcal{E}_\kappa)$; in both cases, the cohomology space is entirely represented by cusp forms [H90, Proposition 5.4.2].

If κ is as above, we define κ^D as in [EHLS, §6.1.3]. Then complex conjugation of functions on the adele group defines an antilinear isomorphism

$$c_B : H_!^0(Sh(V, \Sigma), \mathcal{E}_\kappa) \xrightarrow{\sim} H_!^{\dim Y_V}(Sh(V, \Sigma), \mathcal{E}_{\kappa^D})$$

[EHLS, (6.2.3)]. Moreover, the isomorphism is compatible with the Serre duality pairing

$$\begin{aligned} [\bullet, \bullet]_{Ser} : H_!^0(Sh(V, \Sigma), \mathcal{E}_\kappa) \otimes H_!^{\dim Y_V}(Sh(V, \Sigma), \mathcal{E}_{\kappa^D}) \\ \rightarrow H^{\dim Y_V}(Sh(V, \Sigma), \Omega^{\dim Y_V}_{Sh(V, \Sigma)}) \end{aligned} \qquad (3.7)$$

[EHLS, (6.3.1)]. Note that the line bundle $L(\kappa)$ in *loc. cit.* is trivial for us because it only depends on the similitude factor); in particular we see that we can simply write

$$\mathcal{E}_{\kappa^D} \xrightarrow{\sim} \Omega^{\dim Y_V}_{Sh(V, \Sigma)} \otimes \mathcal{E}_\kappa^\vee.$$

Setting $q = 0$ and bearing in mind that $E(\pi)$ is a CM field, we can take the complex conjugation of both sides of (3.5), we see that, if v_1 corresponds to an $E(\pi)$ rational class in $H_!^0(Sh(V, \Sigma), \mathcal{E}_\kappa)[\pi]$, then, with c denoting complex conjugation,

$$(Q(\pi, c\tau)^{-1}\overline{v}_1|_{\tau:E(\pi)\to\mathbb{C}}) \in H^{\dim Y_V}(Sh(V,\Sigma), \mathcal{E}_\kappa^D)[\pi] \otimes_{E(\pi)} \mathbb{C} \tag{3.8}$$

comes from an $E(\pi)$-rational class.

3.4.1 Change of Polarization

The doubling method naturally gives rise to a different period invariant, which comes from the pairing of a form on $Sh(V, \Sigma)$ with a form on $Sh(-V, \Sigma)$. Namely, with the notation of [EHLS], §6.2, (but with ω replaced by $\mathcal{E}$ as notation for automorphic vector bundles), there is an $E(V, \Sigma) = E(-V, \Sigma)$ linear, $U(V)(\mathbf{A}_f) \xrightarrow{\sim} U(-V)(\mathbf{A}_f)$-equivariant, isomorphism

$$F_\infty : H_!^{\dim Y_V}(Sh(-V,\Sigma), \mathcal{E}_{\kappa^\flat}) \xrightarrow{\sim} H_!^0(Sh(V,\Sigma), \mathcal{E}_{\kappa^D}) \tag{3.9}$$

This isomorphism expresses the $U(V)(\mathbb{R})$-equivariant identity of Y_{-V} with Y_V, endowed with the complex conjugate structure. Now suppose $f \in H_!^{\dim Y_V}(Sh(V,\Sigma), \mathcal{E}_\kappa)[\pi]$ and $f' \in H_!^{\dim Y_V}(Sh(-V,\Sigma), \mathcal{E}_{\kappa^\flat})[\pi^\vee]$ are $E(\pi)$-rational cohomology classes in π and $\pi^\vee$, respectively, for some extension $L \supset E(\pi)$. (In particular, π_∞ is an antiholomorphic discrete series representation, as in [EHLS, §4.5].) Define

$$\begin{aligned} B(\pi)_{f,f'} = [f, F_\infty(f')]_{Ser} &\in H^{\dim Y_V}(Sh(V,\Sigma), \Omega^{\dim Y_V}_{Sh(V,\Sigma)}) \\ &\xrightarrow{\sim} H^0(Sh(V,\Sigma), \mathcal{O}_{Sh(V,\Sigma)}), \end{aligned} \tag{3.10}$$

where the final isomorphism is given by the trace map. It follows from Theorem 3.4 that, for any τ as before, there is a period factor

$$B(\pi) = (B(\pi,\tau), \tau \in \Sigma) \in (E(\pi) \otimes \mathbb{C})^\times$$

such that

$$B(\pi,\tau)^{-1} B(\pi)_{f,f',\tau} \in \tau(E(\pi)),\ \forall f, f', \tau. \tag{3.11}$$

Here we view $B(\pi)_{f,f'}$ as an element of $H^0(Sh(V,\Sigma), \mathcal{O}_{Sh(V,\Sigma)})(E(\pi)) \otimes \mathbb{C} \xrightarrow{\sim} E(\pi) \otimes_{\mathbb{Q}} \mathbb{C}$ and let $B(\pi)_{f,f',\tau}$ denote its projection on the τ-component.

Suppose α is a motivic Hecke character of $\mathbf{A}_F^\times$, with restriction α^U to $U(1)(\mathbf{A})$. Then with f' as above, there is an automorphic line bundle $\Lambda_{\alpha_\infty^U}$ on $Sh(-V, \Sigma)$ such that

$$f' \otimes \alpha^{U,-1} \circ \det \in H_!^{\dim Y_V}(Sh(-V,\Sigma), \mathcal{E}_{\kappa^\flat} \otimes \Lambda_{\alpha_\infty^U})[\pi^\vee \otimes \alpha^{U,-1} \circ \det].$$

There is then a constant $p_U(\alpha, -V) \in (E(\alpha_f) \otimes \mathbb{C})^\times$ such that, letting $E(\pi, \alpha) = E(\pi) \cdot E(\alpha)$, $p_U(\alpha, -V)^{-1} \cdot f' \otimes \alpha^{U,-1} \circ \det$ is an $E(\pi, \alpha)$-rational element of $H_!^{\dim Y_V}(Sh(-V, \Sigma), \mathcal{E}_{\kappa^\flat} \otimes \Lambda_{\alpha_\infty^U})[\pi^\vee \otimes \alpha^{U,-1} \circ \det]$. Put another way, let f'_α be an $E(\pi, \alpha)$-rational element of $H_!^{\dim Y_V}(Sh(-V, \Sigma), \mathcal{E}_{\kappa^\flat} \otimes \Lambda_{\alpha_\infty^U})[\pi^\vee \otimes \alpha^{U,-1} \circ \det]$, and let

$$B(\pi)_{\alpha, f, f'} = [f, F_\infty(f'_\alpha \otimes \alpha^U \circ \det)]_{Ser}.$$

Then for any $\tau : E(\pi, \alpha) \hookrightarrow \mathbb{C}$, there is a constant $B(\pi, \tau)_\alpha \in \mathbb{C}^\times$ such that, with notation as in (3.11)

$$B(\pi, \tau)_\alpha^{-1} B(\pi)_{\alpha, f, f', \tau} \in \tau(E(\pi, \alpha)), \ \forall f, f' \tag{3.12}$$

The formula for $B(\pi)_{\alpha, f, f'}$ may appear artificial but it is what shows up in the analysis of the zeta integral. Of course we have

$$B(\pi)_\alpha \sim p_U(\alpha, -V) \cdot B(\pi). \tag{3.13}$$

We can carry out the same construction with $Sh(GU(-V), X_{-V})$, the Shimura variety attached to the similitude group. If $\tilde{\pi}'$ is an automorphic representation of $GU(-V)(\mathbf{A})$ such that

$$H_!^{\dim Y_V}(Sh(GU(-V), X_{-V}), \mathcal{E}_{\kappa^\flat})[\tilde{\pi}'] \neq 0$$

and if α is as above, there is an automorphic line bundle Λ_{α_∞} on $Sh(GU(-V), X_{-V})$ and a constant $p(\alpha, -V)$ such that, with the analogues of the definitions above (taking into account the similitude factor in the usual way)

$$B(\tilde{\pi})_\alpha \sim p(\alpha, -V) \cdot B(\tilde{\pi}). \tag{3.14}$$

An expression for $p(\alpha, -V)$ can be found in §2.9 of [H97] (the top of p. 138) when $F^+ = \mathbb{Q}$. The relation between $p(\alpha, -V)$ and $p_U(\alpha, -V)$ is given in formula (3.15).

Remark 3.5. The notation F_∞ is not altogether appropriate, because the complex conjugate of $Sh(V, \Sigma)$ is not $Sh(-V, \Sigma)$ but rather $Sh(-V, c\Sigma)$. The difference is reflected in the period invariant. Suppose for simplicity of exposition that $F^+ = \mathbb{Q}$. Because the motive that appears in the π-isotypic component cohomology of $Sh(V, \Sigma)$ is (up to a Tate twist) an exterior power of the motive attached to π, the invariant $B(\pi)$ should be a period of this exterior power. But it is off by an abelian period (denoted $q(M)$ in [GH]), which exactly corresponds to the abelian twist needed to relate rational structures on $Sh(-V, \Sigma)$ and $Sh(-V, c\Sigma)$.

3.5.1 Period Invariants, $n = 1$

We recall the Shimura varieties attached to $H_F := R_{F/\mathbb{Q}}(\mathbb{G}_m)_F$ defined in [H93, §1.1]. For any subset $\Psi \subset \Sigma$ we can define $h_\Psi : \mathbb{S} \to H_{F,\mathbb{R}}$ by the rule that, in the induced Hodge structure on (the $\mathbb{Q}$-vector space) F, the subspace $F_\sigma \subset F \otimes \mathbb{C}$ is of type $(-1, 0)$ (resp. $(0, -1)$, resp. $(0, 0)$) if $\sigma \in \Psi$ (resp. $\sigma \in c\Psi$, where c denotes complex conjugation, resp. if $\sigma \notin \Psi \coprod c\Psi$). If $\Psi = \{\sigma\}$ is a singleton we write h_σ. We define $h_{c\Psi} : \mathbb{S} \to H_{F,\mathbb{R}}$ by exchanging Ψ with $c\Psi$.

For any motivic Hecke character ω of H_F we can thus define the CM periods $p_F(\omega, \Psi)$ and $p_F(\omega, c\Psi)$ as in [H93, Lemma 1.3], with $(H, h) = (H_F, h_\Psi)$ or $(H, h) = (H_F, h_{c\Psi})$ (but see [H97], p. 82 for an explanation of a sign error).

Lemma 3.6. *For any motivic Hecke character ω of H_F we have*

$$p_U(\omega, -V) = p_F(\omega \circ \det, c\Sigma)^{-1} \cdot p(\omega, -V) = p_F(\omega, c\Sigma)^{-n} \cdot p(\omega, -V). \tag{3.15}$$

Proof. By Shimura's product relations ([H93, Corollary 1.5], see [GL], §3.3 for a Galois-equivariant version), this follows by pulling back from $U(-V)$ to $G''(-V) \subset GU(1) \times GU(-V)$ by the map u of (2.2). The ratio between the two invariants is entirely determined by the image in $GU(1)$ of the pullback to $G''(-V)$. □

4 Holomorphic Eisenstein Series

4.1 The Doubled Group and Its Variety

We fix V as in the previous section, with hermitian form $\langle \cdot, \cdot \rangle = \langle \cdot, \cdot \rangle_V$, and write $-V$ to denote the F-vector space V with hermitian form $-\langle \cdot, \cdot \rangle_V$. Let W denote the $2n$-dimensional hermitian space $V \oplus (-V)$, and let $H = H_V = U(W)$ be the corresponding unitary group. Then H is always quasi-split. More precisely, let $V^d = \{(x, x) \in W : x \in V\}$ and $V_d = \{(x, -x) \in W : x \in V\}$, so $W = V_d \oplus V^d$ is a polarization of $\langle\ ,\ \rangle_W$, which is thus a maximally isotropic hermitian form. Projection to the first summand fixes identifications of V^d and V_d with V. Let $P \subset H$ be the stabilizer of V^d; this is a maximal $\mathbb{Q}$-parabolic, the Siegel parabolic. Let $M \subset P$ be the stabilizer of the polarization $W = V_d \oplus V^d$ and $N \subset P$ the group fixing both V^d and W/V^d, so M is a Levi subgroup and N the unipotent radical of P.

Since $\langle\ ,\ \rangle_W$ is maximally isotropic, we know that $H(F_\sigma) \xrightarrow{\sim} U(n, n)$ for any $\sigma : F \hookrightarrow \mathbb{C}$. Let (H, Y_W) be the Shimura datum attached to H and Σ by the procedure described in the previous section. Then $E(W, \Sigma)$ is the reflex field of the CM type Σ. The isomorphism $W = V \oplus (-V)$ determines a morphism of Shimura data

$$(G_V, Y_V) \times (G_{-V}, Y_{-V}) \hookrightarrow (H, Y_W) \tag{4.1}$$

and thus a map of Shimura varieties

$$Sh(V, \Sigma) \times Sh(-V, \Sigma) \rightarrow Sh(W, \Sigma).$$

4.1.1 Tube Domains

As a complex variety, the Shimura variety $Sh(W, \Sigma)$ is a union of arithmetic quotients of the tube domain $\mathfrak{X}_\Sigma$ attached to the group $SU(n, n)^\Sigma$. The tube domain $\mathfrak{X}_\Sigma$, which is isomorphic to a connected component of Y_W, itself factors as a product $\prod_{\sigma \in \Sigma}$ of $|\Sigma|$ copies of the classical tube domain $X_{n,n}^+ \subset Herm_n(\mathbb{C})$, where $Herm_n \subset M(n, \mathbb{C})$ is the space of $n \times n$-hermitian matrices. More precisely, denote by X_σ^+ the copy of $X_{n,n}^+$ corresponding to $\sigma \in \Sigma$, and choose a base point $\daleth_\sigma \in X_\sigma^+$; let $U_\sigma \subset H_\sigma := H(F^+ \times_\sigma \mathbb{R})$ be its stabilizer. Then U_σ is isomorphic to the compact group $U(n) \times U(n)$. Without loss of generality, we may choose $\daleth \in M(n, F)$ to be a diagonal matrix whose entries have trace zero down to F^+, and let $\daleth_\sigma = \sigma(\daleth)$ for $\sigma \in \Sigma$. Then X_σ^+ is identified with the standard tube domain

$$X_{n,n} := \left\{ z \in M_n(\mathbb{C}) \mid \daleth_\sigma \left({}^t\bar{z} - z \right) > 0 \right\}.$$

With respect to this identification, any $h_\sigma = \begin{pmatrix} a_\sigma & b_\sigma \\ c_\sigma & d_\sigma \end{pmatrix} \in H_\sigma$ acts by

$$h_\sigma(z) = (a_\sigma z + b_\sigma)(c_\sigma z + d_\sigma)^{-1}.$$

Here a_σ, b_σ, c_σ, and d_σ are $n \times n$ matrices. We let $y_\daleth = (\sigma(\daleth))_{\sigma \in \Sigma} \in Y_W$.

In particular, the maximal parabolic subgroup $P \subset H$ stabilizes a point boundary component F_P of $\mathfrak{X}_\Sigma$. As we see below in Sects. 4.9 and 4.11, the rationality properties of holomorphic Eisenstein series on H are determined by their restriction to the boundary, which is the $H(\mathbf{A}_f)$-orbit of the Shimura variety attached to the pair (G_P, F_P) for a certain reductive subgroup $G_P \subset P$; the point F_P then corresponds to a homomorphism, also denoted $F_P : \mathbb{S} \rightarrow G_P(\mathbb{R})$.

Remark 4.2. The group U_σ is defined over $\sigma(F)$ and, for all $\gamma \in Gal(\overline{\mathbb{Q}}/\mathbb{Q})$,

$$\gamma(U_\sigma) = U_{\gamma(\sigma)}.$$

We thus have a way of comparing rational representations of the maximal compact subgroups U_σ, and thus the fibers of automorphic vector bundles, as σ varies. Note that, if γ does not fix $E(H, Y_W)$, then

$$U_{\gamma(\sigma)} \subset {}^\gamma H(\mathbb{R}),$$

where ${}^{\gamma}H$ is an inner form of H determined up to isomorphism by Langlands's formula for conjugation of Shimura varieties; see [H13, L79].

4.3 Induced Representations and Eisenstein Series

Denote by Δ the canonical map $\Delta : P \to GL_F(V^d) = GL_F(V)$. Then $M \xrightarrow{\sim} GL_F(V), m \mapsto (\Delta(m))$; the inverse map is

$$A \mapsto m(A) = diag(\left(A^*\right)^{-1}, A).$$

where $A^* = {}^tA^c$ is the transpose of the conjugate under the action of c. Define $\delta_P(\cdot) = |\det \circ \Delta(\cdot)|^n$.

Let $\chi = \otimes \chi_w$ be a character of $F^\times \backslash \mathbf{A}_F^\times$. For $s \in \mathbb{C}$ let

$$I(\chi, s) = Ind_{P(\mathbf{A})}^{H(\mathbf{A})}\left(\chi(\det \circ \Delta(\cdot)) \cdot \delta_P^{-s/n}(\cdot)\right)$$

with the induction smooth and unitarily normalized. This factors as a restricted tensor product

$$I(\chi, s) = \otimes_v I_v(\chi_v, s),$$

with v running over the places of $\mathbb{Q}$, $I_v(\chi_v, s)$ the analogous local induction from $P(\mathbb{Q}_v)$ to $H(\mathbb{Q}_v)$, and $\chi_v = \otimes_{w|v}\chi_w$.

Let $s \mapsto \phi_s \in I(\chi, s)$ be a section, in the sense in which that word is used in the theory of Eisenstein series. We form the *standard (non-normalized)* Eisenstein series,

$$E(\phi_s, h) = \sum_{\gamma \in P(F^+)\backslash H(F^+)} \phi_s(\gamma h). \tag{4.2}$$

If χ is unitary, this series is absolutely convergent, uniformly on compact subsets, for $\mathrm{Re}(s) > \frac{n}{2}$, and defines an automorphic form on $H(\mathbf{A})$. We always assume $\phi_s \in I(\chi, s)$ to be K-finite for a maximal compact subgroup $K \subset H(\mathbf{A})$; then in particular the Eisenstein series $E(\phi_s, h)$ admit a meromorphic continuation in s.

Let $m \geq n$ be a positive integer, which we view as the dimension of a positive-definite hermitian vector space V'. Assume

$$\chi|_{\mathbf{A}_{F^+}^\times} = \varepsilon_{\mathcal{K}}^m \tag{4.3}$$

Then the main result of [Tan] states that the possible poles of $E(\phi_s, h)$ are all simple. Moreover, those poles in the right half plane $Re(s) \geq 0$ can only occur at the points in the set

$$\frac{n-\delta-2r}{2}, \qquad r=0,\ldots,[\frac{n-\delta-1}{2}] \tag{4.4}$$

where $\delta = 0$ if m is even and $\delta = 1$ if m is odd.

Using the theta correspondence between $U(V)$ and $U(V')$, we consider the *Siegel-Weil sections* $\phi_s \in I(\chi, s)$ as in [H08]: these are the sections defined by the functions of the form ϕ_Φ introduced in §2.2.3 of [H08], where V is used to denote what we are now calling V' (and where the presence of the similitude factor introduces an additional complication, irrelevant here). Given a unitary character χ and a Siegel-Weil section $f \in I(\chi, s)$, we put

$$\phi_s := \phi_{\chi,s} := \phi$$
$$E_\phi(s,h) := E(\phi_s, h).$$

4.4 Automorphic Line Bundles on the Doubled Group

Automorphic line bundles on $Sh(W, \Sigma)$ are determined by two sets of parameters $(m_\sigma, k_\sigma)_{\sigma\in\Sigma}$. If we note that, for $U(W)$, $r_\sigma = s_\sigma = n$ for all $\sigma \in \Sigma$, then the parameter $(m_\sigma, k_\sigma)_{\sigma\in\Sigma}$ corresponds to the representations with parameters

$$(a_{1,\sigma} = \cdots = a_{n,\sigma} = m_\sigma;\ a_{n+1,\sigma} = \cdots = a_{2n,\sigma} = k_\sigma) \tag{4.5}$$

in the notation of (3.2). This corresponds to the representation $\otimes_\sigma \det^{m_\sigma} \otimes \otimes_\sigma \det^{k_\sigma}$ of the maximal compact subgroup $U(n)^\Sigma \times U(n)^\Sigma = \prod_{\sigma\in\Sigma} U_\sigma$ of $U(W)(\mathbb{R})$. In applications we only need to consider parameters in which $m_\sigma = m - a_\sigma$ and $k_\sigma = m + b_\sigma$ for all σ, for some Σ-tuple of pairs of integers (a_σ, b_σ) to be introduced below.

4.5 Automorphic Forms on the Point Boundary Shimura Variety

In [Gu16], Guerberoff identifies the Shimura datum attached to the point boundary component of the Shimura variety attached to $GU(W)$. He denotes the datum (G_P, F_P) but we will call it (G'_P, F'_P); then $G'_P = G_h \cdot A_P$, where, for any $\mathbb{Q}$-algebra R,

$$G_h(R) = \{\beta \cdot I_{2n} \mid \beta \in (F \otimes R)^\times, N_{F/F^+}(\beta) \in R^\times\};$$
$$A_P = \{d(\mathbf{a}, \mathbf{d}) := \begin{pmatrix} \mathbf{a}I_n & 0 \\ 0 & \mathbf{d}I_n \end{pmatrix}, \mathbf{a}, \mathbf{d} \in R^\times\}. \tag{4.6}$$

Here N_{F/F^+} is shorthand for the natural map $(F \otimes R)^\times \to (F^+ \otimes R)^\times$. In other words, $G_h = GU(1)$, in the previous notation, and $A_P \xrightarrow{\sim} \mathbb{G}_m \times \mathbb{G}_m$. Moreover,

$$F'_P(z) = \begin{pmatrix} z\bar{z}I_n & 0 \\ 0 & I_n \end{pmatrix},$$

diagonally embedded in $GU(n,n)^\Sigma \cap GU(W)(\mathbb{R})$.

The factor $\mathbf{a}$ in (4.6) is superfluous, because $G_h \cap A_P$ contains all elements of the form $d(\mathbf{a}, \mathbf{a})$. Let $[\beta, \mathbf{a} = 1, \mathbf{d}]$ denote a typical element of G'_P in the coordinates of (4.6). Then $G_P = G'_P \cap U(W)$, in other words is the group of triples $[\beta, 1, \mathbf{d}]$ with $N_{F/F^+}(\beta) \cdot \mathbf{d} = 1$. Since the coordinate $\mathbf{d} \in \mathbb{G}_m$ is thus superfluous, we have an isomorphism

$$GU(1) \xrightarrow{\sim} G_P; \beta \mapsto [\beta, 1, N_{F/F^+}(\beta)^{-1}]. \tag{4.7}$$

Thus we write $[\beta]$ for the typical element of $G_P \xrightarrow{\sim} GU(1)$.

It follows easily from (2.3) and (2.2) that

$$F_P(z) = \begin{pmatrix} zI_n & 0 \\ 0 & \bar{z}^{-1}I_n \end{pmatrix} = [z] \in G_P \subset U(n,n)^\Sigma.$$

Then

Definition 4.5.1. *Let $\alpha : G_P(F^+)\backslash G_P(\mathbf{A}) \to \mathbb{C}^\times$ be a Hecke character of G_P. Let $(a,b) = (a_\sigma, b_\sigma)_{\sigma\in\Sigma})$ be a $|\Sigma|$-tuple of pairs of integers satisfying $a_\sigma + b_\sigma = -\nu$ for some fixed integer ν, called the weight of (a,b). We say α is of type $(m,(a,b)) = (m,(a_\sigma,b_\sigma)_{\sigma\in\Sigma})$ if it is of the form*

$$\alpha = || \bullet ||^m \cdot \alpha_0$$

where the restriction α_∞ of α to $G_P(F^+_\infty)$ satisfies

$$\alpha_{0,\sigma}([\beta]) = \beta^{-a_\sigma} c(\beta)^{-b_\sigma}$$

(see [EHLS], *§4.4). For $h_\infty = (h_\sigma)_{\sigma\in\Sigma} \in U(n,n)^\Sigma$, we define*

$$\begin{aligned} J_{\alpha,\sigma} &:= J_{m,(a_\sigma,b_\sigma)}(h_\sigma) = \det(J(h_\sigma))^{-m+a_\sigma} \cdot \det(J'(h_\sigma))^{-m+b_\sigma}; \\ J_\alpha(h_\infty) &:= J_{m,(a,b)}(h_\infty) = \prod_{\sigma\in\Sigma} J_{\alpha,\sigma}(h_\sigma) \end{aligned} \tag{4.8}$$

Here the automorphy factors $J(\bullet)$ and $J'(\bullet)$ are defined as in [EHLS], §4.4.2, relative to the chosen base point $y_\daleth \in Y_W$ (corresponding to the element $\daleth \in F$ in *loc. cit.*).

Let $M^{der} \subset M$ denote the derived subgroup of M. A function

$$\phi : (P(F^+) \cdot N(\mathbf{A})M^{der}(\mathbf{A}))\backslash H(\mathbf{A}) \to \mathbb{C}$$

is said to be of type $(m, (a, b))$ if it is of the form

$$(h_\infty, h_f) \mapsto J_{m,(a,b)}(h_\infty) \otimes \phi_f(h_f)$$

for some ϕ_f on $(N(\mathbf{A}_f)M^{der}(\mathbf{A}_f))\backslash H(\mathbf{A}_f)$.

Note that we have

$$\alpha \circ \det(F_P(z)) = (z\bar{z})^{[F^+:\mathbb{Q}]nm} \cdot \prod_\sigma z^{-a_\sigma}\bar{z}^{-b_\sigma} \tag{4.9}$$

Lemma 4.6. *The space $I(m, (a, b))$ of functions of type $(m, (a, b)))$ decomposes as the direct sum of subspaces*

$$I(\alpha) = \{J_{m,(a,b))}(h_\infty) \otimes \phi_f(h_f)\}$$

where $\phi_f(h_f) \in I_{P(\mathbf{A}_f)}^{H(\mathbf{A}_f)} \alpha_f \circ \det$.

This follows from the decomposition of $I(m, (a, b))$ under the right action of $M(\mathbf{A}_f)$. We let $I(m, (a, b))(\mathbf{A}_f)$ be the space of $\phi_f(h_f)$ such that $J_{m,(a,b))}(h_\infty) \otimes \phi_f(h_f) \in I(m, (a, b))$, viewed as a space of functions on $(N(\mathbf{A}_f)M^{der}(\mathbf{A}_f))\backslash H(\mathbf{A}_f)$.

For $\sigma \in \Sigma$, let $\sigma^+ : F^+ \to \mathbb{R}$ denote the restriction of σ to F^+; let

$$H_\sigma = H(F^+ \otimes_{\sigma^+} \mathbb{R}), \ \ \mathfrak{h}_\sigma = Lie(H_\sigma), \ \ U(V)_\sigma = U(V \otimes_{F^+,\sigma} \mathbb{R}).$$

Let $U_\sigma = U(r_\sigma) \times U(s_\sigma)$ be the maximal compact subgroup of $U(V)_\sigma$ (with respect to the chosen base point $y \in Y_V$, see the discussion following Lemma 2.3). Let $\alpha = \alpha_0 \cdot || \bullet ||^m$ be an algebraic Hecke character as above, with $m \geq 0$. Define $\mathbb{D}(\alpha_\sigma) = \mathbb{D}(m, \alpha_{0,\sigma})$ to be the holomorphic $(\mathfrak{h}_\sigma, U_\sigma)$-module with highest U_σ-type

$$\begin{aligned}\Lambda(\alpha_\sigma) &= \Lambda(-m, (a_\sigma, b_\sigma)) \\ &= (m - b_\sigma, m - b_\sigma, \ldots, m - b_\sigma; -m + a_\sigma, \ldots, -m + a_\sigma)\end{aligned}$$

in the notation of [H97, (3.3.2)] (with the character on the $\mathbb{R}$-split center omitted). We define a map of $(U(\mathfrak{h}_\sigma), U_\sigma)$-modules

$$\iota(\alpha_\sigma) : \mathbb{D}(\alpha_\sigma) \to C^\infty(H_\sigma) \tag{4.10}$$

as follows. Let $v(\alpha_\sigma)$ be the tautological generator of the $\Lambda(m, \alpha_{0,\sigma})$-isotypic subspace (highest U_σ-type subspace) of $\mathbb{D}(m, \alpha_{0,\sigma})$. Let

$$\iota(\alpha_\sigma)(v(\alpha_\sigma)) = J_{\alpha,\sigma},$$

defined as in (4.8), and extend this to a map of $U(\mathfrak{h}_\sigma)$-modules. Let $C(H_\sigma, \alpha_\sigma)$ denote the image of $\iota(\alpha_\sigma)$.

We let $Sh(W, \Sigma)_{G_P}$ denote the point boundary stratum of the (adèlic) Shimura variety $Sh(W, \Sigma)$, corresponding to the maximal parabolic subgroup $P \subset U(W)$. This is a totally disconnected 0-dimensional proscheme over $E(W, \Sigma)$. As in [H86], §8, there is a line bundle $\mathcal{E}_{m,(a,b)}$ on $Sh(W, \Sigma)_{G_P}$ and an isomorphism (*canonical trivialization*)

$$\begin{aligned} Triv_{m,(a,b)} : \quad & H^0(Sh(W, \Sigma)_{G_P}, \mathcal{E}_{m,(a,b)}) \xrightarrow{\sim} \\ & \{J_{m,(a,b)}(h_\infty) \otimes \phi_f(h_f), \phi_f \in I(m, (a, b))(\mathbf{A}_f)\} \end{aligned} \tag{4.11}$$

normalized as in the discussion in [H08, (2.4)].[2] If $\xi \in H^0(Sh(W, \Sigma)_{G_P}, \mathcal{E}_{m,(a,b)})$ we include ξ in the subscript on the right-hand side:

$$Triv_{m,(a,b)}(\xi) = J_{m,(a,b)}(h_\infty) \otimes \phi_{\xi,f}(h_f).$$

There is a number field $E(m, (a, b)) \supset E(G_P, F_P)$ such that the vector bundle $\mathcal{E}_{m,(a,b)}$ has a canonical model over $E(m, (a, b))$, compatible with the canonical model of $Sh(W, \Sigma)_{G_P}$ over $E(G_P, F_P)$. Moreover, the group $Gal(\overline{\mathbb{Q}}/\mathbb{Q})$ acts on the set of pairs $(Sh(W, \Sigma)_{G_P}, \mathcal{E}_{m,(a,b)})$ through its natural action on the set of CM types Σ and parameters $(m, ((a, b)_\sigma)_{\sigma \in \Sigma})$ (the m is invariant under $Gal(\overline{\mathbb{Q}}/\mathbb{Q})$; see [H13], §4.1 for the action).

Proposition 4.7. *Fix a Σ-tuple of pairs of integers $(a, b) = (a_\sigma, b_\sigma)$ of weight v, and consider characters $\alpha : G_P(F^+)\backslash G_P(\mathbf{A}) \to \mathbb{C}^\times$ of type $(m, (a, b))$. For each such α let $E(\alpha) \supset E(m, (a, b))$ be the field of coefficients of α; it is the field generated by the values of α on $G_P(\mathbf{A}_f)$. Let*

$$H^0(Sh(W, \Sigma)_{G_P}, \mathcal{E}_{m,(a,b)})[\alpha] \subset H^0(Sh(W, \Sigma)_{G_P}, \mathcal{E}_{m,(a,b)})$$

be the subspace corresponding to $I(\alpha)$ with respect to the decomposition of Lemma 4.6 and the isomorphism $Triv_{m,(a,b)}$ (4.11). There is a constant $p_U(\alpha, \Sigma, W) \in \mathbb{C}^\times$ with the property that, for any extension $L/E(\alpha)$, the section $\xi \in H^0(Sh(W, \Sigma)_{G_P}, \mathcal{E}_{m,(a,b)})[\alpha]$ is rational over L if and only if

$$Triv_{m,(a,b)}(\xi) = J_{m,(a,b)}(h_\infty) \otimes \phi_{\xi,f}(h_f) \tag{4.12}$$

where $p_U(\alpha, \Sigma, W)^{-1}\phi_{\xi,f}(h_f)$ takes values in L.

[2] The ϕ_f should be understood as the constant terms of Fourier expansions of the Eisenstein series attached to the functions $J_{m,(a,b)}(h_\infty) \otimes \phi_f(h_f)$.

Proof. This is the analogue of Proposition 4.3.1 of [Gu16] in the setting of the Shimura variety attached to $U(W)$. We may restrict $\mathcal{E}_{m,(a,b)}$ to any $G_P(\mathbf{A}_f)$-orbit in $Sh(W, \Sigma)_{G_P}$ and identify the orbit with an $H(\mathbf{A}_f)$-translate of the Shimura variety $Sh(G_P, F_P)$; then the restriction of $\mathcal{E}_{m,(a,b)}$ to the orbit is an automorphic vector bundle on the toric Shimura variety $Sh(G_P, F_P)$. As such, it has a canonical period invariant $p_U(\alpha, \Sigma, W) \in \mathbb{C}/E(m, (a, b))$ that satisfies the analogue of (4.12) for sections of the bundle over $H(\mathbf{A}_f)$-translate. Since the action of $H(\mathbf{A}_f)$ respects both the property of (4.12) and the rational structure of $\mathcal{E}_{m,(a,b)}$, the Proposition follows from the analogous statement for $Sh(G_P, F_P)$. □

The constant $p_U(\alpha, \Sigma, W)$ is a period attached to automorphic forms on the CM Shimura variety $Sh(G_P, F_P)$, and can thus be related to standard CM periods $p_U(\alpha, \sigma, W)$, defined as follows. We identify G_P with $GU(1)$, as in (4.7). For any $\sigma \in \Sigma$ we let

$$h_\sigma : \mathbb{C}^\times \to GU(1)_\infty = \prod_{\sigma \in \Sigma} \mathbb{C}^\times$$

be the inclusion of $\mathbb{C}^\times$ as the σ-factor of the above product. The pair $(GU(1), h_\sigma)$ is thus a Shimura datum.

On the other hand, as in [Gu16, Prop. 4.3.1] (for α trivial) or [H97, Lemma 3.3.5.3] (for $F^+ = \mathbb{Q}$), there is an analogous invariant $p(\alpha, \Sigma, W)$ attached to the point boundary component of the Shimura variety $Sh(GU(W), \Sigma)$. As in the proof of Lemma 3.6, the ratio of the factors comes from the pullback of $\alpha \circ \det$ to the factor $GU(1)$ of the map u of (2.2)—we use the fact that $u(\mathbb{G}_m \cdot [U(1) \times \{1\}]) \subset G_P$. Here the determinant is taken on the Levi factor $GL(n)$ of $P \subset U(W)$, whereas in Lemma 3.6, the determinant is the map $\det : U(-V) \to U(1)$.

Then Shimura's product relations again imply

Proposition 4.8. *For any motivic Hecke character ω of H_F we have*

$$p_U(\omega, \Sigma, W) = p(\omega \circ \det, c\Sigma)^{-1} \cdot p(\omega, \Sigma, W) = p(\omega, c\Sigma)^{-n} \cdot p(\omega, \Sigma, W). \tag{4.13}$$

4.9 Holomorphic Eisenstein Series: Absolutely Convergent Case

Fix a Σ-tuple of pairs of integers $(a, b) = (a_\sigma, b_\sigma)$ of weight ν. We now fix $\alpha = \alpha_0$ to be a Hecke character of type $(0, (a, b))$, $\alpha_m = ||\bullet||^{m-\frac{n}{2}} \cdot \alpha$.

Let $\phi \in I(\alpha_m) \subset I(m - \frac{n}{2}, (a, b))$. Extend $\phi = \phi_0$ to a section $s \mapsto \phi_s \in I(\alpha, s + m - \frac{n}{2})$, in the sense used in Sect. 4.3, and define $E(\phi, s) := E(\phi_s)$ as in (4.2).

Proposition 4.10. *If $m > n - \frac{\nu}{2}$ then $E(\phi, 0)$ converges absolutely to a holomorphic automorphic form which defines a section*

$$E(\phi, 0) \in H^0(Sh(W, \Sigma), \mathcal{E}_m^{can}).$$

This proposition follows from the analogous statement for the Shimura variety attached to $GU(W)$. Guerberoff ([Gu16], §4.3) treats the case when α is trivial, but the general case is no more difficult. It then follows from Proposition 4.7 and the results of §8 of [H86] that

Corollary 4.10.1. *Let L be a field containing $E(\alpha)$. Then the section $E(\phi, 0) \in H^0(Sh(W, \Sigma), \mathcal{E}_m^{can})$ is rational over L if and only if $p(\alpha, \Sigma)^{-1}\phi_f$ takes values in L.*

Here and in what follows we are thinking of $\phi_f = \phi_{0,f} \in I(\alpha, 0)$ as a complex valued function on $H(\mathbf{A}_f)$, and the condition in the corollary is that the values be L-rational multiples of the period invariant $p(\alpha, \Sigma)$.

4.11 Holomorphic Eisenstein Series: Application of the Siegel-Weil Formula

We continue with the hypotheses of the previous section, but now assume $n \geq m \geq \frac{n-\nu}{2}$. We assume $\chi = \alpha$ satisfies the hypothesis (4.3), with m replaced by $m_0 := 2m + \nu$ (cf. Theorem 4.3 of [H08], where s_0 is what we are calling m, and m is what we are calling m_0, and $\kappa = -\nu$). and we take ϕ to be a *Siegel-Weil section*, in the sense of [H08].

Theorem 4.14. *Suppose $n \geq m \geq \frac{n-\nu}{2}$. Then the function $E(\phi, s)$ has a meromorphic continuation whose value at $s = 0$ is holomorphic and defines a section $E(\phi, 0) \in H^0(Sh(W, \Sigma), \mathcal{E}_m^{can})$. Moreover, with L as in the previous Corollary, the section $E(\phi, 0) \in H^0(Sh(W, \Sigma), \mathcal{E}_m^{can})$ is rational over L if and only if $p(\alpha, \Sigma)^{-1}\phi_f$ takes values in L.*

Proof. This is essentially the main theorem of [H08]. In fact, Corollary 3.3.3 of [H08] is proved in the setting of similitude groups, and in general is only valid on a subgroup of index 2 of $GU(W)(\mathbf{A})$. For the unitary group the method of [H08] works without restriction. □

5 Differential Operators

The method of [H97] and [Gu16] is based on constructing a family of differential operators that take holomorphic Eisenstein series on $Sh(W, \Sigma)$ to holomorphic automorphic forms on $Sh(V, \Sigma) \times Sh(-V, \Sigma)$. The parameters for these differential operators were worked out in §4.2 of [Gu16] and, with different notation, in [EHLS], §4.4. In both cases this was done in the setting of similitude groups, rather than unitary groups, but the differential operators are insensitive to the similitude factor and can be indexed by the same parameters. We follow [EHLS] because it is slightly more general. If κ is a parameter for $Sh(V,\sigma)$, then $\kappa^\flat$ is defined in [EHLS] (6.2.5).

5.1 Parameters for Differential Operators

Theorem 5.1 (Gu16, §4.2). *Let κ be a parameter in* (3.2)*, and let $\mathcal{E}_\kappa$ be the corresponding automorphic vector bundle on $Sh(V,\Sigma)$. Let $\kappa^\flat$ be the corresponding (dual) parameter for $Sh(-V,\Sigma)$ (see* [EHLS, (6.2.5)]*) and define*

$$\mathcal{E}_{\kappa^\flat}((a,b)) = \mathcal{E}_{\kappa^\flat(a,b)}$$

where, if $\kappa^\flat = (\kappa_\sigma^\flat, \sigma \in \Sigma)$, then $\kappa^\flat(a,b) = \kappa_\sigma^\flat \otimes \alpha_{0.\sigma}$ in the notation of [EHLS, (4.4.6)]. *There exists a holomorphic differential operator*

$$\Delta(m,(a,b),\kappa) : \mathcal{E}_{m,(a,b)}|_{Sh(V,\Sigma)\times Sh(-V,\Sigma)} \to \mathcal{E}_\kappa \boxtimes \mathcal{E}_{\kappa^\flat}((a,b)),$$

defined over the reflex field $E(\kappa,(a,b))$ and equivariant with respect to $G_V(\mathbf{A}_f) \times G_{-V}(\mathbf{A}_f)$, if and only if κ has the form $(\kappa_{\sigma,i}, \sigma \in \Sigma, 1 \le i \le n)$ where, for all $\sigma \in \Sigma$,

$$(\kappa_{\sigma,1},\ldots,\kappa_{\sigma,r_\sigma}) = (-m + b_\sigma - c_{r_\sigma},\ldots,-m+b_\sigma - c_1);$$

$$(\kappa_{\sigma,r_\sigma+1},\ldots,\kappa_{\sigma,n}) = (m - a_\sigma + d_1,\ldots,m - a_\sigma + d_{s_\sigma}),$$

with

$$c_1 \ge \cdots \ge c_{r_\sigma} \ge 0;\ d_1 \ge \cdots \ge d_{s_\sigma} \ge 0.$$

Moreover, the space of such differential operators is of dimension 1 *over $E(\kappa,(a,b))$.*

The differential operator $\Delta(m,(a,b),\kappa)$ is obtained by applying a non-holomorphic (Maass) operator to a section of $\mathcal{E}_{m,(a,b)}$, viewed as a C^∞ automorphic form on $U(W)(\mathbf{A})$, and then restricting the result to the subvariety $Sh(V,\Sigma) \times Sh(-V,\Sigma)$. For future use, we denote the Maass operator $D^\infty(m,(a,b),\kappa)$, and we define $G_3 = G_V \times G_{-V}$, as in [EHLS].

Definition 5.2. *Let (a,b) and κ be as in the statement of Theorem 5.1. We say that $m \ge \frac{n-\nu}{2}$ is critical for κ and (a,b) if the above inequalities are satisfied for all σ.*

5.2 Parameters for Nearly Holomorphic Eisenstein Series

Let m be critical for κ and (a,b). Let $\phi \in I(\alpha_m)$ be as in Proposition 4.10 or Theorem 4.14 depending on m. We write $E(m,\phi) = E(\phi, 0, h)$, emphasizing that ϕ is defined relative to the parameter m (although the m is superfluous in the notation). We assume L is a field as in one of those statements, so that

Hypothesis 5.3. *The function $\phi \in I(\alpha_m)$ has the property that the section $E(m,\phi) \in H^0(Sh(W,\Sigma), \mathcal{E}_m^{can})$ is rational over L.*

We also define

$$E^W(m,\phi) = p(\alpha,\Sigma)^{-1} E(m,\phi);$$

this is of course attached to the L-rational function ϕ_f on the finite adèles.

We write

$$\begin{aligned} E(m,\phi,\kappa) &= \Delta(m,(a,b),\kappa)(E(m,\phi)) \\ &\in H^0(Sh(V,\Sigma)\times Sh(-V,\Sigma), \mathcal{E}_\kappa \boxtimes \mathcal{E}_{\kappa^\flat}((a,b))); \end{aligned} \tag{5.1}$$

$$\begin{aligned} E^W(m,\phi,\kappa) &= \Delta(m,(a,b),\kappa)(E^W(m,\phi)) \\ &\in H^0(Sh(V,\Sigma)\times Sh(-V,\Sigma), \mathcal{E}_\kappa \boxtimes \mathcal{E}_{\kappa^\flat}((a,b))). \end{aligned} \tag{5.2}$$

Corollary 5.2.1. *Under Hypothesis 5.3, the sections $E(m,\phi,\kappa)$ are rational over L.*

We also write $E(m,\phi,h)$, $E^W(m,\phi,\kappa,h)$, etc. with $h \in U(W)(\mathbf{A})$, when it is necessary to emphasize that the Eisenstein series are functions on the adèle group. To distinguish the Eisenstein series on $U(W)(\mathbf{A})$ from its restriction to the subgroup $G_3(\mathbf{A})$, we write

$$E^\infty(m,\phi,\kappa,h) = D^\infty(m,(a,b),\kappa)(E(m,\phi))(h).$$

6 The Doubling Integral

6.1 Zeta Integrals

In this section, we briefly summarize key details of the doubling method, which we use to obtain zeta integrals. The doubling method holds for general classes of cuspidal automorphic representations π of $G_V(\mathbf{A})$, but we assume the local factors at Archimedean primes are discrete series representations. By Theorem 3.4, this implies that, for any finite prime v, the component part π_v of π has a model over a number field $E(\pi)$.

Denote by $\mathcal{O}^+$ the ring of integers of F^+. We write $G_V(\mathbf{A}) = \prod_v' G_{V,v}$, with the (restricted) products over all the places of F^+ and $G_{V,v} = G_V(F_v^+)$. Similarly, we write $H(\mathbf{A}) = H_\infty \times \prod_v' H_v$ and $P(\mathbf{A}) = P_\infty \times \prod_v' P_v$.

Let π be an irreducible cuspidal anti-holomorphic automorphic representation of $G_V(\mathbf{A})$, and let $\pi^\vee$ be its contragredient, and set

$$\pi' := \pi^\vee; \;\; \pi'_\alpha = \pi^\vee_\alpha = \pi' \otimes (\alpha \circ \det)^{-1}$$

with α an algebraic Hecke character of weight ν as above. We view $\pi^\vee_\alpha$ as an anti-holomorphic automorphic representation of $G_{-V}(\mathbf{A})$ and as a *holomorphic* automorphic representation of $G_V(\mathbf{A})$. Let S_π be the set of finite primes v of $\mathcal{O}^+$ for which π_v is ramified. Before introducing the zeta integral for π, we need to explain what it means for a function in π to be factorizable over places in F^+. We fix non-zero unramified vectors $\varphi_{w,0}$ and $\varphi'_{w,0}$ in π_w and $\pi^\vee_w$, respectively, for all finite places w outside S_π, and choose factorizations compatible with the unramified choices:

$$\pi \xrightarrow{\sim} \pi_\infty \otimes \pi_f; \;\; \pi_f \xrightarrow{\sim} \pi^{S_\pi} \otimes \pi_{S_\pi}; \tag{6.1}$$

and analogous factorizations for $\pi^\vee$ and $\pi^\vee_\alpha$. Let $\varphi \in \pi^{K_{S_\pi}}$, $\varphi' \in \pi^{\vee, K_{S_\pi}}$; we think of φ and φ' as forms on G_V and G_{-V}, respectively. We suppose they decompose as tensor products with respect to the above factorizations:

$$\varphi = \otimes_v \varphi_v; \;\; \varphi' = \otimes_v \varphi'_v \tag{6.2}$$

with φ_v and φ'_v equal to the chosen $\varphi_{v,0}$ and $\varphi'_{v,0}$ when $v \notin S_\pi$. We write equalities but the formulas we write below depend on the factorizations in (6.1) and its counterpart for $\pi^\vee$ and $\pi^\vee_\alpha$. We write

$$\varphi'_\alpha = \varphi' \otimes \alpha^{-1} \circ \det \in \pi^\vee_\alpha.$$

If $L \supset E(\pi)$, we will call a vector $\varphi_v \in \pi_v$ rational over L if it is rational with respect to the factorization (6.2), with respect to the $E(\pi)$-rational model introduced above.

Remark 6.2. Our conventions are as in [EHLS], except that we are writing $\pi' = \pi^\vee_\alpha$ instead of $\pi^\flat$. The test vectors in π and π' are denoted φ and φ', respectively, whereas the section for the Eisenstein series is denoted $\phi = \phi_s$. In [H97] different choices were made: φ was the datum for the Eisenstein series, whereas f and f' were cusp forms on G_V and G_{-V}, respectively.

We also fix local $G_V(F_v^+)$-invariant pairings $\langle\ ,\ \rangle_{\pi_v} : \pi_v \times \pi_v^\vee \to \mathbb{C}$ for all v such that $\langle \varphi_{v,0}, \varphi'_{v,0}\rangle_{\pi_v} = 1$ for all $v \notin S_\pi$. We assume these pairings are $E(\pi)$-rational for all v.

Normalizations are now as in Sect. 4. Let $\phi = \phi_s(\bullet) \in I(\alpha, s)$. We write $G_3 = G_V \times G_{-V}$ as above. Let $\varphi \in \pi$ and $\varphi' \in \pi'$, be factorizable vectors as above. The zeta integral for ϕ,φ, and φ' is

$$I(\varphi, \varphi', \phi, s) = \int_{G_3(\mathbb{Q})\backslash G_3(\mathbf{A})} E(\phi, s, (g_1, g_2))\varphi(g_1)\varphi'_\alpha(g_2)d(g_1, g_2).$$

Here the measure on G_3 is the Tamagawa measure $dg_V \times dg_{-V}$ discussed in Sect. 2.5. By the cuspidality of φ and φ' this converges absolutely for those values of s at which $E(\phi, s, h)$ is defined and defines a meromorphic function in s (holomorphic wherever $E(\phi, s, h)$ is). Moreover, it follows from the unfolding in [PSR] that $(\varphi, \varphi') \mapsto I(\varphi, \varphi', \phi, s)$ defines a $G_V(\mathbf{A})$-invariant pairing between π and π'. By the multiplicity one property Theorem 3.4, this implies that:

$$\text{If } \langle \varphi, \varphi' \rangle := \int_{G_V(\mathbb{Q})\backslash G_V(\mathbf{A})} \varphi(g)\varphi'(g)dg = 0 \text{ then } I(\varphi, \varphi', \phi, s) = 0 \text{ for all } s.$$

So we suppose

$$\langle \varphi, \varphi' \rangle \neq 0.$$

Then $\langle \varphi_v \otimes \varphi'_v \rangle_{\pi_v} \neq 0$ for all v. For $Re(s)$ sufficiently large, 'unfolding' the Eisenstein series then yields

$$I(\varphi, \varphi', \phi, s) = \int_{G_V(\mathbf{A})} \phi_s(g_V, 1)\langle \pi(g_V)\varphi, \varphi' \rangle_\pi dg_V.$$

Henceforward we assume $\phi(h) = \otimes_v \phi_v(h_v)$ with

$$\phi_v = \phi_{v,s} \in I_v(\alpha_v, s), \alpha_v = \bigotimes_{w|v} \alpha_w.$$

Then bearing in mind the formula (2.6) for the factorization of the measure dg_V, the last expression for $I(\varphi, \varphi', \phi, s)$ factors as

$$\begin{aligned} I(\varphi, \varphi', \phi, s) &= \sqrt{D_{F^+}}^{-n^2} \prod_v I_v(\varphi_v, \varphi_v, \phi_v, s) \cdot \langle \varphi, \varphi' \rangle, \quad \text{where} \\ I_v(\varphi_v, \varphi'_v, \phi_v, s) &= \frac{\int_{G_V(F_v^+)} \phi_{v,s}(g_v, 1)\langle \pi_v(g_v)\varphi_v, \varphi'_v \rangle_{\pi_v} dg_v}{\langle \varphi_v, \varphi'_v \rangle_{\pi_v}}. \end{aligned} \tag{6.3}$$

By hypothesis, the denominator of the above fraction equals 1 whenever $v \notin S_\pi$. We denote the integral in the numerator by $Z_v(s, \varphi_v, \varphi'_v, \phi_v)$.

Theorem 6.4 ([PSR, Li92, LR05]). *We have the identity*

$$I(\varphi, \varphi', \phi, s-\frac{n}{2})$$
$$=\sqrt{D_{F^+}}^{-n^2} B(\pi)_{\alpha,\varphi,\varphi'} \cdot \prod_{v\in S_\pi} Z_v(s-\frac{n}{2}, \varphi_v, \varphi'_v, \phi_v) d^{S_\pi}(s-\frac{n}{2}, \alpha)^{-1} L^{mot,S}(s, \pi, \alpha, St)$$

Here

$$d^{S_\pi}(s, \alpha) = \prod_{0\leq j<n-1} L^{S_\pi}(2s+n-j, \alpha^+ \cdot \varepsilon_{F/F^+}) \tag{6.4}$$

is the product of partial Hecke L-functions of F^+, as in [Gu16, §4.4], and $L^{mot}(s, \pi, \alpha, St)$ denotes the L-function attached to the standard representation of the L-group of G_V, in the motivic normalization. We have written α^+ for the restriction of α to the idèles of F^+. Thus $\alpha^+_\infty(t_\sigma) = t_\sigma^\nu$ for any σ, so that

$$\xi_{\alpha,0} := || \bullet ||^{-\nu} \cdot \xi_\alpha$$

is a Hecke character of F^+ of finite order. We then have (replacing the partial L-function by the complete L-function, since they are equal up to scalars in the field of coefficients)

$$d(s_0, \alpha) = \prod_{0\leq j<n-1} L(2s_0+n-j-\nu, \xi_{\alpha,0} \cdot \varepsilon_{F/F^+})$$
$$\sim \prod_{0\leq j<n-1} (2\pi i)^{(2s_0+n-j-\nu)[F^+:\mathbb{Q}]} \cdot \delta([\xi_{\alpha,0} \cdot \varepsilon_{F/F^+}])^{[\frac{n+1}{2}]} \cdot |D_{F^+}^{\frac{1}{2}}|^{[\frac{n}{2}]}. \tag{6.5}$$

Here $\delta(([\xi_{\alpha,0} \cdot \varepsilon_{F/F^+}])$ is defined as in [Gu16] (2.2.2) (essentially a version of the Gauss sum multiplied by a power of the square root of the discriminant) and $[\bullet]$ denotes the greatest integer function.

Now suppose φ and φ' correspond to elements of

$$[\varphi] \in H_!^{\dim Y_V}(Sh(V, \Sigma), \mathcal{E}_\kappa)[\pi]$$

and

$$[\varphi'] \in H_!^{\dim Y_V}(Sh(-V, \Sigma), \mathcal{E}_{\kappa^\flat} \otimes \Lambda_{\alpha_\infty^U})[\pi^\vee \otimes (\alpha \circ \det)^{-1}].$$

Corollary 6.2.1. *Suppose m is critical for κ and (a,b). Suppose*

$$E(m,\phi,\kappa,h) = \Delta(m,(a,b),\kappa)E(m,\phi)(h)$$

where $\phi \in I(m,\alpha)$. Then

$$I(\varphi,\varphi',\phi,m-\frac{n}{2})$$
$$= \sqrt{D_{F^+}}^{-n^2} B(\pi)_{\alpha,\varphi,\varphi'} \cdot \prod_{v\in S_\pi} Z_v(m-\frac{n}{2},\varphi_v,\varphi'_v,\phi_v)\frac{L^{\mathrm{mot},S}(m,\pi,\alpha,St)}{d^{S_\pi}(m-\frac{n}{2},\alpha)}$$

6.3 Review of the Local Theory

Notation is as in the previous section. We assume φ_v and φ'_v define $\overline{\mathbb{Q}}$-rational vectors for all v (equivalently, we assume φ defines a $\overline{\mathbb{Q}}$-rational section of an appropriate automorphic vector bundle). The following is then proved as in [H97, Lemma 3.5.6] or [Gu16, Lemma 4.5.2].

Lemma 6.4. *For every finite place v, there exists a finite section $\phi_{v,s}(\bullet) \in I_v(\alpha,s)$, rational over $E(\alpha)$ such that*

$$Z_{S_\pi}(m-\frac{n}{2},\varphi,\varphi',\phi) := \prod_{v\in S_\pi} Z_v(m-\frac{n}{2},\varphi_v,\varphi'_v,\phi_v) \in \overline{\mathbb{Q}}^\times.$$

Moreover, for any $\gamma \in Gal(\overline{\mathbb{Q}}/\mathbb{Q})$, we have

$$\gamma(Z_{S_\pi}(m-\frac{n}{2},\varphi,\varphi',\phi)) = Z_{S_{\gamma(\pi)}}(m-\frac{n}{2},\gamma(\varphi),\gamma(\varphi'),\gamma(\phi)).$$

If γ does not fix $E(\pi)\cdot E(\alpha)$ then $\gamma(\pi)$ is not generally an automorphic representation of G_V, but rather of some inner form, as in [H13], which is isomorphic to G_V at all finite places. Nevertheless, $\gamma(\pi)$ is coherent cohomological and the rationality of $\gamma(\varphi)$ and $\gamma(\varphi')$ are determined relative to that structure.

6.5 The Archimedean Theory

The following result of Garrett [H08] has already been used in §3.5 of [H97] and in Lemma 4.5.1 of [Gu16]; we record it here for future reference. Note that Garrett's computation is native to the unitary group and makes no reference to the similitude factor. We recall that the local factors are defined relative to the choice of Haar measure made in [H97].

Lemma 6.6. *For every archimedean place σ, with $\phi_{\sigma,s}(\bullet)$ defined as in Lemma 4.6, the local factor*

$$Z_\sigma(m-\frac{n}{2},\varphi_\sigma,\varphi'_\sigma,\phi_\sigma)\in\sigma(F)^\times.$$

We denote this factor $\mathfrak{z}_\sigma$.

Moreover, for every $\gamma\in Gal(\overline{\mathbb{Q}}/\mathbb{Q})$ and every $\sigma : F\hookrightarrow\mathbb{C}$, we have

$$\gamma(\mathfrak{z}_\sigma)=\mathfrak{z}_{\gamma\circ\sigma},$$

where the latter is viewed as a local zeta integral for the conjugate Shimura variety $\gamma(Sh(V,\Sigma))=Sh(V,\gamma(\Sigma))$.

Proof. We clarify the last assertion. Write $Z_\sigma=Z_\sigma(m-\frac{n}{2},\varphi_\sigma,\varphi'_\sigma,\phi_\sigma)$. Theorem 2.1 of [Ga08] asserts that a zeta integral Z'_σ is a $\sigma(F)$-rational multiple of a constant C, which in turn is a $\sigma(F)$-rational multiple of π^{pq}. As explained in [H08], taking into account the difference between Garrett's normalization of measures and the one given in (2.5), we can conclude that

$$\mathfrak{z}_\sigma:=Z_\sigma(m-\frac{n}{2},\varphi_\sigma,\varphi'_\sigma,\phi_\sigma)\in\sigma(F).$$

It remains to analyze the behavior under $Gal(\overline{\mathbb{Q}}/\mathbb{Q})$ of the $\sigma(F)$-rational constants $\mathfrak{z}_\sigma$ that arise in the computation. Following the argument in §2 of [Ga08], one observes that $\mathfrak{z}_\sigma$ is a product of two factors $\mathfrak{z}'_\sigma\cdot\mathfrak{z}''_\sigma$, determined by a $\sigma(F)$-rational structure on the representation W_κ of U_σ and on $Lie(\mathfrak{g}_\sigma)$. The factor $\mathfrak{z}'_\sigma$ is an element of $End_{\sigma(F)}(W_\kappa)$. Note that (bearing in mind Remarks 3.3 and 4.2) the group U_σ becomes isomorphic to $GL(r_\sigma)\times GL(s_\sigma)$ over F, and the representations of U_σ are thus defined over $\sigma(F)$. The factor $\mathfrak{z}''_\sigma$ is a normalization of the measure which comes down to the choice of ω_σ in Sect. 2.5.

Now it follows from (2.4) that, for $\gamma\in Gal(\overline{\mathbb{Q}}/\mathbb{Q})$,

$$\begin{aligned}\gamma(\omega_\sigma)&=\gamma[\sigma(\daleth)^{-\dim X_\sigma}\bigwedge_{i=1}^{\dim X_\sigma}dz_{\sigma,i}\wedge d\bar{z}_{\sigma,i}]\\&=\gamma[\sigma(\daleth)^{-\dim X_\sigma}\bigwedge_{i=1}^{\dim X_\sigma}dz_{\sigma,i}\wedge dz_{c\circ\sigma,i}]\\&=(\gamma\circ\sigma)(\daleth)^{-\dim X_\sigma}\bigwedge_{i=1}^{\dim X_\sigma}dz_{\gamma\circ\sigma,i}\wedge d\bar{z}_{c\circ\gamma\circ\sigma,i}\\&=\omega_{\gamma\circ\sigma}\end{aligned}$$

Thus

$$\gamma(\bigwedge_{\sigma\in\Sigma}\omega_\sigma) = \bigwedge_{\sigma'\in\gamma(\Sigma)}\omega_{\sigma'}.$$

Thus the collection of factors $\mathfrak{z}''_\sigma$ is equivariant with respect to $Gal(\overline{\mathbb{Q}}/\mathbb{Q})$. Noting that a similar rule applies to $\mathfrak{z}'_\sigma$, the equivariance property follows. □

7 Main Results

Theorem 7.1. *Let π be an automorphic representation of G_V of type κ, and let α be a Hecke character of F of type $(0,(a,b))$, of weight ν. Suppose $m \geq \frac{n-\nu}{2}$ is critical, in the sense of Definition 5.2. Define*

$$P(V,\alpha) = \prod_{\sigma\in\Sigma} p_F(\alpha,\sigma)^{-s_\sigma} p_F(\alpha, c\sigma)^{-r_\sigma}.$$

Then

(i) *If $m > n - \frac{\nu}{2}$, then*

$$L(m,\pi,St,\alpha) \sim_{E(\pi,\alpha,\Sigma)}$$
$$|D_{F^+}|^{\frac{\lfloor\frac{n+1}{2}\rfloor - n^2}{2}} (2\pi i)^{|\Sigma|(mn-\frac{n(n-1)}{2})} \cdot \delta(\varepsilon_F)^{\lfloor\frac{n+1}{2}\rfloor} \cdot P(V,\alpha)\cdot B(\pi)^{-1}. \tag{7.1}$$

Here $\delta(\varepsilon_F)$ is the period of the Artin motive attached to the quadratic character for the extension F/F^+, as in the introduction to [GL].

(ii) *Suppose (a) $n - \frac{\nu}{2} \geq m \geq \frac{n-\nu}{2}$, and (b) $m \neq \frac{n-\nu+1}{2}$. Then* (7.1) *remains true, provided α satisfies* (4.3), *i.e.*

$$\alpha|_{\mathbf{A}^\times_{F^+}} = \varepsilon^\nu_{\mathcal{K}}.$$

(iii) *Suppose $m = \frac{n-\nu+1}{2}$ (the near central point). Suppose that α satisfies* (4.3) *and, for some finite place v of F^+ that does not split in F, α_v does not belong to the finite set A_v of characters introduced in Theorem 3.4 of* [H07]. *Then* (7.1) *is valid for π and α.*

Proof. This is simply an adaptation of the proofs in [H97, H07, H08, Gu16] based on unitary similitude groups. When $F = \mathcal{K}$ is imaginary quadratic then the formula for $P(V,\alpha)$ specializes to the expression $p_{\mathcal{K}}(\alpha,1)^{-s}p_{\mathcal{K}}(\alpha^\iota,1)^{-r}$ at the end of the third line on p. 138 of [H97]. The subsequent lines of that calculation, when $(a,b) = (-\kappa,0)$, yield the expression $(2\pi i)^{\kappa s} g(\alpha_0)^s p_{\mathcal{K}}(\check{\alpha},1)^{r-s}$ that is the contribution of α to the final formula of [H97, Theorem 3.5.13]. In the general case, α's contribution in the computation made using the GU-periods is the expression $P(V,\alpha)$ given above. We thus just need to explain (a) why this doesn't change when we work with the Shimura variety $Sh(V, \Sigma)$; (b) the appearance of the power of the discriminant, and (c) why the period $B(\pi)^{-1}$ is the appropriate replacement for $P^{(s)}(\pi, *, \beta)$.

We interpret Corollary 6.2.1 in terms of our notation. We assume the Eisenstein series that occurs in the integral $I(m, \varphi, \varphi', \phi)$ is the function $E^{\infty}(m, \phi, \kappa, h)$, where

- ϕ_{∞} satisfies the conditions of Lemma 6.6, and
- $p_U(\alpha, \Sigma, W)^{-1}\phi_f$ takes values in $E(\pi, \alpha)$.

Suppose moreover that φ and φ' define L-rational classes in coherent cohomology. Then $I(m, \varphi, \varphi', \phi)$ is the cup product between the L-rational coherent cohomology classes, and thus

$$I(m, \varphi, \varphi', \phi) \in L. \tag{7.2}$$

On the other hand, it follows from Lemmas 6.4 and 6.6, and our hypothesis on ϕ_f, that

$$p(\alpha, \Sigma)^{-1} \prod_{v \in S_\pi} Z_v(s - \frac{n}{2}, \varphi_v, \varphi'_v, \phi_v) \in L. \tag{7.3}$$

Completing the partial Euler products, Theorem 6.4 and (3.13) then yields

$$\begin{aligned} L^{\text{mot}}(s, \pi, \alpha, St) &\sim p_U(\alpha, \Sigma, W) p_U(\alpha, -V)^{-1} B(\pi)^{-1} d(s - \frac{n}{2}, \alpha)^{-1} \sqrt{D_{F^+}}^{n^2} \\ &\sim p(\alpha, \Sigma, W) p(\alpha, -V)^{-1} B(\pi)^{-1} d(s - \frac{n}{2}, \alpha)^{-1} \sqrt{D_{F^+}}^{n^2} \end{aligned} \tag{7.4}$$

where we have used Lemma 3.6 and Proposition 4.8 to simplify the local factors. This explains point (a) above, because the factors related to α are now the same as in the similitude calculation. The discriminant factor in the final formula is the product of the factor $\sqrt{D_{F^+}}^{n^2}$ in (7.4) and the discriminant factor in the expression (6.5) for $d(s - \frac{n}{2}, \alpha)^{-1}$. Finally the power of π is the same as in Theorem 3.6.1 of [GL]. □

Remark 7.1. As far as I can tell, Shimura's proofs in [Sh97], which apply to holomorphic automorphic forms of scalar weight, do not require that α satisfy the hypothesis of (ii) when treating values below the range of absolute convergence. The proofs in [H08] use the Siegel-Weil formula for critical values in this range, whereas Shimura uses an explicit evaluation of the Fourier coefficients of holomorphic Eisenstein series below the range of absolute convergence. The treatment of these values in [EHLS]—where less attention is paid to rationality questions—is based on Shimura's calculations, but it applies as well to vector-valued forms. In particular, [EHLS] implicitly contains a proof of Theorem 7.1 (ii) and (iii) without the indicated restrictions on α.

The reasons for the restriction on α at the near central point are developed at length in [H07]. The point is roughly the following: for each finite place v of F^+ that is not split in F, there are two hermitian spaces $\mathcal{W}_v^+$ and $\mathcal{W}_v^-$ of dimension m. For given π_v and splitting character α_v, the theta lift of π_v to at least one of the $U(\mathcal{W}_v^{\pm})$ is non-trivial. At the central value $m = \frac{n-\nu}{2}$, exactly one of these lifts is non-trivial. Moreover, if v is archimedean then the theta lift of π_v is non-trivial to exactly one of the corresponding unitary groups (a totally definite one). The collection of local $\mathcal{W}_v^{\pm}$ is called *coherent* (following Kudla and Rallis) if and only if there is a global hermitian space $\mathcal{W}$ with the given localizations. Thus at the central value, there is exactly one global unitary group $U(\mathcal{W})$ to which the theta lift of π is non-trivial if and only if the central L-value does not vanish. On the other hand, if $m > \frac{n-\nu+1}{2}$, then it is proved in [H07] that the theta lifts to both $U(\mathcal{W}_v^{\pm})$ are non-trivial as long as π_v is tempered, which is known to be the case in our applications. However, at the near central point $m = \frac{n-\nu+1}{2}$, it is possible that π_v has a theta lift—for the splitting characters α_v—to only one of the $U(\mathcal{W}_v^{\pm})$. If the resulting collection of local hermitian spaces is incoherent then the Rallis inner product formula used in [H08] does not provide an interpretation of the near central value. It is shown in [H07] that this only happens at most for a finite set of α_v at each v; if α_v belongs to this finite set for all v, then π_v is locally a theta lift at all non-split finite v from a $U(\mathcal{W}_v)$ with $\dim \mathcal{W}_v = n - 1$. There seems to be no way a priori to exclude this possibility.

The most comprehensive analysis of the Rallis inner product formula, and of its implications for theta lifts, is contained in [GQT], which contains a complete proof of the second term identity.

8 Unitary and Similitude Periods

The map u of (2.2) is surjective on points over any field; indeed, the restriction of u to $\{1\} \times U(V)$ is just the identity on points. The image of $U(V)(\mathbb{Q})\backslash U(\mathbf{A})$ in $Sh(GU(V), X_V)$ is an open and closed variety that we denote $Sh^U(V)$. It follows that the integral defining $B(\pi)_{f,f'}$ in (3.10) can be viewed as an integral over $Sh^U(V)$. However, the forms f and f' chosen to be rational on $Sh(V, \Sigma)$ over some field L are not rational on $Sh^U(V)$. We assume $f \in \pi$ and $f' \in \pi^\vee$, as before; then there are constants

$$b_U(\pi, V) = (b_U(\pi, V, \tau), \tau \in \Sigma) \in (E(\pi) \otimes \mathbb{C})^\times;$$

$$b_U(\pi^\vee, V^\vee) = b_U(\pi^\vee, -V, \tau), \tau \in \Sigma) \in (E(\pi) \otimes \mathbb{C})^\times$$

such that $b_U(\pi, V)f$ and $b_U(\pi^\vee, V^\vee)f'$ are L-rational elements of

$$H^{\dim Y_V}(Sh^U(V), \mathcal{E}_\kappa) \text{ and } H^{\dim Y_V}(Sh^U(-V), \mathcal{E}_{\kappa^\flat})$$

respectively. Here I use the same notation to denote the automorphic vector bundles on $Sh(V, \Sigma)$ and on $Sh^U(V)$, although this is abusive because the former is attached to (for example) a representation of the stabilizer of $y \in Y_V$ whereas the latter is attached to a representation of the stabilizer of y in X_V.

Define

$$B(\pi)^U_{f,f'} = [b_U(\pi, V)f, F_\infty(b_U(\pi^\vee), V^\vee)f')]_{Ser} = b_U(\pi, V) \cdot b_U(\pi^\vee, V^\vee) B(\pi)_{f,f'},$$

where the second equality is clear because F_∞ and Serre duality are linear.

Lemma 8.1. *With notation as above, we have*

$$b_U(\pi, V) \cdot b_U(\pi^\vee, V^\vee) \sim 1.$$

In particular, $B(\pi) \sim_{E(\pi)} B(\pi)^U$.

Proof. Let ξ_π denote the central character of π; it is an automorphic representation of $U(1)$. Let $\pi_u = \pi \circ u$ be the pullback to G''_V of π, and let $\tilde{\pi}_u$ denote some extension of π_u to $GU(V) \times GU(1)$. Then we can factor $\tilde{\pi}_u$ as $\tilde{\pi} \otimes \tilde{\xi}_\pi^{-1}$, where $\tilde{\pi}$ is an extension of π to $GU(V)$ and $\tilde{\xi}_\pi$ is an extension of ξ_π from $U(1)$ to $GU(1)$. Define $\tilde{\pi}^\vee_u$, $\tilde{\pi}^\vee$, and $\tilde{\xi}^{-1}_{\pi^\vee}$ likewise. We can and do assume that $\tilde{\xi}_\pi$ is a motivic Hecke character and that

$$\tilde{\xi}^{-1}_{\pi^\vee} = \tilde{\xi}_\pi.$$

Now the statement is clear because the constants $b_U(\pi, V)$ and $b_U(\pi^\vee, V^\vee)$ depend only on the restrictions of $\tilde{\pi}_u$ and $\tilde{\pi^\vee}_u$ to the factor $(GU(1), X_1)$ of π and $\pi^\vee$, and these are $\tilde{\xi}_\pi^{-1}$ and $\tilde{\xi}_\pi$, respectively. More precisely, we view the Shimura variety $Sh(G''_V, X''_V)$ as an open and closed subvariety of the product $Sh(GU(V), X_V) \times Sh(GU(1), X_1)$ and one checks that the constants $b_U(\pi, V)$ and $b_U(\pi^\vee, V^\vee)$ are determined by viewing the pullbacks via u of f and f' as cohomology classes of this open and closed subvariety of the product $Sh(GU(V), X_V) \times Sh(GU(1), X_1)$, with coefficients in the bundles $\mathcal{E}_\kappa$ and $\mathcal{E}_{\kappa^\flat}$. Using the product structure, the result follows. □

Remark 8.2. Note that the coefficient field $E(\pi)$ in Lemma 8.1 is defined relative to the rational structure on $Sh(V, \Sigma)$, and in particular is an extension of the reflex field $E(V, \Sigma)$. This is in general strictly larger than the reflex field $E(GU(V), X_V)$. Thus if π extends to an automorphic representation $\tilde{\pi}$ of $GU(V)$, the fields $E(\pi)$ and $E(\tilde{\pi})$ are in general not the same. We use the same $E(\pi)$ in Proposition 8.3 below.

Proposition 8.3. *Suppose there is a character α and a critical value m of $L(s, \pi, \alpha, St)$ such that $L(m, \pi, \alpha, St) \neq 0$. Then*

$$B(\pi)^U \sim_{E(\pi)} Q^*(\pi)$$

where $Q^*(\pi)$ is the period that occurs in Corollary 3.7.2 of [GL].

Proof. We compare two formulas for the non-vanishing critical value $L(m, \pi, \alpha, St)$. The one in [GL] is of the form

$$L(m, \pi, \alpha, St) \sim_{E(\pi,\alpha)} A(\alpha, m) \cdot Q^*(\pi)^{-1}$$

and the one in Theorem 7.1 is of the form

$$L(m, \pi, \alpha, St) \sim_{E(\pi,\alpha)} A(\alpha, m) \cdot B(\pi)^{-1},$$

for the same non-vanishing $A(\alpha, m)$. Thus, letting α vary among its Galois conjugates, we have

$$B(\pi) \sim_{E(\pi)} Q^*(\pi),$$

and the Proposition follows from Lemma 8.1. □

Acknowledgments The immediate motivation for this paper was to provide a convenient comparison between periods computed on similitude groups, as in [H97, Gu16, GL], and periods computed on unitary groups, which appear in the Ichino-Ikeda and N. Harris conjectures. The comparison is applied to factorization of automorphic periods in [GHL]. I thank my co-authors Harald Grobner and Lin Jie for discussions of this point that made clear the need for a paper like this one. I also thank Wee Teck Gan for an exchange that helped to clarify the limits of the approach to special values based on the Rallis inner product formula. Finally, I thank the organizers of the Simons Symposium on Relative Trace Formulas—Werner Müller, Sug Woo Shin, and Nicolas Templier—for their invitation, for their eclectic choice of topics, and for encouraging me to contribute to the proceedings.

This work was partially supported by NSF Grant DMS-1701651. This work was also supported by the National Science Foundation under Grant No. DMS-1440140 while the author was in residence at the Mathematical Sciences Research Institute in Berkeley, California, during the Spring 2019 semester.

References

[B] G. Boxer, Torsion in the coherent cohomology of Shimura varieties and Galois representations. https://arxiv.org/abs/1507.05922

[Ca] A. Caraiani, Local-global compatibility and the action of monodromy on nearby cycles. Duke Math. J. **161**, 2311–2413 (2012)

[CHLN] L. Clozel, M. Harris, J.-P. Labesse, B.-C. Ngô, *The Stable Trace Formula, Shimura Varieties, and Arithmetic Applications*, vol. I. Stabilization of the Trace Formula (International Press, Boston, 2011)

[D79] P. Deligne, Valeurs de fonctions L et périodes d'intégrales, in *Automorphic Forms, Representations and L-Functions*, ed. by A. Borel, W. Casselman, vol. 33. Proceedings of the Symposium in Pure Mathematics of the American Mathematical Society (American Mathematical Society, Providence, 1979)

[EHLS] E. Eischen, M. Harris, J.-S. Li, C. Skinner, p-adic L-functions for unitary groups, Forum Math. Pi 8 (2020), e9

[GGP] W.T. Gan, B. Gross, D. Prasad, Symplectic local root numbers, central critical L values, and restriction problems in the representation theory of classical groups. Sur les conjectures de Gross et Prasad. I. Astérisque **346**, 1–109 (2012)

[GQT] W.T. Gan, Y. Qiu, S. Takeda, The regularized Siegel-Weil formula (the second term identity) and the Rallis inner product formula. Invent. Math. **198**, 739–831 (2014)

[Ga84] P. Garrett, Pullbacks of Eisenstein series; applications, in *Automorphic Forms of Several Variables (Katata, 1983)*. Progress in Mathematics, vol. 46 (Birkhuser, Boston, 1984), pp. 114–137

[Ga08] P. Garrett, Values of Archimedean zeta integrals for unitary groups, in *Eisenstein Series and Applications*. Progr. Math., vol. 258 (Birkhäuser, Boston, 2008), pp. 125–148

[GH] H. Grobner, M. Harris, Whittaker periods, motivic periods, and special values of tensor product L-functions. J. Ins. Math. Jussieu **15**, 711–769 (2016)

[GHL] H. Grobner, M. Harris, J. Lin, Deligne's conjecture for automorphic motives over CM-fields, Part I: factorization. arXiv:1802.02958 [math.NT] (2018)

[Gu16] L. Guerberoff, Period relations for automorphic forms on unitary groups and critical values of L-functions. Doc. Math. **21**, 1397–1458 (2016)

[GL] L. Guerberoff, J. Lin, Galois equivariance of critical values of L-functions for unitary groups. Preprint (2016)

[H86] M. Harris, Arithmetic vector bundles and automorphic forms on Shimura varieties, II. Compos. Math. **60**, 323–378 (1986)

[H90] M. Harris, Automorphic forms of $\bar{\partial}$-cohomology type as coherent cohomology classes. J. Diff. Geom. **32**, 1–63 (1990)

[H93] M. Harris, L-functions of 2×2 unitary groups and factorization of periods of Hilbert modular forms. J. Amer. Math. Soc. **6**, 637–719 (1993)

[H97] M. Harris, L-functions and periods of polarized regular motives. J. Reine Angew. Math. **483**, 75–161 (1997)

[H07] M. Harris, Cohomological automorphic forms on unitary groups. II. Period relations and values of L-functions, in *Harmonic Analysis, Group Representations, Automorphic Forms and Invariant Theory*. Lecture Notes Series, Institute for Mathematical Sciences National University of Singapore, vol. 12 (World Scientific Publishing, Hackensack, 2007), pp. 89–149

[H08] M. Harris, A simple proof of rationality of Siegel-Weil Eisenstein series, in *Eisenstein Series and Applications*. Progr. Math., vol. 258 (Birkhäuser, Boston, 2008), pp. 149–185

[H13] M. Harris, Beilinson-Bernstein localization over $\mathbb{Q}$ and periods of automorphic forms. Int. Math. Res. Notices **2013**, 2000–2053 (2013). Erratum. https://doi.org/10.1093/imrn/rny043

[HT01] M. Harris, R. Taylor, *The Geometry and Cohomology of Some Simple Shimura Varieties*. Annals of Mathematics Studies, vol. 151 (Princeton University Press, Princeton, 2001)

[HLTT] M. Harris, K.W. Lan, R. Taylor, J. Thorne, Res. Math. Sci. **3**, 37 (2016). https://doi.org/10.1186/s40687-016-0078-5

[KMSW] T. Kaletha, A. Minguez, S.W. Shin, P.-J. White, Endoscopic classification of representations: inner forms of unitary groups. Preprint (2014). arXive: 1409.3731

[KSZ] M. Kisin, S.W. Shin, Y. Zhu, The stable trace formula for certain Shimura varieties of abelian type. In preparation (2021)

[L79] R.P. Langlands, Automorphic representations, Shimura varieties, and motives, in *Proceedings of Symposia in Pure Mathematics*, ed. by E. Märchen, vol. XXXIII, part II (AMS, Providence, 1979), pp. 205–246

[LR05] E. Lapid, S. Rallis, On the local factors of representations of classical groups, in *Automorphic Representations, L-functions and Applications: Progress and Prospects*, ed. by J.W. Cogdell et al. (de Gruyter, Berlin, 2005), pp. 309–360

[Li92] J.-S. Li, Nonvanishing theorems for the cohomology of certain arithmetic quotients. J. Reine Angew. Math. **428**, 177–217 (1992)

[M10] S. Morel, *On the Cohomology of Certain Non-compact Shimura Varieties*. With an appendix by Robert Kottwitz. Annals of Mathematics Studies, vol. 173 (Princeton University Press, Princeton, 2010)

[PSR] S. Rallis, I.I. Piatetski-Shapiro, L-functions for the classical groups, in *Explicit Constructions of Automorphic L-functions*. Lecture Notes in Mathematics, vol. 1254 (Springer, Berlin, 1987), pp. 1–52

[Sch] P.Scholze, On torsion in the cohomology of locally symmetric varieties. Ann. Math. **182**, 945–1066 (2015)

[Sh64] G. Shimura, On the field of definition for a field of automorphic functions. Ann. Math. **80**, 160–189 (1964)

[Sh97] G. Shimura, *Euler Products and Eisenstein Series*. CBMS Regional Conference Series in Mathematics, vol. 93 (1997)

[Sh11] S.W. Shin, Galois representations arising from some compact Shimura varieties. Ann. Math. **173**, 1645–1741 (2011)

[Tan] V. Tan, Poles of Siegel Eisenstein series on $U(n, n)$. Can. J. Math. **51**, 164-175 (1999)

Bessel Descents and Branching Problems

Dihua Jiang and Lei Zhang

Abstract We discuss the theory of automorphic descents of Bessel type and its relation to automorphic version of branching problem and its relevant reciprocal branching problem.

2000 Mathematics Subject Classification Primary 11F67, 11F70, 22E55; Secondary 11F30, 11F66

1 Introduction

Automorphic descent is a method to construct automorphic forms on a smaller group from these on a larger group. The automorphic descent of Bessel type is a construction of such descents using the Bessel-Fourier coefficients of automorphic forms on a larger group, which is simply called the *Bessel descent* as in the title of this paper. There is another type of the automorphic descent, which may be called *Fourier-Jacobi descent*, using Fourier-Jacobi coefficients of automorphic forms. The idea goes back to the construction of the Saito-Kurokawa cusp forms of Sp_4, which are non-tempered at almost all local places and hence are counter-examples to the generalized Ramanujan conjecture [2, 26], and [7].

The more recent development of the automorphic descent was re-initiated by the work of D. Ginzburg, S. Rallis and D. Soudry in the 1990s in order to construct the backwards global Langlands functorial transfers from the general linear groups to quasi-split classical groups. The idea is to construct explicit cuspidal automorphic modules on classical groups by means of their global Arthur parameters. Their work has been well presented in their book [13]. Note that the automorphic descents in

D. Jiang (✉)
School of Mathematics, University of Minnesota, Minneapolis, MN, USA
e-mail: dhjiang@math.umn.edu

L. Zhang
Department of Mathematics, National University of Singapore, Singapore, Singapore
e-mail: matzhlei@nus.edu.sg

W. Müller et al. (eds.), *Relative Trace Formulas*, Simons Symposia,
https://doi.org/10.1007/978-3-030-68506-5_7

[13] require that the classical groups are quasi-split and the cuspidal automorphic representations on quasi-split classical groups must be generic in the sense that they admit a non-zero Whittaker-Fourier coefficient.

The notion of twisted automorphic descents was introduced by the authors in their previous work ([23], and also [21]), which can be viewed as a natural extension of the automorphic descents of Ginzburg-Rallis-Soudry. The twisted automorphic descent is to take, as source representations, square-integrable automorphic representations π of classical group G_n with global Arthur parameters of type

$$\psi^{(2,1)} = \phi_\tau^{(2)} \boxplus \phi_\delta.$$

Here ϕ_τ and ϕ_δ are generic global Arthur parameters of the form in (2.3) with all $b_j = 1$, and $\phi_\tau^{(2)}$ is defined to be the following non-generic global Arthur parameter:

$$\phi_\tau^{(2)} = (\tau_1, 2) \boxplus \cdots \boxplus (\tau_r, 2), \tag{1.1}$$

if $\phi_\tau = (\tau_1, 1) \boxplus \cdots \boxplus (\tau_r, 1)$ with $\tau = \tau_1 \boxplus \cdots \boxplus \tau_r$. We refer to Sect. 2.1 for details on the discrete spectrum of classical groups and their Arthur parameters.

The square-integrable residual representations of quasi-split classical group G_n^* with global Arthur parameter $\psi^{(2,1)}$ are explicitly determined in [18]. Note that the automorphic descent of Ginzburg-Rallis-Soudry [13] only considers the residual representations with global Arthur parameters of the form $\phi_\tau^{(2)}$ and only for quasi-split classical groups. The key point in the theory of twisted automorphic descents is to introduce a generic global Arthur parameter ϕ_δ, with the property that the central L-value $L(\frac{1}{2}, \tau \times \delta)$ of the twisted L-functions is non-zero. This property plays an indispensable role in the construction of the twisted automorphic descents for general classical groups.

The twisted automorphic descents have been proved to be fruitful and productive in the endoscopy theory of automorphic forms and its applications. We list the following important development in the theory of twisted automorphic descents:

(1) In [17], a special case of the twisted automorphic descents was explicitly studied, which leads to a new proof of the classical Jacquet-Langlands correspondence between PGL_2 and $\mathrm{PD}^\times$, where D is a quaternion algebra defined over a number field k.
(2) In [23], the twisted automorphic descents were systematically developed in order to construct explicit cuspidal automorphic modules for general classical groups by means of their global Arthur parameters, with assumption that the global Arthur parameters are generic, and with the assumption of the *Generic Summand Conjecture* (Conjecture 5.4).
(3) In [23], the twisted automorphic descents (ideas and methods) are used to prove one direction of the global Gan–Gross–Prasad conjecture for unitary groups and special orthogonal groups in a uniform way, and prove the other direction of the global Gan–Gross–Prasad conjecture with a global assumption that can only be verified in some special cases, currently.

(4) In [22], the idea and method of the twisted automorphic descents were used to prove the non-vanishing of certain tensor product L-functions for classical groups.
(5) In [16], the twisted automorphic descents were used to test the *Branching Problem* and the *Reciprocal Branching Problem* for cuspidal automorphic representations of classical groups.

We will discuss in this paper the theory of twisted automorphic descents and its relation to the automorphic version of the branching problem and its relevant reciprocal branching problem. Refer to [14, 15], and [21] for further discussions on other relevant topics. Instead of treating the theory for general classical groups, we mainly discuss the theory of Bessel descents for special orthogonal groups and leaves more detailed treatment of other classical groups in our future work.

The branching decomposition problem roots in the classical theory of representations and invariants. Let G be a group and H be a subgroup of G. The classical *Branching Decomposition Problem* is to ask: For a given representation π of G, what is the spectrum of the restriction of π to H as a representation of H? The *Reciprocal Branching Problem* is then to ask: For a given representation σ of H, how to find a representation π of G, such that the restriction of π to H contains σ? It is clear that without any extra conditions or requirements on π and/or σ, the two problems are related by the well-known *Frobenius Reciprocity*. However, with some extra requirements on π and σ, it could become an interesting and possibly difficult problem.

Let G_n be a special orthogonal group over a number field k as defined in Sect. 2, although the discussion here makes sense for general classical groups, or even for general reductive algebraic groups. Let H_m be a special orthogonal group over a number field k, which has a natural embedding into G_n as a subgroup, so that the pair (G_n, H_m) is relevant in the sense of the global Gan–Gross–Prasad conjecture [8]. For any irreducible unitary representation π of $G_n(\mathbb{A})$ (where $\mathbb{A}$ is the ring of adeles of k), which has a realization $\mathcal{C}_\pi$ in the space of the automorphic discrete spectrum of G_n (see Sect. 2.1 for definition), the automorphic version of *branching problem* may ask: what is the spectral decomposition of the space $\mathcal{C}_\pi$ restricted to $H_m(\mathbb{A})$ as an automorphic $H_m(\mathbb{A})$-module? It is in general a complicated problem if H_m is small comparing to G_n, since the decomposition may not be discrete, and the multiplicity may be infinite, for instance. From representation-theoretic point of view, it will be reasonable to consider first the case of multiplicity one situation. To this end, one may enlarge the subgroup H_m in a reasonable way.

One of the natural ways is to consider the centralizer $C_{G_n}(H_m)$ of the subgroup H_m in G_n, which is a reductive subgroup of G_n. It makes sense to consider the natural embedding:

$$C_{G_n}(H_m) \times H_m \to G_n.$$

It is usually expected that one may consider the branching decomposition of the space $\mathcal{C}_\pi$ restricted to $C_{G_n}(H_m)(\mathbb{A}) \times H_m(\mathbb{A})$ as an intermediate step towards the eventual understanding of the branching decomposition problem from G_n to the subgroup H_m. This leads to an interesting topic, but beyond the reach of this paper.

In this paper, we consider another way to enlarge the subgroup H_m. Namely, we consider, when $m = \ell^-$, the Bessel subgroup of G_n

$$R_\ell := H_{\ell^-}^{\mathcal{O}_\ell} \rtimes V_\ell, \tag{1.2}$$

where $H_m := H_{\ell^-}^{\mathcal{O}_\ell}$ and a unipotent subgroup V_ℓ of G_n are defined in Sect. 2.2. The Bessel-Fourier coefficient of $\mathcal{C}_\pi$ attached to the Bessel subgroup produces an automorphic Bessel module $\mathcal{F}^{\mathcal{O}_\ell}(\mathcal{C}_\pi)$ of $H_m(\mathbb{A})$ as defined in (2.10). The question is: how to understand the spectral structure of the automorphic Bessel module $\mathcal{F}^{\mathcal{O}_\ell}(\mathcal{C}_\pi)$ as a representation of $H_m(\mathbb{A})$? In particular, we are interested in the following branching problem.

Problem 1.1 (L^2-Automorphic Branching Problem). *Let π, $\mathcal{C}_\pi$, G_n, H_m, and $\mathcal{F}^{\mathcal{O}_\ell}(\mathcal{C}_\pi)$ be as given above. Find irreducible unitary representations σ of $H_m(\mathbb{A})$, which has a realization $\mathcal{C}_\sigma$ in the space of discrete spectrum of $H_m(\mathbb{A})$, so that the Bessel period $\mathcal{B}^{\mathcal{O}_\ell}(\varphi_\pi, \varphi_\sigma)$, as defined in* (2.11), *is non-zero for some $\varphi_\pi \in \mathcal{C}_\pi$ and $\varphi_\sigma \in \mathcal{C}_\sigma$.*

Note that the Bessel period $\mathcal{B}^{\mathcal{O}_\ell}(\varphi_\pi, \varphi_\sigma)$, if it is non-zero, defines a non-zero linear functional in

$$\mathrm{Hom}_{H_{\ell^-}^{\mathcal{O}_\ell}(\mathbb{A})}(\mathcal{F}^{\mathcal{O}_\ell}(\mathcal{C}_\pi) \otimes \sigma^\vee, 1).$$

From the uniqueness of local Bessel models [1, 8, 29], and [19], for any $\sigma \in \mathcal{A}_{\mathrm{cusp}}(H_{\ell^-}^{\mathcal{O}_\ell})$, the Hom-space is at most one-dimensional. Hence if the Bessel period $\mathcal{B}^{\mathcal{O}_\ell}(\varphi_\pi, \varphi_\sigma)$ is non-zero, σ occurs in the spectral decomposition of the automorphic Bessel module $\mathcal{F}^{\mathcal{O}_\ell}(\mathcal{C}_\pi)$ with multiplicity one.

The *Reciprocal Branching Problem* can be formulated as follows.

Problem 1.2 (L^2-Automorphic Reciprocal Branching Problem). *Let H_m be a special orthogonal group defined over a number field k. Given an irreducible unitary representation σ of $H_m(\mathbb{A})$, which has a realization $\mathcal{C}_\sigma$ in the space of discrete spectrum of $H_m(\mathbb{A})$, find a special orthogonal group G_n containing H_m as a subgroup such that (G_n, H_m) is a relevant pair in the sense of the global Gan–Gross–Prasad conjecture, and find an irreducible unitary representation π of $G_n(\mathbb{A})$, which has a realization $\mathcal{C}_\pi$ in the space of discrete spectrum of $G_m(\mathbb{A})$, such that the Bessel period $\mathcal{B}^{\mathcal{O}_\ell}(\varphi_\pi, \varphi_\sigma)$, as defined in* (2.11), *is non-zero for some $\varphi_\pi \in \mathcal{C}_\pi$ and $\varphi_\sigma \in \mathcal{C}_\sigma$.*

The objective of this paper is to understand these two problems by means of the theory of twisted automorphic descents, as developed in [23]. We expect the discussion in this paper will be applicable to other classical groups, according to our work [14, 15, 23], and [21].

The paper is organized as follows. We recall in Sect. 2 the basic statement of endoscopy classification of the discrete spectrum for special orthogonal groups [3], the basic definition of Bessel-Fourier coefficients and Bessel periods of automorphic forms on special orthogonal groups [23], and a family of global zeta integrals that built from Bessel periods and the basic identity of such global zeta integral with tensor product L-functions (Theorem 2.2). By applying Theorem 2.2 in various situations, we are able to control the basic structure of global Arthur parameters of π and σ when they have a non-zero Bessel periods in Sect. 3. Due to technical issues as explained in Sect. 3, we are able to prove Theorems 3.1 and 3.3 when one of π and σ has a generic global Arthur parameter. Based on these results, we go further to prove Theorem 3.4 that the Bessel period of π and σ is non-zero implies the central value ($s = \frac{1}{2}$) of a quotient of relevant L-functions is non-zero. According to the work of Gan–Gross–Prasad [9], this can be viewed as one direction of the global Gan–Gross–Prasad conjecture for non-generic global Arthur parameters. As explained in [23], the theory of twisted automorphic descents is naturally connected to the global Gan–Gross–Prasad conjecture for generic global Arthur parameters. Hence as expected, the further refinement of the theory of twisted automorphic descents will be a powerful method to treat the global Gan–Gross–Prasad conjecture in general. After recalling the theory of twisted automorphic descents in Sect. 4, we go on in Sect. 5 with a discussion of the automorphic version *Branching Problem* for cuspidal automorphic representations with generic global Arthur parameters. This can be viewed as reformulation of some main results in [23] on construction of cuspidal automorphic modules for special orthogonal groups via the twisted automorphic descents. In the last section (Sect. 6), we consider the automorphic version of the *Reciprocal Branching Problem* for cuspidal automorphic representations with a given generic global Arthur parameter (Theorems 6.1 and 6.2). At the end of Sect. 6, the connection of the results here with the global Gan–Gross–Prasad conjecture in general is briefly discussed.

The research work of this paper started during the visit in January, 2018, of the first named author to National University of Singapore. He would like to thank Math. Department for providing productive research environment and hospitality. The first named author would like to thank the organizers: W. Mueller, S. W. Shin and N. Templier for invitation to the wonderful Simons Symposium 2018 on Relative Trace Formulas and thank the Simons Foundation for hospitality. We would like to thank D. Prasad for sending us a preliminary version of the reference [9] when we were working on this paper, and thank Wee Teck Gan for his interests and comments on some aspects of the paper. We finally would like to thank the referee for very helpful comments on an early version of this paper.

2 Bessel Modules and Bessel Periods

Let k be a number field and $\mathbb{A}$ the ring of adeles of k. As in [23] and [3, Chapter 9], we may use $G_n^* = \mathrm{SO}(V^*, q_{V^*}^*)$ for a k-quasisplit special orthogonal group that is defined by a non-degenerate, $\mathfrak{n}$-dimensional quadratic space (V^*, q^*) over k with $n = [\frac{\mathfrak{n}}{2}]$ and use $G_n = \mathrm{SO}(V, q)$ to denote a pure inner k-form of G_n^*. This means that both quadratic spaces (V^*, q^*) and (V, q) have the same dimension and the same discriminant, as discussed in [8] and [23], for instance.

Let (V_0, q) be the k-anisotropic kernel of (V, q) with dimension $\mathfrak{d}_0 = \mathfrak{n} - 2\mathfrak{r}$, where the k-rank $\mathfrak{r} = \mathfrak{r}_\mathfrak{n} = \mathfrak{r}(G_n)$ of G_n is equal to the Witt index of (V, q). Let V^+ be a maximal totally isotropic subspace of (V, q), with a basis $\{e_1, \dots, e_\mathfrak{r}\}$. Then (V, q) has the following polar decomposition

$$V = V^+ \oplus V_0 \oplus V^-,$$

where $V_0 = (V^+ \oplus V^-)^\perp$. Let $\{e_{-1}, \dots, e_{-\mathfrak{r}}\}$ be a basis of V^- that is dual to the basis $\{e_1, \dots, e_\mathfrak{r}\}$, i.e. $q(e_i, e_{-j}) = \delta_{i,j}$ for all $1 \le i, j \le \mathfrak{r}$. We may choose an orthogonal basis $\{e'_1, \dots, e'_{\mathfrak{d}_0}\}$ of V_0 with the property that $q(e'_i, e'_i) = d_i$, where d_i is nonzero for all $1 \le i \le \mathfrak{d}_0$. Hence we obtain the following basis for (V, q):

$$e_1, \dots, e_\mathfrak{r}, e'_1, \dots, e'_{\mathfrak{d}_0}, e_{-\mathfrak{r}}, \dots, e_{-1}. \tag{2.1}$$

We may fix a full isotropic flag in (V, q):

$$\mathrm{Span}\{e_1\} \subset \mathrm{Span}\{e_1, e_2\} \subset \cdots \subset \mathrm{Span}\{e_1, \dots, e_\mathfrak{r}\},$$

which defines a minimal parabolic k-subgroup P_0. Moreover, P_0 contains a maximal k-split torus S, consisting of elements

$$\mathrm{diag}\{t_1, \dots, t_\mathfrak{r}, 1, \dots, 1, t_\mathfrak{r}^{-1}, \dots, t_1^{-1}\},$$

with $t_i \in k^\times$ for $i = 1, 2, \cdots, \mathfrak{r}$. Then the centralizer $Z(S)$ in G_n is $S \times G_{d_0}$, the Levi subgroup of P_0, where G_{d_0} is defined by the k-anisotropic (V_0, q_0), which has dimension $\mathfrak{d}_0$, and $d_0 = [\frac{\mathfrak{d}_0}{2}]$. Then P_0 has the Levi decomposition: $P_0 = (S \times G_{d_0}) \ltimes N_0$ where N_0 is the unipotent radical of P_0. Also, with respect to the order of the basis in (2.1), the group G_n is also defined by the following symmetric matrix:

$$J_\mathfrak{r}^\mathfrak{n} = \begin{pmatrix} & & 1 \\ & J_{\mathfrak{r}-1}^{\mathfrak{n}-2} & \\ 1 & & \end{pmatrix}_{\mathfrak{n}\times\mathfrak{n}} \quad \text{and } J_0^{\mathfrak{d}_0} = \mathrm{diag}\{d_1, \dots, d_{\mathfrak{d}_0}\} \tag{2.2}$$

as defined inductively.

2.1 Discrete Spectrum

Following [3], we denote by $\mathcal{A}_2(G_n)$ the set of equivalence classes of irreducible unitary representations π of $G_n(\mathbb{A})$, where $\mathbb{A}$ is the ring of adeles of the number field k, occurring in the discrete spectrum $L^2_{\mathrm{disc}}(G_n)$ of $L^2(G_n(F)\backslash G_n(\mathbb{A}))$. Also denote by $\mathcal{A}_{\mathrm{cusp}}(G_n)$ for the subset of $\mathcal{A}_2(G)$, whose elements occur in the cuspidal spectrum $L^2_{\mathrm{cusp}}(G_n)$. The theory of endoscopic classification for G_n is to parameterize the set $\mathcal{A}_2(G_n)$ by means of the global Arthur parameters, which can be realized as certain automorphic representations of general linear group GL_{2n}.

Let $G_n^\vee$ be the complex dual group of G_n, which is $\mathrm{SO}_{2n}(\mathbb{C})$ if $\mathfrak{n} = 2n$ and is Sp_{2n} if $\mathfrak{n} = 2n+1$. Following [3], we denote by $\widetilde{\mathcal{E}}_{\mathrm{sim}}(\mathrm{GL}_{2n})$ the set of the equivalence classes of simple twisted endoscopic data of GL_{2n}. Each member in $\widetilde{\mathcal{E}}_{\mathrm{sim}}(\mathrm{GL}_{2n})$ is represented by a triple (G_n^*, s, ξ), where G_n^* is a k-quasisplit classical group, s is a semi-simple element as described in [3, Page 11], and ξ is the L-embedding: ${}^LG_n \to {}^L\mathrm{GL}_{2n}$. The set of global Arthur parameters for G_n^* is denoted by $\widetilde{\Psi}_2(G_n^*, \xi)$, or simply by $\widetilde{\Psi}_2(G_n^*)$ if the L-embedding ξ is well understood in the discussion. The elements of $\widetilde{\Psi}_2(G_n^*)$ can be formally written as

$$\psi = (\tau_1, b_1) \boxplus \cdots \boxplus (\tau_r, b_r), \tag{2.3}$$

where for each j, τ_j belongs to $\mathcal{A}_{\mathrm{cusp}}(\mathrm{GL}_{a_j})$ and is self-dual, and b_j is a positive integer. Moreover, for ψ to be a global Arthur parameter in $\widetilde{\Psi}_2(G_n^*)$, one must require that $2n = \sum_{j=1}^r a_j b_j$, and the pairs (τ_j, b_j) are mutually distinct and satisfy the parity condition, which will be explained as follows.

A self-dual $\tau \in \mathcal{A}_{\mathrm{cusp}}(\mathrm{GL}_a)$ is called *of symplectic type* if the (partial) exterior square L-function $L^S(s, \tau, \wedge^2)$ has a (simple) pole at $s = 1$; otherwise, τ is called *of orthogonal type*. In the latter case, the (partial) symmetric square L-function $L^S(s, \tau, \mathrm{sym}^2)$ has a (simple) pole at $s = 1$. When G_n is an odd special orthogonal group, the pairs (τ_j, b_j) are of symplectic type. This means that if b_j is odd, then τ_j must be of symplectic type; and if b_j is even, then τ_j must be of orthogonal type. Similarly, when G_n is an even special orthogonal group, the pairs (τ_j, b_j) are of orthogonal type. This means that if b_j is odd, then τ_j must be of orthogonal type; and if b_j is even, then τ_j must be of symplectic type.

A global parameter ψ as in (2.3) is called *generic* if $b_i = 1$ for $i = 1, 2, \cdots, r$. The set of generic, elliptic, global Arthur parameters is denoted by $\widetilde{\Phi}_2(G_n^*)$. When $r = 1$, the parameter $\psi = (\tau, b)$ is called *simple*. The following is an overly simplified version of Theorem 1.5.2 of [3], which is enough for the purpose of this paper. We refer to [3, Section 1.5] for unexplained notions in the statement, including the notion of global Arthur-packets.

Theorem 2.1 (Endoscopic Classification [3]). *For each $\pi \in \mathcal{A}_2(G_n)$, there is a G_n-relevant global Arthur parameter $\psi \in \widetilde{\Psi}_2(G_n^*, \xi)$, such that π belongs to the global Arthur packet, $\widetilde{\Pi}_\psi(G_n)$, attached to the global Arthur parameter ψ.*

Following [3, Theorem 1.5.2], when G_n is an odd special orthogonal group, i.e. $\mathfrak{n} = 2n + 1$, the multiplicity of $\pi \in \mathcal{A}_2(G_n)$ realizing in the discrete spectrum $L^2_{\mathrm{disc}}(G_n)$ is expected to be one. However, when G_n is an even special orthogonal group, the discrete multiplicity of $\pi \in \mathcal{A}_2(G_n)$ could be two. In this case, we need to fix a realization of $\pi \in \mathcal{A}_2(G_n)$ in the discrete spectrum $L^2_{\mathrm{disc}}(G_n)$, which will be denoted by $\mathcal{C}_\pi$, especially when the discrete multiplicity of π is two.

Recall from [3, Chapter 8] (see also [23]) that

$$\widetilde{\mathrm{O}}(G_n) := \widetilde{\mathrm{Out}}_N(G_n) := \widetilde{\mathrm{Aut}}_N(G_n)/\widetilde{\mathrm{Int}}_N(G_n) \tag{2.4}$$

is regarded as a diagonal subgroup of $\widetilde{\mathrm{Out}}_N(G_n(\mathbb{A}))$. When G_n is an even special orthogonal group, one may take $\varepsilon \in \mathrm{O}_{2n}(k)$ with $\det \varepsilon = -1$ and $\varepsilon^2 = I_{2n}$, such that the action of $\widetilde{\mathrm{O}}(G_n)$ on π can be realized as the ε-conjugate on π, i.e., $\pi^\varepsilon(g) = \pi(\varepsilon g \varepsilon^{-1})$. Hence the $\widetilde{\mathrm{O}}(G_n)$-orbit of π has one or two elements. If $\widetilde{\mathrm{O}}(G_n)$ acts freely on π, following the notation in [3], we denote the $\widetilde{\mathrm{O}}(G_n)$-torsor of π by $\{\pi, \pi_\star\}$. When G_n is not an even special orthogonal group, the group $\widetilde{\mathrm{O}}(G_n)$ is trivial, so is its action. Hence in this case, the $\widetilde{\mathrm{O}}(G_n)$-orbit of π contains only π itself.

When G_n is an even special orthogonal group, an elliptic global Arthur parameter ψ as in (2.3) may be conjugated under the action of $\widetilde{\mathrm{O}}(G_n)$ to a different global Arthur parameter $\psi_\star$ for G_n, which form an $\widetilde{\mathrm{O}}(G_n)$-orbit $\{\psi, \psi_\star\}$. If the $\widetilde{\mathrm{O}}(G_n)$-orbit is an $\widetilde{\mathrm{O}}(G_n)$-torsor $\{\psi, \psi_\star\}$, then they define different global Arthur packets and different global Vogan packets. In the rest of this paper, when we say that ψ is a global Arthur parameter of an even special orthogonal group G_n, we really mean that ψ is identified with either ψ or $\psi_\star$, through a specific twisted endoscopic datum.

2.2 Bessel Modules and Bessel Periods

The general notion of Fourier coefficients of automorphic forms on $G_n(\mathbb{A})$ associated to nilpotent orbits was discussed in previous work (see [14, Section 4] for instance). The Bessel-Fourier coefficient is a special type of Fourier coefficients associated to nilpotent orbits that are parameterized by the following partition

$$\underline{p}_\ell = [(2\ell + 1)1^{\mathfrak{n}-2\ell-1}]. \tag{2.5}$$

This partition is G_n-relevant if $0 \le \ell \le \mathfrak{r}$. In this case, we may choose the unipotent subgroup $V_\ell := V_{\underline{p}_\ell}$ of G_n to consist of all unipotent elements of the form:

$$V_\ell = \left\{ v = \begin{pmatrix} z & y & x \\ & I_{\mathfrak{n}-2\ell} & y' \\ & & z^* \end{pmatrix} \in G_n \mid z \in Z_\ell \right\}, \tag{2.6}$$

where Z_ℓ is the standard maximal (upper-triangular) unipotent subgroup of GL_ℓ. As discussed in length in [23, Subsection 2.3], the partition $\underline{p}_\ell$ is in one-to-one correspondence with the k-stable unipotent orbit $\mathcal{O}^{\mathrm{st}}_{\underline{p}_\ell}$ in $G_n(k)$. Here two k-unipotent orbits in $G_n(k)$ belong to the same k-stable unipotent orbit if they belong to the same unipotent orbit under the adjoint action of $G_n(\overline{k})$, where $\overline{k}$ is the algebraic closure of k. As explained in [23, Subsection 2.4], the k-rational unipotent orbits $\mathcal{O}_\ell$ in the k-stable unipotent orbit $\mathcal{O}^{\mathrm{st}}_{\underline{p}_\ell}$ are in one-to-one correspondence with the $\mathrm{GL}_1 \times G_{n-\ell}$-orbits of k-anisotropic vectors in $(k^{\mathfrak{n}-2\ell}, q)$, which can be viewed as a subspace of (V, q) and consists of the last row of y in (2.6). It follows that the generic character $\psi_{\mathcal{O}_\ell}$ of $V_\ell(\mathbb{A})$, whose restriction to $V_\ell(k)$ is trivial, may also be explicitly defined as follows. Fix a nontrivial character ψ_k of $k\backslash\mathbb{A}$ and consider the following identification:

$$V_\ell/[V_\ell, V_\ell] \cong \oplus_{i=1}^{\ell-1}\mathfrak{g}_{\alpha_i} \oplus k^{\mathfrak{n}-2\ell},$$

where $\mathfrak{g}_{\alpha_i}$ is the one-dimensional k-vector space associated to the simple root α_i, which is the position $(i, i+1)$ in z, as displayed in (2.6). Let w_0 be an anisotropic vector in $(k^{\mathfrak{n}-2\ell}, q)$ and define a character ψ_{ℓ,w_0} of $V_\ell(\mathbb{A})$ by

$$\psi_{\mathcal{O}_\ell}(v) = \psi_{\ell,w_0}(v) := \psi_k(\sum_{i=1}^{\ell-1} z_{i,i+1} + q(y_\ell, w_0)), \tag{2.7}$$

where y_ℓ is the last row of y as defined in (2.6). We denote by $H^{\mathcal{O}_\ell}_{\ell^-} = H^{w_0}_{\ell^-}$ is the identity connected component of the stabilizer of ψ_{ℓ,w_0} in $\mathrm{GL}_1 \times G_{n-\ell}$, which is given by

$$\left\{\begin{pmatrix} I_\ell & & \\ & \gamma & \\ & & I_\ell \end{pmatrix} \in G_n \mid \gamma J_{\mathfrak{n}-2\ell} w_0 = J_{\mathfrak{n}-2\ell} w_0\right\}, \tag{2.8}$$

where $\ell^- = [\frac{\mathfrak{l}^-}{2}]$ with $\mathfrak{l}^- := \mathfrak{n}-2\ell-1$. We may write $(k^{\mathfrak{n}-2\ell}, q) = (k^{\mathfrak{n}-2\ell-1}, q) \perp kw_0$. By [23, Proposition 2.5], the k-anisotropic kernel of the quadratic space $(k^{\mathfrak{n}-2\ell-1}, q)$ has dimension equal to $\mathfrak{d}_0 - 1$ if w_0 belongs to the $\mathrm{GL}_1 \times G_{n-\ell}$-orbit of a non-zero vector in the k-anisotropic kernel (V_0, q) of (V, q); otherwise, it has dimension equal to $\mathfrak{d}_0 + 1$. This determines the structure of the group $H^{\mathcal{O}_\ell}_{\ell^-} = H^{w_0}_{\ell^-}$.

Let φ be an automorphic form on $G_n(\mathbb{A})$. We define the ψ_{ℓ,w_0}-Bessel-Fourier coefficient of φ by the following integral:

$$\mathcal{F}^{\psi_{\ell,w_0}}(\varphi)(g) := \int_{V_\ell(k)\backslash V_\ell(\mathbb{A})} \varphi(vg)\psi^{-1}_{\ell,w_0}(v)dv. \tag{2.9}$$

It is clear that $\mathcal{F}^{\psi_{\ell,w_0}}(\varphi)(g)$ is a smooth automorphic function on $H_{\ell-}^{w_0}$ with moderate growth. For any $\pi \in \mathcal{A}_2(G_n)$, which has a realization $\mathcal{C}_\pi$ in the discrete spectrum $L^2_{\mathrm{disc}}(G_n)$, we denote by

$$\mathcal{F}^{\psi_{\ell,w_0}}(\mathcal{C}_\pi) = \mathcal{F}^{\mathcal{O}_\ell}(\mathcal{C}_\pi) \tag{2.10}$$

the space generated by all $\mathcal{F}^{\psi_{\ell,w_0}}(\varphi_\pi)(g)$ with all $\varphi_\pi \in \mathcal{C}_\pi$, and call it an ℓ-th *Bessel module of* π.

Let $\sigma \in \mathcal{A}_2(H_m)$ have a realization $\mathcal{C}_\sigma$ in the space of the discrete spectrum of $H_m(\mathbb{A})$. The Bessel period of $\mathcal{C}_\pi$ and $\mathcal{C}_\sigma$ is defined to be

$$\mathcal{B}^{\mathcal{O}_\ell}(\varphi_\pi, \varphi_\sigma) := \int_{H_{\ell-}^{\mathcal{O}_\ell}(k)\backslash H_{\ell-}^{\mathcal{O}_\ell}(\mathbb{A})} \mathcal{F}^{\mathcal{O}_\ell}(\varphi_\pi)(h)\overline{\varphi_\sigma(h)}dh. \tag{2.11}$$

Note that if both π and σ are not cuspidal, then the Bessel period $\mathcal{B}^{\mathcal{O}_\ell}(\varphi_\pi, \varphi_\sigma)$ needs regularization to define (see [10, 11] and [12] for instance). In spirit of the relation between parabolic induction and Bessel periods, we may only consider in this paper the situation that at least one of π, σ is cuspidal. Hence the Bessel periods are well defined.

In order to understand Question 1.1, the L^2-*Automorphic Branching Problem*, and Question 1.2, the L^2-*Automorphic Reciprocal Branching Problem*, we would like to understand the Bessel periods $\mathcal{B}^{\mathcal{O}_\ell}(\varphi_\pi, \varphi_\sigma)$ in terms of other invariants attached to π and σ. One natural invariant attached to π and σ are the relevant automorphic L-functions; and another natural invariant attached to π and σ is the global Arthur parameters. We will discuss the relation between the Bessel periods $\mathcal{B}^{\mathcal{O}_\ell}(\varphi_\pi, \varphi_\sigma)$ with the relevant automorphic L-functions in Sect. 2.3, and the relation between the Bessel periods $\mathcal{B}^{\mathcal{O}_\ell}(\varphi_\pi, \varphi_\sigma)$ with the relevant global Arthur parameters in Sect. 3.

2.3 A Family of Global Zeta Integrals

We recall from [23, Section 4] a family of global zeta integrals that represent tensor product L-functions for the product of orthogonal groups with general linear groups. The global zeta integrals are nothing but the Bessel periods as defined in (2.11) with one of the cuspidal automorphic forms replaced by an Eisenstein series.

Let τ be an irreducible unitary automorphic representation of $\mathrm{GL}_a(\mathbb{A})$ of the following isobaric type:

$$\tau = \tau_1 \boxplus \tau_2 \boxplus \cdots \boxplus \tau_r, \tag{2.12}$$

where $\tau_i \in \mathcal{A}_{\mathrm{cusp}}(\mathrm{GL}_{a_i})$, $\sum_{i=1}^r a_i = a$, and $\tau_i \not\cong \tau_j$ if $i \neq j$. As explained in [20] and in [23, Section 4], for the purpose of the global zeta integrals considered below, we do not have to assume that τ is associated to a generic global Arthur parameter of G_n. Instead, we only need to assume that τ is generic, i.e. has a non-zero Whittaker-Fourier coefficient.

Let H_{a+m} be the special orthogonal group with a Levi subgroup $M_a = \mathrm{GL}_a \times H_m$. Since (G_n, H_m) is a relevant pair in the sense of the Gan–Gross–Prasad conjecture [8], the pair (G_n, H_{a+m}) or (H_{a+m}, G_n) (depending on the size of n and $a+m$) is also relevant. Let $P_a = M_a U_a$ be the standard parabolic k-subgroup of H_{a+m} with the Levi subgroup $M_a = \mathrm{GL}_a \times H_m$. For $\sigma \in \mathcal{A}_{\text{cusp}}(H_m)$, we obtain the datum $(M_a, \tau \otimes \sigma)$, which is going to define an Eisenstein series, following [24] and [25]. For any

$$\phi = \phi_{\tau\otimes\sigma} \in \mathcal{A}(U_a(\mathbb{A})M_a(F)\backslash H_{a+m}(\mathbb{A}))_{\tau\otimes\sigma}, \tag{2.13}$$

set $\phi_s := |\det|_{\mathbb{A}}^s \cdot \phi$, where $|\det|_{\mathbb{A}}$ is first defined on $M_a(\mathbb{A})$, through the projection: $M_a \to \mathrm{GL}_a$, and then defined on $H_{a+m}(\mathbb{A})$ through the Iwasawa decomposition with respect to $P_a(\mathbb{A})$ and the standard maximal compact subgroup of $H_{a+m}(\mathbb{A})$. The associated Eisenstein series is defined to be

$$E(h, \phi, s) = E(h, \phi_{\tau\otimes\sigma}, s) = \sum_{\delta\in P_a(F)\backslash H_{a+m}(F)} \phi_s(\delta g). \tag{2.14}$$

The theory of Langlands on Eisenstein series [24] and [25] shows that $E(h, \phi, s)$ converges absolutely for $\mathrm{Re}(s)$ large, has meromorphic continuation to the complex plane $\mathbb{C}$, and defines an automorphic form on $H_{a+m}(\mathbb{A})$ when s is not a pole.

As in (2.5), we take a family of H_{a+m}-relevant partitions

$$\underline{p}_\kappa = [(2\kappa+1)1^{\mathfrak{m}+2a-2\kappa-1}]$$

for H_{a+m} with $\kappa \leq a + \mathfrak{r}_{\mathfrak{m}}$, where $\mathfrak{r}_{\mathfrak{m}} := \mathfrak{r}(H_m)$ is the k-rank of H_m. As in (2.9), we define the Bessel-Fourier coefficient of the Eisenstein series $E(h, \phi, s)$ on $H_{a+m}(\mathbb{A})$:

$$\mathcal{F}^{\psi_{\mathcal{O}_\kappa}}(E(\cdot, \phi, s))(h) := \int_{N_\kappa(k)\backslash N_\kappa(\mathbb{A})} E(nh, \phi, s)\psi_{\mathcal{O}_\kappa}^{-1}(n)\, dn, \tag{2.15}$$

where the unipotent subgroup N_κ of H_{a+m} determined by the partition $\underline{p}_\kappa$ is similar to the unipotent subgroup V_ℓ of G_n as in (2.6), and the character $\psi_{\mathcal{O}_\kappa}(n)$ is defined as in (2.7). Hence its centralizer $G_{m^-}^{\mathcal{O}_\kappa}$ is similar to the subgroup as described in (2.8), with $m^- = [\frac{\mathfrak{m}^-}{2}]$ and $\mathfrak{m}^- := 2a + \mathfrak{m} - 2\kappa - 1$ and $\kappa \leq a + \mathfrak{r}_{\mathfrak{m}}$. By [23, Proposition 2.6], the stabilizer $G_{m^-}^{\mathcal{O}_\kappa}$ and the subgroup H_m form a relevant pair in the sense of the Gan–Gross–Prasad conjecture [8]. In this case, the Bessel subgroup of H_{a+m} is defined to be

$$R_{\mathcal{O}_\kappa}^{a+m} := G_{m^-}^{\mathcal{O}_\kappa} \ltimes N_\kappa. \tag{2.16}$$

For $\pi \in \mathcal{A}_{\mathrm{cusp}}(G_{m^-}^{\mathcal{O}_\kappa})$, the *global zeta integral* $\mathcal{Z}(s, \phi_{\tau\otimes\sigma}, \varphi_\pi, \psi_{\mathcal{O}_\kappa})$, as in [23, (4.9)], is defined by the following Bessel period:

$$\mathcal{Z}(s, \varphi_\pi, \phi_{\tau\otimes\sigma}, \psi_{\mathcal{O}_\kappa}) := \mathcal{B}^{\mathcal{O}_\kappa}(E(\phi_{\tau\otimes\sigma}, s), \varphi_\pi). \tag{2.17}$$

As given in Proposition 2.1 of [20], $\mathcal{Z}(s, \phi_{\tau\otimes\sigma}, \varphi_\pi, \psi_{\mathcal{O}_\kappa})$ converges absolutely and hence is holomorphic at s where the Eisenstein series $E(h, \phi, s)$ has no poles. The main result of [23, Section 4] is the following theorem.

Theorem 2.2 (Basic Identity). *With the notation given as above, the global zeta integral $\mathcal{Z}(s, \phi_{\tau\otimes\sigma}, \varphi_\pi, \psi_{\mathcal{O}_\kappa})$ can be expressed as an eulerian product with factorizable data, and can be expressed by the following formula*

$$\mathcal{Z}(s, \phi_{\tau\otimes\sigma}, \varphi_\pi, \psi_{\mathcal{O}_\kappa}) = \mathcal{Z}_S(s, \cdot)\frac{L^S(s+\frac{1}{2}, \tau\times\pi)}{L^S(s+1, \tau\times\sigma)L^S(2s+1, \tau, \rho)}, \tag{2.18}$$

where $\mathcal{Z}_S(s, \cdot)$ is a finite eulerian product of the local zeta integrals at archimedean local places and ramified finite local places, $L^S(\cdot)$ denotes the partial L-function with unramified data, and $\rho = \wedge^2$ if H_{m+a} is an even orthogonal group; and $\rho =$ sym^2 *if H_{m+a} is an odd orthogonal group.*

Note that in the proof of the formula in (2.18), it is proved in [23, Section 4], especially in Theorems 4.5, 4.7 and 4.8 of [23], that the Bessel period for π and σ is not zero for a certain choice of data if and only if the global zeta integral is not zero as function in s for a corresponding choice of data. By establishing some refined properties of the global intertwining operators and the local zeta integrals, we obtain the reciprocal relation of the relevant Bessel periods, as in [23, Section 5], which will be discussed in the next section.

3 Bessel Periods and Global Arthur Parameters

We are going to understand the relation between the Bessel periods for π and σ and the structure of the global Arthur parameters. The main tool is the basic identity (2.18) in Theorem 2.2. As explained in [23, Section 5], in order to find fruitful applications of the basic identity (2.18), one needs to have a good control of the analytic properties of the finite eulerian product of local zeta integrals, which is $\mathcal{Z}_S(s, \cdot)$ on the right hand side of the basic identity; meanwhile, one needs to have a good control of the analytic properties of the Eisenstein series on the left hand side of the basic identity. Up to date, we are able to handle these two technical points when one of the $\sigma \in \mathcal{A}_{\mathrm{cusp}}(H_m)$ and $\pi \in \mathcal{A}_{\mathrm{cusp}}(G_n)$ has a generic global Arthur parameter. Hence we have to limit our discussion with such an *assumption*, and hope to get back to those technical issues in our future work.

3.1 π Has a Generic Global Arthur Parameter

Let (G_n, H_m) be a relevant pair. Assume that $\pi \in \mathcal{A}_{\mathrm{cusp}}(G_n)$ has a generic global Arthur parameter. For any $\sigma \in \mathcal{A}_{\mathrm{cusp}}(H_m)$, assume that for some ℓ with $0 \le \ell \le \mathfrak{r}_n = \mathfrak{r}(G_n)$, the Bessel period $\mathcal{B}^{\mathcal{O}_\ell}(\varphi_\pi, \varphi_\sigma)$ is non-zero for a choice of $\varphi_\pi \in \mathcal{C}_\pi$ and $\varphi_\sigma \in \mathcal{C}_\sigma$. What can one say about the structure of the global Arthur parameter of σ?

By assumption, we write the generic global Arthur parameter of π by

$$\phi_n = (\eta_1, 1) \boxplus \cdots \boxplus (\eta_t, 1). \tag{3.1}$$

We may write the global Arthur parameter of σ by

$$\psi_m = (\zeta_1, 2b_1 + 1) \boxplus \cdots \boxplus (\zeta_l, 2b_l + 1) \boxplus (\xi_1, 2a_1) \boxplus \cdots \boxplus (\xi_k, 2a_k). \tag{3.2}$$

Note that η_i and ξ_j are in the same parity as irreducible cuspidal automorphic representations of general linear groups, i.e. either they are all of symplectic type or they are all of orthogonal type; and η_i and ζ_j are in different parity, as irreducible cuspidal automorphic representations of general linear groups. Also the global Arthur parameter ψ_m in (3.2) includes the following three cases that $l = 0$, but $k \neq 0$; $l \neq 0$, but $k = 0$; and $l \times k \neq 0$. Note that $k = 0$ is the same as $a_1 = a_2 = \cdots = a_k = 0$.

Theorem 3.1 (Structure of ψ_m). *With π and σ as given above, and with the assumption that π has a generic global Arthur parameter ϕ_n as in* (3.1), *if the Bessel period $\mathcal{B}^{\mathcal{O}_\ell}(\varphi_\pi, \varphi_\sigma)$ is non-zero for a choice of $\varphi_\pi \in \mathcal{C}_\pi$ and $\varphi_\sigma \in \mathcal{C}_\sigma$, then the global Arthur parameter ψ_m of σ as given in* (3.2) *must be of the form:*

$$\psi_m = (\zeta_1, 1) \boxplus \cdots \boxplus (\zeta_l, 1) \boxplus (\xi_1, 2) \boxplus \cdots \boxplus (\xi_k, 2), \tag{3.3}$$

and $\{\xi_1, \xi_2, \ldots, \xi_k\}$ is a subset of $\{\eta_1, \eta_2, \ldots, \eta_t\}$. In other words, in this case, the global Arthur parameters ϕ_n of π and ψ_m of σ must be relevant in the sense of Gan–Gross–Prasad in [9, Conjecture 9.1].

Note that the forms of ϕ_n in (3.1) and ψ_m in (3.3) force that both $\{\xi_1, \xi_2, \ldots, \xi_k\}$ and $\{\eta_1, \eta_2, \ldots, \eta_t\}$ are sets instead of multiset. A special case when π has a non-zero Whittaker-Fourier coefficient, which implies that G_n and H_m must be quasi-split, was treated in [27] and [28] by the local unramified argument.

Proof. Without loss of generality, we may assume that

$$0 \le a_1 \le a_2 \le \cdots \le a_k \text{ and } 0 \le b_1 \le b_2 \le \cdots \le b_l \text{ in (3.2)}, \tag{3.4}$$

and if $k \neq 0$, then $a_1 \ge 1$. To prove this theorem, we consider two cases separately: *Case: $0 \le k$, $a_k \le b_l$, and $l \neq 0$* and *Case: $a_k \ge b_l + 1$; or $k \neq 0$ but $l = 0$.*

Case: $0 \le k$, $a_k \le b_l$, and $l \ne 0$ In this case, we may assume that $\zeta_l \in \mathcal{A}_{\mathrm{cusp}}(\mathrm{GL}_{\alpha_l})$. Note ζ_l is in different parity with η_i. Let G_{α_l+n} be the special orthogonal group containing a standard Levi subgroup $M_{\alpha_l} = \mathrm{GL}_{\alpha_l} \times G_n$. Then associated to the cuspidal datum $(M_{\alpha_l}, \zeta_l \otimes \pi)$, we define an Eisenstein series $E(g, \phi_{\zeta_l\otimes\pi}, s)$ as in (2.14). By Proposition 2.6 of [23], we are able to choose an integer κ with $0 \le \kappa \le \alpha_l + \mathfrak{r}_n = \mathfrak{r}(G_n)$, and find a k-rational unipotent orbit $\mathcal{O}_\kappa$ such that $\mathfrak{k}^- = 2\alpha_l + \mathfrak{n} - 2\kappa - 1 = \mathfrak{m}$ with $\kappa^- = [\frac{\mathfrak{k}^-}{2}] = m$, and the stabilizer $G_{\kappa^-}^{\mathcal{O}_\kappa} = H_m$. As in (2.17), the Bessel period of $E(g, \phi_{\zeta_l\otimes\pi}, s)$ and φ_σ gives a global zeta integral $\mathcal{Z}(s, \phi_{\zeta_l\otimes\pi}, \varphi_\sigma, \psi_{\mathcal{O}_\kappa})$. By Theorem 2.2, we have the following identity:

$$\mathcal{Z}(s, \phi_{\zeta_l\otimes\pi}, \varphi_\sigma, \psi_{\mathcal{O}_\kappa}) = \mathcal{Z}_S(s, \cdot) \cdot \frac{L^S(s+\frac{1}{2}, \zeta_l \times \sigma)}{L^S(s+1, \zeta_l \times \pi) L^S(2s+1, \zeta_l, \rho)}. \tag{3.5}$$

Because the Bessel period $\mathcal{B}^{\mathcal{O}_\ell}(\varphi_\pi, \varphi_\sigma)$ is non-zero for a choice of $\varphi_\pi \in \mathcal{C}_\pi$ and $\varphi_\sigma \in \mathcal{C}_\sigma$, the identity in (3.5) holds non-trivially, according to Theorems 4.5 and 4.7 in [23].

We first consider the right-hand side of the basic identity (3.5). By (3.2), the partial L-function $L^S(s+\frac{1}{2}, \zeta_l \times \sigma)$ can be written as

$$L^S(s+\frac{1}{2}, \zeta_l \times \sigma) = \prod_{j=1}^{l} L^S(s+\frac{1}{2}, \zeta_l \times (\zeta_j, 2b_j+1)) \cdot \prod_{i=1}^{k} L^S(s+\frac{1}{2}, \zeta_l \times (\xi_i, 2a_i)). \tag{3.6}$$

Then we look at each factors in the expression above, and have

$$L^S(s+\frac{1}{2}, \zeta_l \times (\zeta_j, 2b_j+1)) = \prod_{i=0}^{2b_j} L^S(s+\frac{1}{2}+b_j-i, \zeta_l \times \zeta_j), \tag{3.7}$$

and

$$L^S(s+\frac{1}{2}, \zeta_l \times (\xi_i, 2a_i)) = \prod_{j=0}^{2a_i-1} L^S(s+a_i-j, \zeta_l \times \xi_i). \tag{3.8}$$

We *claim* that the partial L-function $L^S(s+\frac{1}{2}, \zeta_l \times \sigma)$ has a right-most simple pole at $s = b_l + \frac{1}{2}$, which can be proved as follows.

Let us look at the global Arthur parameter ψ_m of σ in (3.2) with the order in (3.4). From the definition of global Arthur parameters, the pair $(\zeta_j, 2b_j+1)$ and the pair $(\zeta_l, 2b_l+1)$ are distinct if $j \ne l$. This means that if $b_j = b_l$, then $\zeta_j \not\cong \zeta_l$. When ζ_l is not equivalent to ζ_j, the partial L-function $L^S(s+\frac{1}{2}+b_j-i, \zeta_l \times \zeta_j)$ in (3.7) has no poles. When ζ_l is equivalent to ζ_j for some j, the partial L-function $L^S(s + \frac{1}{2} + b_j - i, \zeta_l \times \zeta_j)$ has poles. But in this case $b_j < b_l$ by the order in (3.4). Hence the right hand side of (3.7) must have a right-most (simple) pole at $s = b_l + \frac{1}{2}$.

If $k = 0$, then the partial L-function $L^S(s + \frac{1}{2}, \zeta_l \times \sigma)$ has a right-most simple pole at $s = b_l + \frac{1}{2}$. This proves the *claim* for the case of $k = 0$.

Now we consider the case of $k \neq 0$. By assumption in this case, we have

$$1 \leq a_1 \leq a_k \leq b_l.$$

As irreducible cuspidal automorphic representations of general linear groups, ζ_j and ξ_i are in different parity. It follows that the partial L-functions $L^S(s+a_i-j, \zeta_l \times \xi_i)$ has no poles. However, they may have zeros at $\mathrm{Re}(s) + a_i - j < 1$. We want to show that there are no zeros of the partial L-functions $L^S(s + a_i - j, \zeta_l \times \xi_i)$ at $\mathrm{Re}(s) + a_i - j < 1$ that can cancel the right-most pole at $s = b_l + \frac{1}{2}$ of (3.7) for all $i = 1, 2, \cdots, k$.

In fact, if $\mathrm{Re}(s) + a_i - j < 1$, then $\mathrm{Re}(s) < 1 + j - a_i$. Hence if the right-most simple pole at $s = b_l + \frac{1}{2}$ of (3.7) can be cancelled by the possible zero of $L^S(s + a_i - j, \zeta_l \times \xi_i)$ at $\mathrm{Re}(s) + a_i - j < 1$ for some i, we must have that

$$s = b_l + \frac{1}{2} < 1 + j - a_i.$$

It follows that

$$a_i = 1 + (2a_i - 1) - a_i \geq 1 + j - a_i > b_l + \frac{1}{2}.$$

From $a_i > b_l + \frac{1}{2}$, we must have, by (3.4), that $a_k \geq b_l + 1$. But this is impossible because the situation under consideration here assumes that $a_k \leq b_l$.

This proves that the right-most simple pole at $s = b_l + \frac{1}{2}$ of the partial L-function in (3.7) can not be cancelled by any possible zeros of the partial L-functions in (3.8). Hence $L^S(s + \frac{1}{2}, \zeta_l \times \sigma)$ has a right-most simple pole at $s = b_l + \frac{1}{2}$. This completes the proof of the *claim*.

As for the denominators at the right-hand side of (3.5), we write

$$L^S(s + 1, \zeta_l \times \pi) = \prod_{j=1}^{t} L^S(s + 1, \zeta_l \times \eta_j).$$

Since ζ_l and η_j are in different parity, the partial L-functions $L^S(s + 1, \zeta_l \times \eta_j)$ are holomorphic and non-zero for $\mathrm{Re}(s) > 0$. The same holds for the partial L-function $L^S(2s + 1, \zeta_l, \rho)$.

On the right-hand side, we remain to discuss the finite eulerian product of local zeta integrals $\mathcal{Z}_S(s, \cdot)$. By Proposition 5.5 of [23], when the Bessel period $\mathcal{B}^{\mathcal{O}_\ell}(\varphi_\pi, \varphi_\sigma)$ is non-zero for some choice of global data, $\mathcal{Z}_S(s, \cdot)$ can be made a non-zero constant at $s = b_l + \frac{1}{2}$ for a corresponding choice of local data. Therefore, the right-hand side of (3.5) has a right-most simple at $s = b_l + \frac{1}{2}$ for some choice of data.

Next we consider the left-hand side of (3.5). The global zeta integral $\mathcal{Z}(s, \phi_{\zeta_l\otimes\pi}, \varphi_\sigma, \psi_{\mathcal{O}_\kappa})$ is given by the Bessel period $\mathcal{B}^{\mathcal{O}_\kappa}(E(\cdot, \phi_{\zeta_l\otimes\pi}, s), \varphi_\sigma)$, as in (2.17). According to the existence of the simple pole at $s = b_l + \frac{1}{2}$ on the right-hand side, the Bessel period $\mathcal{B}^{\mathcal{O}_\kappa}(E(\cdot, \phi_{\zeta_l\otimes\pi}, s), \varphi_\sigma)$ has a simple pole at $s = b_l + \frac{1}{2}$ for some choice of data given from the choice on the right-hand side of (3.5). Since φ_σ is cuspidal, we obtain that the Eisenstein series $E(\cdot, \phi_{\zeta_l\otimes\pi}, s)$ has a right-most simple pole at $s = b_l + \frac{1}{2}$ for the chosen data. By Proposition 5.2 of [23], if this Eisenstein series has a pole, then the pole must be at $s = \frac{1}{2}$. Hence we must have that $\frac{1}{2} = \frac{1}{2} + b_l$. This implies that $b_l = 0$. Since the case under consideration assumes that $a_k \leq b_l$, we also have that $a_k = 0$. Therefore, in this case, we must have that

$$a_1 = a_2 = \cdots = a_k = b_1 = b_2 = \cdots = b_l = 0,$$

which of course includes the case that $k = 0$, and the global Arthur parameter of σ by

$$\psi_m = (\zeta_1, 1) \boxplus \cdots \boxplus (\zeta_l, 1).$$

Case: $a_k \geq b_l + 1$; **or** $k \neq 0$ **but** $l = 0$ In this case, we consider an Eisenstein series $E(\cdot, \phi_{\xi_k\otimes\pi}, s)$. To set up the notation, we assume that ξ_k is an irreducible cuspidal automorphic representation of $\mathrm{GL}_{\beta_k}(\mathbb{A})$. It follows that this Eisenstein series is defined on $G_{\beta_k+n}(\mathbb{A})$ with the defining cuspidal datum $(\mathrm{GL}_{\beta_k} \times G_n, \xi_k\otimes\pi)$. By Proposition 2.6 of [23], there is an integer γ with $0 \leq \gamma \leq \beta_k + \mathfrak{r}_n$, such that $H_m = G^{\mathcal{O}_\gamma}_{\mathfrak{q}^-}$, with $\mathfrak{q}^- = 2\beta_k + \mathfrak{n} - 2\gamma - 1$. Then we consider the Bessel period $\mathcal{B}^{\mathcal{O}_\gamma}(E(\cdot, \phi_{\xi_k\otimes\pi}, s), \varphi_\sigma)$. As before, this Bessel period, which defines the global zeta integral for this case, produces the following basic identity:

$$\mathcal{B}^{\mathcal{O}_\gamma}(E(\cdot, \phi_{\xi_k\otimes\pi}, s), \varphi_\sigma) = \mathcal{Z}_S(s, \cdot) \cdot \frac{L^S(s + \frac{1}{2}, \xi_k \times \sigma)}{L^S(s + 1, \xi_k \times \pi)L^S(2s + 1, \xi_k, \rho)}. \tag{3.9}$$

In the right-hand side of (3.9), by (3.2), $L^S(s + \frac{1}{2}, \xi_k \times \sigma)$ can be written as

$$L^S(s + \frac{1}{2}, \xi_k \times \sigma) = \prod_{j=1}^{l} L^S(s + \frac{1}{2}, \xi_k \times (\zeta_j, 2b_j + 1))$$
$$\cdot \prod_{i=1}^{k} L^S(s + \frac{1}{2}, \xi_k \times (\xi_i, 2a_i)). \tag{3.10}$$

Each factors in the expression above can be further expressed as follows:

$$L^S(s+\frac{1}{2},\xi_k\times(\zeta_j,2b_j+1))=\prod_{i=0}^{2b_j}L^S(s+\frac{1}{2}+b_j-i,\xi_k\times\zeta_j),\tag{3.11}$$

and

$$L^S(s+\frac{1}{2},\xi_k\times(\xi_i,2a_i))=\prod_{j=0}^{2a_i-1}L^S(s+a_i-j,\xi_k\times\xi_i).\tag{3.12}$$

We *claim* that the partial L-function $L^S(s+\frac{1}{2},\xi_k\times\sigma)$ has a right-most simple pole at $s=a_k$, which can be proved as follows.

We first look at the possible right-most pole on the right-hand side of (3.12). From the definition of global Arthur parameters, the pair $(\xi_i,2a_i)$ and the pair $(\xi_k,2a_k)$ are distinct if $i\neq k$. This means that if $a_i=a_k$, then $\xi_i\not\cong\xi_k$. When ξ_k is not equivalent to ξ_i, the partial L-function $L^S(s+a_i-j,\xi_k\times\xi_i)$ has no poles. When ξ_k is equivalent to ξ_i for some i, the partial L-function $L^S(s+a_i-j,\xi_k\times\xi_i)$ has poles. But in this case $a_i<a_k$ by the order in (3.4). Hence the right-hand side of (3.12) must have a right-most (simple) pole at $s=a_k$.

Therefore, if $k\neq 0$ but $l=0$, the partial L-function $L^S(s+\frac{1}{2},\xi_k\times\sigma)$ has a right-most (simple) pole at $s=a_k$. This proves the *claim* in the case of $k\neq 0$ but $l=0$.

Now we consider the case of $l\neq 0$, where we also have $a_k\geq b_l+1\geq 1$. As irreducible cuspidal automorphic representations of general linear groups, ζ_j and ξ_i are in different parity. It follows that the partial L-functions $L^S(s+\frac{1}{2}+b_j-i,\xi_k\times\zeta_j)$ in (3.11) has no poles for all $j=1,2,\cdots,l$. However, they may have zeros at $\mathrm{Re}(s)+\frac{1}{2}+b_j-i<1$. We want to show that there are no zeros of the partial L-functions $L^S(s+\frac{1}{2}+b_j-i,\xi_k\times\zeta_j)$ at $\mathrm{Re}(s)+\frac{1}{2}+b_j-i<1$ that can cancel the right-most pole at $s=a_k$ of (3.12) for all $j=1,2,\cdots,l$.

In fact, from $\mathrm{Re}(s)+\frac{1}{2}+b_j-i<1$, one has $\mathrm{Re}(s)<\frac{1}{2}+i-b_j$. If such a zero will cancel the pole at $s=a_k$, then we have

$$a_k<\frac{1}{2}+i-b_j\leq\frac{1}{2}+b_j\leq\frac{1}{2}+b_l.$$

But from the assumption of this case, we have $a_k\geq b_l+1$. This contradiction shows that there are no zeros of the partial L-functions $L^S(s+\frac{1}{2}+b_j-i,\xi_k\times\zeta_j)$ at $\mathrm{Re}(s)+\frac{1}{2}+b_j-i<1$ that can cancel the right-most pole at $s=a_k$ of (3.12) for all $j=1,2,\cdots,l$. Therefore the partial L-function $L^S(s+\frac{1}{2},\xi_k\times\sigma)$ must have a right-most simple pole at $s=a_k$ and we prove the *claim* in the this case.

As discussed in the previous case, the denominators $L^S(s+1,\xi_k\times\pi)$ and $L^S(2s+1,\xi_k,\rho)$ in the right-hand side of (3.9) are holomorphic and non-zero for $\mathrm{Re}(s)>0$. We remain to discuss the finite eulerian product of local zeta integrals

$\mathcal{Z}_S(s,\cdot)$. As discussed in the previous case, by Proposition 5.5 of [23], when the Bessel period $\mathcal{B}^{\mathcal{O}_\ell}(\varphi_\pi, \varphi_\sigma)$ is non-zero for some choice of global data, $\mathcal{Z}_S(s,\cdot)$ can be made a non-zero constant at $s = b_l + \frac{1}{2}$ for a corresponding choice of local data. Therefore, the right-hand side of (3.9) has a right-most simple at $s = a_k$ for some choice of data.

On the left-hand side of (3.9), it follows that the Bessel period $\mathcal{B}^{\mathcal{O}_\gamma}(E(\cdot, \phi_{\xi_k\otimes\pi}, s), \varphi_\sigma)$ has a right-most simple pole at $s = a_k$ for some choice of data, and so does the Eisenstein series $E(\cdot, \phi_{\xi_k\otimes\pi}, s)$, because φ_σ is cuspidal. From the generic global Arthur parameter ϕ_n of π given in (3.1), the cuspidal automorphic representations η_j and ξ_k are in the same parity. It follows from [18, Theorem 1.2] that the right-most possible pole of $E(\cdot, \phi_{\xi_k\otimes\pi}, s)$ is at $s = 1$. Hence, we must have that $a_k = 1$, which forces that $b_l = 0$. Therefore, we obtain that

$$a_1 = a_2 = \cdots = a_k = 1 \text{ and } b_1 = b_2 = \cdots = b_l = 0,$$

and the global Arthur parameter of σ is of the form

$$\psi_m = (\zeta_1, 1) \boxplus \cdots \boxplus (\zeta_l, 1) \boxplus (\xi_1, 2) \boxplus \cdots \boxplus (\xi_k, 2). \tag{3.13}$$

Of course, if $l = 0$, we must have that the global Arthur parameter of σ is of the form

$$\psi_m = (\xi_1, 2) \boxplus \cdots \boxplus (\xi_k, 2).$$

Finally, let us assume that ψ_m is of form (3.13) and show that the set $\{\xi_1, \xi_2, \ldots, \xi_k\}$ is a subset of $\{\eta_1, \eta_2, \ldots, \eta_s\}$.

For each irreducible cuspidal automorphic representation ξ_i with $1 \le i \le k$, we consider the Bessel period $\mathcal{B}^{\mathcal{O}_\gamma}(E(\cdot, \phi_{\xi_i\otimes\pi}, s), \varphi_\sigma)$ and we have the same basic identity as (3.9):

$$\mathcal{B}^{\mathcal{O}_\gamma}(E(\cdot, \phi_{\xi_i\otimes\pi}, s), \varphi_\sigma) = \mathcal{Z}_S(s,\cdot) \cdot \frac{L^S(s+\frac{1}{2}, \xi_i \times \sigma)}{L^S(s+1, \xi_i \times \pi) L^S(2s+1, \xi_i, \rho)}. \tag{3.14}$$

Similar as above, $\mathcal{Z}_S(s,\cdot)$ at the right-hand side of (3.14) is nonzero at $s = 1$ for some choice of data. Following (3.10), (3.11) and (3.12), $L^S(s+\frac{1}{2}, \xi_i \times \sigma)$ has a right-most simple pole at $s = 1$. Then the right-hand side of (3.14) has a right-most simple pole at $s = 1$, because the denominator is holomorphic and non-zero for $\mathrm{Re}(s) > 0$ as discussed before.

It follows that the Bessel period $\mathcal{B}^{\mathcal{O}_\gamma}(E(\cdot, \phi_{\xi_i\otimes\pi}, s), \varphi_\sigma)$ on the left-hand side of (3.14) has a right-most simple pole at $s = 1$ for some choice of data, so does the Eisenstein series $E(\cdot, \phi_{\xi_i\otimes\pi}, s)$. Recall that $E(\cdot, \phi_{\xi_i} \otimes \pi, s)$ has a simple pole at $s = 1$ if and only if $L^S(s, \xi_i \times \pi)$ has a simple pole at $s = 1$. Thus the existence of the simple pole of $L^S(s, \xi_i \times \pi)$ at $s = 1$ is equivalent to $\xi_i = \eta_j$ for some η_j. Hence, we obtain that $\{\xi_1, \xi_2, \ldots, \xi_k\}$ is a subset of $\{\eta_1, \eta_2, \ldots, \eta_s\}$.

This finishes the proof. □

Based on the above discussion, we make the following conjecture, which can be viewed as a reformulation of the Conjecture 2.3 in [23].

Conjecture 3.2. *Assume that $\pi \in \mathcal{A}_{\mathrm{cusp}}(G_n)$ has a generic global Arthur parameter ϕ_n as in* (3.1). *For any $\sigma \in \mathcal{A}_{\mathrm{cusp}}(H_m)$, assume that the Bessel period $\mathcal{B}^{\mathcal{O}_\ell}(\varphi_\pi, \varphi_\sigma)$ is non-zero for a choice of $\varphi_\pi \in \mathcal{C}_\pi$ and $\varphi_\sigma \in \mathcal{C}_\sigma$. If the global Arthur parameter ψ_m of the σ has the part $(\xi_1, 2) \boxplus \cdots \boxplus (\xi_k, 2)$ with non-zero k, then the integer ℓ with $0 \le \ell \le \mathfrak{r}_n = \mathfrak{r}(G_n)$ is not the first occurrence index of π.*

The *tower property* of automorphic descents of Bessel type and the notion of the *first occurrence index* of Bessel periods will be reviewed with details in Sect. 5, and more explanations of Conjecture 2.3 can found in Section 2 of [23].

3.2 σ Has a Generic Global Arthur Parameter

Let (G_n, H_m) be a relevant pair. Assume that $\sigma \in \mathcal{A}_{\mathrm{cusp}}(H_m)$ has a generic global Arthur parameter. For any $\pi \in \mathcal{A}_{\mathrm{cusp}}(G_n)$, assume that for some ℓ with $0 \le \ell \le \mathfrak{r}_n = \mathfrak{r}(G_n)$, the Bessel period $\mathcal{B}^{\mathcal{O}_\ell}(\varphi_\pi, \varphi_\sigma)$ is non-zero for a choice of $\varphi_\pi \in \mathcal{C}_\pi$ and $\varphi_\sigma \in \mathcal{C}_\sigma$. What can one say about the structure of the global Arthur parameter of π? We may regard this question reciprocal to that in Sect. 3.1. Indeed, we are going to apply the same theory developed in [23, Sections 4,5], but to different situation. For completeness, we include the details for this case.

In this case, we may write the global Arthur parameters of π and σ as follows:

$$\psi_n = (\eta_1, 2a_1 + 1) \boxplus \cdots \boxplus (\eta_k, 2a_k + 1) \boxplus (\delta_1, 2b_1) \boxplus \cdots \boxplus (\delta_l, 2b_l) \tag{3.15}$$

for π, with the order:

$$0 \le a_1 \le a_2 \le \cdots \le a_k \text{ and } 0 \le b_1 \le b_2 \le \cdots \le b_l, \tag{3.16}$$

and if $l \neq 0$, then $b_1 \ge 1$; and

$$\phi_m = (\zeta_1, 1) \boxplus \cdots \boxplus (\zeta_t, 1) \tag{3.17}$$

for σ. Note that δ_i and ζ_j are in the same parity, while η_i and ζ_j are in different parity, as irreducible cuspidal automorphic representations of general linear groups. Also the global Arthur parameter ψ_n in (3.15) includes the following three cases that $l = 0$, but $k \neq 0$; $l \neq 0$, but $k = 0$; and $l \times k \neq 0$. Note that $l = 0$ is the same as $b_1 = \cdots = b_l = 0$.

Cases: $0 \le l, b_l \le a_k$ and $k \neq 0$ We consider the Eisenstein series $E(\cdot, \phi_{\eta_k \otimes \sigma}, s)$ of $H_{a_k+m}(\mathbb{A})$, where we may assume that $\eta_k \in \mathcal{A}_{\mathrm{cusp}}(\mathrm{GL}_{a_k})$. It is clear that the

Eisenstein series $E(\cdot, \phi_{\eta_k\otimes\sigma}, s)$ is associated to the cuspidal datum $(M_{\alpha_k}, \eta_k \otimes \sigma)$, with $M_{\alpha_k} = \mathrm{GL}_{\alpha_k} \times H_m$ being a standard Levi subgroup of H_{α_k+m}. The global zeta integral that is defined to be the Bessel period $\mathcal{B}^{\mathcal{O}_\kappa}(E(\cdot, \phi_{\eta_k\otimes\sigma}, s), \varphi_\pi)$ has the following identity, by Theorem 2.2 and an argument similar to that of Theorem 3.1.

$$\mathcal{B}^{\mathcal{O}_\kappa}(E(\cdot, \phi_{\eta_k\otimes\sigma}, s), \varphi_\pi) = \mathcal{Z}_S(s, \cdot) \cdot \frac{L^S(s+\frac{1}{2}, \eta_k \times \pi)}{L^S(s+1, \eta_k \times \sigma)L^S(2s+1, \eta_k, \rho)}. \tag{3.18}$$

We consider the partial L-function $L^S(s+\frac{1}{2}, \eta_k\times\pi)$ on the right-hand side of (3.18):

$$L^S(s+\frac{1}{2}, \eta_k \times \pi) = \prod_{j=1}^{k} L^S(s+\frac{1}{2}, \eta_k \times (\eta_j, 2a_j+1)) \cdot \prod_{i=1}^{l} L^S(s+\frac{1}{2}, \eta_k \times (\delta_i, 2b_i)). \tag{3.19}$$

As before, we write

$$L^S(s+\frac{1}{2}, \eta_k \times (\eta_j, 2a_j+1)) = \prod_{i=0}^{2a_j} L^S(s+\frac{1}{2}+a_j-i, \eta_k \times \eta_j), \tag{3.20}$$

and

$$L^S(s+\frac{1}{2}, \eta_k \times (\delta_i, 2b_i)) = \prod_{j=0}^{2b_i-1} L^S(s+b_i-j, \eta_k \times \delta_i). \tag{3.21}$$

We *claim* that the partial L-function $L^S(s+\frac{1}{2}, \eta_k \times \pi)$ has a right-most simple pole at $s = a_k + \frac{1}{2}$, which can be proved as follows.

Let us look at the global Arthur parameter ψ_n of π in (3.15) with the order in (3.16). From the definition of global Arthur parameters, the pair $(\eta_j, 2a_j+1)$ and the pair $(\eta_k, 2a_k+1)$ are distinct if $j \neq k$. This means that if $a_j = a_k$, then $\eta_j \not\equiv \eta_k$. When η_k is not equivalent to η_j, the partial L-function $L^S(s+\frac{1}{2}+a_j-i, \eta_k \times \eta_j)$ in (3.20) has no poles. When η_k is equivalent to η_j for some j, the partial L-function $L^S(s+\frac{1}{2}+a_j-i, \eta_k \times \eta_j)$ has poles. But in this case $a_j < a_k$ by the order in (3.16). Hence the right hand side of (3.20) must have a right-most (simple) pole at $s = a_k + \frac{1}{2}$.

If $l = 0$, then the partial L-function $L^S(s+\frac{1}{2}, \eta_k \times \pi)$ has a right-most simple pole at $s = a_k + \frac{1}{2}$. This proves the *claim* for the case of $l = 0$.

Now we consider the case of $l \neq 0$. By assumption in this case, we have

$$1 \le b_1 \le b_l \le a_k.$$

As irreducible cuspidal automorphic representations of general linear groups, η_j and δ_i are in different parity. It follows that the partial L-functions $L^S(s+b_i-j, \eta_k \times \delta_i)$ has no poles. However, they may have zeros at $\mathrm{Re}(s) + b_i - j < 1$. We want to show that there are no zeros of the partial L-functions $L^S(s + b_i - j, \eta_k \times \delta_i)$ at $\mathrm{Re}(s) + b_i - j < 1$ that can cancel the right-most pole at $s = a_k + \frac{1}{2}$ of (3.20) for all $i = 1, 2, \cdots, l$.

In fact, if $\mathrm{Re}(s) + b_i - j < 1$, then $\mathrm{Re}(s) < 1 + j - b_i$. Hence if the right-most simple pole at $s = a_k + \frac{1}{2}$ of (3.20) can be cancelled by the possible zero of $L^S(s + b_i - j, \eta_k \times \delta_i)$ at $\mathrm{Re}(s) + b_i - j < 1$ for some i, we must have that

$$s = a_k + \frac{1}{2} < 1 + j - b_i.$$

It follows that

$$b_i = 1 + (2b_i - 1) - b_i \geq 1 + j - b_i > a_k + \frac{1}{2}.$$

From $b_i > a_k + \frac{1}{2}$, we must have, by (3.16), that $b_l \geq a_k + 1$. But this is impossible because the situation under consideration here assumes that $b_l \leq a_k$.

This proves that the right-most simple pole at $s = a_k + \frac{1}{2}$ of the partial L-function in (3.20) can not be cancelled by any possible zeros of the partial L-functions in (3.21). Hence $L^S(s + \frac{1}{2}, \eta_k \times \pi)$ has a right-most simple pole at $s = a_k + \frac{1}{2}$. The *claim* is completely proved.

As for the denominators at the right-hand side of (3.18), we write

$$L^S(s + 1, \eta_k \times \sigma) = \prod_{j=1}^{t} L^S(s + 1, \eta_k \times \zeta_j).$$

Since η_k and ζ_j are in different parity, the partial L-functions $L^S(s + 1, \eta_k \times \zeta_j)$ are holomorphic and non-zero for $\mathrm{Re}(s) > 0$. The same holds for the partial L-function $L^S(2s + 1, \eta_k, \rho)$.

On the right-hand side, we remain to discuss the finite eulerian product of local zeta integrals $\mathcal{Z}_S(s, \cdot)$. By Proposition 5.5 of [23], when the Bessel period $\mathcal{B}^{\mathcal{O}_\ell}(\varphi_\pi, \varphi_\sigma)$ is non-zero for some choice of global data, $\mathcal{Z}_S(s, \cdot)$ can be made a non-zero constant at $s = a_k + \frac{1}{2}$ for a corresponding choice of local data. Therefore, the right-hand side of (3.18) has a right-most simple at $s = a_k + \frac{1}{2}$ for some choice of data.

It follows that the left-hand side of the identity in (3.18) has a right-most simple pole at $s = a_k + \frac{1}{2}$ for some choice of data, and so is the Eisenstein series $E(\cdot, \phi_{\eta_k \otimes \sigma}, s)$. Finally by Proposition 5.2 of [23], we obtain that the Eisenstein series can have the right-most simple pole at $s = \frac{1}{2}$. Therefore, we must have that $a_k = 0$, which implies that $b_l = 0$. This of course implies that

$$a_1 = \cdots = a_k = b_1 = \cdots = b_l = 0.$$

In particular, the global Arthur parameter ψ_n of π must be of the form:

$$\psi_n = (\eta_1, 1) \boxplus \cdots \boxplus (\eta_k, 1). \tag{3.22}$$

Case: $b_l \geq a_k + 1$; or $l \neq 0$ but $k = 0$ In this case, we consider the Eisenstein series $E(\cdot, \phi_{\delta_l \otimes \sigma}, s)$, where we may assume that $\delta_l \in \mathcal{A}_{\text{cusp}}(\mathrm{GL}_{\beta_l})$. This Eisenstein series is defined over $H_{\beta_l+m}(\mathbb{A})$, with its cuspidal datum $(M_{\beta_l}, \delta_l \otimes \sigma)$, where $M_{\beta_l} = \mathrm{GL}_{\beta_l} \times H_m$ is a standard Levi subgroup of H_{β_l+m}. The Bessel period $\mathcal{B}^{\mathcal{O}_\gamma}(E(\cdot, \phi_{\delta_l \otimes \sigma}, s), \varphi_\pi)$ defines the global zeta integral and has the following identity:

$$\mathcal{B}^{\mathcal{O}_\gamma}(E(\cdot, \phi_{\delta_l\otimes\sigma}, s), \varphi_\pi) = \mathcal{Z}_S(s, \cdot) \cdot \frac{L^S(s + \frac{1}{2}, \delta_l \times \pi)}{L^S(s+1, \delta_l \times \sigma) L^S(2s+1, \delta_l, \rho)}, \tag{3.23}$$

by Theorem 2.2 and an argument similar to that of Theorem 3.1. Following the same idea, we consider the partial L-functions on the right-hand side of the identity in (3.23) and have

$$L^S(s + \frac{1}{2}, \delta_l \times \pi) = \prod_{j=1}^{k} L^S(s + \frac{1}{2}, \delta_l \times (\eta_j, 2a_j + 1))$$
$$\cdot \prod_{i=1}^{l} L^S(s + \frac{1}{2}, \delta_l \times (\delta_i, 2b_i)). \tag{3.24}$$

As before, we write

$$L^S(s + \frac{1}{2}, \delta_l \times (\eta_j, 2a_j + 1)) = \prod_{i=0}^{2a_j} L^S(s + \frac{1}{2} + a_j - i, \delta_l \times \eta_j), \tag{3.25}$$

and

$$L^S(s + \frac{1}{2}, \delta_l \times (\delta_i, 2b_i)) = \prod_{j=0}^{2b_i - 1} L^S(s + b_i - j, \delta_l \times \delta_i). \tag{3.26}$$

By the same proof for the corresponding case of Theorem 3.1, we deduce, without further details, that the right-most simple pole of (3.26) is at $s = b_l$. If $k = 0$, then the partial L-function $L^S(s + \frac{1}{2}, \delta_l \times \pi)$ has a right-most simple pole at $s = b_l$.

If $k \neq 0$, the partial L-functions $L^S(s + \frac{1}{2} + a_j - i, \delta_l \times \eta_j)$ has no poles, since δ_l and η_j are in different parity. But they might have a zero at $\mathrm{Re}(s) + \frac{1}{2} + a_j - i < 1$. If this possible zero is going to cancel the right-most simple pole at $s = b_l$, we must have that $b_l + \frac{1}{2} + a_j - i < 1$, which implies that $b_l - a_k < \frac{1}{2}$. But this is impossible since in this case, we assume that $b_l \geq a_k + 1$.

This proves that the partial L-function $L^S(s + \frac{1}{2}, \delta_l \times \pi)$ has a right-most simple pole at $s = b_l$. As discussed in the proof of Theorem 3.1, by Proposition 5.5 of [23], when the Bessel period $\mathcal{B}^{\mathcal{O}_\ell}(\varphi_\pi, \varphi_\sigma)$ is non-zero for some choice of global data, $\mathcal{Z}_S(s, \cdot)$ can be made a non-zero constant at $s = b_l$ for a corresponding choice of local data. Therefore, the right-hand side of (3.23) has a right-most simple at $s = b_l$ for some choice of data.

It follows that on the left-hand side of (3.23), the Bessel period $\mathcal{B}^{\mathcal{O}_\gamma}(E(\cdot, \phi_{\delta_l \otimes \sigma}, s), \varphi_\pi)$ has a right-most simple pole at $s = b_l$ for some choice of data, based on the choice of the data on the right-hand side, and so does the Eisenstein series $E(\cdot, \phi_{\delta_l \otimes \sigma}, s)$. By [18], this Eisenstein series can have a right-most simple pole at $s = 1$. It follows that $b_l = 1$ and $a_k = 0$. It implies that if $k = 0$, then the global Arthur parameter of π is of the form:

$$\psi_n = (\delta_1, 2) \boxplus \cdots \boxplus (\delta_l, 2);$$

and if $k \neq 0$, then the global Arthur parameter of π is of the form:

$$\psi_n = (\eta_1, 1) \boxplus \cdots \boxplus (\eta_k, 1) \boxplus (\delta_1, 2) \boxplus \cdots \boxplus (\delta_l, 2). \tag{3.27}$$

Let us assume that ψ_n is of form (3.27). Analogous to Theorem 3.1, we will show that $\{\delta_1, \delta_2, \ldots, \delta_l\}$ is a subset of $\{\zeta_1, \zeta_2, \ldots, \zeta_s\}$.

The proof is completely the same as the corresponding part (**Case:** $a_k \geq b_l + 1$**; or** $k \neq 0$ **but** $l = 0$) in the proof of Theorem 3.1. We omit the details, but only give a sketch of the argument. We consider the Bessel period $\mathcal{B}^{\mathcal{O}_\gamma}(E(\cdot, \phi_{\delta_i \otimes \sigma}, s), \varphi_\pi)$ for $1 \leq i \leq l$, and the following identity:

$$\mathcal{B}^{\mathcal{O}_\gamma}(E(\cdot, \phi_{\delta_i \otimes \sigma}, s), \varphi_\pi) = \mathcal{Z}_S(s, \cdot) \cdot \frac{L^S(s + \frac{1}{2}, \delta_i \times \pi)}{L^S(s + 1, \delta_i \times \sigma) L^S(2s + 1, \delta_i, \rho)}, \tag{3.28}$$

by Theorem 2.2 and an argument similar to that of Theorem 3.1. By computing the partial L-function $L^S(s + \frac{1}{2}, \delta_i \times \pi)$ on the right-hand side of (3.28), we deduce that the Bessel period on the left-hand side must have a right-most simple pole at $s = 1$. It follows that the Eisenstein series $E(\cdot, \phi_{\delta_i \otimes \sigma}, s)$ has a simple pole at $s = 1$, which implies that $L^S(s, \delta_i \times \sigma)$ has a simple pole at $s = 1$. Hence $\delta_i = \zeta_j$ for some $1 \leq j \leq s$.

We summarize the above discussion as a theorem.

Theorem 3.3 (Structure of ψ_n). *With π and σ as given above, and with the assumption that σ has a generic global Arthur parameter ϕ_m as in* (3.17)*, if the Bessel period $\mathcal{B}^{\mathcal{O}_\ell}(\varphi_\pi, \varphi_\sigma)$ is non-zero for a choice of $\varphi_\pi \in \mathcal{C}_\pi$ and $\varphi_\sigma \in \mathcal{C}_\sigma$, then the global Arthur parameter ψ_n of π as given in* (3.15) *must be of the form:*

$$\psi_n = (\eta_1, 1) \boxplus \cdots \boxplus (\eta_k, 1) \boxplus (\delta_1, 2) \boxplus \cdots \boxplus (\delta_l, 2)$$

and $\{\delta_1, \delta_2, \ldots, \delta_l\}$ *is a subset of* $\{\zeta_1, \zeta_2, \ldots, \zeta_s\}$. *In other words, in this case, the global Arthur parameters* ψ_n *of* π *and* ϕ_m *of* σ *must be relevant in the sense of Gan–Gross–Prasad in* [9, Conjecture 9.1].

3.3 Reformulation in Terms of L-Functions

We continue our discussion for the pair (π, σ) with one of them having a generic global Arthur parameter. We may first consider the situation that π has a generic global Arthur parameter $\phi_n = (\eta_1, 1) \boxplus \cdots \boxplus (\eta_s, 1)$ as in (3.1), and σ has a global Arthur parameter ψ_m as in (3.2). In the spirit of the global Gan–Gross–Prasad conjecture for non-generic global Arthur parameters [9], we prove the following result.

Theorem 3.4 (Bessel Period and Central L-Value). *Assume that* π *and* σ *are as given in Theorem 3.1, with* π *having a generic global Arthur parameter* ϕ_n *as in* (3.1) *and* σ *having a global Arthur parameter* ψ_m *as* (3.2), *which is of the form:*

$$\psi_m = (\zeta_1, 2b_1 + 1) \boxplus \cdots \boxplus (\zeta_l, 2b_l + 1) \boxplus (\xi_1, 2a_1) \boxplus \cdots \boxplus (\xi_k, 2a_k).$$

Assume that the following non-vanishing

$$\prod_{1\le i\le l,\ 1\le j\le k} L(\frac{1}{2}, \zeta_i \times \xi_j) \neq 0, \tag{3.29}$$

holds, and for irreducible cuspidal automorphic representations ζ *of* $\mathrm{GL}_a(\mathbb{A})$ *the following normalized finite eulerian product of local zeta integrals in* (2.18)

$$\frac{\mathcal{Z}_S(s, \phi_{\zeta\otimes\pi}, \varphi_\sigma)}{\prod_{\nu\in S} L(s + \frac{1}{2}, \zeta_\nu \times \psi_{m,\nu})} \tag{3.30}$$

is not identically zero at $s = \frac{1}{2}$.

If the Bessel period $\mathcal{B}^{\mathcal{O}_\ell}(\varphi_\pi, \varphi_\sigma)$ *is non-zero for a choice of* $\varphi_\pi \in \mathcal{C}_\pi$ *and* $\varphi_\sigma \in \mathcal{C}_\sigma$ *then*

$$\frac{L(s + \frac{1}{2}, \phi_n \times \psi_m)}{L(s + 1, \phi_n, \rho_G^-)L(s + 1, \psi_m, \rho_H^-)} \tag{3.31}$$

is holomorphic and nonzero at $s = 0$.

Here ρ_G^- (resp. ρ_H^-) is either a symmetric square or exterior square representation of a general linear group, depending on that G_n (resp. H_m) is an even orthogonal group or odd orthogonal group, which is opposite to the one defined in Theorem 2.2.

Note that here we use the complete (resp. local) L-functions of the Arthur parameters, which are defined in terms of the relevant complete (resp. local) L-functions associated to irreducible cuspidal automorphic (resp. smooth admissible) representations of general linear groups.

Proof. Since the assumption of the theorem is taken from Theorem 3.1, it is clear by Theorem 3.1 that the global Arthur parameters ϕ_n and ψ_m are of the forms:

$$\phi_n = (\eta_1, 1) \boxplus \cdots \boxplus (\eta_t, 1),$$

and

$$\psi_m = (\zeta_1, 1) \boxplus \cdots \boxplus (\zeta_l, 1) \boxplus (\xi_1, 2) \boxplus \cdots \boxplus (\xi_k, 2)$$

with the property that $\{\xi_1, \xi_2, \ldots, \xi_k\}$ is a subset of $\{\eta_1, \eta_2, \ldots, \eta_t\}$.

Now, we may write

$$L(s, \phi_n, \rho_{\bar{G}}) = \prod_{i=1}^{t} L(s, \eta_i, \rho_{\bar{G}}) \prod_{1 \le i < j \le s} L(s, \eta_i \times \eta_j), \tag{3.32}$$

$$\begin{aligned} & L(s, \psi_m, \rho_{\bar{H}}) \\ = & \prod_{i=1}^{l} L(s, \zeta_i, \rho_{\bar{H}}) \prod_{1 \le i < j \le s} L(s, \zeta_i \times \zeta_j) \\ & \prod_{\substack{1 \le i \le l \\ 1 \le j \le k}} L(s + \frac{1}{2}, \zeta_i \times \xi_j) L(s - \frac{1}{2}, \zeta_i \times \xi_j) \\ & \prod_{1 \le i < j \le k} L(s, \xi_i \times \xi_j)^2 L(s+1, \xi_i \times \xi_j) L(s-1, \xi_i \times \xi_j) \\ & \prod_{j=1}^{k} L(s, \xi_j \times \xi_j) L(s+1, \xi_j, \rho_{\bar{H}}) L(s-1, \xi_j, \rho_{\bar{H}}), \end{aligned} \tag{3.33}$$

and

$$L(s, \phi_n \times \psi_m) = \prod_{\substack{1 \le i \le l \\ 1 \le j \le t}} L(s, \zeta_i \times \eta_j) \prod_{\substack{1 \le i \le t \\ 1 \le j \le k}} L(s + \frac{1}{2}, \eta_i \times \xi_j) L(s - \frac{1}{2}, \eta_i \times \xi_j). \tag{3.34}$$

We are going to show that

$$\frac{L(s+\frac{1}{2},\phi_n\times\psi_m)}{L(s+1,\phi_n,\rho_G^-)L(s+1,\psi_m,\rho_H^-)}$$

is holomorphic and nonzero at $s=0$.

Since η_i are distinct to each other and of the type different with ρ_G^-, by (3.32), $L(s+1,\phi_n,\rho_G^-)$ is holomorphic and non-zero at $s=0$. It reduces to show that

$$\frac{L(s+\frac{1}{2},\phi_n\times\psi_m)}{L(s+1,\psi_m,\rho_H^-)} \tag{3.35}$$

is holomorphic and nonzero at $s=0$.

To consider $L(s+1,\psi_m,\rho_H^-)$ at $s=0$, we study the equation in (3.33). It is clear that $L(s+1,\zeta_i,\rho_H^-)$, $L(s+\frac{3}{2},\zeta_i\times\xi_j)$, $L(s+2,\xi_j,\rho_H^-)$, and $L(s+2,\xi_i\times\xi_j)$ are holomorphic and non-zero at $s=0$. In addition, $L(s+1,\zeta_i\times\zeta_j)$, $L(s+1,\xi_i\times\xi_j)$ and $L(s,\xi_i\times\xi_j)$ are holomorphic and non-zero at $s=0$ as $\xi_i\not\cong\xi_j$ and $\zeta_i\not\cong\zeta_j$ for $i\neq j$. Following the non-vanishing assumption $\prod_{1\le i\le l,\ 1\le j\le k}L(\frac{1}{2},\zeta_i\times\xi_j)\neq 0$ in (3.29), we have that the non-vanishing of (3.35) at $s=0$ is equivalent to the non-vanishing of

$$\frac{\prod_{\substack{1\le i\le l\\ 1\le j\le t}}L(s+\frac{1}{2},\zeta_i\times\eta_j)}{\prod_{\substack{1\le i\le l\\ 1\le j\le k}}L(s+\frac{1}{2},\zeta_i\times\xi_j)}\times\frac{\prod_{\substack{1\le i\le t\\ 1\le j\le k}}L(s+1,\eta_i\times\xi_j)L(s,\eta_i\times\xi_j)}{\prod_{j=1}^{k}L(s+1,\xi_j\times\xi_j)L(s,\xi_j,\rho_H^-)} \tag{3.36}$$

at $s=0$.

Recall that $\{\xi_1,\xi_2,\ldots,\xi_k\}$ is a subset of $\{\eta_1,\eta_2,\ldots,\eta_t\}$ and denote by $\{\xi_1',\xi_2',\ldots,\xi_{k'}'\}$ its complementary subset in $\{\eta_1,\eta_2,\ldots,\eta_t\}$. Note that by definition (3.2), ξ_j and ψ_m are in different parity, that is, $L(s,\xi_j,\rho_H^-)$ has a pole at $s=1$. Then $L(s,\xi_j,\rho_H)$ is holomorphic and nonzero at $s=1$. We may decompose $L(s,\xi_j\times\xi_j)=L(s,\xi_j,\rho_H)L(s,\xi_j,\rho_H^-)$. Then the following factor in (3.36)

$$\frac{\prod_{\substack{1\le i\le t\\ 1\le j\le k}}L(s+1,\eta_i\times\xi_j)L(s,\eta_i\times\xi_j)}{\prod_{j=1}^{k}L(s+1,\xi_j\times\xi_j)L(s,\xi_j,\rho_H^-)}$$

is equal to

$$\prod_{\substack{1\le i\le k'\\ 1\le j\le k}}L(s+1,\xi_i'\times\xi_j)L(s,\xi_i'\times\xi_j)$$

$$\times\prod_{\substack{1\le i,j\le k\\ i\neq j}}L(s+1,\xi_i\times\xi_j)L(s,\xi_i\times\xi_j)\times\prod_{j=1}^{k}L(s,\xi_j,\rho_H),$$

which is holomorphic and nonzero at $s = 0$.

Hence, it is sufficient to show that the remaining factor in (3.36)

$$\frac{\prod_{\substack{1\le i\le l\\ 1\le j\le t}} L(s+\frac{1}{2}, \zeta_i \times \eta_j)}{\prod_{\substack{1\le i\le l\\ 1\le j\le k}} L(s+\frac{1}{2}, \zeta_i \times \xi_j)} = \prod_{1\le i\le l,\ 1\le j\le k'} L(s+\frac{1}{2}, \zeta_i \times \xi'_j) \tag{3.37}$$

is holomorphic and nonzero at $s = 0$.

Under the assumption (3.29), the non-vanishing of (3.37) is equivalent to

$$\prod_{1\le i\le l,\ 1\le j\le t} L(\frac{1}{2}, \zeta_i \times \eta_j) \neq 0.$$

We are going to show that for each ζ_i $(1 \le i \le l)$

$$L(\frac{1}{2}, \zeta_i \times \pi) = \prod_{1\le j\le t} L(\frac{1}{2}, \zeta_i \times \eta_j)$$

is nonzero.

Consider the basic identity as (3.9) for the Bessel period

$$\begin{aligned} &\mathcal{B}^{\mathcal{O}_\gamma}(E(\cdot, \phi_{\zeta_i\otimes\pi}, s), \varphi_\sigma) \\ =&\frac{\mathcal{Z}_S(s, \phi_{\zeta_i\otimes\pi}, \varphi_\sigma)}{\prod_{v\in S} L(s+\frac{1}{2}, \zeta_{i,v} \times \psi_{m,v})} \times \frac{L(s+\frac{1}{2}, \zeta_i \times \psi_m)}{L^S(s+1, \zeta_i \times \pi) L^S(2s+1, \zeta_i, \rho_G)}. \end{aligned} \tag{3.38}$$

The L-function in the numerator of the right hand side equals

$$\begin{aligned} &L(s+\frac{1}{2}, \zeta_i \times \psi_m) \\ =&\prod_{j=1}^{l} L(s+\frac{1}{2}, \zeta_i \times \zeta_j) \prod_{r=1}^{k} L(s+1, \zeta_i \times \xi_r) L(s, \zeta_i \times \xi_r). \end{aligned}$$

By the assumption (3.29), it follows that $L(s+\frac{1}{2}, \zeta_i \times \psi_m)$ has a simple pole at $s = \frac{1}{2}$. Since the normalized finite eulerian product (3.30) is not identically zero, the right hand side of (3.38) for ζ_i has a simple pole at $s = \frac{1}{2}$ for some choice of data, so does the left hand side of (3.38) for ζ_i. It implies that the Eisenstein series $E(\cdot, \phi_{\zeta_i\otimes\pi}, s)$ has a simple pole at $s = \frac{1}{2}$. Since π has a generic global Arthur parameter ϕ_n, by Proposition 5.2 of [23],

$$L(2s, \zeta_i, \rho) L(s, \zeta_i \times \pi)$$

must have a simple pole at $s = \frac{1}{2}$. Hence we must have that

$$\prod_{1\le j\le t} L(s, \zeta_i \times \eta_j) \neq 0$$

for each ζ_i. This completes the proof.

□

It is clear that the same result holds when σ has a generic global Arthur parameter ϕ_m and π has a global Arthur parameter ψ_n, by the same argument. We state the result and omit the proof here.

Theorem 3.5. *Assume that π and σ are given as in Theorem 3.3, with σ having a generic global Arthur parameter ψ_m as* (3.17) *and π having a global Arthur parameter ϕ_n as in* (3.15)*, which is of the form:*

$$\psi_n = (\eta_1, 2a_1 + 1) \boxplus \cdots \boxplus (\eta_k, 2a_k + 1) \boxplus (\delta_1, 2b_1) \boxplus \cdots \boxplus (\delta_l, 2b_l).$$

Assume that the following non-vanishing

$$\prod_{1\le i\le l,\ 1\le j\le k} L(\frac{1}{2}, \delta_i \times \eta_j) \neq 0,$$

holds, and for irreducible cuspidal automorphic representations η of $\mathrm{GL}_a(\mathbb{A})$ the following normalized finite eulerian product of local zeta integrals in (2.18)

$$\frac{\mathcal{Z}_S(s, \phi_{\eta\otimes\sigma}, \varphi_\pi)}{\prod_{v\in S} L(s+\frac{1}{2}, \eta_v \times \psi_{n,v})}$$

is not identically zero at $s = \frac{1}{2}$.

If the Bessel period $\mathcal{B}^{\mathcal{O}_\ell}(\varphi_\pi, \varphi_\sigma)$ is non-zero for a choice of $\varphi_\pi \in \mathcal{C}_\pi$ and $\varphi_\sigma \in \mathcal{C}_\sigma$ then

$$\frac{L(s+\frac{1}{2}, \psi_n \times \phi_m)}{L(s+1, \psi_n, \rho_G^-)L(s+1, \phi_m, \rho_H^-)}$$

is holomorphic and nonzero at $s = 0$.

4 Theory of Twisted Automorphic Descents

We are going to explain the theory of twisted automorphic descents that are mainly developed by the authors in [23], as part of the general theory of explicit constructions of integral transforms with automorphic kernel functions to realize the endoscopic transfers and endoscopic descents as outlined in [14]. This theory is part of the effort to develop refined structures of the discrete spectrum of classical groups based on the theory of endoscopic classification that has been achieved through the Arthur-Selberg trace formula approach ([3] for orthogonal groups as considered in this paper).

Although the idea of construction of automorphic descents goes back to [2, 26] and [7], or maybe even earlier, the recent development of the automorphic descents was re-initiated by the work of Ginzburg-Rallis-Soudry, which has been well presented in their book [13]. The motivation of the work of Ginzburg-Rallis-Soudry is to construct the backwards Langlands functorial transfer from the general linear group GL_N to a quasi-split classical group G, corresponding to the standard embedding of the complex dual groups: $G^\vee(\mathbb{C}) \to \mathrm{GL}_N(\mathbb{C})$, where N depends on G. It is well-known that the existence of such Langlands functorial transfers can be proved either through the Arthur-Selberg trace formula approach [3], or through the approach that is based on the converse theorem of Cogdell and Piatetski-Shapiro [5] and [6].

However, in order to develop a refined theory of discrete spectrum in general, it may be more important to understand the automorphic descent of Ginzburg-Rallis-Soudry as an explicit construction of cuspidal automorphic representations for classical groups [13]. The explicit cuspidal modules of quasi-split classical groups that have been constructed through the Ginzburg-Rallis-Soudry descent have a non-zero Whittaker-Fourier coefficient. Hence it remains very interesting to ask:

Question 4.1. *For quasi-split classical groups, how to develop a more general version of the Ginzburg-Rallis-Soudry automorphic descents in order to construct explicit cuspidal modules for cuspidal automorphic representations that have no non-zero Whittaker-Fourier coefficients?*

and even more generally, to ask:

Question 4.2. *How to develop a more general theory of automorphic descents to construct cuspidal modules for the whole cuspidal spectrum when the classical groups are not quasi-split?*

In order to understand these questions, it is better to understand the construction theory as generally formulated in [14] in terms of the endoscopic classification of the discrete spectrum of classical groups. By Theorem 2.1, any $\pi \in \mathcal{A}_2(G)$ belongs to a global Arthur packet $\widetilde{\Pi}_\psi(G)$ for some global Arthur parameter $\psi \in \widetilde{\Psi}_2(G)$. The following is a question of W. Schmid, which was posed and discussed by Arthur in [4].

Question 4.3 (Arthur-Schmid). *How to construct an explicit cuspidal module for π if assume that $\pi \in \widetilde{\Pi}_\psi(G)$?*

Guided by Questions 4.1 and 4.2, and motivated by the Arthur-Schmid question (Question 4.3), the authors formulated a general principle [23, Principle 1.1] for construction of cuspidal automorphic modules. Instead of getting into the details here, we would like to recall what we have established in the theory of twisted automorphic descents.

For more precise discussions, we only consider special orthogonal group G_n. The Ginzburg-Rallis-Soudry descent [13] constructs a generic cuspidal automorphic module for a quasi-split G_n^*, which is the quasi-split inner form of G_n, by means of

$$\phi_\tau = (\tau_1, 1) \boxplus \cdots \boxplus (\tau_r, 1) \tag{4.1}$$

which is a generic global Arthur parameter of G_n^*. Their construction uses the source automorphic representations of a quasi-split group H_{2n}^*. Here the group H_{2n}^* has a parity different from that of G_n^* and forms with G^* a relevant pair in the sense of the Gan–Gross–Prasad conjecture [8], with global Arthur parameters:

$$\phi_\tau^{(2)} = (\tau_1, 2) \boxplus \cdots \boxplus (\tau_r, 2). \tag{4.2}$$

For any $\Sigma \in \mathcal{A}_2(H_{2n}^*)$ belonging to the global Arthur packet $\widetilde{\Pi}_{\phi_\tau^{(2)}}(H_{2n}^*)$, a Bessel module $\mathcal{F}^{\mathcal{O}_\kappa}(\mathcal{C}_\Sigma)$ with $\kappa = n$ or $n-1$, depending on the type of H_{2n}^*, is cuspidal on $H_{2n}^{\mathcal{O}_\kappa} \cong G_n^*$. One of the technical difficulties in the theory is to prove that the Bessel module $\mathcal{F}^{\mathcal{O}_\kappa}(\mathcal{C}_\Sigma)$ is in fact non-zero. This global non-vanishing property was established if one takes Σ to be the residual representation $\mathcal{E}_\tau$ in $\widetilde{\Pi}_{\phi_\tau^{(2)}}(H_{2n}^*)$. In this case, the Bessel module $\mathcal{F}^{\mathcal{O}_\kappa}(\mathcal{C}_{\mathcal{E}_\tau})$ turns out to be an irreducible generic cuspidal representation of $G_n^*(\mathbb{A})$ with its global Arthur parameter ϕ_τ.

The theory of twisted automorphic descents is to take automorphic representations in the global Arthur packet $\widetilde{\Pi}_{\psi^{(2,1)}}(H_{2n+m})$ as the source representations, where H_{2n+m} is a pure inner form of a quasi-split orthogonal group H_{2n+m}^*, and

$$\psi^{(2,1)} = \phi_\tau^{(2)} \boxplus \phi_\delta,$$

with $\phi_\delta = (\delta_1, 1) \boxplus \cdots \boxplus (\delta_s, 1)$ a generic global Arthur parameters of H_m. The key point in the theory of twisted automorphic descents is the input of the generic global Arthur parameter ϕ_δ. We fix $\phi_\tau^{(2)}$-part and let ϕ_δ vary. The theory predicts the construction of all cuspidal members in the global Arthur packet $\widetilde{\Pi}_{\phi_\tau}(G_n)$ for pure inner forms G_n of G_n^*. The global non-vanishing of the descent $\mathcal{F}^{\mathcal{O}_\ell}(\mathcal{C}_{\mathcal{E}_{\tau\otimes\sigma}})$ for σ varying in the global Arthur packet $\widetilde{\Pi}_{\phi_\delta}(H_m)$ is directly related to the non-vanishing of the central value of the tensor product L-function $L(s, \tau \times \sigma)$. This has been extensively discussed in [23], especially in Sections 5 and 6 of [23], with connection to the global Gan–Gross–Prasad conjecture [8]. Since τ is fixed, but σ varying, this L-function is also commonly called the standard L-function of τ twisted by σ. This is where the name of twisted automorphic descents comes from. We refer to [15, 23], and [21] for further discussions of the theory, and its applications.

5 Branching Problem: Cuspidal Case

For $\pi \in \mathcal{A}_{\mathrm{cusp}}(G_n)$ with a cuspidal realization $\mathcal{C}_\pi$, the *Branching Problem* is to consider the spectral structure of the Bessel module $\mathcal{F}^{\mathcal{O}_\ell}(\mathcal{C}_\pi)$ as a representation of $G_{\ell^-}^{\mathcal{O}_\ell}(\mathbb{A})$. We first recall Proposition 2.2 of [23], which establishes the tower property and the cuspidality of Bessel modules $\mathcal{F}^{\mathcal{O}_\ell}(\mathcal{C}_\pi)$ of π.

Proposition 5.1 (Tower Property and Cuspidality). *Assume that $\pi \in \mathcal{A}_{\mathrm{cusp}}(G_n)$ has a cuspidal realization $\mathcal{C}_\pi$. Then the ℓ-th Bessel module $\mathcal{F}^{\mathcal{O}_\ell}(\mathcal{C}_\pi)$ of π enjoys the following property: There exists an integer ℓ_0 in $\{0, 1, \cdots, \mathfrak{r}\}$, where $\mathfrak{r}$ is the k-rank of G_n, such that*

(1) *the ℓ_0-th Bessel module $\mathcal{F}^{\mathcal{O}_{\ell_0}}(\mathcal{C}_\pi)$ of $\mathcal{C}_\pi$ is nonzero, but for any $\ell \in \{0, 1, \cdots, \mathfrak{r}\}$ with $\ell > \ell_0$, the ℓ-th Bessel module $\mathcal{F}^{\mathcal{O}_\ell}(\mathcal{C}_\pi)$ is identically zero; and*
(2) *the ℓ_0-th Bessel module $\mathcal{F}^{\mathcal{O}_{\ell_0}}(\mathcal{C}_\pi)$ is cuspidal in the sense that its constant terms along all the parabolic subgroups of $H_{\ell_0^-}^{\mathcal{O}_{\ell_0}}$ are zero.*

We call $\ell_0 = \ell_0(\mathcal{C}_\pi)$ in Proposition 5.1 the *first occurrence index* of $\mathcal{C}_\pi$. When the realization $\mathcal{C}_\pi$ is unique, we may write $\ell_0 = \ell_0(\pi)$. The tower property and the cuspidality of the ℓ-th Bessel modules as in Proposition 5.1 lead naturally to the following questions.

Question 5.2 (First Occurrence Index). *For a given $\pi \in \mathcal{A}_{\mathrm{cusp}}(G_n)$ with a cuspidal realization $\mathcal{C}_\pi$, how to determine the first occurrence index $\ell_0 = \ell_0(\mathcal{C}_\pi)$ in terms of basic invariants of π?*

In terms of the global Arthur parameters of $\pi \in \mathcal{A}_{\mathrm{cusp}}(G_n)$, Question 5.2 is closely related to Conjecture 4.2 of [14]. It will be very interesting to understand the relations between the first occurrence index $\ell_0 = \ell_0(\mathcal{C}_\pi)$ and the other basic invariants of π. Note that Conjecture 4.2 of [14] has been intensively studied by B. Liu and the first named author of this papers for symplectic groups and the metaplectic double cover of symplectic groups, but it is much less known for orthogonal groups.

Question 5.3 (Spectrum). *For a given $\pi \in \mathcal{A}_{\mathrm{cusp}}(G_n)$ with a cuspidal realization $\mathcal{C}_\pi$, what can one say about the spectrum of the ℓ_0-th Bessel module $\mathcal{F}^{\mathcal{O}_{\ell_0}}(\mathcal{C}_\pi)$ as a representation of $H_{\ell_0^-}^{\mathcal{O}_{\ell_0}}(\mathbb{A})$?*

This can be regarded as a cuspidal version of the branching problem (Question 1.1) and may be quite difficult. A preliminary understanding to Question 5.3 is the *Generic Summand Conjecture* (Conjecture 2.3 of [23]). For completeness, we state it here.

Conjecture 5.4 (Generic Summand). *Assume that $\pi \in \mathcal{A}_{\mathrm{cusp}}(G_n)$ has a G_n-relevant, generic global Arthur parameter $\phi \in \widetilde{\Phi}_2(G_n^*)$. Then there exists a cuspidal* *realization $\mathcal{C}_\pi$ of π in $L^2_{\mathrm{cusp}}(G_n)$ with the first occurrence index $\ell_0 = \ell_0(\mathcal{C}_\pi)$, such that there exists a k-rational orbit $\mathcal{O}_{\ell_0} = \mathcal{O}_{\underline{p}_{\ell_0}}$ in the k-stable orbits $\mathcal{O}^{\mathrm{st}}_{\underline{p}_{\ell_0}}$ associated to the partition $\underline{p}_{\ell_0}$ with the* **Generic Summand Property:** *There exists at least one σ in $\mathcal{A}_{\mathrm{cusp}}(H^{\mathcal{O}_{\ell_0}}_{\ell_0^-})$ with an $H^{\mathcal{O}_{\ell_0}}_{\ell_0^-}$-relevant, generic global Arthur parameter ϕ_σ in $\widetilde{\Phi}_2(H^*_{\ell_0^-})$, and with a cuspidal realization $\mathcal{C}_\sigma$ of σ in $L^2_{\mathrm{cusp}}(H^{\mathcal{O}_{\ell_0}}_{\ell_0^-})$, such that the L^2-inner product*

$$\left\langle \mathcal{F}^{\psi_{\mathcal{O}_{\ell_0}}}(\varphi_\pi), \varphi_\sigma \right\rangle_{H^{\mathcal{O}_{\ell_0}}_{\ell_0^-}}$$

in the Hilbert space $L^2_{\mathrm{cusp}}(H^{\mathcal{O}_{\ell_0}}_{\ell_0^-})$ is nonzero for some $\varphi_\pi \in \mathcal{C}_\pi$ and $\varphi_\sigma \in \mathcal{C}_\sigma$.

It is well-known that one may explain the existence of such a generic summand in the ℓ-Bessel modules in terms of Vogan packets and of the non-vanishing of the central value of the tensor product L-function $L(s, \pi \times \sigma)$, according to the global Gan–Gross–Prasad conjecture. However, Conjecture 5.4 asserts that if an irreducible cuspidal automorphic representation π has a generic global Arthur parameter, then for one of its automorphic realizations $\mathcal{C}_\pi$, the Bessel module $\mathcal{F}^{\mathcal{O}_{\ell_0}}(\mathcal{C}_\pi)$ at the first occurrence index $\ell_0 = \ell_0(\mathcal{C}_\pi)$, which is cuspidal, has at least one irreducible summand σ, which has a generic global Arthur parameter. The proof of Conjecture 5.4 for general situation is still out of reach, although one may find the proof of some special cases in Section 7 of [23].

According to Theorem 3.1, with the assumption in Conjecture 5.4, the global Arthur parameter of any cuspidal summand σ in the Bessel module $\mathcal{F}^{\psi_{\mathcal{O}_\ell}}(\mathcal{C}_\pi)$ is of the form:

$$\psi_m = (\zeta_1, 1) \boxplus \cdots \boxplus (\zeta_l, 1) \boxplus (\xi_1, 2) \boxplus \cdots \boxplus (\xi_k, 2),$$

with $l, k \geq 0$. The reason, with which Conjecture 5.4 should be true, is that Conjecture 5.4 only consider the spectrum of the Bessel module $\mathcal{F}^{\psi_{\mathcal{O}_\ell}}(\mathcal{C}_\pi)$ at the first occurrence index ℓ_0. Hence Conjecture 3.2 can be viewed as a reformulation of Conjecture 5.4 (Conjecture 2.3 of [23]) with help of Theorem 3.1.

One may consider a further refinement of the branching problem by requiring that both π and σ have generic global Arthur parameters. This leads to a close connection with the global Gan–Gross–Prasad conjecture.

Assume that $\pi \in \mathcal{A}_{\mathrm{cusp}}(G_n)$ has a generic global Arthur parameter $\phi_n \in \widetilde{\Phi}_2(G_n^*)$, which is G_n-relevant, and $\sigma \in \mathcal{A}_{\mathrm{cusp}}(H_m)$, with $H_m = H^{\mathcal{O}_\ell}_{\ell^-}$, has a generic global Arthur parameter $\phi_m \in \widetilde{\Phi}_2(H_m^*)$, which is H_m-relevant. The global Gan–Gross–Prasad conjecture implies that if the Bessel period $\mathcal{B}^{\mathcal{O}_\ell}(\varphi_\pi, \varphi_\sigma)$ in (2.11) is non-zero for some $\varphi_\pi \in \mathcal{C}_\pi$ and $\varphi_\sigma \in \mathcal{C}_\sigma$, then the central L-value $L(\frac{1}{2}, \phi_n \times \phi_m)$ is non-zero. This is Theorem 5.7 of [23]. Moreover, in this case there exists one and at most one pair (π, σ) in the global Vogan packet $\widetilde{\Pi}_{\phi_n \times \phi_m}[G_n^* \times H_m^*]$, such that the Bessel period $\mathcal{B}^{\mathcal{O}_\ell}(\varphi_\pi, \varphi_\sigma)$ is non-zero for some $\varphi_\pi \in \mathcal{C}_\pi$ and $\varphi_\sigma \in \mathcal{C}_\sigma$. Of course, this global uniqueness is a consequence of the local Gan–Gross–Prasad conjecture over all local fields, as explained in Section 3 of [23]. Hence by means of the formulation of the global Gan–Gross–Prasad conjecture, the non-vanishing of the central L-value $L(\frac{1}{2}, \phi_n \times \phi_m)$, plus the compatibility of the sign of the local symplectic root numbers for the pair (π_ν, σ_ν) at all local places ν of k will detect the automorphic branching property that the Bessel period $\mathcal{B}^{\mathcal{O}_\ell}(\varphi_\pi, \varphi_\sigma)$ is non-zero for some data with the unique choice of the pair (π, σ) in the global Vogan packet $\widetilde{\Pi}_{\phi_n \times \phi_m}[G_n^* \times H_m^*]$. Therefore, in order to understand the branching decomposition problem in the framework of the global Gan–Gross–Prasad conjecture, it remains interesting to ask the following question.

Question 5.5 (Generic Automorphic Branching Problem). *For a $\pi \in \mathcal{A}_{\text{cusp}}(G_n)$ with a generic global Arthur parameter ϕ_n of G_n^*, which is G_n-relevant, and for an integer $\ell \in \{0, 1, \cdots, \mathfrak{r}\}$ with $m = \ell^-$, find a special orthogonal group H_m, such that $G_n \times H_m$ is relevant, and find a $\sigma \in \mathcal{A}_{\text{cusp}}(H_m)$ with a generic global Arthur parameter ϕ_m of H_m^*, which is H_m-relevant, such that*

1. *the central L-value $L(\frac{1}{2}, \phi_n \times \phi_m)$ is non-zero for certain choice of ϕ_m; and*
2. *the compatibility of the sign of the local symplectic root numbers for the pair (π_v, σ_v) at all local places v of the number field k.*

Following from the global Gan–Gross–Prasad conjecture, the representations σ that satisfy the two conditions must occur in the automorphic branching decomposition of π, that is, the Bessel period $\mathcal{B}^{\mathcal{O}_\ell}(\varphi_\pi, \varphi_\sigma)$ is non-zero for some $\varphi_\pi \in \mathcal{C}_\pi$ and $\varphi_\sigma \in \mathcal{C}_\sigma$. It is a very interesting problem to construct explicitly such representations σ via a variant of the twisted automorphic descent method.

6 Reciprocal Branching Problem

Let $\sigma \in \mathcal{A}_2(H_m)$ with an automorphic realization $\mathcal{C}_\sigma$. According to Question 1.2, the L^2-*automorphic reciprocal branching problem* is to find a special orthogonal group G_n, such that (G_n, H_m) is a relevant pair in the sense of global Gan–Gross–Prasad conjecture, and find $\pi \in \mathcal{A}_2(G_n)$ such that the Bessel period $\mathcal{B}^{\mathcal{O}_\ell}(\varphi_\pi, \varphi_\sigma)$ is non-zero for some $\varphi_\pi \in \mathcal{C}_\pi$ and $\varphi_\sigma \in \mathcal{C}_\sigma$. Note that when both σ and π are not cuspidal, the Bessel period $\mathcal{B}^{\mathcal{O}_\ell}(\varphi_\pi, \varphi_\sigma)$ needs to be defined with regularization. In this case, one may follow the compatibility of Bessel periods with parabolic induction as indicated in [10, 11], and [12], for instance, which can also be viewed as the global unfolding of the global zeta integrals for tensor product L-functions for the product of the classical groups and general linear groups (as in [20] and [23, Section 4], for example). Hence we may, without loss of generality, only consider the case when σ is cuspidal and $\pi \in \mathcal{A}_2(G_n)$.

6.1 σ Has a Generic Global Arthur Parameter

We may consider first the case that σ has a generic global Arthur parameter

$$\phi_\zeta = (\zeta_1, 1) \boxplus \cdots \boxplus (\zeta_l, 1) \tag{6.1}$$

where $\zeta = \zeta_1 \boxplus \cdots \boxplus \zeta_l$ is the generic isobaric sum of irreducible cuspidal automorphic representations ζ_j of $\mathrm{GL}_{\alpha_j}(\mathbb{A})$ with $\zeta_j \ncong \zeta_i$ if $j \neq i$. By Theorem 3.3, if there is a special orthogonal group G_n such that (G_n, H_m) is relevant, and there is a $\pi \in \mathcal{A}_{\mathrm{cusp}}(G_n)$ such that the Bessel period $\mathcal{B}^{\mathcal{O}_\ell}(\varphi_\pi, \varphi_\sigma)$ is non-zero for some $\varphi_\pi \in \mathcal{C}_\pi$ and $\varphi_\sigma \in \mathcal{C}_\sigma$, then the global Arthur parameter of π must be of the form

$$\psi_n = (\eta_1, 1) \boxplus \cdots \boxplus (\eta_k, 1) \boxplus (\delta_1, 2) \boxplus \cdots \boxplus (\delta_l, 2). \tag{6.2}$$

We are going to show how to refine this structure via the construction in the theory of twisted automorphic descents.

Take $\tau = \tau_1 \boxplus \cdots \boxplus \tau_r$ to be an isobaric automorphic representation of $\mathrm{GL}_a(\mathbb{A})$ with $\tau_i \ncong \tau_j$ if $i \neq j$. Let $\mathrm{GL}_a \times H_m$ be a standard Levi subgroup of H_{a+m}. With $(\mathrm{GL}_a \times H_m, \tau \otimes \sigma)$, we may define an Eisenstein series $E(\cdot, \phi_{\tau\otimes\sigma}, s)$. By [18], this Eisenstein series has a pole at $s = \frac{1}{2}$ if and only if the L-function $L(s, \tau, \rho)$ has a pole at $s = 1$ and the central value $L(\frac{1}{2}, \tau \otimes \sigma)$ is non-zero. The condition that $L(s, \tau, \rho)$ has a pole at $s = 1$ is equivalent to say that τ generates a generic global Arthur parameter for an even special orthogonal group of the same type as G_n. This implies in particular that a must be even, because for the special orthogonal groups considered in this paper, their complex dual groups are naturally embedded into even degree general linear groups. We may assume that $a \geq 2m$ if H_m is an even special orthogonal group; and $a > 2m$ if H_m is an odd special orthogonal group, in the following discussion. We also assume that the central value $L(\frac{1}{2}, \tau\otimes\sigma)$ is non-zero. Under these two assumptions, we obtain that the Eisenstein series $E(\cdot, \phi_{\tau\otimes\sigma}, s)$ has a non-zero residue $\mathcal{E}_{\tau\otimes\sigma}$ at $s = \frac{1}{2}$, which is square-integrable and has a global Arthur parameter of the form $\phi_\tau^{(2)} \boxplus \phi_\zeta$, where $\phi_\tau^{(2)}$ is as defined in (4.2).

We may consider the global Arthur packet $\widetilde{\Pi}_{\phi_\tau^{(2)}\boxplus\phi_\zeta}(H_{a+m})$. The automorphic members Σ in $\widetilde{\Pi}_{\phi_\tau^{(2)}\boxplus\phi_\zeta}(H_{a+m})$ could be cuspidal or residual representations. We may define $\mathcal{A}_{\phi_\tau^{(2)}\boxplus\phi_\zeta}(H_{a+m})$ to be the set of all $\Sigma \in \mathcal{A}_2(H_{a+m})$ that also belongs to $\widetilde{\Pi}_{\phi_\tau^{(2)}\boxplus\phi_\zeta}(H_{a+m})$. The tower of the Bessel modules $\mathcal{B}^{\mathcal{O}_\kappa}(\Sigma)$ for all $\Sigma \in \mathcal{A}_{\phi_\tau^{(2)}\boxplus\phi_\zeta}(H_{a+m})$ has been studied in [23, Section 6.1]. In particular, Proposition 6.2 of [23] proves that for every $\Sigma \in \mathcal{A}_{\phi_\tau^{(2)}\boxplus\phi_\zeta}(H_{a+m})$, the Bessel modules $\mathcal{B}^{\mathcal{O}_\kappa}(\Sigma)$ are zero if $\kappa > \frac{a}{2} + m - 1$ when H_m is an even special orthogonal group; and if $\kappa > \frac{a}{2} + m$ when H_m is an odd special orthogonal group, respectively. However, it remains mysterious about the first occurrence index $\ell_0(\mathcal{C}_\Sigma)$ of each member Σ in $\mathcal{A}_{\phi_\tau^{(2)}\boxplus\phi_\zeta}(H_{a+m})$. In the spirit of the general conjecture [14, Conjecture 4.2], or more specifically, Conjecture 6.7 of [23], there should be at least one member Σ in

$\mathcal{A}_{\phi_\tau^{(2)}\boxplus\phi_\zeta}(H_{a+m})$, such that the first occurrence index $\ell_0(\mathcal{C}_\Sigma) = \frac{a}{2}+m-1$ when H_m is an even special orthogonal group; and $\ell_0(\mathcal{C}_\Sigma) = \frac{a}{2}+m$ when H_m is an odd special orthogonal group, respectively. This expectation has been verified in Proposition 6.8 of [23] for $m = 0$ and $m = 1$. The close relation of this expectation to the global Gan–Gross–Prasad conjecture has been explained in Theorem 6.9 and its proof in [23].

We define $\ell_0(\phi_\tau^{(2)} \boxplus \phi_\zeta)$ to be the largest possible first occurrence index of all Σ in $\mathcal{A}_{\phi_\tau^{(2)}\boxplus\phi_\zeta}(H_{a+m})$ and call it the *first occurrence index* of the global Arthur packet $\widetilde{\Pi}_{\phi_\tau^{(2)}\boxplus\phi_\zeta}(H_{a+m})$, which implies, by means of Conjecture 6.7 of [23], that $\ell_0(\phi_\tau^{(2)} \boxplus \phi_\zeta) = \frac{a}{2} + m - 1$ when H_m is an even special orthogonal group; and $\ell_0(\phi_\tau^{(2)} \boxplus \phi_\zeta) = \frac{a}{2}+m$ when H_m is an odd special orthogonal group. For $\Sigma = \mathcal{E}_{\tau\otimes\sigma}$ with $\sigma \in \mathcal{A}_{\phi_\zeta}(H_m)$, which belongs to $\mathcal{A}_{\phi_\tau^{(2)}\boxplus\phi_\zeta}(H_{a+m})$, we consider the Bessel modules $\mathcal{F}^{\mathcal{O}_\kappa}(\mathcal{E}_{\tau\otimes\sigma})$ with $0 \le \kappa \le \ell_0(\phi_\tau^{(2)} \boxplus \phi_\zeta)$. Assume that $\kappa_0 = \ell_0(\mathcal{E}_{\tau\otimes\sigma})$ is the first occurrence index of the residual representation $\mathcal{E}_{\tau\otimes\sigma}$. Then by the same argument of the proof of Proposition 6.3 of [23], the Bessel module $\mathcal{F}^{\mathcal{O}_{\kappa_0}}(\mathcal{E}_{\tau\otimes\sigma})$ is non-zero and cuspidal, and hence can be written as

$$\mathcal{F}^{\mathcal{O}_{\kappa_0}}(\mathcal{E}_{\tau\otimes\sigma}) = \pi_1 \oplus \pi_2 \oplus \cdots \oplus \pi_j \oplus \cdots \tag{6.3}$$

as a representation of $G_{\kappa_0^-}^{\mathcal{O}_{\kappa_0}}(\mathbb{A})$, where every π_j is an irreducible cuspidal automorphic representation of $G_{\kappa_0^-}^{\mathcal{O}_{\kappa_0}}(\mathbb{A})$. Since the residual representation $\mathcal{E}_{\tau\otimes\sigma}$ is irreducible, by the uniqueness of local Bessel models at all local places of k, the decomposition in (6.3) is of multiplicity free. It is clear that for each π_i, the Bessel period $\mathcal{B}^{\mathcal{O}_{\kappa_0}}(\mathcal{E}_{\tau\otimes\sigma}, \varphi_{\pi_i})$ is non-zero for some choice of data. By the reciprocal non-vanishing of Bessel periods as established in Theorem 5.3 of [23], there exists an integer ℓ_* such that the Bessel period $\mathcal{B}^{\mathcal{O}_{\ell_*}}(\varphi_{\pi_i}, \varphi_\sigma)$ is non-zero. According to the discussion in Section 5 of [23], the index ℓ_* is completely determined by (m, a, κ_0), but may not be the first occurrence index of π. The above discussion can be summarized as the following theorem, which can be viewed as an answer to the *Cuspidal Reciprocal Branching Problem*.

Theorem 6.1 (Cuspidal Reciprocal Branching). *Assume that $\sigma \in \mathcal{A}_{\mathrm{cusp}}(H_m)$ has a generic global Arthur parameter ϕ_ζ as in (6.1). For any even integer a, which is greater than or equal to 2m if H_m is an even special orthogonal group; and is greater than 2m if H_m is an odd special orthogonal group, assume that there exists an isobaric automorphic representation $\tau = \tau_1 \boxplus \cdots \boxplus \tau_r$ of $\mathrm{GL}_a(\mathbb{A})$ with the properties that the global Arthur parameter ϕ_τ is of the type opposite to that of ϕ_δ, and the central value $L(\frac{1}{2}, \tau \times \sigma)$ is not zero. Then there exists a special orthogonal group G_n, which is of the type $G_{\kappa_0^-}^{\mathcal{O}_{\kappa_0}}$ with $n = \kappa_0^-$, such that (G_n, H_m) is a relevant pair, and there is a $\pi \in \mathcal{A}_{\mathrm{cusp}}(G_n)$ with an automorphic realization $\mathcal{C}_\pi$, whose global Arthur parameter ψ_n is of the form as in (6.2), such that the Bessel period*

$\mathcal{B}^{\mathcal{O}_{\ell_*}}(\varphi_\pi, \varphi_\sigma)$ *is non-zero for some choice of* $\varphi_\sigma \in \mathcal{C}_\sigma$ *and* $\varphi_\pi \in \mathcal{C}_\pi$, *where* ℓ_* *is an integer that is uniquely determined by* (κ_0^-, m) *or by* (m, a, κ_0).

It may be worthwhile to mention that the index ℓ_* in Theorem 6.1 can be calculated by means of the global unfolding of the global zeta integrals of Bessel type for the tensor product L-function $L(s, \tau \times \pi)$ as in [20] and [23], in particular, in Sections 4 and 5 in [23].

Let us discuss a special case when the first occurrence index $\kappa_0 = \ell_0(\mathcal{E}_{\tau\otimes\sigma})$ is equal to $\ell_0(\phi_\tau^{(2)} \boxplus \phi_\delta)$, the first occurrence index of the global Arthur packet $\widetilde{\Pi}_{\phi_\tau^{(2)}\boxplus\phi_\zeta}(H_{a+m})$, then $a = 2n$ and $\kappa_0^- = n$. In this case, by Theorem 6.5 of [23], the irreducible summands π_i of $\mathcal{F}^{\mathcal{O}_{\kappa_0}}(\mathcal{E}_{\tau\otimes\sigma})$ in (6.3) belonging to $\mathcal{A}_{\mathrm{cusp}}(G_n^{\mathcal{O}_{\kappa_0}})$ have ϕ_τ as their global Arthur parameter. Again, by the reciprocal non-vanishing of Bessel Periods (Theorem 5.3 of [23]), the Bessel periods $\mathcal{B}^{\mathcal{O}_{\ell_0}}(\varphi_{\pi_i}, \varphi_\sigma)$ are non-zero for some choice of data. Assuming the local Gan–Gross–Prasad conjecture for special orthogonal groups over all local fields, we deduce that $\mathcal{F}^{\mathcal{O}_{\kappa_0}}(\mathcal{E}_{\tau\otimes\sigma}) = \pi$ is irreducible if $G_n^{\mathcal{O}_{\kappa_0}}$ is an odd special orthogonal group; and $\mathcal{F}^{\mathcal{O}_{\kappa_0}}(\mathcal{E}_{\tau\otimes\sigma}) = \pi \oplus \pi^\star$ if $G_n^{\mathcal{O}_{\kappa_0}}$ is an even special orthogonal group. The details of the proof for this last assertion are the same as the proof of Theorem 7.1 of [23].

Theorem 6.2. *Take the same assumption as in Theorem 6.1. If the first occurrence index* $\kappa_0 = \ell_0(\mathcal{E}_{\tau\otimes\sigma})$ *is equal to* $\ell_0(\phi_\tau^{(2)} \boxplus \phi_\delta)$, *the first occurrence index of the global Arthur packet* $\widetilde{\Pi}_{\phi_\tau^{(2)}\boxplus\phi_\zeta}(H_{a+m})$, *then there exists a* $\pi \in \mathcal{A}_{\mathrm{cusp}}(G_n^{\mathcal{O}_{\kappa_0}})$ *with* $2n = a$ *with* ϕ_τ *as its generic global Arthur parameter, such that the Bessel period* $\mathcal{B}^{\mathcal{O}_{\ell_0}}(\varphi_\pi, \varphi_\sigma)$ *is non-zero for some choice of* $\varphi_\sigma \in \mathcal{C}_\sigma$ *and* $\varphi_\pi \in \mathcal{C}_\pi$.

6.2 Relations with the Global Gan–Gross–Prasad Conjecture

We are going to discuss the relation between the *Cuspidal Reciprocal Branching Problem* and the global Gan–Gross–Prasad conjecture.

Given $\sigma \in \mathcal{A}_{\mathrm{cusp}}(H_m)$ with an H_m-relevant, generic global Arthur parameter ϕ_δ of H_m^*, as in (6.1), it is not hard to find a family of special orthogonal groups G_n such that (G_n, H_m) is a relevant pair in the sense of global Gan–Gross–Prasad conjecture. It is itself a hard problem to find a G_n-relevant, generic global Arthur parameter ϕ_τ of G_n^* with $a = 2n$, such that the central L-value $L(\frac{1}{2}, \phi_\tau \times \phi_\delta)$ is non-zero. We take it as an assumption. The global Gan–Gross–Prasad conjecture asserts that there exists a unique pair (π_0, σ_0) in the global Vogan packet $\widetilde{\Pi}_{\phi_G\times\phi_H}[G_n^* \times H_m^*]$ with the property that the Bessel period $\mathcal{B}^{\mathcal{O}_{\ell_0}}(\varphi_{\pi_0}, \varphi_{\sigma_0})$, with $m = \ell_0^-$, is non-zero for some $\varphi_{\pi_0} \in \mathcal{C}_{\pi_0}$ and $\varphi_{\sigma_0} \in \mathcal{C}_{\sigma_0}$.

If the given σ, sharing the global Vogan packet with σ_0, is happened to be σ_0, then the cuspidal automorphic module π as constructed by the twisted automorphic descent in Theorem 6.2 is equal to π_0.

If σ shares a global Vogan packet with σ_0, but they are not equivalent to each other, the global Gan–Gross–Prasad conjecture asserts that the first occurrence index $\kappa_0 = \ell_0(\mathcal{E}_{\tau\otimes\sigma})$ must be less than $\ell_0(\phi_\tau^{(2)} \boxplus \phi_\delta)$. This goes to the situation considered in Theorem 6.1. In this case, we may still apply the twisted automorphic descent method to construct an irreducible cuspidal automorphic representation π of G_n, such that the Bessel period $\mathcal{B}^{\mathcal{O}_{\ell_0}}(\varphi_\pi, \varphi_\sigma)$ is non-zero for some choice of $\varphi_\sigma \in \mathcal{C}_\sigma$ and $\varphi_\pi \in \mathcal{C}_\pi$. However, we do not have precise information about the global Arthur parameter ψ_n of π, except that ψ_n is of the form in (6.2). Hence it is a desirable question to find the precise information about the global Arthur parameter of π that gives an answer the *Cuspidal Reciprocal Branching Problem*! In a recent work of Jiang-Liu-Xu [16], a special situation when H_m is SO(3) is explicitly studied. In this case, we are able to prove in some situation that the twisted automorphic descent method may still construct a cuspidal π with generic global Arthur parameter (Theorem 6.3 of [16]). We will leave further details of discussion about this mysterious question to our future work.

On the other hand, Theorem 6.1 can be viewed as examples of the global Gan–Gross–Prasad conjecture for non-generic global Arthur parameters, which has been recently reformulated in [9].

Acknowledgments The research of the first named author "Dihua Jiang" is supported in part by the NSF Grant DMS–1600685 and DMS-1901802, and that of the second named author "Lei Zhang" is supported in part by AcRF Tier 1 grant R-146-000-237-114 and R-146-000-277-114 of National University of Singapore.

References

1. A. Aizenbud, D. Gourevitch, S. Rallis, G. Schiffmann, Multiplicity one theorems. Ann. Math. **172**(2), 1407–1434 (2010)
2. A.N. Andrianov, Modular descent and the Saito-Kurokawa conjecture. Invent. Math. **53**(3), 267–280 (1979)
3. J. Arthur, *The Endoscopic Classification of Representations. Orthogonal and Symplectic Groups*. American Mathematical Society Colloquium Publications, vol. 61 (American Mathematical Society, Providence, 2013)
4. J. Arthur, Characters and modules, in *Representation Theory, Automorphic Forms and Complex Geometry, a Tribute to W. Schmid* (International Press, Vienna, 2020), pp. 15–16
5. J.W. Cogdell, I.I. Piatetski-Shapiro, Converse theorems for GL_n. Inst. Hautes Études Sci. Publ. Math. **79**, 157–214 (1994)
6. J.W. Cogdell, H.H. Kim, I.I. Piatetski-Shapiro, F. Shahidi, Functoriality for the classical groups. Publ. Math. Inst. Hautes Études Sci. **99**, 163–233 (2004)
7. M. Eichler, D. Zagier, *The Theory of Jacobi Forms*. Progress in Mathematics, vol. 55 (Birkhäuser Boston, Boston, 1985)
8. W.T. Gan, B.H. Gross, D. Prasad, Symplectic local root numbers, central critical L-values, and restriction problems in the representation theory of classical groups. Astérisque **346**, 1–109, (2012)
9. W.T. Gan, B.H. Gross, D. Prasad, Branching laws for Classical Groups: The non-tempered case. Compos. Math. **156**(11), 2298–2367 (2020)

10. D. Ginzburg, D. Jiang, S. Rallis, On the nonvanishing of the central value of the Rankin-Selberg L-functions. J. Amer. Math. Soc. **17**(3), 679–722 (2004)
11. D. Ginzburg, D. Jiang, S. Rallis, On the nonvanishing of the central value of the Rankin-Selberg L-functions. II, in *Automorphic Representations, L-Functions and Applications: Progress and Prospects*. Ohio State University Mathematics Research Institute Publication, vol. 11 (de Gruyter, Berlin, 2005), pp. 157–191
12. D. Ginzburg, D. Jiang, S. Rallis, Models for certain residual representations of unitary groups, in *Automorphic Forms and L-Functions I. Global Aspects*. Contemporary Mathematics, vol. 488 (American Mathematical Society, Providence 2009), pp. 125–146
13. D. Ginzburg, S. Rallis, D. Soudry, *The Descent Map from Automorphic Representations of* GL(n) *to Classical Groups* (World Scientific, Singapore, 2011)
14. D. Jiang, Automorphic integral transforms for classical groups I: Endoscopy correspondences, *Automorphic Forms: L-Functions and Related Geometry: Assessing the Legacy of I.I. Piatetski-Shapiro*. Contemporary Mathematics, vol. 614 (American Mathematical Society, Providence, 2014), pp. 179–242
15. D. Jiang, On tensor product L-functions and Langlands functoriality. Bull. Iranian Math. Soc. **43**(4), 169–189 (2017)
16. D. Jiang, B. Liu, B. Xu, A reciprocal branching problem for automorphic representations and global Vogan packets. J. Reine Angew. Math. **765**, 249–277 (2020)
17. D. Jiang, B. Liu, B. Xu, L. Zhang, The Jacquet-Langlands correspondence via twisted descent. Int. Math. Res. Not. **2016**(18), 5455–5492 (2016)
18. D. Jiang, B. Liu, L. Zhang, Poles of certain residual Eisenstein series of classical groups. Pacific J. Math. **264**(1), 83–123 (2013)
19. D. Jiang, B. Sun, C.-B. Zhu, Uniqueness of Bessel models: the archimedean case. Geom. Funct. Anal. **20**, 690–709 (2010)
20. D. Jiang, L. Zhang, A product of tensor product L-functions for classical groups of hermitian type. Geom. Funct. Anal. **24**(2), 552–609 (2014)
21. D. Jiang, L. Zhang, Automorphic integral transforms for classical groups II: Twisted descents, in *Representation Theory, Number Theory, and Invariant Theory*. Progress in Mathematics, vol. 323 (Birkhäuser/Springer, Cham, 2017), pp. 303–335
22. D. Jiang, L. Zhang, On the nonvanishing of the central value of certain L-functions: unitary groups. J. Eur. Math. Soc. (JEMS) **22**(6), 1759–1783 (2020)
23. D. Jiang, L. Zhang, Arthur parameters and cuspidal automorphic modules of classical groups. Ann. of Math. **191**(3), 739–827 (2020)
24. R.P. Langlands, *On the Functional Equations Satisfied by Eisenstein Series*. Springer Lecture Notes in Mathematics, vol. 544 (Springer, Berlin, 1976)
25. C. Mœglin, J.-P. Waldspurger, *Spectral Decomposition and Eisenstein Series*. Cambridge Tracts in Mathematics, vol. 113 (Cambridge University Press, Cambridge, 1995)
26. I.I. Piatetski-Shapiro, On the Saito-Kurokawa lifting. Invent. Math. **71**(2), 309–338 (1983)
27. D. Soudry, The unramified computation of Rankin-Selberg integrals expressed in terms of Bessel models for split orthogonal groups: Part I. Israel J. Math. **222**(2), 711–786 (2017)
28. D. Soudry, The unramified computation of Rankin-Selberg integrals expressed in terms of Bessel models for split orthogonal groups: Part II. J. Number Theory **186**, 62–102 (2018)
29. B. Sun, C. Zhu, Multiplicity one theorems: the Archimedean case. Ann. Math. **175**(1), 23–44 (2012)

Distinguished Representations of $SO(n+1,1) \times SO(n,1)$, Periods and Branching Laws

Toshiyuki Kobayashi and Birgit Speh

Abstract Given irreducible representations Π and π of the rank one special orthogonal groups $G = SO(n+1,1)$ and $G' = SO(n,1)$ with nonsingular integral infinitesimal character, we state in terms of θ-stable parameter necessary and sufficient conditions so that

$$\mathrm{Hom}_{G'}(\Pi|_{G'}, \pi) \neq \{0\}.$$

In the special case that both Π and π are tempered, this implies the Gross–Prasad conjectures (Gross and Prasad, Canad J Math 44:974–1002, 1992) for tempered representations of $SO(n+1,1) \times SO(n,1)$ which are nontrivial on the center.

We apply these results to construct nonzero periods and distinguished representations. If both Π and π have the trivial infinitesimal character ρ then we use a theorem that the periods are nonzero on the minimal K-type to obtain a nontrivial bilinear form on the $(\mathfrak{g}, K)$-cohomology of the representations.

Mathematic Subject Classification (2010) Primary 22E30; Secondary 11F70, 22E45, 22E46, 53A30, 58J70

1 Introduction

In the book [We46] *The Classical Groups: Their Invariants and Representations* published in 1939, Hermann Weyl discusses the restriction of irreducible finite-dimensional representations of the orthogonal group $O(n)$ to an orthogonal sub-

T. Kobayashi (✉)
Graduate School of Mathematical Sciences, The University of Tokyo and Kavli IPMU (WPI), Tokyo, Japan
e-mail: toshi@ms.u-tokyo.ac.jp

B. Speh
Department of Mathematics, Cornell University, Ithaca, NY, USA
e-mail: bes12@cornell.edu

W. Müller et al. (eds.), *Relative Trace Formulas*, Simons Symposia,
https://doi.org/10.1007/978-3-030-68506-5_8

group $O(n-1)$. The restriction splits into a direct sum of irreducible representations and can be roughly described by the following *branching law*. Suppose $n \geq 2$. Denote by $[x]$ the greatest integer that does not exceed x. To each irreducible representation of $O(n)$ is assigned a highest weight

$$\mu = (\mu_1, \mu_2, \ldots, \mu_{[\frac{n}{2}]})$$

satisfying

$$\mu_1 \geq \mu_2 \geq \cdots \geq \mu_{[\frac{n}{2}]}$$

and a character ε of $O(n)/SO(n)$. We denote by this representation $F^{O(n)}(\mu)_\varepsilon$ where $\varepsilon \in \{+, -\}$ (ε is not unique when n is even and $\mu_{\frac{n}{2}} \neq 0$). H. Weyl obtained the "branching law" as

$$F^{O(n)}(\mu_1, \ldots, \mu_{[\frac{n}{2}]})_\varepsilon|_{O(n-1)} = \bigoplus F^{O(n-1)}(\nu_1, \ldots, \nu_{[\frac{n-1}{2}]})_\varepsilon \tag{1.1}$$

where the summation is taken over $(\nu_1, \ldots, \nu_{[\frac{n-1}{2}]}) \in \mathbb{Z}^{[\frac{n-1}{2}]}$ subject to

$$\mu_1 \geq \nu_1 \geq \mu_2 \geq \cdots \geq \nu_{\frac{n-1}{2}} \geq 0 \qquad \text{for } n \text{ odd,}$$

$$\mu_1 \geq \nu_1 \geq \mu_2 \geq \cdots \geq \nu_{\frac{n-2}{2}} \geq \mu_{\frac{n}{2}} \geq 0 \quad \text{for } n \text{ even.}$$

In this article we present similar branching laws for the restriction of irreducible *infinite-dimensional* representations of $SO(n+1, 1)$ to the subgroup $SO(n, 1)$. Since the restriction of an infinite-dimensional representation Π of $SO(n+1, 1)$ to $SO(n, 1)$ is not a direct sum of irreducible representations [Ko98], we consider as in [KS15, KS18b] the representations Π and π of $SO(n+1, 1)$, respectively $SO(n, 1)$, realized as smooth representations of moderate growth [Wa92, Chap. 11] and define the *multiplicity* by

$$m(\Pi, \pi) := \dim_{\mathbb{C}} \operatorname{Hom}_{SO(n,1)}(\Pi|_{SO(n,1)}, \pi).$$

The multiplicity is either 0 or 1 [SZ12].

We consider in the article only representations of the special orthogonal group which have the same infinitesimal character as an irreducible finite-dimensional representation F of $SO(n+1)$. To simplify the notation and presentation we assume in this article that F is "self-dual", i.e., assume that the highest weight $\mu = (\mu_1, \cdots, \mu_{m+1})$ satisfies $\mu_{m+1} = 0$ when $n = 2m+1$. We do not impose any assumption on F when n is even. See Assumption A in Sect. 2.2.1. For the general case, see [KS18b] and [KS].

For every irreducible representation π of $SO(n, 1)$ we define in Sect. 2.3 a *height*

$$h_\pi \in \{0, \ldots m\} \quad \text{if } n = 2m \text{ or } 2m+1$$

and in Sect. 2.4 a *signature* $\delta \in \{+,-\}$. The signature is unique except for discrete series representations. If π has the same infinitesimal character as $F(\mu)$ we say that (μ, h_π, δ) are the *enhanced* θ*-stable parameters* of π. The representations with *enhanced* θ*-stable parameters* $((0,\ldots,0), i, \delta)$ are representations with trivial infinitesimal character ρ and are denoted by $\Pi_{i,\delta}$. See [KS18b, Chap. 2, Sect. 4] for a description of the representations of $O(n+1,1)$ with trivial infinitesimal character ρ and [KS18b, Chap. 14, Sect. 9] for their enhanced θ-parameters.

Remark 1.1. In [KS18b] we have treated mainly the full group $O(n+1,1)$ rather than the special orthogonal group $SO(n+1,1)$, and stated results for $SO(n+1,1)$ in Chapter 11 with "bar" for the corresponding objects. The relation between branching laws for $O(n+1,1) \downarrow O(n,1)$ and $SO(n+1,1) \downarrow SO(n,1)$ is discussed in [KS18b, Chap. 15 (Appendix II)]. In this article, we treat mainly the special orthogonal group $SO(n+1,1)$, and use different convention in the point that we omit the "bar" for representations of $SO(n+1,1)$.

All results in this article are based on the following **branching theorem**:

Theorem 1.2 (Branching Law). *Let Π and π be irreducible representations of $SO(n+1,1)$ respectively $SO(n,1)$ with enhanced θ-stable parameters $(\mu, h_\Pi, \varepsilon)$, respectively (ν, h_π, δ).*

(1) *Suppose that $n = 2m$. Then*

$$\dim_{\mathbb{C}} \mathrm{Hom}_{SO(n,1)}(\Pi|_{SO(n,1)}, \pi) = 1$$

if and only if the enhanced θ-stable parameters of the representations Π and π satisfy

(a) $\varepsilon = \delta$,
(b) $h_\pi \in \{h_\Pi, h_\Pi - 1\}$ *when* $h_\pi < m$ *and* $h_\pi = h_\Pi$ *when* $h_\pi = m$,
(c) $\mu_0 \geq \nu_0 \geq \mu_1 \geq \cdots \geq \mu_m \geq \nu_m \geq 0$.

(2) *Suppose that $n = 2m - 1$. Then*

$$\dim_{\mathbb{C}} \mathrm{Hom}_{SO(n,1)}(\Pi|_{SO(n,1)}, \pi) = 1$$

if and only if the enhanced θ-stable parameters of the representations satisfy

(a) $\varepsilon = \delta$,
(b) $h_\pi \in \{h_\Pi, h_\Pi - 1\}$,
(c) $\mu_0 \geq \nu_0 \geq \mu_1 \geq \cdots \geq \mu_m \geq \nu_{m-1} \geq \mu_m = 0$.

A detailed proof will be given in [KS].

Remark 1.3. A similar theorem was proved in [KS18b, Thms. 4.1 and 4.2] for the restriction of irreducible representations of $O(n+1,1)$ to $O(n,1)$ with trivial infinitesimal character ρ.

Applications of the Branching Theorem

Gross–Prasad Conjectures The discussion in [KS18b, Chap. 13, Sect. 3.3] shows that Theorem 1.2 in the special case where $h_\Pi = \left[\frac{n+1}{2}\right]$ and $h_\pi = \left[\frac{n}{2}\right]$ implies the following.

Theorem 1.4. *The Gross–Prasad conjectures are valid for all tempered representations with nonsingular infinitesimal character of $SO(n+1,1)$ and $SO(n,1)$, which are nontrivial on the center.*

Remark 1.5.

(1) For tempered principal series representations Π of $SO(n+1,1)$, and π of $SO(n,1)$ which are nontrivial on each center, it was proved in [KS18b, Chap. 11, Sect. 4]. For irreducible tempered representations with trivial infinitesimal character ρ, this was announced in [KS18a] and proved in [KS18b, Chap. 11, Sect. 5].
(2) The branching law for nontempered representations (Theorem 1.2) interpolates between the classical branching laws of finite-dimensional representations (1.1) and the branching laws of the conjecture by Gross and Prasad for tempered representations [GGP12, GP92].

Periods For representations Π, π of a real reductive Lie group G, respectively of a reductive subgroup G', the space of symmetry breaking operators

$$\mathrm{Hom}_{G'}(\Pi|_{G'}, \pi^\vee)$$

and the space of G'-invariant continuous linear functionals

$$\mathrm{Hom}_{G'}(\Pi \otimes \pi, \mathbb{C})$$

are naturally isomorphic to each other [KS18b, Thm. 5.4], where $\pi^\vee$ denotes the contragredient representation of π in the category of admissible smooth representations (see Sect. 3.2), and $\Pi \otimes \pi$ denotes the representation of G' acting on the outer tensor product representation $\Pi \boxtimes \pi$ of $G \times G'$ diagonally. Thus we may use symmetry breaking operators to determine G'-invariant continuous linear functionals on $\Pi \otimes \pi$, i.e., periods. Hence, the branching theorem implies the following.

Theorem 1.6. *Suppose that the representations Π and π of G, respectively of G' satisfy Assumption A (see Sect. 2.2.1) with height i respectively j. Then the following statements on the pair (Π, π) are equivalent:*

(i) *The representation $\Pi \boxtimes \pi$ has a nontrivial G'-period;*
(ii) *Π and π have the same signature, $j = i$ or $i - 1$ and their enhanced θ-stable parameters satisfy the interlacing conditions of Theorem 1.2.*

For representations with infinitesimal character ρ we proved in [KS18b, Chap. 10] furthermore the following:

Theorem 1.7. *If Π and π are representations with trivial infinitesimal character ρ, then any nonzero period does not vanish on the minimal K-type of the outer tensor product representation $\Pi \boxtimes \pi$.*

In a special case we determine in [KS18b, Chap. 12] the value of the period on vectors in the minimal K-type of a representation with infinitesimal character ρ. See also Theorem 5.11.

Distinguished Representations Let G be a reductive group and H a reductive subgroup. We regard H as a subgroup of the direct product group $G \times H$ via the diagonal embedding $H \hookrightarrow G \times H$.

Definition 1.8. Let ψ be a one-dimensional representation of H. We say an admissible smooth representation Π of G is *(H, ψ)-distinguished* if

$$(\mathrm{Hom}_H(\Pi \boxtimes \psi^\vee, \mathbb{C}) \simeq)\, \mathrm{Hom}_H(\Pi|_H, \psi) \neq \{0\}.$$

If the character ψ is trivial, we say Π is *H-distinguished.*

Let $G = SO(n+1,1)$, and $\mathfrak{g} \simeq \mathfrak{so}(n+1,1)$ its Lie algebra. We fix a fundamental Cartan subalgebra $\mathfrak{h}$ of $\mathfrak{g}$. For $0 \leq i \leq [\frac{n+1}{2}]$, there are θ-stable parabolic subalgebras $\mathfrak{q}_i \equiv \mathfrak{q}_i^+ = (\mathfrak{l}_i)_\mathbb{C} + \mathfrak{u}_i$ and $\mathfrak{q}_i^- = (\mathfrak{l}_i)_\mathbb{C} + \mathfrak{u}_i^-$ in $\mathfrak{g}_\mathbb{C} = \mathrm{Lie}(G) \otimes_\mathbb{R} \mathbb{C} = \mathfrak{u}_i^- + (\mathfrak{l}_i)_\mathbb{C} + \mathfrak{u}_i$ such that $\mathfrak{q}_i$ and $\mathfrak{q}_i^-$ contain $\mathfrak{h}_\mathbb{C}$ and that both the Levi subgroups L_i of $\mathfrak{q} = \mathfrak{q}_i$ and $\mathfrak{q}_i^-$ are given by $L_i \simeq SO(2)^i \times SO(n-2i+1,1)$, see [KS18b, Lem. 14.38].

There are two one-dimensional representations of $SO(n-2i+1,1)$ when $n \neq 2i-1$: we denote by χ_+ the trivial representation $\mathbf{1}$ and by χ_- the nontrivial one. For $n = 2i-1$, we consider only χ_+. We consider the differential $\lambda = (\lambda_1, \cdots, \lambda_i, 0, \ldots 0)$ of a one-dimensional representation of L_i and assume that λ satisfies the conditions in [KS18b, Chap. 14, Sect. 9] and $\delta \in \{+,-\}$. We consider an irreducible one-dimensional L_i-module

$$\mathbb{C}_\lambda \boxtimes \chi_\delta$$

and define an admissible smooth representation of G of moderate growth denoted by $A_{\mathfrak{q}_i}(\lambda)_\delta$. Its underlying $(\mathfrak{g}, K)$-module is given by the cohomological parabolic induction from $\mathfrak{q}_i$, see [KV95, VZ84]. For $\delta = +$ we often omit the subscript.

Theorem 1.9. *Suppose that $\Pi \in \mathcal{A}$ (see Sect. 2.2.1 below for definition) is a representation of $SO(n+1,1)$ cohomologically induced from a one-dimensional representation of a θ-stable parabolic subalgebra $\mathfrak{q}_i$, i.e. that $\Pi = A_{\mathfrak{q}_i}(\lambda)$. Then the height (see Sect. 2.3) of Π is i and Π is $SO(n+1-i,1)$-distinguished.*

Remark 1.10. The proof of this theorem will be given in a subsequent paper, based on the work [KS18b]. For a different proof and perspective of this theorem, see [Ko92].

The irreducible representations with trivial infinitesimal character ρ are obtained through cohomological induction. We set $\Pi_i := A_{\mathfrak{q}_i}(0)$. Then Theorem 1.9 may be regarded as a generalization of the following results in [KS18b, Thm. 12.4 and Lem. 15.10]:

Theorem 1.11. *Let $0 \leq i \leq n+1$. Then the representations Π_i of the group $G = SO(n+1,2)$ are $SO(n+1-i,1)$-distinguished.*

For details see [KS18b, Chap. 12].

A Bilinear Form on the $(\mathfrak{g}, K)$-cohomology In [KS18b, Chap. 12, Sect. 3], we considered the morphism on $(\mathfrak{g}, K)$-cohomologies of representations induced by a symmetry breaking operator: Let $(G, G') = (SO(n+1,1), SO(n,1))$ and Π, π be irreducible representations and V, V' irreducible finite-dimensional representations of G and G', respectively. By abuse of notation, we use the same symbols Π and π to denote their underlying $(\mathfrak{g}, K)$-module and $(\mathfrak{g}', K')$-module, respectively, when we take $(\mathfrak{g}, K)$-cohomologies and $(\mathfrak{g}', K')$-cohomologies. Suppose that there exists a symmetry breaking operator $T : \Pi \otimes V \to \pi \otimes V'$. Then the symmetry breaking operator $T : \Pi \otimes V \to \pi \otimes V'$ induces for every j a morphism

$$T^j : H^j(\mathfrak{g}, K; \Pi \otimes V) \to H^j(\mathfrak{g}', K'; \pi \otimes V')$$

on the $(\mathfrak{g}, K)$-cohomologies and a bilinear form

$$B_T : H^j(\mathfrak{g}, K; \Pi \otimes V) \times H^{n-j}(\mathfrak{g}', K'; (\pi \otimes V')^\vee \otimes \chi_{(-1)^{n+1}}) \to \mathbb{C}$$

where $(\pi \otimes V')^\vee$ denotes the contragredient representation of $\pi \otimes V'$.

The induced morphism on the $(\mathfrak{g}, K)$-cohomologies may be zero, but in some special cases we conclude that it is nonzero.

For $0 \leq \ell \leq [\frac{n+1}{2}]$ and $\delta \in \{+, -\}$, we denote by $\Pi_{\ell,\delta}$ the irreducible admissible smooth representation of $G = SO(n+1,1)$ with underlying $(\mathfrak{g}, K)$-module $A_{\mathfrak{q}_\ell}(0)_\delta$ (see Theorem 1.9 for notation). We also write simply Π_ℓ for $\Pi_{\ell,\delta}$ if $\delta = +$. We recall from [KS18b, Thm. 2.20] that $\Pi_{\ell,\delta}$ is the unique submodule of the principal series representation $I_\delta(\Lambda^\ell(\mathbb{C}^n), \ell))$ for $\ell \neq \frac{n}{2}$ (see (2.5) below), and is isomorphic to $I_\delta(V_+, \frac{n}{2}) \simeq I_\delta(V_-, \frac{n}{2})$ for $\ell = \frac{n}{2}$ where V_+ and V_- are irreducible $SO(n)$-modules such that $\Lambda^{\frac{n}{2}}(\mathbb{C}^n) = V_+ \oplus V_-$. We also recall from [KS18b, Prop. 15.11] that the set of irreducible admissible representations of $G = SO(n+1,1)$ with the trivial $\mathfrak{Z}(\mathfrak{g})$-infinitesimal character ρ is classified as follows:

$$\{\Pi_{\ell,\delta} : 0 \leq \ell \leq \frac{n-1}{2},\ \delta \in \{+,-\}\} \cup \{\Pi_{\frac{n+1}{2},+}\} \quad \text{if } n \text{ is odd,}$$

$$\{\Pi_{\ell,\delta} : 0 \leq \ell \leq \frac{n}{2},\ \delta \in \{+,-\}\} \quad \text{if } n \text{ is even.}$$

Analogous notation $\pi_{j,\varepsilon}$ is applied to the subgroup $G' = SO(n,1)$.

Then the following theorem follows from [KS18b, Cor. 12.19 and Lem. 15.10] for the nonvanishing of B_T and [KS18b, Thm. 15.19] for the uniqueness of T.

Theorem 1.12. *Let* $(G, G') = (SO(n+1,1), SO(n,1))$, $0 \leq i \leq \frac{n}{2}$, *and* $\delta \in \{\pm\}$. *Let* T *be a nontrivial symmetry breaking operator* $\Pi_{i,\delta} \to \pi_{i,\delta}$.

(1) T *induces bilinear forms*

$$B_T \colon H^j(\mathfrak{g}, K; \Pi_{i,\delta}) \times H^{n-j}(\mathfrak{g}', K'; \pi_{i,(-1)^{n-1}\delta}) \to \mathbb{C} \quad \textit{for all } j.$$

(2) *The bilinear form* B_T *is nonzero if and only if* $j = i$ *and* $\delta = (-1)^i$.

In Sect. 6, we also state a nonvanishing theorem for bilinear forms on the $(\mathfrak{g}, K)$-cohomologies of principal series representations, see Theorem 6.1.

Remark 1.13. If Π and π have trivial infinitesimal character ρ then V and V' are the trivial representations and the theorem follows from [KS18b, Chaps. 9 and 12].

Detailed proofs of the results will be published elsewhere [KS].

2 Representations with Nonsingular Integral Infinitesimal Character

In this section we recall from [KS18b] the results about principal series representations and irreducible representations of $G = SO(n+1,1)$ and introduce their height and signature.

2.1 Notation

We are using the same notation and assumptions as in the book [KS18b] except that we do not use "bar" for subgroups of $SO(n+1,1)$, see Remark 1.1. The proofs of most of the results stated in this section can be found in Chapter 2 and Appendices I, II, and III therein.

We first recall some notation which is up to small changes (see the comments in Remark 1.1) the same as the notation in the Memoir article [KS15].

Consider the standard quadratic form

$$x_0^2 + x_1^2 + \cdots + x_n^2 - x_{n+1}^2 \tag{2.1}$$

of signature $(n+1,1)$. We define G to be the indefinite special orthogonal group $SO(n+1,1)$ that preserves the quadratic form (2.1) and the orientation. Let G' be the stabilizer of the vector $e_n = {}^t(0, 0, \cdots, 0, 1, 0)$. Then G' is isomorphic to $SO(n,1)$. We set

$$K := O(n+2) \cap G = \{\begin{pmatrix} B & \\ & \det B \end{pmatrix} : B \in O(n+1)\} \simeq O(n+1), \tag{2.2}$$

$$K' := K \cap G' \qquad = \{\begin{pmatrix} B & & \\ & 1 & \\ & & \det B \end{pmatrix} : B \in O(n)\} \simeq O(n).$$

Then K and K' are maximal compact subgroups of G and G', respectively.

2.2 Principal Series Representations

Let $\mathfrak{g} = \mathfrak{so}(n+1, 1)$ and $\mathfrak{g}' = \mathfrak{so}(n, 1)$ be the Lie algebras of G and G', respectively. We take a hyperbolic element H as

$$H := E_{0,n+1} + E_{n+1,0} \in \mathfrak{g}', \tag{2.3}$$

and set

$$\mathfrak{a} := \mathbb{R}H \qquad \text{and } A := \exp \mathfrak{a}.$$

Then the centralizers of H in G and G' are given by MA and $M'A$, respectively, where

$$M := \left\{\begin{pmatrix} \varepsilon & & \\ & B & \\ & & \varepsilon \end{pmatrix} : B \in SO(n),\ \varepsilon = \pm 1\right\} \qquad \simeq SO(n) \times O(1),$$

$$M' := \left\{\begin{pmatrix} \varepsilon & & & \\ & B & & \\ & & 1 & \\ & & & \varepsilon \end{pmatrix} : B \in SO(n-1) : \varepsilon = \pm 1\right\} \quad \simeq SO(n-1) \times O(1).$$

We observe that $\operatorname{ad}(H) \in \operatorname{End}_{\mathbb{R}}(\mathfrak{g})$ has eigenvalues -1, 0, and $+1$. Let

$$\mathfrak{g} = \mathfrak{n}_- + (\mathfrak{m} + \mathfrak{a}) + \mathfrak{n}_+$$

be the corresponding eigenspace decomposition, and P a minimal parabolic subgroup with Langlands decomposition $P = MAN_+$. Likewise, $P' := M'AN'_+$ is a compatible Langlands decomposition of a minimal (also maximal) parabolic subgroup P' of G' with Lie algebra

$$\mathfrak{p}' = \mathfrak{m}' + \mathfrak{a} + \mathfrak{n}'_+ = (\mathfrak{m} \cap \mathfrak{g}') + (\mathfrak{a} \cap \mathfrak{g}') + (\mathfrak{n}_+ + \mathfrak{g}'). \tag{2.4}$$

We note that we have chosen $H \in \mathfrak{g}'$ so that $P' = P \cap G'$ and $A = \exp(\mathbb{R}H)$ is a common maximally split abelian subgroup in P' and P.

The character group of $O(1)$ consists of two characters. We write $+$ for the trivial character $\mathbf{1}$, and $-$ for the nontrivial character. Since $M \simeq SO(n) \times O(1)$, any irreducible representation of M is the outer tensor product of an irreducible representation (σ, V) of $SO(n)$ and a character δ of $O(1)$.

Given $(\sigma, V) \in \widehat{SO(n)}$, $\delta \in \{\pm\} \simeq \widehat{O(1)}$, and a character $e_\lambda(\exp(tH)) = e^{\lambda t}$ of A for $\lambda \in \mathbb{C}$, we define the (unnormalized) principal series representation

$$I_\delta(V, \lambda) = \mathrm{Ind}_P^G(V \otimes \delta, \lambda) \tag{2.5}$$

of $G = SO(n+1,1)$ on the Fréchet space of smooth maps $f\colon G \to V$ subject to

$$f(gmm'e^{tH}n) = \sigma(m)^{-1}\delta(m')e^{-\lambda t} f(g)$$
$$\text{for all } g \in G,\ mm' \in M \simeq SO(n) \times O(1),\ t \in \mathbb{R},\ n \in N_+.$$

By a result of R. Langlands [L89] every irreducible nontempered representation with nonsingular integral infinitesimal character is isomorphic to the unique subrepresentation of a principal series representation $I_\delta(V, \lambda)$ with $\lambda < \frac{n}{2}$. We denote it by $\Pi_\delta(V, \lambda)$, see also [KS18b, Chap. 15, Sect. 7].

Definition 2.1. We call the triple (V, δ, λ) the *Langlands parameter* of the irreducible nontempered representation $\Pi_\delta(V, \lambda)$.

2.2.1 The Set $\mathcal{A}$

Since we highlight in the article representations which are of interest in number theory, we consider from now on a subset of irreducible representations of special orthogonal groups. For results about irreducible representation of $SO(n+1,1)$ in the general case see [KS18b, Chap. 15].

We start with irreducible finite-dimensional representations of $SO(n+1,1)$ that are self-dual, or equivalently, that are obtained as the restriction of irreducible representations of $O(n+1,1)$. So for $n = 2m$ we assume that the highest weight $(\mu_1, \ldots, \mu_{m+1})$ of an irreducible finite-dimensional representation F of $G = SO(2m+1,1)$ is of the form $(\mu_1, \ldots, \mu_m, 0)$.

Assumption A. Suppose that Π is an irreducible representation of the group $G = SO(n+1,2)$ with regular integral infinitesimal character, see [KS18b, Chap. 2, Sect. 1.4]. We say that a representation Π of G satisfies *Assumption* A if it has the same infinitesimal character as a self-dual irreducible finite-dimensional representation of G. When n is odd, Assumption A is automatically satisfied.

Notation 2.2. The set of irreducible representations of $G = SO(n+1,1)$ satisfying Assumption A is denoted by $\mathcal{A}$.

For the convenience of the reader, we give a description of irreducible admissible representations of $G = SO(n+1, 1)$ satisfying Assumption A in Sect. 2.2.2 for n even and in Sect. 2.2.3 for n odd.

2.2.2 Classification of the Set $\mathcal{A}$ for n Even

Suppose $n = 2m$. By using the highest weight, we write $V \in \widehat{SO(n)}$ as $V = F^{SO(n)}(\sigma)$ with $\sigma = (\sigma_1, \cdots, \sigma_m) \in \mathbb{Z}^m$ satisfying $\sigma_1 \geq \cdots \geq \sigma_{m-1} \geq |\sigma_m|$. Then the contragredient representation of V is given as

$$V^\vee \simeq F^{SO(n)}(\sigma_1, \cdots, \sigma_{m-1}, -\sigma_m).$$

Hence V is self-dual if and only if $\sigma_m = 0$.

Proposition 2.3. *For $V = F^{SO(2m)}(\sigma)$, we consider the following condition on λ:*

$$\lambda \in \mathbb{Z}, \ \lambda < m, \ \lambda \notin \{1 - \sigma_1, \cdots, m - \sigma_m\}. \tag{2.6}$$

Then irreducible admissible representations of $SO(2m+1, 1)$ in $\mathcal{A}$ are classified as

- (nontempered case)

$$\{\Pi_\delta(V, \lambda) : \delta = \pm, \sigma_m = 0, \ \textit{and}\ \lambda\ \textit{satisfies}\ (2.6)\},$$

or

- (tempered case)

$$\{I_\delta(V, m) : \delta = \pm, \sigma_m > 0\}.$$

We note that in the tempered case $V \not\simeq V^\vee$ as $SO(n)$-modules because $\sigma_m > 0$, whereas there is a G-isomorphism, see [KS18b, Prop.15.5]:

$$I_\delta(V, m) \simeq I_\delta(V^\vee, m).$$

2.2.3 Classification of the Set $\mathcal{A}$ for n Odd

Suppose $n = 2m - 1$. We write $V \in \widehat{SO(n)}$ as $V = F^{SO(n)}(\sigma)$ with $\sigma = (\sigma_1, \cdots, \sigma_{m-1}) \in \mathbb{Z}^{m-1}$ satisfying $\sigma_1 \geq \cdots \geq \sigma_{m-1} \geq 0$.

Proposition 2.4. *For $V = F^{SO(2m-1)}(\sigma)$, we consider the following conditions on λ:*

$$\lambda \in \mathbb{Z}, \ \lambda \leq m - 1, \ \ \textit{and}\ \lambda \notin \{1 - \sigma_1, \cdots, m - 1 - \sigma_{m-1}\}. \tag{2.7}$$

$$\lambda \in \mathbb{Z}, \ m \leq \lambda \leq m-1+\sigma_{m-1}. \tag{2.8}$$

All irreducible admissible representations of $SO(2m,1)$ *belong to* $\mathcal{A}$*, which are classified as*

- (nontempered case)

$$\{\Pi_\delta(V,\lambda) : \delta = \pm, \ \lambda \textit{ satisfies } (2.7)\},$$

  or
- (discrete series)

$$\{\Pi_+(V,\lambda) : \sigma_{m-1} > 0, \ \lambda \textit{ satisfies } (2.8)\}.$$

We note that in the discrete series case there is a G-isomorphism

$$\Pi_+(V,\lambda) \simeq \Pi_-(V,\lambda).$$

2.3 *The Height of Representations in* $\mathcal{A}$

In this section, we define a *height*

$$h \colon \mathcal{A} \to \{0, 1, \cdots, [\frac{n+1}{2}]\}$$

for irreducible representations of $G = SO(n+1,1)$ that belong to $\mathcal{A}$ (see Sect. 2.2.1).

We recall that the i-th exterior representation $\bigwedge^i(\mathbb{C}^n)$ of $SO(n)$ is irreducible if $2i \neq n$, and splits into two irreducible representations if $2i = n$, which may be written as

$$\bigwedge^{\frac{n}{2}}(\mathbb{C}^n) \simeq \bigwedge^{\frac{n}{2}}(\mathbb{C}^n)^{(+)} \oplus \bigwedge^{\frac{n}{2}}(\mathbb{C}^n)^{(-)}.$$

We define a finite family of principal series representations of $G = SO(n+1,1)$ by

$$I_\delta(i,i) := \begin{cases} I_\delta(\bigwedge^i(\mathbb{C}^n), i) & \text{for } i \neq \frac{n}{2}, \\ I_\delta(\bigwedge^i(\mathbb{C}^n)^{(\varepsilon)}, i) & \text{for } i = \frac{n}{2}. \end{cases}$$

Then $I_\delta(i,i)$ has the trivial infinitesimal character ρ, and it does not depend on the choice $\varepsilon = +$ or $-$ when $i = \frac{n}{2}$, see [KS18b, (15.5)].

Suppose that Π is an irreducible nontempered representation of $SO(n+1,1)$ in $\mathcal{A}$ with Langlands parameter (V,λ,δ). The principal series representation $I_\delta(V,\lambda)$

can be obtained by using a translation functor from exactly one principal series representation $I_\delta(i, i)$, $i \in \{0, 1, \ldots, [\frac{n+1}{2}]\}$ without crossing a wall. See [KS18b, Thm. 16.24].

Following [KS18b, Chap. 14, Sect. 5 and Thm. 16.17], we say that the principal series representation $I_\delta(V, \lambda)$ has *height* i if it can be obtained from a principal series representations $I_\delta(i, i)$ by a translation functor without crossing walls.

Definition 2.5 (Height).

(1) Suppose that the representation $\Pi \in \mathcal{A}$ is not tempered and has Langlands parameter (V, λ, δ). If $I_\delta(V, \lambda)$ has *height* i, we say that Π has height i.
(2) If $n = 2m - 1$ we say that a discrete series representation $\Pi \in \mathcal{A}$ has *height* m.
(3) If $n = 2m$ and $\Pi \in \mathcal{A}$ is a tempered representation, then we say that it has *height* m.

An explicit formula of the height can be derived from the case for $O(n+1, 1)$ in [KS18b, Def. 14.26 and Chap. 15]:

Proposition 2.6 (Height for Nontempered Representation).

(1) *Suppose* $n = 2m$. *With notation as in Proposition 2.3, the height* i *of the nontempered representation* $\Pi_\delta(V, \lambda)$ *takes the value in* $\{0, 1, \cdots, m-1\}$, *and is determined by the following inequalities:*

$$i - \sigma_i < \lambda < i + 1 - \sigma_{i+1}.$$

(2) *Suppose* $n = 2m - 1$. *With notation as in Proposition 2.4, the height of the nontempered representation* $\Pi_\delta(V, \lambda)$ *takes the value in* $\{0, 1, \cdots, m-1\}$, *and is determined by the following condition:*

- *for* $\lambda < m - 1 - \sigma_{m-1}$, *we have* $0 \leq i \leq m-2$ *and*

$$i - \sigma_i < \lambda < i + 1 - \sigma_{i+1};$$

- *for* $m - 1 - \sigma_{m-1} < \lambda < m$, *we have* $i = m - 1$.

Example 2.7. For $\Pi \in \mathcal{A}$, the height of Π is zero if and only if Π is finite-dimensional.

2.4 Signatures of Representations in $\mathcal{A}$

In this section, we define a *signature*

$$\mathrm{sgn}\colon \mathcal{A} \to \{+, -, \pm\}$$

for irreducible representations of $G = SO(n+1,1)$ that belong to $\mathcal{A}$. We shall impose the condition

$$\operatorname{sgn}(\Pi \otimes \chi_-) = -\operatorname{sgn}\Pi. \tag{2.9}$$

2.4.1 Tempered Representations

We recall from Propositions 2.3 and 2.4 (see also [KS18b, Thms. 13.7 and 13.9]) that for every irreducible finite-dimensional representation F in $\mathcal{A}$ there exist irreducible tempered representations with the same infinitesimal character.

Suppose $G = SO(2m+1,1)$ and $\lambda = m$ $(= \frac{n}{2})$. The unitary principal series representation $I_\delta(V,m)$ is tempered, and it has the same infinitesimal character as an irreducible finite-dimensional representation of G if and only if V is not self-dual. In this case, there is an isomorphism $I_\delta(V,m) \simeq I_\delta(V^\vee,m)$ as $SO(2m+1,1)$-modules as we saw in Sect. 2.2.2. We define the signature of $I_\delta(V,m)$ to be δ.

For $G = SO(2m,1)$, there is exactly one discrete series representation Π of G having the same infinitesimal character as F (see Proposition 2.4). Furthermore there is an isomorphism

$$\Pi \otimes \chi_- \simeq \Pi$$

as G-modules, hence we define the signature of a discrete series representation Π to be $\pm$.

2.4.2 Nontempered Representations

An irreducible representation Π in $\mathcal{A}$ of $G = SO(n+1,1)$, which is not tempered, is isomorphic to a subrepresentation of a principal series representation $I_\delta(V,\lambda)$ of height $i \in \{0,1,\cdots,[\frac{n-1}{2}]\}$, see Proposition 2.6. It has a Langlands parameter (V,δ,λ) where V is a self-dual representation of $SO(n)$, $\lambda < \frac{n}{2}$ and $\delta \in \{+,-\}$. We refer to $\delta(-1)^{i-\lambda}$ as the signature of the representation Π. Since the Langlands parameter of a nontempered representation is unique, the signature is unique as far as Π is not tempered.

Remark 2.8. If a representation Π has signature δ, then one sees $\Pi \otimes \chi_-$ has signature $-\delta$.

Example 2.9 (One-Dimensional Representations). There are two one-dimensional representations of $G = SO(n+1,1)$: we denote the trivial representation $\mathbf{1}$ by χ_+ and the nontrivial one by χ_-. The Langlands parameter of χ_+ and χ_- is $(+,\mathbf{1},0)$ and $(-,\mathbf{1},0)$, respectively, and their height is 0. Hence the representation χ_+ has signature $+$ and the representation χ_- has signature $-$.

Example 2.10 (Irreducible Representations with Trivial Infinitesimal Character ρ). If V is the representation of $SO(n)$ on the i-th exterior tensor space $\bigwedge^i(\mathbb{C}^n)$ ($2i \neq n$), we write for simplicity $I_\delta(i, \lambda)$ instead of $I_\delta(V, \lambda)$. Then the $SO(n)$-isomorphism on the exterior representations $\bigwedge^i(\mathbb{C}^n) \simeq \bigwedge^{n-i}(\mathbb{C}^n)$ leads us to the following G-isomorphism:

$$I_\delta(i, \lambda) \simeq I_\delta(n-i, \lambda).$$

If n is even and $n = 2i$, the exterior representation $\bigwedge^i(\mathbb{C}^n)$ splits into two irreducible representations of $SO(n)$:

$$\bigwedge^{\frac{n}{2}}(\mathbb{C}^n) \simeq \bigwedge^{\frac{n}{2}}(\mathbb{C}^n)_+ \oplus \bigwedge^{\frac{n}{2}}(\mathbb{C}^n)_-$$

with highest weights $(1, \cdots, 1, 1)$ and $(1, \cdots, 1, -1)$, respectively, with respect to a fixed positive system for $\mathfrak{so}(n, \mathbb{C})$. Accordingly, we have a direct sum decomposition of the induced representation:

$$\mathrm{Ind}_P^G(\bigwedge^{\frac{n}{2}}(\mathbb{C}^n) \otimes \delta, \lambda) = I_\delta(\bigwedge^{\frac{n}{2}}(\mathbb{C}^n)_+, \lambda) \oplus I_\delta(\bigwedge^{\frac{n}{2}}(\mathbb{C}^n)_-, \lambda),$$

which we shall write as

$$I_\delta\left(\frac{n}{2}, \lambda\right) = I_\delta^{(+)}\left(\frac{n}{2}, \lambda\right) \oplus I_\delta^{(-)}\left(\frac{n}{2}, \lambda\right). \tag{2.10}$$

The representations $I_\delta^{(+)}\left(\frac{n}{2}, \lambda\right)$ and $I_\delta^{(-)}\left(\frac{n}{2}, \lambda\right)$ are isomorphic to each other and have the signature δ.

Let $\mathfrak{Z}(\mathfrak{g})$ be the center of the enveloping algebra $U(\mathfrak{g})$ of the complexified Lie algebra $\mathfrak{g}_\mathbb{C} = \mathfrak{g} \otimes_\mathbb{R} \mathbb{C} \simeq \mathfrak{so}(n+2, \mathbb{C})$. Via the Harish-Chandra isomorphism, the $\mathfrak{Z}(\mathfrak{g})$-infinitesimal character of the trivial one-dimensional representation $\mathbf{1}$ is given by

$$\rho = \left(\frac{n}{2}, \frac{n}{2} - 1, \cdots, \frac{n}{2} - [\frac{n}{2}]\right)$$

in the standard coordinates of the Cartan subalgebra of $\mathfrak{g}_\mathbb{C} = \mathfrak{so}(n+2, \mathbb{C})$, whereas up to conjugation by the Weyl group the infinitesimal character of $I_\delta^{(\pm)}(i, \lambda)$ (when $2i \leq n$) is given by

$$\left(\frac{n}{2}, \frac{n}{2} - 1, \cdots, \frac{n}{2} - i + 1, \widehat{\frac{n}{2} - i}, \frac{n}{2} - i - 1, \cdots, \frac{n}{2} - [\frac{n}{2}], \lambda - \frac{n}{2}\right). \tag{2.11}$$

The irreducible representation $\Pi_{i,\delta}$ has height $\min(i, n-i)$. Since $\lambda = i$, the signature is equal to δ if $2i \leq n$ and to $\delta(-1)^n$ if $2i \geq n$. The irreducible tempered representations are denoted by Π_m if $n = 2m - 1$ and $\Pi_{m,\delta}$ if $n = 2m$. See [KS18b, Chap. 2, Sect. 4.5] for $O(n+1,1)$ and [KS18b, Chap. 15, Sect. 5] for $SO(n+1,1)$ in detail.

For the group $G' = SO(n,1)$, we shall use the notation $J_\varepsilon(j,\nu)$ for the unnormalized parabolic induction $\mathrm{Ind}_{P'}^{G'}(\bigwedge^j(\mathbb{C}^{n-1}) \otimes \varepsilon, \nu)$ for $0 \leq j \leq n-1$, $\varepsilon \in \{+,-\}$, and $\nu \in \mathbb{C}$. The irreducible representations are denoted by $\pi_{j,\varepsilon}$ respectively π_j.

2.4.3 Hasse and Standard Sequences

The notion of the height of representation in $\mathcal{A}$ is motivated by the Hasse sequences in [KS18b, Chap. 13], which were defined for the full orthogonal group $O(n+1,1)$. We adapt the definition for the special orthogonal group $G = SO(n+1,1)$ as follows. Let $n = 2m$ or $2m - 1$. For every irreducible finite-dimensional representation F of the group G, there exists a unique sequence

$$U_0\ ,\ \ldots\ ,\ U_{m-1}\ ,\ U_m$$

of irreducible admissible smooth representations $U_i \equiv U_i(F)$ of G such that

1. $U_0 \simeq F$;
2. consecutive representations are composition factors of a principal series representation;
3. U_i $(0 \leq i \leq m)$ are pairwise inequivalent as G-modules.

Definition 2.11 (Hasse Sequence and Standard Sequence). We refer to the sequence

$$U_0\ ,\ \ldots\ ,\ U_{m-1}\ ,\ U_m$$

as the *Hasse sequence* of irreducible representations starting with the finite-dimensional representation $U_0 = F$. We shall write $U_j(F)$ for U_j if we emphasize the sequence $\{U_j(F)\}$ starts with $U_0 = F$, and we refer to

$$\Pi_0 := U_0\ ,\ \ldots\ ,\ \Pi_{m-1} := U_{m-1} \otimes (\chi_-)^{m-1}\ ,\ \Pi_m := U_m \otimes (\chi_-)^m$$

as the *standard sequence* of irreducible representations $\Pi_i = \Pi_i(F)$ starting with $\Pi_0 = U_0 = F$, where χ_- is the nontrivial one-dimensional representation of G defined in Example 2.9.

More details about the standard sequence for $O(n+1,1)$ can be found in [KS18b, Chap. 13], from which the case for the normal subgroup $G = SO(n+1,1)$ is derived as follows.

Theorem 2.12. *Let $G = SO(n+1,1)$.*

(1) *Suppose that $\Pi \in \mathcal{A}$ is not a discrete series representation of G, having height j and signature δ. Then there exists exactly one irreducible finite-dimensional representation F of G with signature δ so that Π is the j-th representation in the standard sequence starting with F.*
(2) *Suppose that $\Pi \in \mathcal{A}$ is a discrete series representation with signature $\delta \in \{\pm\}$. Then n is odd, and there exists a unique irreducible finite-dimensional representation F of G with signature +, so that Π is the $\frac{n+1}{2}$-th representation in the standard sequence starting with F and with $F \otimes \chi_{-}$.*

Remark 2.13. The last representation in the standard sequence starting at an irreducible finite-dimensional representation $F \in \mathcal{A}$ is tempered.

2.4.4 The θ-stable and Enhanced θ-stable Parameters of Irreducible Representations

We summarize the results in [KS18b, Chap. 14] and give a parametrization of irreducible subquotients of the principal series representations $I_\delta(V, \lambda)$ of the group $G = SO(n+1, 1)$ in terms of cohomological parabolic induction.

We recall quickly cohomological parabolic induction. A basic reference is Vogan [Vo81] and Knapp–Vogan [KV95]. We begin with a *connected* real reductive Lie group G. Let K be a maximal compact subgroup, and θ the corresponding Cartan involution. Given an element $X \in \mathfrak{k}$, the complexified Lie algebra

$$\mathfrak{g}_{\mathbb{C}} = \mathrm{Lie}(G) \otimes_{\mathbb{R}} \mathbb{C}$$

is decomposed into the eigenspaces of $\sqrt{-1}\,\mathrm{ad}(X)$, and we write

$$\mathfrak{g}_{\mathbb{C}} = \mathfrak{u}_{-} + \mathfrak{l}_{\mathbb{C}} + \mathfrak{u}$$

for the sum of the eigenspaces with negative, zero, and positive eigenvalues. Then $\mathfrak{q} := \mathfrak{l}_{\mathbb{C}} + \mathfrak{u}$ is a θ-stable parabolic subalgebra with Levi subgroup

$$L = \{g \in G : \mathrm{Ad}(g)\mathfrak{q} = \mathfrak{q}\}. \tag{2.12}$$

The homogeneous space G/L is endowed with a G-invariant complex manifold structure such that its holomorphic cotangent bundle is given as $G \times_L \mathfrak{u}$. As an algebraic analogue of Dolbeault cohomology groups for G-equivariant holomorphic vector bundle over G/L, Zuckerman introduced a cohomological parabolic induction functor $\mathcal{R}^j_{\mathfrak{q}}(\cdot \otimes \mathbb{C}_{\rho(\mathfrak{u})})$ $(j \in \mathbb{N})$ from the category of $(\mathfrak{l}, L \cap K)$-modules to the category of $(\mathfrak{g}, K)$-modules. We adopt here the normalization of the cohomological parabolic induction $\mathcal{R}^j_{\mathfrak{q}}$ from a θ-stable parabolic subalgebra $\mathfrak{q} = \mathfrak{l}_{\mathbb{C}} + \mathfrak{u}$ so that the $\mathfrak{Z}(\mathfrak{g})$-infinitesimal character of the $(\mathfrak{g}, K)$-module $\mathcal{R}^j_{\mathfrak{q}}(F)$ equals

the $\mathfrak{Z}(\mathfrak{l})$-infinitesimal character of the $\mathfrak{l}$-module F

modulo the Weyl group via the Harish-Chandra isomorphism.

For each i with $0 \leq i \leq [\frac{n+1}{2}]$, there are θ-stable parabolic subalgebras $\mathfrak{q}_i \equiv \mathfrak{q}_i^+ = (\mathfrak{l}_i)_{\mathbb{C}} + \mathfrak{u}_i$ and $\mathfrak{q}_i^- = (\mathfrak{l}_i)_{\mathbb{C}} + \mathfrak{u}_i^-$ in $\mathfrak{g}_{\mathbb{C}} = \mathrm{Lie}(G) \otimes_{\mathbb{R}} \mathbb{C}$ such that $\mathfrak{q}_i$ and $\mathfrak{q}_i^-$ contain a fundamental Cartan subalgebra $\mathfrak{h}$. The Levi subgroup $L_i = N_G(\mathfrak{q}_i)$ of the θ-stable parabolic subalgebra $\mathfrak{q}_i$ and $\mathfrak{q}_i^-$ is isomorphic to $L_i = SO(2)^i \times SO(n-2i+1, 1)$.

We set

$$\Lambda^+(N) := \{(\lambda_1, \cdots, \lambda_N) \in \mathbb{Z}^N : \lambda_1 \geq \lambda_2 \geq \cdots \geq \lambda_N \geq 0\}.$$

For $\nu = (\nu_1, \cdots, \nu_i) \in \mathbb{Z}^i$, $\mu \in \Lambda^+([\frac{n}{2}] - i + 1)$, and $\delta \in \{+, -\}$, we consider an irreducible finite-dimensional L_i-module

$$F^{O(n-2i+1,1)}(\mu)_\delta \otimes \mathbb{C}_\nu$$

and define an admissible smooth representation of G of moderate growth, whose underlying $(\mathfrak{g}, K)$-module is given by the cohomological parabolic induction

$$\mathcal{R}_{\mathfrak{q}_i}^{S_i}(F^{SO(n-2i+1,1)}(\mu)_\delta \otimes \mathbb{C}_{\nu+\rho(\mathfrak{u}_i)}) \tag{2.13}$$

of degree S_i, where we set

$$S_i := \dim_{\mathbb{C}}(\mathfrak{u}_i \cap \mathfrak{k}_{\mathbb{C}}) = i(n-i). \tag{2.14}$$

It is denoted by

$$(\nu_1, \cdots, \nu_i \,||\, \mu_1, \cdots, \mu_{[\frac{n}{2}]-i+1})_\delta.$$

We note that if $i = 0$ then $(||\, \mu_1, \cdots, \mu_{[\frac{n}{2}]+1})_\delta$ is finite-dimensional.

Definition 2.14. We call $(\nu_1, \cdots, \nu_i \,||\, \mu_1, \cdots, \mu_{[\frac{n}{2}]-i+1})_\delta$ the **θ-stable parameter** of the representation $\mathcal{R}_{\mathfrak{q}_i}^{S_i}(F^{O(n-2i+1,1)}(\mu)_\delta \otimes \mathbb{C}_{\nu+\rho(\mathfrak{u}_i)})$.

Remark 2.15. By [KS18b, Chaps. 14 and 16], the double bars $||$ in the θ-stable parameter of a representation in $\mathcal{A}$ of height i are before the $i+1$-th entry.

Remark 2.16. In the introduction we refer to (μ, i, δ) as the **enhanced θ-stable parameter** of the representation with θ-stable parameter $(\mu_1, \cdots, \mu_i \,||\, \mu_{i+1}, \cdots, \mu_{[\frac{n}{2}]+1})_\delta$.

Example 2.17.

(1) An irreducible finite-dimensional representation $F^G(\mu)_\delta$ has the θ-stable parameter $(||\mu_1, \mu_2, \ldots, \mu_{[\frac{n}{2}]+1})_\delta$.
(2) The θ-stable parameter of a representation of height i with trivial infinitesimal character ρ is

$$(0, 0, \ldots, 0 \,||\, 0, \ldots, 0)_\delta$$

where the double bars $||$ are before the $i + 1$-th zero (see [KS18b, Chap. 14, Sect. 9.3]).

(3) The representations Π in $\mathcal{A}$ with θ-stable parameter $(\lambda_1, \lambda_2, \ldots, \lambda_i \,||\, 0, \ldots, 0)_\delta$ are unitary and are often referred to as $A_\mathfrak{q}(\lambda)_\delta$. There exists a finite-dimensional representation V of G so that $H^*(\mathfrak{g}, K; \Pi \otimes V) \neq \{0\}$, see [VZ84].

2.4.5 The Hasse and Standard Sequences in θ-stable Parameters

We set $m := [\frac{n+1}{2}]$, namely $n = 2m - 1$ or $2m$. Let $F = F^G(s_0, \ldots, s_{[\frac{n}{2}]})_\delta$ be an irreducible finite-dimensional representation of $G = SO(n+1, 1)$, and $U_i \equiv U_i(F)$ $(0 \leq i \leq [\frac{n+1}{2}])$ be the Hasse sequence with $U_0 \simeq F$. In [KS18b, Chap. 14] we show:

Theorem 2.18. *Let* $n = 2m$ *and* $0 \leq i \leq m$.

(1) (Hasse sequence) $U_i(F) \simeq (s_0, \cdots, s_{i-1} || s_i, \cdots, s_m)_{(-1)^{i-s_i}\delta}$.
(2) (standard sequence) $U_i(F) \otimes \chi_-^i \simeq (s_0, \cdots, s_{i-1} || \, s_i, \cdots, s_m)_{(-1)^{s_i}\delta}$.

The case n odd is given similarly as follows.

Theorem 2.19. *Let* $n = 2m - 1$, *and* $0 \leq i \leq m - 1$.

(1) (Hasse sequence) $U_i(F) \simeq (s_0, \cdots, s_{i-1} || \, s_i, \cdots, s_{m-1})_{(-1)^{i-s_i}\delta}$.
(2) (standard sequence) $U_i(F) \otimes \chi_-^i \simeq (s_0, \cdots, s_{i-1} || \, s_i, \cdots, s_{m-1})_{(-1)^{s_i}\delta}$.

3 The Restriction of Representations of $SO(n + 1, 1)$ in $\mathcal{A}$ to the Subgroup $SO(n, 1)$

In this section we discuss the branching law for the restriction of irreducible representations $\Pi \in \mathcal{A}$ of $SO(n + 1, 1)$ to the subgroup $SO(n, 1)$. We state it for infinite-dimensional representations in Langlands parameter and θ-stable parameters as well in the language of height and signature. A branching law for irreducible representations without the assumption $\Pi \in \mathcal{A}$ will appear in [KS].

3.1 *Branching Laws for Finite-Dimensional Representations*

We first recall the branching laws for finite-dimensional representations. As in the classical branching law for $SO(N) \downarrow SO(N - 1)$ the irreducible decomposition of finite-dimensional representations of $SO(N, 1)$ when restricted to the subgroup $SO(N - 1, 1)$ is as follows:

Theorem 3.1 (Branching Rule for $SO(N,1) \downarrow SO(N-1,1)$). *Let $N \geq 2$. Suppose that $(\lambda_1, \cdots, \lambda_{[\frac{N+1}{2}]}) \in \Lambda^+([\frac{N+1}{2}])$ and $\delta \in \{+,-\}$. Then the irreducible finite-dimensional representation $F^{O(N,1)}(\lambda_1, \cdots, \lambda_{[\frac{N+1}{2}]})_\delta$ of $SO(N,1)$ decomposes into a multiplicity-free sum of irreducible representations of $SO(N-1,1)$ as follows:*

$$F^{SO(N,1)}(\lambda_1, \cdots, \lambda_{[\frac{N+1}{2}]})_\delta|_{SO(N-1,1)} \simeq \bigoplus F^{SO(N-1,1)}(\nu_1, \cdots, \nu_{[\frac{N}{2}]})_\delta,$$

where the summation is taken over $(\nu_1, \cdots, \nu_{[\frac{N}{2}]}) \in \mathbb{Z}^{[\frac{N}{2}]}$ subject to

$$\lambda_1 \geq \nu_1 \geq \lambda_2 \geq \cdots \geq \nu_{\frac{N}{2}} \geq 0 \qquad \textit{for } N \textit{ even},$$

$$\lambda_1 \geq \nu_1 \geq \lambda_2 \geq \cdots \geq \nu_{\frac{N-1}{2}} \geq \lambda_{\frac{N+1}{2}} \quad \textit{for } N \textit{ odd}.$$

Example 3.2. We proved the branching rule for the restriction of the representations of $SO(n,1)$ on the space of harmonic polynomials in [KS15, Prop. 2.3].

3.2 *Symmetry Breaking Operators*

Irreducible infinite-dimensional representations of G typically do not decompose into a direct sum of irreducible representations of G when restricted to a noncompact subgroup G', see [Ko98] for details. To obtain information about the restriction and the branching laws we have to proceed differently.

For a continuous representation Π of G on a complete, locally convex topological vector space $\mathcal{H}$, the space $\mathcal{H}^\infty$ of C^∞-vectors of $\mathcal{H}$ is naturally endowed with a Fréchet topology, and $(\Pi, \mathcal{H})$ induces a continuous representation Π^∞ of G on $\mathcal{H}^\infty$. If Π is an admissible representation of finite length on a Banach space $\mathcal{H}$, then the Fréchet representation $(\Pi^\infty, \mathcal{H}^\infty)$, which we refer to as an *admissible smooth representation*, depends only on the underlying $(\mathfrak{g}, K)$-module $\mathcal{H}_K$. In the context of asymptotic behaviour of matrix coefficients, these representations are also referred to as an admissible representations of moderate growth [Wa92, Chap. 11]. We shall work with these representations and write simply Π for Π^∞. We denote by $\mathrm{Irr}(G)$ the set of equivalence classes of irreducible admissible smooth representations. We also sometimes call these representations "irreducible admissible representations" for simplicity.

Given another admissible smooth representation π of a reductive subgroup G', we consider the space of continuous G'-intertwining operators (*symmetry breaking operators*)

$$\mathrm{Hom}_{G'}(\Pi|_{G'}, \pi).$$

If $G = G'$ then these operators include the Knapp–Stein operators [KS71] and the differential intertwining operators studied by B. Kostant [Ks74]. Including the general case where $G \neq G'$, we define now the multiplicity of π occurring in the restriction $\Pi|_{G'}$ as follows.

Definition 3.3 (Multiplicity). For $G \supset G'$, we say

$$m(\Pi, \pi) := \dim_{\mathbb{C}} \operatorname{Hom}_{G'}(\Pi|_{G'}, \pi)$$

the *multiplicity* of π occurring in the restriction $\Pi|_{G'}$.

A finiteness criterion and a uniformly boundedness criterion are proved in [KO13, Thms. C and D]. Moreover, by a result of B. Sun and C. B. Zhu [SZ12], our assumptions imply that the multiplicities are either 0 or 1. These multiplicities yield important information of the restriction of Π to G', as we will see in the applications in the next part of the article.

3.3 *Branching Laws for Representations in $\mathcal{A}$: First Formulation*

Theorem 3.4. *Let F be an irreducible finite-dimensional representations of $G = SO(n+1, 1)$, and $\{\Pi_i(F)\}$ be the standard sequence starting at $\Pi_0(F) = F$. Let F' be an irreducible finite-dimensional representation of the subgroup $G' = SO(n, 1)$, and $\{\pi_j(F')\}$ the standard sequence starting at $\pi_0(F') = F'$. Assume that*

$$\operatorname{Hom}_{G'}(F|_{G'}, F') \neq \{0\}.$$

Then symmetry breaking for the representations $\Pi_i(F)$, $\pi_j(F')$ in the standard sequences is presented graphically in Figs. 1 and 2. In the first row are representations of G, in the second row are representations of G'. Nontrivial symmetry breaking operators are represented by arrows, namely, there exist nonzero symmetry breaking operators between 2 representations if and only if there are arrows in Figs. 1 and 2.

Fig. 1 Symmetry breaking for $SO(2m+1, 1) \downarrow SO(2m, 1)$

$$\begin{array}{ccccccccc}
\Pi_0(F) & & \Pi_1(F) & & \cdots & & \Pi_{m-1}(F) & & \Pi_m(F) \\
\downarrow & \swarrow & \downarrow & \swarrow & & \swarrow & \downarrow & \swarrow & \downarrow \\
\pi_0(F') & & \pi_1(F') & & \cdots & & \pi_{m-1}(F') & & \pi_m(F')
\end{array}$$

$$\begin{array}{ccccccccc}
\Pi_0(F) & & \Pi_1(F) & & \dots & & \Pi_{m-1}(F) & & \Pi_m(F) & & \Pi_{m+1}(F) \\
\downarrow & \swarrow & \downarrow & \swarrow & & \swarrow & \downarrow & \swarrow & \downarrow & \swarrow & \\
\pi_0(F') & & \pi_1(F') & & \dots & & \pi_{m-1}(F') & & \pi_m(F') & &
\end{array}$$

Fig. 2 Symmetry breaking for $SO(2m+2,1) \downarrow SO(2m+1,1)$

3.4 Branching Laws for Representations in $\mathcal{A}$: Second Formulation

Let $F^G(\mu)_\delta$ and $F^{G'}(\nu)_\delta$ be irreducible finite-dimensional representations in $\mathcal{A}$ of $G = SO(n+1,1)$, respectively of the subgroup $G' = SO(n,1)$, where $\mu \in \Lambda^+([\frac{n+2}{2}])$, $\nu \in \Lambda^+([\frac{n+1}{2}])$, and $\delta \in \{+,-\}$.

Suppose that

$$\mathrm{Hom}_{G'}(F^G(\mu)_\delta|_{G'}, F^{G'}(\nu)_\delta) \neq \{0\}.$$

If $n = 2m$, then $\mu = (\mu_0, \cdots, \mu_{m+1}) \in \Lambda^+(m+1)$ and $\nu = (\nu_0, \cdots, \nu_m) \in \Lambda^+(m)$ and

$$\mu_0 \geq \nu_1 \geq \mu_2 \geq \cdots \geq \mu_n \geq \nu_n \geq \mu_{n+1} = 0. \tag{3.1}$$

If $n = 2m+1$, then $\mu = (\mu_0, \cdots, \mu_{m+1}) \in \Lambda^+(m+1)$ and $\nu = (\nu_0, \cdots, \nu_m) \in \Lambda^+(m+1)$ and

$$\mu_0 \geq \nu_1 \geq \mu_2 \geq \cdots \geq \mu_n \geq \nu_n \geq 0. \tag{3.2}$$

We represent the result graphically in the following theorem by representing nontrivial symmetry breaking operators by arrows connecting the θ-stable parameters of the representations.

Theorem 3.5. *Two representations in the standard sequences of $F^G(\mu)_\delta$ respectively $F^{G'}(\nu)_\delta$ have a nontrivial symmetry breaking operator if and only if the θ-stable parameters of the representations satisfy one of the following conditions: First case: $n = 2m$.*

Case A

$$(\mu_0, \ldots, \mu_i \,||\, \mu_{i+1}, \ldots, \mu_{m+1})_\delta$$

$$\Downarrow$$

$$(\nu_0, \ldots, \nu_i \,||\, \nu_{i+1}, \ldots, \nu_m)_\delta$$

or Case B

$$(\mu_0, \ldots, \mu_i \mid\mid \mu_{i+1}, \ldots, \mu_{m+1})_\delta$$
$$\Downarrow$$
$$(\nu_0, \ldots, \nu_{i-1} \mid\mid \nu_i, \nu_{i+1}, \ldots, \nu_m)_\delta$$

Second case: $n = 2m + 1$.

Case A

$$(\mu_0, \ldots, \mu_i \mid\mid \mu_{i+1}, \ldots, \mu_{m+1})_\delta$$
$$\Downarrow$$
$$(\nu_0, \ldots, \nu_i \mid\mid \nu_{i+1}, \ldots, \nu_{m+1})_\delta$$

or Case B

$$(\mu_0, \ldots, \mu_i \mid\mid \mu_{i+1}, \ldots, \mu_{m+1})_\delta$$
$$\Downarrow$$
$$(\nu_0, \ldots, \nu_{i-1} \mid\mid \nu_i, \ldots, \nu_{m+1})_\delta$$

3.5 *Branching Laws for Representations in $\mathcal{A}$: Third Formulation*

We summarize the results as follows:

Theorem 3.6 (Branching Law). *Let Π and π be irreducible representations in $\mathcal{A}$ of $SO(n+1, 1)$ respectively $SO(n, 1)$.*

(1) *Suppose first that $n = 2m$. Then*

$$\mathrm{Hom}_{SO(n,1)}(\Pi|_{SO(n,1)}, \pi) \neq \{0\}$$

if and only if the enhanced θ-stable parameters of the representations satisfy the following conditions:

(a) *Π and π have the same signature δ;*
(b) *$h_\pi \in \{h_\Pi, h_\Pi - 1\}$;*
(c) $\mu_0 \geq \nu_0 \geq \mu_1 \geq \cdots \geq \mu_n \geq \nu_n \geq \mu_{n+1}$.

(2) *Suppose that $n = 2m + 1$. Then*

$$\mathrm{Hom}_{SO(n,1)}(\Pi|_{SO(n,1)}, \pi) \neq \{0\}$$

if and only if the enhanced θ-stable parameters of the representations satisfy the following conditions:

(a) *they have the same signature δ;*
(b) $h_\pi \in \{h_\Pi, h_\Pi - 1\}$;
(c) $\mu_0 \geq \nu_0 \geq \mu_1 \geq \cdots \geq \mu_n \geq \nu_n \geq 0.$

4 Gross–Prasad Conjectures for Tempered Representations of $(SO(n+1,1), SO(n,1))$

In this section we discuss the Gross–Prasad conjectures for irreducible tempered representations in $\mathcal{A}$ which are nontrivial on the center. This is a generalization of the results for irreducible tempered representations with infinitesimal character ρ which are nontrivial on the center in [KS18b, Chap. 11]. For simplicity we discuss here only the case $n = 2m$.

Recall that for $n = 2m$, tempered representations Π_m of $G = SO(n+1,1)$ in $\mathcal{A}$ are irreducible unitary principal series representations [KS18b, Prop. 15.5], with height m, signature δ and θ-stable parameter

$$(\mu_1, \mu_2, \ldots, \mu_m \mid\mid 0)_\delta.$$

The tempered representations π_m of $G' = SO(n,1)$ in $\mathcal{A}$ are discrete series representations with height m. Their signature is not unique and their θ-stable parameter is

$$(\nu_1, \nu_2, \ldots, \nu_m \mid\mid)_\delta.$$

We now assume that the representation $\Pi = \Pi_m$ is nontrivial on the center. These data determine Vogan packets $VP(\Pi_m)$ and $VP(\pi_m)$ of the representations Π of G, respectively π of the subgroup G' [Vo93]. By Theorem 3.6, there is a nontrivial symmetry breaking operator

$$B \colon \Pi \to \pi$$

if and only if the interlacing conditions are satisfied. Following exactly the steps of the algorithm by Gross and Prasad outlined in [KS18a], see also [KS18b, Chap. 11], we conclude that the Gross–Prasad conjecture predicts correctly that the pair (Π, π) in $(VP(\Pi_m), VP(\pi_m))$ has a nontrivial symmetry breaking operator.

Together with the results about tempered principal series representations in [KS18b, Chap. 11, Sect. 4] this completes the proof of the following:

Theorem 4.1. *The Gross–Prasad conjectures are correct for tempered representations of the pair* $(G, G') = (SO(n+1, 1), SO(n, 1))$ *which are nontrivial on the center of G and G'.*

Remark 4.2. The third formulation of the branching laws (Theorem 3.6) shows that branching for representations in $\mathcal{A}$ "interpolate" between the classical branching laws of finite-dimensional representations and the branching laws of Gross–Prasad for tempered representations.

5 Distinguished Representations and Periods

We discuss periods of a pair of irreducible representations Π of $G = SO(n+1, 1)$ and π of the subgroup $G' = SO(n, 1)$. Using the branching law (Theorem 3.4), we see in this section that we can prove that the representations $A_{\mathfrak{q}}(\lambda) \in \mathcal{A}$ are distinguished for some orthogonal group H.

5.1 Periods

We recall from [KS18b, Thm. 5.4] that for representations Π, π of a real reductive Lie group G, respectively of a reductive subgroup G', the space of symmetry breaking operators

$$\mathrm{Hom}_{G'}(\Pi|_{G'}, \pi^{\vee})$$

and the space of G'-invariant continuous linear functionals

$$\mathrm{Hom}_{G'}(\Pi \boxtimes \pi, \mathbb{C})$$

are naturally isomorphic to each other. Thus we may use symmetry breaking operators to construct G'-invariant continuous linear functionals. This technique allows us to obtain G'-invariant continuous linear functionals not only for unitary representations but also for nonunitary representations.

Definition 5.1. A nontrivial linear functional $\mathcal{F}$ on $\Pi \boxtimes \pi$ is called a *period* of $\Pi \boxtimes \pi$ if $\mathcal{F}$ is invariant under the diagonal G'-action, i.e., $\mathcal{F} \in \mathrm{Hom}_{G'}(\Pi \boxtimes \pi, \mathbb{C})$.

We say that vector $\Phi \otimes \phi \in \Pi \boxtimes \pi$ is a *test vector* for the period $\mathcal{F}$ if $\Phi \otimes \phi$ is not in the kernel of $\mathcal{F}$. If the period is nontrivial on a test vector $\Phi \otimes \phi$, we refer to its image as the *value of the period* on $\Phi \otimes \phi$.

Remark 5.2. If (Π, π) is a pair of discrete series representations for the symmetric pairs $(G_1(\mathbb{R}), G_2(\mathbb{R}))$ we may consider a realization of the outer tensor product representation $\Pi \boxtimes \pi$ in $L^2(G_1(\mathbb{R}) \times G_2(\mathbb{R}))$. The integral

$$\int_{G_2(\mathbb{R})} \Phi(h)\phi(h)dh$$

converges for some smooth vectors $(\Phi, \phi) \in \Pi \boxtimes \pi$ in the minimal K-types and so if it is nonzero it defines a period integral for the discrete series representations $\Pi \boxtimes \pi$ [Va01]. If the representation $\Pi \boxtimes \pi$ is not tempered the integral usually does not converge, but nevertheless we can consider periods via symmetry breaking operators.

The next theorem describes for the pair $(G, G') = (SO(n+1,1), SO(n,1))$ the θ-stable parameters of the representations in $\mathcal{A}$ which have a nontrivial period $\mathrm{Hom}_{G'}(\Pi \boxtimes \pi, \mathbb{C})$. Recall that the θ-stable parameter of a representation $\Pi \in \mathcal{A}$ of height i is of the form

1. $(\mu_1, \ldots, \mu_i \mid\mid \mu_{i+1}, \ldots, 0)_\delta$ if $G = SO(2m+1, 1)$,
2. $(\mu_1, \ldots, \mu_i \mid\mid \mu_{i+1}, \ldots, \mu_m)_\delta$ if $G' = SO(2m, 1)$.

Theorem 5.3. *Suppose that Π and π are representations of G, respectively G' in $\mathcal{A}$ of height i respectively j. The following statements are equivalent:*

(i) *The representation $\Pi \boxtimes \pi$ has a nontrivial G'-period;*
(ii) *Π and π have the same signature, $j = i$ or $i - 1$ and their θ-stable parameters satisfy the interlacing conditions of the branching result in Theorem 3.6.*

In [KS18b, Props. 10.12 and 10.30] we proved furthermore for the pair $(O(n+1,1), O(n,1))$.

Theorem 5.4. *If Π and π are representations of $O(n+1,1)$ respectively $O(n,1)$ with trivial infinitesimal character ρ such that $\Pi \otimes \pi$ has a nontrivial G'-period. Then there is a test vector for a nonzero period in the minimal K-type $\Pi \boxtimes \pi$.*

Remark 5.5. We expect that Theorem 5.4 also holds for unitary representations in $\mathcal{A}$.

Remark 5.6. Similar results for cohomologically induced representations of other pairs (G, G') of reductive groups were obtain by B. Sun [S19].

5.2 *Distinguished Representations*

Let G be a reductive group, and H a reductive subgroup. We regard H as a subgroup of the direct product group $G \times H$ via the diagonal embedding $H \hookrightarrow G \times H$.

Definition 5.7. Let ψ be a one-dimensional representation of H. We say an admissible smooth representation Π of G is *(H, ψ)-distinguished* if

$$\mathrm{Hom}_H(\Pi \boxtimes \psi^\vee, \mathbb{C}) \simeq \mathrm{Hom}_H(\Pi|_H, \psi) \neq \{0\}.$$

If the character ψ is trivial, we say Π is *H-distinguished.*

Repeated application of Theorem 3.6 proves

Theorem 5.8. *Suppose that $\Pi = A_{\mathfrak{q}}(\lambda) \in \mathcal{A}$ is a representation of $SO(n+1,1)$ of height h. Then Π is $SO(n+1-h,1)$-distinguished.*

Remark 5.9. For a different proof and perspective of this theorem see [Ko92].

Since the representations $\Pi_{i,+}$ have height i (see Example 2.10), this generalizes the following theorem proved in [KS18b, Thm. 12.4].

Theorem 5.10. *Let $0 \le i \le \frac{n}{2}$. Then the representations $\Pi_{i,\delta}$ ($\delta \in \{+,-\}$) of $G = SO(n+1,1)$ are $SO(n+1-i,1)$-distinguished.*

In the remainder of the section we recall a formula for the period of representations with trivial infinitesimal character ρ of the pair

$$(G, H) = (SO(n+1,1), SO(m+1,1)) \quad \text{for } m \le n.$$

We use here the notation of Example 2.10 in Sect. 2.4.2.

The period can be computed by applying the composition of the regular symmetry breaking operators that we constructed in [KS18b, Chap. 12, Sect. 1] with respect to the chain of subgroups

$$G = SO(n+1,1) \supset SO(n,1) \supset SO(n-1,1) \supset \cdots \supset SO(m+1,1) = H, \tag{5.1}$$

to test vectors. We write simply Π_i for $\Pi_{i,+}$. We recall from [KS18b, Prop. 14.44] that $\Pi_i \equiv \Pi_{i,+}$ has a minimal K-type $K_{min}(i,+) = \bigwedge^i(\mathbb{C}^{n+1}) \boxtimes \mathbf{1}$.

Let $v \in \bigwedge^i(\mathbb{C}^{n+1})$ be the image of $1 \in \mathbb{C}$ via the following successive inclusions:

$$\bigwedge^i(\mathbb{C}^{n+1}) \supset \bigwedge^{i-1}(\mathbb{C}^{n}) \supset \cdots \supset \bigwedge^{i-\ell}(\mathbb{C}^{n+1-\ell}) \supset \cdots \supset \bigwedge^{0}(\mathbb{C}^{n+1-i}) \simeq \mathbb{C} \ni 1,$$

and we regard v as an element of the minimal K-type $K_{min}(i,+)$ of Π_i.

Theorem 5.11 ([KS18b, Thm. 12.5]). *Let Π_i be the irreducible representation of $G = SO(n+1,1)$, with infinitesimal character ρ, height i and signature $+$. Let v be the normalized element of its minimal K-type as above. For $0 \le i \le n$, the value $F(v)$ of the $SO(n+1-i,1)$-period F on $v \in \Pi_i$ is*

$$\frac{\pi^{\frac{1}{4}i(2n-i-1)}}{((n-i)!)^{i-1}} \times \begin{cases} \frac{1}{(n-2i)!} & \text{if } 2i < n+1, \\ (-1)^{n+1}(2i-n-1)! & \text{if } 2i \ge n+1. \end{cases}$$

6 Bilinear Forms on $(\mathfrak{g}, K)$-cohomologies Induced by Symmetry Breaking Operators

Consider now the induced map by a symmetry breaking operator

$$T : \Pi \to \pi$$

on $(\mathfrak{g}, K)$-cohomologies of a pair of representations Π and π. In what follows, by abuse of notation, we denote an admissible smooth representation and its underlying $(\mathfrak{g}, K)$-module by the same letter when we discuss their $(\mathfrak{g}, K)$-cohomologies.

Recall that a theorem of Vogan–Zuckerman [VZ84] states that every irreducible *unitary* representation Π of $SO(n+1,1)$ with

$$H^*(\mathfrak{g}, K; \Pi \otimes V) \neq \{0\}$$

for a finite-dimensional representation V is of the form $\Pi = A_{\mathfrak{q}}(\lambda)$. If we assume that

$$H^*(\mathfrak{g}, K; \Pi) \neq \{0\}$$

then Π is isomorphic to a unitary irreducible representation with infinitesimal character ρ i.e., it is of the form $A_{\mathfrak{q}}$. See [KS18b, Chap. 14, Sect. 9.4] for $O(n+1,1)$.

Note also that an irreducible representation Π with

$$H^*(\mathfrak{g}, K; \Pi \otimes V) \neq \{0\}$$

for some finite-dimensional representation V is not always unitarizable.

Suppose that Π is a principal series representation of a connected reductive Lie group with nonsingular integral infinitesimal character. If the $(\mathfrak{g}, K)$-cohomology of $\Pi \otimes V$ is nonzero the highest weight of V satisfies the conditions in [BW80, Chap. III, Thm. 3.3]. For representations of $O(n+1,1)$ the situation is more complicated and the finite-dimensional representation V is also described in [KS18b, Chap. 16, Sect. 4]. Using the results in [KS18b, Chap. 15] about the restriction of representations of $O(n+1,1)$ to $SO(n+1,1)$ we obtain a formula for the representation V for $SO(n+1,1)$.

Let $(G, G') = (SO(n+1,1), SO(n,1))$, and Π, π be representations of G and G', respectively with

$$H^*(\mathfrak{g}, K; \Pi \otimes V) \neq \{0\}$$

and

$$H^*(\mathfrak{g}', K'; \pi \otimes V') \neq \{0\},$$

where V and V' are irreducible finite-dimensional representations of G and G', respectively. Suppose in addition that

1. $\mathrm{Hom}_{G'}(V|_{G'}, V') \neq \{0\}$;
2. Π and π have the same height i;
3. Π and π have the same signature δ.

A nontrivial symmetry breaking operator $T\colon \Pi \otimes V \to \pi \otimes V'$ induces a canonical homomorphism

$$T^*\colon H^j(\mathfrak{g}, K; \Pi \otimes V) \to H^j(\mathfrak{g}', K'; \pi \otimes V') \tag{6.1}$$

and a bilinear form

$$B_T\colon H^j(\mathfrak{g}, K; \Pi \otimes V) \times H^{n-j}(\mathfrak{g}', K'; (\pi \otimes V')^\vee \otimes \chi_{(-1)^n}) \to \mathbb{C} \quad \text{for all } j,$$

where $(\pi \otimes V')^\vee$ denotes the contragredient representation of $\pi \otimes V'$.

The formulas in [KS18b, Chap. 16, Sect. 3] and [BW80, Chap. III, Thm. 3.3] imply

Theorem 6.1. *Suppose that Π and π are principal series representations and that*

1. $H^*(\mathfrak{g}, K, \Pi \otimes V) \neq \{0\}$ *and* $H^*(\mathfrak{g}', K', \pi \otimes V') \neq \{0\}$;
2. $\mathrm{Hom}_{G'}(V|_{G'}, V') \neq \{0\}$;
3. Π *and* π *have the same height* i;
4. Π *and* π *have the same signature* δ;
5. *there exists a nontrivial symmetry breaking operator* $T\colon \Pi \to \pi$.

Then the symmetry breaking operator T induces a nontrivial homomorphism

$$T^i\colon H^i(\mathfrak{g}, K; \Pi \otimes V) \to H^i(\mathfrak{g}', K'; \pi \otimes V')$$

and hence a nontrivial bilinear form

$$B_T\colon H^i(\mathfrak{g}, K; \Pi \otimes V) \times H^{n-i}(\mathfrak{g}', K'; (\pi \otimes V')^\vee \chi_{(-1)^n}) \to \mathbb{C}.$$

The following theorem provides a criterium for the nonvanishing of this bilinear form on the $(\mathfrak{g}, K)$-cohomology of representations with trivial infinitesimal character.

Theorem 6.2 ([KS18b, Thm. 12.11]). *Let $T\colon X \to Y$ be a $(\mathfrak{g}', K')$-homomorphism, where X is a $(\mathfrak{g}, K)$-module $A_\mathfrak{q}$ and Y is a $(\mathfrak{g}', K')$-module $A_{\mathfrak{q}'}$. Let U be the representation space of the minimal K-type μ in X, and U' that of the minimal K'-type μ' in Y. We define a K'-homomorphism by*

$$\varphi_T := \mathrm{pr} \circ T|_U\colon U \to U'. \tag{6.2}$$

(1) *If φ_T is zero, then the homomorphisms $T_*\colon H^j(\mathfrak{g}, K; X) \to H^j(\mathfrak{g}', K'; Y)$ and the bilinear form B_T vanish for all degrees $j \in \mathbb{N}$.*
(2) *If φ_T is $\mathfrak{p}$-nonvanishing at degree j, then T_* and the bilinear forms B_T are nonzero for this degree j.*

This theorem together with our results [KS18b] implies

Theorem 6.3 (cf. [KS18b, Thm. 12.13]). *Let $(G, G') = (SO(n+1, 1), SO(n, 1))$, $0 \le i \le n$, and $\delta \in \{+, -\}$. Let T be the symmetry breaking operator $\Pi_{i,\delta} \to \pi_{i,\delta}$ given in Theorem 3.4*

(1) *T induces bilinear forms*

$$B_T\colon H^j(\mathfrak{g}, K; \Pi_{i,\delta}) \times H^{n-j}(\mathfrak{g}', K'; \pi_{n-i,(-1)^n\delta}) \to \mathbb{C} \quad \textit{for all } j.$$

(2) *The bilinear form B_T is nonzero if and only if $j = i$ and $\delta = (-1)^i$.*

Remark 6.4. A theorem similar to Theorem 6.3 was proved by B. Sun [S17] for the $(\mathfrak{g}, K)$-cohomology with nontrivial coefficients of irreducible tempered representations of the pair

$$(GL(n, \mathbb{R}), GL(n-1, \mathbb{R})).$$

Acknowledgments The first author was partially supported by Grant-in-Aid for Scientific Research (A) (JP18H03669), Japan Society for the Promotion of Science. The second author was partially supported by NSF grant DMS-1500644.

References

[BW80] A. Borel, N.R. Wallach, *Continuous Cohomology, Discrete Subgroups, and Representations of Reductive Groups*. Math. Surveys Monogr., vol. 67, 2nd edn. (Amer. Math. Soc., Providence, 2000), xviii+260 pp.; Ann. of Math. Stud., vol. 94, 1st edn. (Princeton Univ. Press, Princeton, 1980), xvii+388 pp.

[GGP12] W.T. Gan, B.H. Gross, D. Prasad, J.-L. Waldspurger, *Sur les conjectures de Gross et Prasad. I.* Astérique, vol. 346 (Soc. Math. France, Paris, 2012), xi+318 pp.

[GP92] B.H. Gross, D. Prasad, On the decomposition of a representations of SO_n when restricted to SO_{n-1}. Canad. J. Math. **44**, 974–1002 (1992)

[KS71] A.W. Knapp, E.M. Stein, Intertwining operators for semisimple groups. Ann. Math. (2) **93**, 489–578 (1971)

[KV95] A.W. Knapp, D.A. Vogan Jr., *Cohomological Induction and Unitary Representations*. Princeton Math. Ser., vol. 45 (Princeton Univ. Press, Princeton, 1995), xx+948 pp., ISBN: 978-0-691-03756-6

[Ko92] T. Kobayashi, *Singular Unitary Representations and Discrete Series for Indefinite Stiefel Manifolds $U(p, q; \mathbb{F})/U(p-m, q; \mathbb{F})$*. Mem. Amer. Math. Soc., vol. 462 (Amer. Math. Soc., Providence, 1992), v+106 pp., ISBN-10:0-8218-2524-0; ISBN-13:978-0-8218-2524-2

[Ko98] T. Kobayashi, Discrete decomposability of the restriction of $A_{\mathfrak{q}}(\lambda)$ with respect to reductive subgroups III—restriction of Harish-Chandra modules and associated varieties. Invent. Math. **131**, 229–256 (1998)

[KO13] T. Kobayashi, T. Oshima, Finite multiplicity theorems for induction and restriction. Adv. Math. **248**, 921–944 (2013)

[KS15] T. Kobayashi, B. Speh, *Symmetry Breaking for Representations of Rank One Orthogonal Groups*. Mem. Amer. Math. Soc., vol. 238 (Amer. Math. Soc., Providence, 2015), v+112 pp. https://doi.org/10.1090/memo/1126, ISBNs: 978-1-4704-1922-6 (print); 978-1-4704-2615-6 (online)

[KS18a] T. Kobayashi, B. Speh, Symmetry breaking for orthogonal groups and a conjecture by B. Gross and D. Prasad, in *Geometric Aspects of the Trace Formula*, Simons Symposia (Springer, Berlin, 2018), pp. 245–266

[KS18b] T. Kobayashi, B. Speh, *Symmetry Breaking for Representations of Rank One Orthogonal Groups II*. Lecture Notes in Math., vol. 2234 (Springer, Berlin, 2018), xv + 342 pp., ISBN: 978-981-13-2900-5

[KS] T. Kobayashi, B. Speh, Branching laws for irreducible representations of rank one orthogonal groups (in preparation)

[Ks74] B. Kostant, Verma modules and the existence of quasi-invariant differential operators, in *Non-commutative Harmonic Analysis, Marseille–Luminy, 1974*. Lecture Notes in Math., vol. 466 (Springer, Berlin, 1975), pp. 101–128

[L89] R.P. Langlands, On the classification of irreducible representations of real algebraic groups, in *Representation Theory and Harmonic Analysis on Semisimple Lie Groups*, Math. Surveys Monogr., vol. 31 (Amer. Math. Soc., Providence, 1989), pp. 101–170

[S17] B. Sun, The nonvanishing hypothesis at infinity for Rankin–Selberg convolutions. J. Am. Math. Soc. **30**, 1–25 (2017)

[S19] B. Sun, Cohomologically induced distinguished representations and cohomological test vectors. Duke Math. J. **168**, 85–126 (2019)

[SZ12] B. Sun, C.-B. Zhu, Multiplicity one theorems: the Archimedean case. Ann. Math. (2), **175**, 23–44 (2012). https://doi.org/10.4007/annals.2012.175.1.2

[Va01] J.A. Vargas, Restriction of some discrete series representations. Algebras Groups Geom. **18**, 85–99 (2001)

[Vo81] D.A. Vogan Jr., *Representations of Real Reductive Lie Groups*. Progr. Math., vol. 15 (Birkhäuser, Boston, 1981), xvii+754 pp.

[Vo93] D.A. Vogan Jr., The local Langlands conjecture, in *Representation Theory of Groups and Algebras*. Contemp. Math., vol. 145 (Amer. Math. Soc., Providence, 1993), pp. 305–379

[VZ84] D.A. Vogan Jr., G.J. Zuckerman, Unitary representations with nonzero cohomology. Compos. Math. **53**, 51–90 (1984)

[Wa92] N.R. Wallach, *Real Reductive Groups*. I, Pure Appl. Math., vol. 132 (Academic, Boston, 1988), xx+412 pp., ISBN: 0-12-732960-9; II, ibid, vol. 132 (Academic, Boston, 1992), xiv+454 pp., ISBN: 978-0127329611

[We46] H. Weyl, *The Classical Groups. Their Invariants and Representations*. Princeton Landmarks Math. (Princeton Univ. Press, Princeton, 1997), xiv+320 pp. A reprint of the second edition (1946)

Explicit Decomposition of Certain Induced Representations of the General Linear Group

Erez Lapid

Abstract We provide some experimental results on the decomposition of the parabolic induction of $\pi \otimes \pi$ in the Grothendieck group where π is an irreducible representation of GL_n over a local non-archimedean field.

1 Introduction

The purpose of this short note is to report on some experimental results in the representation theory of the general linear group over a non-archimedean local field F. This theory was developed in the fundamental work of Bernstein–Zelevinsky [2, 3].

The direct sum over $n \geq 0$ of the categories of finite length, smooth, complex representations of $\mathrm{GL}_n(F)$, forms a ring category[1] $\mathcal{C}$ with respect to (normalized) parabolic induction, which we denote by $\times$. (The unit object is the one-dimensional representation of the trivial group $\mathrm{GL}_0(F)$.)

For any $a, b \in \mathbb{Z}$ with $a \leq b$ we call $[a, b]$ a *segment* and denote by $Z([a, b])$ the character $|\det|^{(a+b)/2}$ of $\mathrm{GL}_{b-a+1}(F)$. By definition, a *multisegment* is a formal finite sum of segments $\mathfrak{m} = \sum_{i=1}^{k} \Delta_i$. The set of multisegments forms a commutative monoid. Writing $\Delta_i = [a_i, b_i]$ with $b_1 \geq \cdots \geq b_k$, the induced representation

$$Z(\Delta_1) \times \cdots \times Z(\Delta_k)$$

admits an irreducible socle which depends only on $\mathfrak{m}$, up to equivalence. We denote it by $Z(\mathfrak{m})$. By Zelevinsky classification [13], the map $\mathfrak{m} \mapsto Z(\mathfrak{m})$ is injective. Its image consists of the irreducible representations of $\mathrm{GL}_n(F)$, $n \geq 0$ whose

[1] I.e., a locally finite abelian monoidal category with bilinear and biexact tensor product.

E. Lapid (✉)
Department of Mathematics, Weizmann Institute of Science, Rehovot, Israel

W. Müller et al. (eds.), *Relative Trace Formulas*, Simons Symposia,
https://doi.org/10.1007/978-3-030-68506-5_9

(normalized) Jacquet module with respect to the Borel subgroup contains a character of the diagonal torus of the form $\mathrm{diag}(t_1, \ldots, t_n) \mapsto \prod_{i=1}^{n} |t_i|^{\lambda_i}$ where $\lambda_i \in \mathbb{Z}$ for all i. One of the nice properties of this classification is that for any multisegments $\mathfrak{m}_1$ and $\mathfrak{m}_2$, the representation $Z(\mathfrak{m}_1 + \mathfrak{m}_2)$ occurs with multiplicity one in the Jordan–Hölder sequence of $Z(\mathfrak{m}_1) \times Z(\mathfrak{m}_2)$. In particular, if $Z(\mathfrak{m}_1) \times Z(\mathfrak{m}_2)$ is irreducible, then it is necessarily equivalent to $Z(\mathfrak{m}_1 + \mathfrak{m}_2)$.

The irreducible representations of the form $Z(\mathfrak{m})$ generate a Serre subcategory[2] $\mathcal{C}'$ of $\mathcal{C}$ that is closed under parabolic induction. In fact, one can decompose $\mathcal{C}$ as a direct sum of categories, each of which is equivalent (as a ring category) to $\mathcal{C}'$. Thus, as far as understanding the decomposition of parabolic induction of objects in $\mathcal{C}$ is concerned, the restriction to $\mathcal{C}'$ entails no loss in generality.

2 Basics

Let V be a finite-dimensional $\mathbb{Z}$-graded vector space over $\mathbb{C}$. The group $\mathrm{Aut}(V)$ of grading preserving automorphisms of V acts with finitely many orbits on the space $E_+(V)$ of degree 1 (nilpotent) graded endomorphisms of V. By a graded version of the Jordan normal form, the orbits are indexed by multisegments whose support (in the obvious sense) is the graded dimension of V [12]. Similarly, we may consider the space $E_-(V)$ of degree -1 graded endomorphisms of V. It is the dual space to $E_+(V)$ under the trace pairing, which is $\mathrm{Aut}(V)$-invariant. By a general principle [11], the orbits in $E_+(V)$ (as well as those of $E_-(V)$) correspond bijectively to the irreducible components of the commuting variety

$$\mathcal{X}(V) = \{(A, B) \in E_+(V) \times E_-(V) : AB = BA\}.$$

We denote by $\mathcal{C}_\mathfrak{m}$ the irreducible component corresponding to (the orbit indexed by) a multisegment $\mathfrak{m}$. Note that $\mathrm{Aut}(V)$ acts on $\mathcal{X}(V)$ and preserves its irreducible components.

An interesting class of irreducible representations of $\mathrm{GL}_n(F)$ are those π for which $\pi \times \pi$ is irreducible. They were called *□-irreducible* in [10] and studied there following the work of Kang–Kashiwara–Kim–Oh [6, 7] (in a broader context which will not be discussed here). In particular, the reducibility properties of $\pi_1 \times \pi_2$ are more tractable if at least one of π_1 or π_2 is □-irreducible. (We refer to [10] for more details.)

In their work (again, in a much more general context), Geiß–Leclerc–Schröer highlighted the property that $\mathcal{C}_\mathfrak{m}$ admits an open $\mathrm{Aut}(V)$-orbit [5]. (We will refer to it as the *GLS condition*.) A variant of their conjecture, formulated in [10], states that for a multisegment $\mathfrak{m}$, $Z(\mathfrak{m})$ is □-irreducible if and only if the GLS condition is satisfied.

[2] I.e., a full subcategory closed under subobjects, quotients and extensions.

3 Permutations

Let S_k denote the symmetric group on k elements. For any permutation $\sigma \in S_k$ consider the multisegment

$$\mathfrak{m}_\sigma = \sum_{i=1}^{k} [\sigma(i), 2k - i].$$

A special case of the main result of [10] is that the representation $Z(\mathfrak{m}_\sigma)$ is $\square$-irreducible if and only if σ is smooth, namely σ is 3412 and 4231 avoiding. The terminology is motivated by a result of Lakshmibai–Sandhya which asserts that σ is smooth if and only if the corresponding Schubert variety of type A_{k-1} is smooth [8]. Another equivalent condition is that the Kazhdan–Lusztig polynomial $P_{e,\sigma}(q)$ with respect to S_k is 1 [4]. It turns out that this is also equivalent to the GLS condition in the case at hand, although we are unaware of a direct relation between the smoothness condition for Schubert varieties and the GLS condition.

4 Calculation

Suppose that $\mathfrak{m}_1$ and $\mathfrak{m}_2$ are two multisegments. Using the Arakawa–Suzuki functor [1], it is in principle possible to compute the decomposition of $Z(\mathfrak{m}_1) \times Z(\mathfrak{m}_2)$ in the Grothendieck group. The computation involves Kazhdan–Lusztig polynomials with respect to S_k where k is the number of segments in $\mathfrak{m}_1 + \mathfrak{m}_2$. We refer to [10, §10] and the references therein for more details.

In December 2017 I computed all Kazhdan–Lusztig polynomials for each pair of permutations in S_{12}. (In reality, "only" about 46×10^9 pairs need to be considered.) This computation was carried out on a 1 terabyte RAM machine at the Weizmann Institute and resulted in about 4.3×10^9 different polynomials. More details about this month-long computation (on a single core) can be found in [9].

Following up on this computation, I calculated the decomposition of $\pi \times \pi$ in the Grothendieck group for any $\pi = Z(\mathfrak{m})$ where $\mathfrak{m}$ consists of at most 6 segments. It would be unwieldy to fully reproduce here the entire computation, but let me provide some details about the upshot.

The most striking feature of the calculation, which was discovered following a question by David Kazhdan, is that there appears to be a strong correlation between the length l of $\pi \times \pi$ and the minimal codimension d of an orbit in $C_\mathfrak{m}$. We will call the latter the *deficiency* of $\mathfrak{m}$. Note that the GLS condition precisely means that $d = 0$. Our empirical findings are:

(1) $d = 0 \iff l = 1$ (i.e., if and only if π is $\square$-irreducible), in accordance with [10, Conjecture 4.4].
(2) $d = 1 \iff l = 2 \iff \pi \times \pi$ is multiplicity free.
(3) $d = 2 \iff 7 \le l \le 9$
(4) $d = 3 \iff 39 \le l \le 53$
(5) $d = 4 \iff 251 \le l \le 257$.

(In all cases checked, $d \le 4$.)

In the case $\mathfrak{m} = \mathfrak{m}_\sigma$ there also seems to be a correlation, although not as clear cut, between d and $P_{e,\sigma}(1)$, i.e., the sum of the coefficients of $P_{e,\sigma}$. (Recall that in general, we know, albeit indirectly, that $d = 0 \iff P_{e,\sigma}(1) = 1$.)

For simplicity, let us say that $\sigma \in S_k$ is *almost smooth* if $Z(\mathfrak{m}_\sigma) \times Z(\mathfrak{m}_\sigma)$ has length two. We say that σ is *wild* if it is neither smooth nor almost smooth.

Among the 354 non-smooth permutations in S_6, 230 are almost smooth. For 224 of these, $P_{e,\sigma}(1) = 2$, the 6 exceptions being

$$\sigma = (463152), (465132), (526413), (546213), (632541), (653421)$$

where $P_{e,\sigma}(1) = 3$. In the wild case, $P_{e,\sigma}(1) > 2$.

5 Tables

In Table 1 we list for each almost smooth permutations σ in S_k ($k = 6$) the w such that

$$Z(\mathfrak{m}_\sigma) \times Z(\mathfrak{m}_\sigma) = Z(\mathfrak{m}_\sigma + \mathfrak{m}_\sigma) \oplus Z(\mathfrak{n}_w).$$

Here w is a two-to-one function from $\{1, \dots, 2k\}$ to $\{1, \dots, k\}$ and

$$\mathfrak{n}_w = \sum_{i=1}^{2k} [w(i), 2k - \lceil \frac{i}{2} \rceil]. \tag{1}$$

In Table 2 we list for each wild $\sigma \in S_6$, the length l of $\Pi = Z(\mathfrak{m}_\sigma) \times Z(\mathfrak{m}_\sigma)$, the number t of distinct constituents in the Jordan–Hölder decomposition of Π, the deficiency d of $\mathfrak{m}$ and the value $p = P_{e,\sigma}(1)$ of the Kazhdan–Lusztig polynomial. The deficiency is calculated as the dimension of the null space of the matrix given by formula (6) of [10, p. 138] for generic λ. (In practice, this is computed by taking several random values of λ. While this does not provide a rigorous proof for the value of d, the veracity of the calculation is beyond reasonable doubt.)

Table 1 Decomposition of $Z(\mathfrak{m}_\sigma) \times Z(\mathfrak{m}_\sigma)$ in the almost smooth case

σ	w	σ	w	σ	w
125634	112235463546	126453	112246354635	135624	113325462546
136452	113346254625	145236	112435243566	145263	112435246635
146253	112436254635	146352	114436253625	146523	112466352435
153426	113524352466	153462	113524356624	153624	113524362546
153642	113524663524	154623	115524362436	156423	112546352436
163542	113625463524	164352	114635243625	164532	114635463522
215634	221135463546	216453	221146354635	235614	223315461546
236451	223346154615	245163	221435146635	246135	221436143655
246153	221436154635	246351	224436153615	246513	221466351435
251634	223511463546	253461	223514356614	253614	223514361546
253641	223514663514	254613	225514361436	256143	221546154633
256314	223546351146	256413	221546351436	261453	224611354635
263415	223614361455	263514	223615361544	263541	223615463514
264153	224635114635	264351	224635143615	264531	224635463511
265341	226635143514	325614	332215461546	326451	332246154615
346251	334426152615	351642	132514663524	352641	332514662514
356214	332546251146	361542	132615463524	362514	332615261544
362541	332615462514	364251	334625142615	364521	334625462511
365142	136625143524	365241	336625142514	365412	136655241324
426153	241326154635	426513	241366251435	426531	241366552413
436512	441366251325	463152	443625113625	463251	443625132615
463521	443625362511	465132	146635143522	465213	246635241135
465312	146635241325	564312	154635241326	625431	261546352413
632541	362513462514	635421	362546352411	645321	463546352211
653421	663524352411				

In order to avoid repetition, in both tables we only list one representative for each equivalence class of the equivalence relation $\sigma \sim \sigma^{-1} \sim \sigma^{w_0} \sim (\sigma^{-1})^{w_0}$ where σ^{w_0} is the conjugation of σ by the longest permutation $w_0 \in S_k$. (The representation $Z(\mathfrak{m}_{\sigma^{-1}})$ is essentially the contragredient of $Z(\mathfrak{m}_\sigma)$. The relation between the decompositions of $Z(\mathfrak{m}_\sigma) \times Z(\mathfrak{m}_\sigma)$ and $Z(\mathfrak{m}_{\sigma^{w_0}}) \times Z(\mathfrak{m}_{\sigma^{w_0}})$ is more mysterious but follows from the abovementioned computation using Kazhdan–Lusztig polynomials.)

In all the cases where $d < 4$, i.e., when $\sigma \neq (456123), (562341), (634512)$, any constituent of Π occurs with multiplicity one or two. For $d = 2$ exactly one irreducible representation occurs with multiplicity two in Π, but for $d = 3$ there are at least 10 such representations in each case. For $\sigma = (456123)$ where $l = 251$, there are 59 constituents occurring with multiplicity one, 81 with multiplicity two, 4 with multiplicity three and 2 with multiplicity 9. For $\sigma = (562341)$ or (634512) where the length is 257, there are 61 constituents occurring with multiplicity one, 83 with multiplicity two, 4 with multiplicity three and 2 with multiplicity 9.

Table 2 Length, number of distinct constituents of $Z(\mathfrak{m}_\sigma) \times Z(\mathfrak{m}_\sigma)$, deficiency and $P_{e,\sigma}(1)$ in the wild case

σ	l	t	d	p	σ	l	t	d	p	σ	l	t	d	p
145623	8	7	2	3	156342	8	7	2	3	163452	9	8	2	4
245613	8	7	2	3	256134	8	7	2	3	256341	8	7	2	3
263451	9	8	2	4	264513	8	7	2	3	345612	36	26	3	5
346152	7	6	2	3	346512	8	7	2	3	351624	9	8	2	4
354612	9	8	2	4	356124	39	28	3	5	356142	7	6	2	3
356241	8	7	2	3	356412	7	6	2	3	361452	8	7	2	3
362451	9	8	2	4	364152	9	8	2	4	364512	44	31	3	6
426351	7	6	2	4	456123	251	146	4	10	456132	9	8	2	4
456231	44	31	3	6	456312	7	6	2	3	462351	46	33	3	6
462513	7	6	2	3	462531	7	6	2	3	463512	9	8	2	4
465231	9	8	2	4	562341	257	150	4	9	562431	7	6	2	3
563412	53	37	3	8	563421	9	8	2	4	564231	7	6	2	3
623451	48	35	3	8	623541	7	6	2	4	624351	7	6	2	4
624531	8	7	2	4	634521	9	8	2	4	635241	9	8	2	4
645231	53	37	3	8										

6 Supplement

In the aforementioned tables, we have only listed (some of) the data pertaining to the multisegments of the form $\mathfrak{m}_\sigma$. However, for any multisegment $\mathfrak{m}$ consisting of k segments, one can easily read off the decomposition of $Z(\mathfrak{m}) \times Z(\mathfrak{m})$ from the decomposition of $Z(\mathfrak{m}_\sigma) \times Z(\mathfrak{m}_\sigma)$ for a suitable $\sigma \in S_k$. (While we get more values for the length l of $Z(\mathfrak{m}) \times Z(\mathfrak{m})$ than in the case $\mathfrak{m} = \mathfrak{m}_\sigma$, the correlation between l and the deficiency of $\mathfrak{m}$ continues to hold for all multisegments consisting of up to 6 segments.)

More precisely, any multisegment $\mathfrak{m}$ consisting of k segments can be written as $\mathfrak{m} = [a_{\sigma(1)}, b_1] + \cdots + [a_{\sigma(k)}, b_k]$ where $a_1 \le \cdots \le a_k$, $b_1 \ge \cdots \ge b_k$ and $\sigma \in S_k$. We may (and shall) assume that $\sigma^{-1}(i) < \sigma^{-1}(i+1)$ whenever $a_i = a_{i+1}$ and that $\sigma(j) < \sigma(j+1)$ whenever $b_j = b_{j+1}$. Suppose that the decomposition of $Z(\mathfrak{m}_\sigma) \times Z(\mathfrak{m}_\sigma)$ in the Grothendieck group is given by

$$Z(\mathfrak{m}_\sigma) \times Z(\mathfrak{m}_\sigma) = \sum \mu_w \, Z(\mathfrak{n}_w)$$

where $\mu_w \in \mathbb{Z}_{\ge 0}$ and the sum is over the two-to-one functions $w : \{1, \dots, 2k\} \to \{1, \dots, k\}$ such that $w(2i-1) < w(2i)$ for all $i = 1, \dots, k$ and $\mathfrak{n}_w$ is as in (1). Then,

$$Z(\mathfrak{m}) \times Z(\mathfrak{m}) = \sum{}' \mu_w \, Z(\sum_{i=1}^{2k} [a_{w(i)}, b_{\lceil \frac{i}{2} \rceil}]) \tag{1}$$

where the prime indicates that we only sum over the w's above such that

(1) $w(2j) < w(2j+1)$ whenever $b_j = b_{j+1}$.
(2) Whenever $w(r) = i$, $w(s) = i+1$ and $a_i = a_{i+1}$, we have $r < s$.
(3) $a_{w(i)} \leq b_{\lceil \frac{i}{2} \rceil} + 1$ for all $i = 1, \ldots, 2k$. (Whenever $a_{w(i)} = b_{\lceil \frac{i}{2} \rceil} + 1$ we ignore the i-th summand, which is an empty segment, on the right-hand side of (1).)

Note that the irreducible representations on the right-hand side of (1) are distinct.

Acknowledgement I am indebted to David Kazhdan for his insightful advice. I would also like to thank the referee for making useful suggestions.

References

1. T. Arakawa, T. Suzuki, Duality between $\mathfrak{sl}_n(\mathbf{C})$ and the degenerate affine Hecke algebra. J. Algebra **209**(1), 288–304 (1998). MR1652134
2. I.N. Bernšteĭn, A.V. Zelevinskiĭ, Induced representations of the group GL(n) over a p-adic field, Funkcional. Anal. i Priložen. **10**(3), 74–75 (1976). MR0425031 (54 #12989)
3. I.N. Bernstein, A.V. Zelevinsky, Induced representations of reductive $\mathfrak{p}$-adic groups. I. Ann. Sci. École Norm. Sup. **10**(4), 441–472 (1977). MR0579172 (58 #28310)
4. V.V. Deodhar, Local Poincaré duality and nonsingularity of Schubert varieties. Comm. Algebra **13**(6), 1379–1388 (1985). MR788771 (86i:14015)
5. C. Geiß, B. Leclerc, J. Schröer, Kac-Moody groups and cluster algebras. Adv. Math. **228**(1), 329–433 (2011). MR2822235
6. S.-J. Kang, M. Kashiwara, M. Kim, S.-j. Oh, Simplicity of heads and socles of tensor products. Compos. Math. **151**(2), 377–396 (2015). MR3314831
7. S.-J. Kang, M. Kashiwara, M. Kim, S.-j. Oh, Monoidal categorification of cluster algebras. J. Amer. Math. Soc. **31**(2), 349–426 (2018). MR3758148
8. V. Lakshmibai, B. Sandhya, Criterion for smoothness of Schubert varieties in $SL(n)/B$. Proc. Indian Acad. Sci. Math. Sci. **100**(1), 45–52 (1990). MR1051089
9. E. Lapid, Conjectures about certain parabolic Kazhdan–Lusztig polynomials, in *Geometric Aspects of the Trace Formula* (Springer, Cham, 2018), pp. 267–297. MR3969878
10. E. Lapid, A. Mínguez, Geometric conditions for □-irreducibility of certain representations of the general linear group over a non-archimedean local field. Adv. Math. **339**, 113–190 (2018). MR3866895
11. V.S. Pjaseckiĭ, Linear Lie groups that act with a finite number of orbits. Funkcional. Anal. i Priložen. **9**(4), 85–86 (1975). MR0390138 (52 #10964)
12. A.V. Zelevinskiĭ, The p-adic analogue of the Kazhdan-Lusztig conjecture. Funktsional. Anal. i Prilozhen. **15**(2), 9–21, 96 (1981). MR617466 (84g:22039)
13. A.V. Zelevinsky, Induced representations of reductive $\mathfrak{p}$-adic groups. II. On irreducible representations of GL(n). Ann. Sci. École Norm. Sup. **13**(2), 165–210 (1980). MR584084 (83g:22012)

Mixed Arithmetic Theta Lifting for Unitary Groups

Yifeng Liu

Abstract In this article, we construct mixed generating functions of special cycles and mixed arithmetic theta lifting, and prove the product relation. We then formulate a conjecture on the arithmetic inner product formula for odd rank unitary groups, complementary to the even rank case previously formulated in the author's thesis.

2010 Mathematics Subject Classification 11G18, 11G40, 14G40

1 Introduction

Let F be a totally real number field. For every integer $r \geq 1$, let Sp_{2r} be the symplectic group of rank $2r$. Kudla in [Kud97] defined the arithmetic generating functions Z_r^{ϕ}, which are functions on $\mathrm{Sp}_{2r}(\mathbf{A}_F)$ valued in the Chow groups of certain orthogonal (more precisely, GSpin) Shimura varieties, associated to Schwartz functions ϕ. Recall that the classical theta series can be viewed as functions on $\mathrm{Sp}_{2r}(\mathbf{A}_F)$ valued in the space of automorphic forms on orthogonal groups. Therefore, one may regard Z_r^{ϕ} as the arithmetic analogue of theta series. As theta series play an important role in the doubling method, which relates to the (central) L-value via Rallis' inner product formula [Ral87, KR94], it is expected that the arithmetic generating functions are closely related to the (central) L-derivatives—this is known as (part of) the Kudla program [Kud02, Kud04].

Kudla's construction can be easily carried over to the case of unitary groups. Let E/F be a totally imaginary quadratic extension. For $r \geq 1$, let W_r be the split skew-Hermitian space over E of dimension $2r$. Then one can construct arithmetic generating functions $Z_{r;\nu}^{\phi}$, which are functions on $\mathrm{U}(\mathrm{W}_r)(\mathbf{A}_F)$ valued in Chow groups of certain unitary Shimura varieties, where ν is some auxiliary automorphic character (see [Liu11a]). However, in the unitary case, $\{\mathrm{W}_r\}_{r\geq 1}$ are

Y. Liu (✉)
Department of Mathematics, Yale University, New Haven, CT, USA
e-mail: yifeng.liu@yale.edu

W. Müller et al. (eds.), *Relative Trace Formulas*, Simons Symposia,
https://doi.org/10.1007/978-3-030-68506-5_10

not the only skew-Hermitian spaces. Thus, it is interesting to ask whether one can define arithmetic generating functions for any skew-Hermitian space W over E. In this article, we will perform the construction for W that are split of *odd* dimensions when the unitary Shimura varieties come from complex ball quotients (that is, those attached to totally positive definite incoherent Hermitian spaces over $\mathbf{A}_E$), based on our previous results in [Liu] and the idea of mixed models of Weil representations. Moreover, as suggested in Remark 5.6, one shall be able to define arithmetic generating functions on U(W) as long as $\mathrm{W} \otimes_{\mathbb{Q}} \mathbb{R}$ is split.

To explain in more details, we need some notation. For every integer $r \geq 0$, we denote by $(\mathrm{W}_r, \langle\ ,\ \rangle)$ the skew-Hermitian space over E (with respect to E/F) spanned by $\{e_{\pm 1}, \dots, e_{\pm r}\}$ with $\langle e_i, e_{-j}\rangle = \delta_{ij}$ for $0 \leq i, j \leq r$. For every nonzero purely imaginary element $\varepsilon \in E^{\times -}$, we put $\mathrm{W}_{0,\varepsilon} := Ee_0$ where $\langle e_0, e_0\rangle = \varepsilon$, and $\mathrm{W}_{r,\varepsilon} := \mathrm{W}_r \oplus \mathrm{W}_{0,\varepsilon}$ in general. Denote by G_r and $\mathrm{G}_{r,\varepsilon}$ the unitary groups of W_r and $\mathrm{W}_{r,\varepsilon}$, respectively, which are quasi-split reductive groups over F. Let $\mu_{E/F}\colon F^\times\backslash\mathbf{A}_F^\times \to \{\pm 1\}$ be the quadratic character associated to E/F.

Let $\mathbf{V}$ be a totally positive definite incoherent Hermitian space over $\mathbf{A}_E$ of rank $n \geq 2$, with the associated unitary Shimura varieties $\{\mathrm{Sh}(\mathbf{V})_K\}_K$ smooth of dimension $n-1$ over E. Since we will consider $\mathrm{W}_{r,\varepsilon}$ whose dimension is odd, we need to choose a certain splitting character $\mu\colon E^\times\backslash\mathbf{A}_E^\times \to \mathbb{C}^\times$ satisfying $\mu|_{\mathbf{A}_F^\times} = \mu_{E/F}$ (see Sect. 3 for more details). We also choose another character $\nu\colon E^\times\backslash\mathbf{A}_E^\times \to \mathbb{C}^\times$ satisfying $\nu|_{\mathbf{A}_F^\times} = \mu_{E/F}^n$. The character μ will determine a CM abelian variety A_μ over E, unique up to isogeny. We construct for certain "Schwartz function" ϕ, a (mixed) arithmetic generating function $Z^{\natural,\phi}_{r,\varepsilon;\mu,\nu}$ on $\mathrm{G}_{r,\varepsilon}(\mathbf{A}_F^\infty)$ as long as ε is μ-admissible (basically a sign condition at infinity), valued in a certain variant of $\varinjlim_K \mathrm{CH}^{r+\dim A_\mu}(\mathrm{Sh}(\mathbf{V})_K \times A_\mu)_{\mathbb{C}}$. We then discuss its relation with Kudla's arithmetic generating functions $Z^{\phi}_{r;\nu}$ on $\mathrm{G}_r(\mathbf{A}_F^\infty)$ via a series of product relations in Sect. 5. In Sect. 6, we formulate the modularity conjecture of (mixed) arithmetic generating functions, assuming which we define the (mixed) arithmetic theta lifting, and in Sect. 7 formulate the conjecture on *arithmetic inner product formula* when n is odd (and at least 3) using the Beilinson–Bloch–Poincaré height pairing introduced in [Liu]. Note that the arithmetic inner product formula has already been formulated in [Liu11a] when n is even, and confirmed when $n = 2$ in [Liu11b].

The (conjectural) arithmetic inner product formula is the first-derivative counterpart of Rallis' inner product formula (for the central L-value); it computes the central derivative of a certain L-function of degree n over E in terms of the height of the arithmetic theta lifting which is a cycle. Note that when $\mathrm{Sh}(\mathbf{V})_K$ has even dimension, that is, when n is odd, the Beilinson–Bloch height pairing of any cycle on $\mathrm{Sh}(\mathbf{V})_K$ with itself would be zero. This is the main reason we assumed n even in our earlier work [Liu11a]. Thus in this article, we complement the picture by formulating the arithmetic inner product formula for odd n (if not 1) using these new (mixed) arithmetic generating functions and arithmetic theta lifting.

Notation and Conventions

- For an abelian group A and a ring R, we put $A_R := A \otimes_{\mathbb{Z}} R$ as an R-module.
- A reflex field is always a *subfield* of $\mathbb{C}$.
- Denote by $\mathbf{A}^\infty := \widehat{\mathbb{Z}}_{\mathbb{Q}}$ the ring of finite adèles of $\mathbb{Q}$, and put $\mathbf{A} := \mathbb{R} \times \mathbf{A}^\infty$.
- For a number field k, we put $\mathbf{A}_k^\infty := \mathbf{A}^\infty \otimes_{\mathbb{Q}} k$ and $\mathbf{A}_k := \mathbf{A} \otimes_{\mathbb{Q}} k$.
  - Denote by $|\ |_{\mathbb{Q}} \colon \mathbb{Q}^\times \backslash \mathbf{A}^\times \to \mathbb{R}_{>0}^\times$ the character, uniquely determined by the properties that $|x|_{\mathbb{Q}} = |x|$ for $x \in \mathbb{R}^\times$ and that $|\ |_{\mathbb{Q}}$ is trivial on $\widehat{\mathbb{Z}}^\times$. For every $s \in \mathbb{C}$, Put $|\ |_k^s := |\ |_{\mathbb{Q}}^s \circ \mathrm{N}_{k/\mathbb{Q}} \colon k^\times \backslash \mathbf{A}_k^\times \to \mathbb{R}_{>0}^\times$.
  - Denote by $\psi_{\mathbb{Q}} \colon \mathbb{Q} \backslash \mathbf{A} \to \mathbb{C}^\times$ the character, uniquely determined by the properties that $\psi_{\mathbb{Q}}(x) = \exp(2\pi i x)$ for $x \in \mathbb{R}$ and that $\psi_{\mathbb{Q}}$ is trivial on $\widehat{\mathbb{Z}}$. Put $\psi_k := \psi_{\mathbb{Q}} \circ \mathrm{Tr}_{k/\mathbb{Q}} \colon k \backslash \mathbf{A}_k \to \mathbb{C}^\times$, which we call the *standard additive character* for k.
- Let E/F be a totally imaginary quadratic extension of a totally real number field.
  - We denote by Φ_F and Φ_E the set of all archimedean places of F and E, respectively.
  - We denote by $\mathtt{c} \in \mathrm{Gal}(E/F)$ the nontrivial Galois involution.
  - We let $\mu_{E/F} \colon F^\times \backslash \mathbf{A}_F^\times \to \{\pm 1\}$ be the quadratic character associated to E/F.
  - For every integer $r \geq 0$, we denote by Herm_r the F-scheme of Hermitian matrices of size r with respect to E/F.
  - For a place v of F and a character $\chi \colon E_v^\times \to \mathbb{C}^\times$ satisfying $\chi|_{F_v^\times} = 1$, we denote by $\chi^\flat \colon E_v^1 \to \mathbb{C}^\times$ the unique character such that $\chi(x) = \chi^\flat(x/x^{\mathtt{c}})$ for every $x \in E_v^\times$.

2 Special Cycles via Incoherent Spaces

We first review the notion of unitary Shimura varieties attached to incoherent Hermitian spaces. The readers can consult [Gro] in this volume for more details. Let $(\mathbf{V}, (\ ,\ )_{\mathbf{V}})$ be a totally positive definite incoherent Hermitian space over $\mathbf{A}_E$ of rank $n \geq 1$. Attached to $\mathbf{V}$, we have a projective system $\{\mathrm{Sh}(\mathbf{V})_K\}_K$ of smooth quasi-projective schemes over E of dimension $n-1$, indexed by (sufficiently small) open compact subgroups K of $\mathrm{U}(\mathbf{V}^\infty)$, where $\mathbf{V}^\infty := \mathbf{V} \otimes_{\mathbf{A}_E} \mathbf{A}_E^\infty$. The relation between $\{\mathrm{Sh}(\mathbf{V})_K\}_K$ with the classical notion of Shimura varieties is recalled as follows: Given a τ-nearby[1] Hermitian space V of $\mathbf{V}$ for some element $\tau \in \Phi_F$, a Deligne homomorphism $\mathrm{h}_{\mathrm{V},\hat{\tau}}$ (depending on the choice of $\hat{\tau} \in \Phi_E$ lifting τ), and an isomorphism $\iota \colon \mathbf{V}^\infty \xrightarrow{\sim} \mathrm{V} \otimes_E \mathbf{A}_E^\infty$ of Hermitian spaces over $\mathbf{A}_E^\infty$, we have an isomorphism

[1]It means that V has signature $(n-1, 1)$ at τ and $(n, 0)$ at other archimedean places, and satisfies $\mathrm{V} \otimes_E \mathbf{A}_E^\infty \simeq \mathbf{V}^\infty$.

$$\kappa_{\mathrm{V},\mathrm{h}_{\mathrm{V},\hat{\tau}},\iota}\colon \{\mathrm{Sh}(\mathbf{V})_K \otimes_{E,\hat{\tau}} \hat{\tau}(E)\}_K \xrightarrow{\sim} \{\mathrm{Sh}(\mathrm{U}(\mathrm{V}),\mathrm{h}_{\mathrm{V},\hat{\tau}})_{\iota(K)}\}_K$$

of projective systems of schemes over $\hat{\tau}(E)$. We also recall that $\mathrm{Sh}(\mathbf{V})_K$ has a canonical (smooth) toroidal compactification, which we denote by $\widetilde{\mathrm{Sh}}(\mathbf{V})_K$. In particular, $\mathrm{Sh}(\mathbf{V})_K = \widetilde{\mathrm{Sh}}(\mathbf{V})_K$ if $[F : \mathbb{Q}] > 1$. Finally, we recall that there is a Hodge divisor D_K in $\mathrm{CH}^1(\mathrm{Sh}(\mathbf{V})_K)_{\mathbb{Q}}$ and its canonical extension $\widetilde{D}_K$ in $\mathrm{CH}^1(\widetilde{\mathrm{Sh}}(\mathbf{V})_K)_{\mathbb{Q}}$.

Definition 2.1. An E-subspace $\mathrm{V}_\circ$ of $\mathbf{V}^\infty$ is *admissible* if the restriction of the Hermitian form $(\ ,\ )_{\mathbf{V}}$ to $\mathrm{V}_\circ$ is non-degenerate, takes values in E, and is totally positive definite.

For every admissible E-subspace $\mathrm{V}_\circ$ of $\mathbf{V}^\infty$, we let $\mathbf{V}_\bullet$ be the totally definite Hermitian space over $\mathbf{A}_E$ such that $\mathbf{V}_\bullet^\infty = (\mathrm{V}_\circ)^\perp \subseteq \mathbf{V}^\infty$. Then $\mathbf{V}_\bullet$ is necessarily incoherent. By the proof of Langlands' conjecture on the conjugation of Shimura varieties by Milne–Shih for those of abelian type in [DMOS82], we see that there is a unique finite morphism

$$\varsigma_{\mathrm{V}_\circ}\colon \mathrm{Sh}(\mathbf{V}_\bullet)_{K_\bullet} \to \mathrm{Sh}(\mathbf{V})_K \tag{2.1}$$

where $K_\bullet := K \cap \mathrm{U}(\mathbf{V}_\bullet^\infty)$, such that for every τ-nearby Hermitian space $\mathrm{V}_\bullet$ of $\mathbf{V}_\bullet$ for some $\tau \in \Phi_F$, a Deligne homomorphism $\mathrm{h}_{\mathrm{V}_\bullet,\hat{\tau}}$, and an isomorphism $\iota_\bullet\colon \mathbf{V}_\bullet^\infty \xrightarrow{\sim} \mathrm{V}_\bullet \otimes_E \mathbf{A}_E^\infty$, the following diagram

$$\begin{array}{ccc} \mathrm{Sh}(\mathbf{V}_\bullet)_{K_\bullet} \otimes_{E,\hat{\tau}} \hat{\tau}(E) & \xrightarrow{\varsigma_{\mathrm{V}_\circ}\otimes_{E,\hat{\tau}}\hat{\tau}(E)} & \mathrm{Sh}(\mathbf{V})_K \otimes_{E,\hat{\tau}} \hat{\tau}(E) \\ {\scriptstyle \kappa_{\mathrm{V}_\bullet,\mathrm{h}_{\mathrm{V}_\bullet,\hat{\tau}},\iota_\bullet}} \downarrow \simeq & & \simeq \downarrow {\scriptstyle \kappa_{\mathrm{V},\mathrm{h}_{\mathrm{V},\hat{\tau}},\iota}} \\ \mathrm{Sh}(\mathrm{U}(\mathrm{V}_\bullet),\mathrm{h}_{\mathrm{V}_\bullet,\hat{\tau}})_{\iota_\bullet(K_\bullet)} & \longrightarrow & \mathrm{Sh}(\mathrm{U}(\mathrm{V}),\mathrm{h}_{\mathrm{V},\hat{\tau}})_{\iota(K)} \end{array}$$

commutes, where $\mathrm{V} := \mathrm{V}_\bullet \oplus \mathrm{V}_\circ$; $\mathrm{h}_{\mathrm{V},\hat{\tau}}$ is the direct sum of $\mathrm{h}_{\mathrm{V}_\bullet,\hat{\tau}}$ and the trivial homomorphism; $\iota = \iota_\bullet \oplus \mathrm{id}_{\mathrm{V}_\circ \otimes_E \mathbf{A}_E^\infty}$; and the lower horizontal arrow is the morphism induced by the map of Shimura data.

Definition 2.2. Let $\mathrm{V}_\circ$ be an admissible E-subspace of $\mathbf{V}^\infty$. We define the *special cycle* associated to $\mathrm{V}_\circ$ to be

$$Z(\mathrm{V}_\circ)_K := (\varsigma_{\mathrm{V}_\circ})_*[\mathrm{Sh}(\mathbf{V}_\bullet)_{K_\bullet}]$$

as an element in $\mathrm{Z}^{\dim \mathrm{V}_\circ}(\mathrm{Sh}(\mathbf{V})_K)$. We also define the *compactified special cycle* associated to $\mathrm{V}_\circ$ to be the Zariski closure of $Z(\mathrm{V}_\circ)_K$, denoted by $\widetilde{Z}(\mathrm{V}_\circ)_K$, as an element in $\mathrm{Z}^{\dim \mathrm{V}_\circ}(\widetilde{\mathrm{Sh}}(\mathbf{V})_K)$.

Now let $r \geq 0$ be an integer. For an element $x = (x_1, \dots, x_r) \in (\mathbf{V}^\infty)^r$, we denote by V_x the E-subspace of $\mathbf{V}^\infty$ spanned by $\{x_1, \dots, x_r\}$.

Definition 2.3. We define the *special (Chow) cycle* associated to x to be

$$Z(x)_K := \begin{cases} Z(\mathrm{V}_x)_K \cdot (-D_K)^{r-\dim \mathrm{V}_x}, & \text{if } \mathrm{V}_x \text{ is admissible,} \\ 0, & \text{if } \mathrm{V}_x \text{ is not admissible,} \end{cases}$$

as an element in $\mathrm{CH}^r(\mathrm{Sh}(\mathbf{V})_K)_{\mathbb{Q}}$, and the *compactified special (Chow) cycle* associated to x to be

$$\widetilde{Z}(x)_K := \begin{cases} \widetilde{Z}(\mathrm{V}_x)_K \cdot (-\widetilde{D}_K)^{r-\dim \mathrm{V}_x}, & \text{if } \mathrm{V}_x \text{ is admissible,} \\ 0, & \text{if } \mathrm{V}_x \text{ is not admissible,} \end{cases}$$

as an element in $\mathrm{CH}^r(\widetilde{\mathrm{Sh}}(\mathbf{V})_K)_{\mathbb{Q}}$.

3 Mixed Arithmetic Generating Functions: the $r = 0$ Case

From now on, we assume that n, the rank of $\mathbf{V}$, is at least 2. We will work in the category of E-schemes.

We first recall some key constructions from [Liu, Section 4], which contribute to the initial case of our mixed arithmetic theta series, corresponding to $r = 0$.

Recall that an automorphic character $\mu\colon E^\times \backslash \mathbf{A}_E^\times \to \mathbb{C}^\times$ is a *conjugate symplectic character of weight one* if

- the restriction of μ to $F^\times \backslash \mathbf{A}_F^\times$ coincides with $\mu_{E/F}$;
- there exists a CM type Φ_μ of E, necessarily unique, such that for every $\tau \in \Phi_F$, the component $\mu_\tau\colon (E \otimes_{F,\tau} \mathbb{R})^\times \to \mathbb{C}^\times$ is the character

$$z \mapsto \left(\frac{z}{\overline{z}}\right)^{-\frac{1}{2}} = \frac{\sqrt{z\overline{z}}}{z}$$

if we have identified $(E \otimes_{F,\tau} \mathbb{R})^\times$ with $\mathbb{C}^\times$ via the unique element $\hat{\tau} \in \Phi_\mu$ above τ.

Note that the character $\mu^{\mathrm{alg}} := \mu \cdot |\ |_E^{-1/2}$ is algebraic. Denote by $M_\mu \subseteq \mathbb{C}$ the subfield generated by values $\mu^{\mathrm{alg}}(x)$ for $x \in (\mathbf{A}_E^\infty)^\times$, which is a number field containing the reflex field of (E, Φ_μ).

Before going to the details, let us explain the main idea behind the construction. The character μ gives rise to a family of irreducible admissible representations $\omega(\mu, \varepsilon, \chi)$ of $\mathrm{U}(\mathbf{V})(\mathbf{A}_F^\infty)$, where $\omega(\mu, \varepsilon, \chi)$ is just the theta lifting of a finite character χ of the unitary group of the one-dimensional skew-Hermitian space defined by a purely imaginary element $\varepsilon \in E^{\times -}$, via the splitting character μ. The main ingredient in the construction of the mixed arithmetic generating functions in the $r = 0$ case is to find algebraic cycles whose cycle classes live in the representations

$\omega(\mu, \varepsilon, \chi)$ with ε and χ varying. We will later see that such representations appear naturally in the first cohomology $\mathrm{H}^1(\widetilde{\mathrm{Sh}}(\mathbf{V}))$ of the compactified family $\{\widetilde{\mathrm{Sh}}(\mathbf{V})_K\}$, for some appropriate cohomology theory. Now we have two problems: first, we have to indicate the character μ in the cohomology; second, since cohomology classes in degree one cannot be realized as cycle classes, we need to find a suitable replacement. It turns out that the two problems are solved simultaneously, by introducing an auxiliary CM abelian variety (A_μ, i_μ) over E determined by μ, which is unique up to isogeny. Instead of considering $\mathrm{H}^1(\widetilde{\mathrm{Sh}}(\mathbf{V}))$, we consider $\mathrm{H}^{2\dim A_\mu}(\widetilde{\mathrm{Sh}}(\mathbf{V}) \times A_\mu)$, which canonically contains $\mathrm{H}^1(\widetilde{\mathrm{Sh}}(\mathbf{V})) \otimes \mathrm{H}_1(A_\mu)$ as a direct summand. Then the homology $\mathrm{H}_1(A_\mu)$ will serve as an indicator for the μ-part of $\mathrm{H}^1(\widetilde{\mathrm{Sh}}(\mathbf{V}))$, as well as we can realize classes in $\mathrm{H}^{2\dim A_\mu}(\widetilde{\mathrm{Sh}}(\mathbf{V}) \times A_\mu)$ by a family of algebraic cycles indexed by homomorphisms from the Albanese of $\{\widetilde{\mathrm{Sh}}(\mathbf{V})_K\}$ to A_μ.

Now we proceed with more details. To every conjugate symplectic character μ of weight one, we associate a category $\mathcal{A}(\mu)$ whose objects are essentially pairs (A_μ, i_μ) in which A_μ is an abelian variety over E and $i_\mu\colon M_\mu \to \mathrm{End}_E(A_\mu)_\mathbb{Q}$ is a certain CM structure, and whose morphisms are isogenies (see [Liu, Definition 4.5] for more details). Then the category $\mathcal{A}(\mu)$ is nonempty and connected [Liu, Proposition 4.6]. Moreover, we have $2\dim A_\mu = [M_\mu : \mathbb{Q}]$.

Let $A(\mathbf{V})_K$ be the Albanese variety [Liu, Definition 2.3] of $\widetilde{\mathrm{Sh}}(\mathbf{V})_K$; and put

$$A(\mathbf{V})_\infty := \varprojlim_K A(\mathbf{V})_K,$$

which is an abelian group pro-object in the category of E-schemes. For $(A_\mu, i_\mu) \in \mathcal{A}(\mu)$, put

$$\Omega(A_\mu, i_\mu) := \mathrm{Hom}_E(A(\mathbf{V}), A_\mu).$$

Then $\Omega(A_\mu, i_\mu)_\mathbb{Q}$ does not depend on the choice of $(A_\mu, i_\mu) \in \mathcal{A}(\mu)$, which we denote by $\Omega(\mu)$ (see [Liu, Definition 4.16] for a more precise definition). It is an M_μ-vector space, on which $\mathrm{U}(\mathbf{V})(\mathbb{A}_F^\infty)$ acts M_μ-linearly via its action on $A(\mathbf{V})_\infty$.

Definition 3.1. We collect the following notations for later use.

(1) Let X be a scheme of finite type over E and $\mathbb{L}$ an abelian group. For every rational prime ℓ, the Hochschild–Serre spectral sequence on the target of the absolute ℓ-adic cycle class map

$$\mathrm{cl}_\ell\colon \mathrm{CH}^i(X)_\mathbb{L} \to \mathrm{H}^{2i}_{\text{ét}}(X, \mathbb{L} \otimes_\mathbb{Z} \mathbb{Z}_\ell(i))$$

induces a decreasing filtration $\mathcal{F}^\bullet_\ell \mathrm{CH}^i(X)_\mathbb{L}$ on $\mathrm{CH}^i(X)_\mathbb{L}$ such that $\mathcal{F}^0_\ell \mathrm{CH}^i(X)_\mathbb{L} = \mathrm{CH}^i(X)_\mathbb{L}$ and $\mathcal{F}^1_\ell \mathrm{CH}^i(X)_\mathbb{L}$ is the subgroup of cohomologically trivial cycles (which is independent of ℓ). We put

$$\mathrm{CH}^i(X)^\natural_\mathbb{L} := \mathrm{CH}^i(X)_\mathbb{L} / \bigcap_\ell \mathcal{F}^2_\ell \mathrm{CH}^i(X)_\mathbb{L}.$$

(2) For an E-scheme X of finite type, $i \in \mathbb{Z}$, and an M_μ-module $\mathbb{L}$, we denote by $\mathrm{CH}^i_\mu(X)^\natural_{\mathbb{L}}$ the submodule[2] of $\mathrm{CH}^{i+\dim A_\mu}(X \times A_\mu)^\natural_{\mathbb{L}}$ on which $(\mathrm{id}_X \times \varphi)_*$ acts by $i_\mu^{-1}(\varphi)$ for every $\varphi \in \mathrm{End}_E(A_\mu)$.
(3) Whenever there is a contravariant functor F on the category of finite type E-schemes, we put

$$\mathsf{F}(\mathrm{Sh}(\mathbf{V})) := \varinjlim_K \mathsf{F}(\mathrm{Sh}(\mathbf{V})_K), \quad \mathsf{F}(\widetilde{\mathrm{Sh}}(\mathbf{V})) := \varinjlim_K \mathsf{F}(\widetilde{\mathrm{Sh}}(\mathbf{V})_K),$$

as long as the colimit exists.
(4) For various Chow groups, we add a superscript 0 to denote the subgroup of cohomologically trivial cycles.

Remark 3.2. Beilinson's conjecture predicts that when X is smooth and projective and $\mathbb{L}$ is a $\mathbb{Q}$-module, we have $\mathrm{CH}^i(X)^\natural_{\mathbb{L}} = \mathrm{CH}^i(X)_{\mathbb{L}}$.

We have the following result from [Liu] concerning $\Omega(\mu)$ as an $M_\mu[\mathrm{U}(\mathbf{V})(\mathbf{A}_F^\infty)]$-module. First recall the following definition.

Definition 3.3. An element $\varepsilon \in E^{\times -}$ is *μ-admissible* if $\hat{\tau}(\varepsilon)$ has negative imaginary part for every $\hat{\tau} \in \Phi_\mu$. We denote by $E^{\times -}_\mu \subseteq E^{\times -}$ the subset of μ-admissible elements, which is stable under the multiplication by $\mathrm{N}_{E/F}E^\times$.

Proposition 3.4 ([Liu, Theorem 4.18]). *There is a unique decomposition*

$$\Omega(\mu) = \bigoplus_{\varepsilon \in E^{\times -}_\mu / \mathrm{N}_{E/F}E^\times} \Omega(\mu, \varepsilon)$$

of $M_\mu[\mathrm{U}(\mathbf{V})(\mathbf{A}_F^\infty)]$-modules such that

$$\Omega(\mu, \varepsilon) \otimes_{M_\mu} \mathbb{C} \simeq \bigoplus_\chi \omega(\mu, \varepsilon, \chi),$$

where the direct sum is taken over all finite characters $\chi : E^1 \backslash (\mathbf{A}_E^\infty)^1 \to \mathbb{C}^\times$. Here, $\omega(\mu, \varepsilon, \chi)$ is the representation of $\mathrm{U}(\mathbf{V})(\mathbf{A}_F^\infty)$ obtained as a theta lifting of χ, which we have mentioned previously.

Moreover, we have $\Omega(\mu)^K = \mathrm{Hom}_E(A(\mathbf{V})_K, A_\mu)_{\mathbb{Q}}$ for every sufficiently small open compact subgroup $K \subseteq \mathrm{U}(\mathbf{V})(\mathbf{A}_F^\infty)$ and every $(A_\mu, i_\mu) \in \mathcal{A}(\mu)$.

Remark 3.5. For $\varepsilon \in E^{\times -}_\mu$, if we let $\mathrm{H}_\varepsilon(\mathbf{V})$ be the Heisenberg group of $\mathrm{Res}_{\mathbf{A}_E/\mathbf{A}_F}\mathbf{V}$ equipped with the symplectic form $\mathrm{Tr}_{E/F}\,\varepsilon \cdot (\ ,\ )_{\mathbf{V}}$, then we have a unique (irreducible) action of $\mathrm{H}_\varepsilon(\mathbf{V})(\mathbf{A}_F^\infty)$ on $\Omega(\mu, \varepsilon) \otimes_{M_\mu} \mathbb{C}$ with central character ψ_F^∞, where we recall that ψ_F is the standard additive character for F.

[2] It is independent of the choice of $(A_\mu, i_\mu) \in \mathcal{A}(\mu)$; hence the notation is justified.

In what follows, we choose a pair (μ, ε) with $\varepsilon \in E_{\mu}^{\times -}$. For later use, we put

$$\Omega(A_{\mu}, i_{\mu}, \varepsilon)^K := \Omega(A_{\mu}, i_{\mu}) \cap \Omega(\mu, \varepsilon)^K \subseteq \mathrm{Hom}_E(A(\mathbf{V})_K, A_{\mu}),$$

which is an $i_{\mu}^{-1}\,\mathrm{End}_E(A_{\mu})$-lattice in $\Omega(A_{\mu}, i_{\mu}, \varepsilon)^K$, and

$$\Omega(A_{\mu}, i_{\mu}, \varepsilon) := \bigcup_K \Omega(A_{\mu}, i_{\mu}, \varepsilon)^K.$$

Definition 3.6. Let $\nu \colon E^{\times}\backslash\mathbf{A}_E^{\times} \to \mathbb{C}^{\times}$ be an automorphic character satisfying $\nu|_{\mathbf{A}_F^{\times}} = \mu_{E/F}^n$. We define an action of $\mathrm{G}_{0,\varepsilon}(\mathbf{A}_F^{\infty}) = (\mathbf{A}_E^{\infty})^1$ on $\Omega(\mu, \varepsilon) \otimes_{M_{\mu}} \mathbb{C}$, denoted by $\omega_{0,\varepsilon;\mu,\nu}$, such that it acts on $\omega(\mu, \varepsilon, \chi)$ by the character $\chi \cdot (\nu\mu^{-n})^{\flat}$.

For every $\phi \in \Omega(A_{\mu}, i_{\mu}, \varepsilon)$, we will construct elements

$$\widetilde{Z}(\phi)_K \in \mathrm{CH}^{\dim A_{\mu}}(\widetilde{\mathrm{Sh}}(\mathbf{V})_K \times A_{\mu})_{M_{\mu}}$$

for every K under which ϕ is invariant, that are compatible with respect to K under pullbacks, such that if we denote by $\widetilde{Z}(\phi)_K^{\natural}$ the image of $\widetilde{Z}(\phi)_K$ in $\mathrm{CH}^{\dim A_{\mu}}(\widetilde{\mathrm{Sh}}(\mathbf{V})_K \times A_{\mu})_{M_{\mu}}^{\natural}$, then it belongs to $\mathrm{CH}^0_{\mu}(\widetilde{\mathrm{Sh}}(\mathbf{V})_K)_{M_{\mu}}^{\natural}$. Then by linearity, we obtain a complex linear map

$$\widetilde{Z}(-)^{\natural} \colon \Omega(\mu, \varepsilon) \otimes_{M_{\mu}} \mathbb{C} \to \mathrm{CH}^0_{\mu}(\widetilde{\mathrm{Sh}}(\mathbf{V}))_{\mathbb{C}}^{\natural}.$$

Definition 3.7. For every $\phi \in \Omega(\mu, \varepsilon) \otimes_{M_{\mu}} \mathbb{C}$, we define the *(mixed) arithmetic generating function* associated to ϕ to be the function $\widetilde{Z}_{0,\varepsilon;\mu,\nu}^{\natural,\phi}$ on $\mathrm{G}_{0,\varepsilon}(\mathbf{A}_F^{\infty})$ valued in $\mathrm{CH}^0_{\mu}(\widetilde{\mathrm{Sh}}(\mathbf{V}))_{\mathbb{C}}^{\natural}$ such that

$$\widetilde{Z}_{0,\varepsilon;\mu,\nu}^{\natural,\phi}(g) = \widetilde{Z}(\omega_{0,\varepsilon;\mu,\nu}(g)\phi)^{\natural}.$$

Remark 3.8. The function $\widetilde{Z}_{0,\varepsilon;\mu,\nu}^{\natural,\phi}$ extends uniquely to an automorphic form on $\mathrm{G}_{0,\varepsilon}(\mathbf{A}_F)$ via the assignment $g = g_{\infty}g^{\infty} \mapsto (\nu_{\infty}\mu_{\infty}^{-n})^{\flat}(g_{\infty}) \cdot \widetilde{Z}_{0,\varepsilon;\mu,\nu}^{\natural,\phi}(g^{\infty})$.

In the remaining part of this section, we will construct $\{\widetilde{Z}(\phi)_K\}_K$. The process is similar to the construction of the Fourier–Jacobi cycles in [Liu, Section 4.3]. We realize ϕ (uniquely) as a homomorphism $\phi \colon A(\mathbf{V})_K \to A_{\mu}$. By [Liu, Definition 2.3 and (4.1)], we have the subscheme $\nabla\widetilde{\mathrm{Sh}}(\mathbf{V})_K \subseteq \widetilde{\mathrm{Sh}}(\mathbf{V})_K \times \widetilde{\mathrm{Sh}}(\mathbf{V})_K$ and the Albanese morphism $\alpha_K \colon \nabla\widetilde{\mathrm{Sh}}(\mathbf{V})_K \to A(\mathbf{V})_K$.

Step 1: We start from the cycle

$$\Delta\widetilde{\mathrm{Sh}}(\mathbf{V})_K \times \widetilde{D}_K^{n-1} \in \mathrm{CH}^{2(n-1)}(\widetilde{\mathrm{Sh}}(\mathbf{V})_K \times \widetilde{\mathrm{Sh}}(\mathbf{V})_K \times \widetilde{\mathrm{Sh}}(\mathbf{V})_K)_{\mathbb{Q}},$$

where $\widetilde{D}_K$ is actually denoted by D_K in [Liu, Definition 4.22]. Put

$$(\Delta\widetilde{\mathrm{Sh}}(\mathbf{V})_K \times \widetilde{D}_K^{n-1})^\nabla := \Delta\widetilde{\mathrm{Sh}}(\mathbf{V})_K \times \widetilde{D}_K^{n-1} \cap \widetilde{\mathrm{Sh}}(\mathbf{V})_K \times \nabla\widetilde{\mathrm{Sh}}(\mathbf{V})_K$$

as an element in $\mathrm{CH}^{2(n-1)}(\widetilde{\mathrm{Sh}}(\mathbf{V})_K \times \nabla\widetilde{\mathrm{Sh}}(\mathbf{V})_K)_{\mathbb{Q}}$.

Step 2: We push the above cycle along the morphism

$$\mathrm{id}_{\widetilde{\mathrm{Sh}}(\mathbf{V})_K} \times (\phi \circ \alpha_K) \colon \widetilde{\mathrm{Sh}}(\mathbf{V})_K \times \nabla\widetilde{\mathrm{Sh}}(\mathbf{V})_K \to \widetilde{\mathrm{Sh}}(\mathbf{V})_K \times A_\mu$$

to obtain a cycle

$$(\mathrm{id}_{\widetilde{\mathrm{Sh}}(\mathbf{V})_K} \times (\phi \circ \alpha_K))_*(\Delta\widetilde{\mathrm{Sh}}(\mathbf{V})_K \times \widetilde{D}_K^{n-1})^\nabla \in \mathrm{CH}^{\dim A_\mu}(\widetilde{\mathrm{Sh}}(\mathbf{V})_K \times A_\mu)_{\mathbb{Q}}.$$

Step 3: To proceed, we use the canonical projector $\mathrm{T}_\mu^{\mathrm{can}} \in \mathrm{CH}^{\dim A_\mu}(A_\mu \times A_\mu)_{M_\mu}$ on A_μ in [Liu, Definition 4.9]. Define

$$\widetilde{Z}(\phi)_K := (\deg \widetilde{D}_K^{n-1})^{-1}(\mathrm{id}_{\widetilde{\mathrm{Sh}}(\mathbf{V})_K} \otimes \mathrm{T}_\mu^{\mathrm{can}})^*(\mathrm{id}_{\widetilde{\mathrm{Sh}}(\mathbf{V})_K} \times (\phi \circ \alpha_K))_*(\Delta\widetilde{\mathrm{Sh}}(\mathbf{V})_K \times D_K^{n-1})^\nabla$$

as an element in $\mathrm{CH}^{\dim A_\mu}(\widetilde{\mathrm{Sh}}(\mathbf{V})_K \times A_\mu)_{M_\mu}$, where $\deg \widetilde{D}_K^{n-1}$ is the degree of $\widetilde{D}_K^{n-1}$ [Liu, Definition 4.22]. The introduction of the projector $\mathrm{T}_\mu^{\mathrm{can}}$ ensures that $\widetilde{Z}(\phi)_K^\natural$ belongs to $\mathrm{CH}_\mu^0(\widetilde{\mathrm{Sh}}(\mathbf{V})_K)_{M_\mu}^\natural$.

It is a direct consequence from the definition that for another smaller open compact subgroup $K' \subseteq K$, we have $(u_K^{K'})^*\widetilde{Z}(\phi)_K = \widetilde{Z}(\phi)_{K'}$ in $\mathrm{CH}^{\dim A_\mu}(\widetilde{\mathrm{Sh}}(\mathbf{V})_{K'} \times A_\mu)_{M_\mu}$, where $u_K^{K'} : \widetilde{\mathrm{Sh}}(\mathbf{V})_{K'} \to \widetilde{\mathrm{Sh}}(\mathbf{V})_K$ is the transition morphism. Thus, we have completed the construction of $\{\widetilde{Z}(\phi)_K\}_K$.

4 Mixed Arithmetic Generating Functions: the $r > 0$ Case

For an integer $r > 0$, it is well-known that there is a mixed model of the Weil representation $\omega_{r,\varepsilon;\mu,\nu}$ of $\mathrm{G}_{r,\varepsilon}(\mathbf{A}_F^\infty)$ realized on $\mathscr{S}((\mathbf{V}^\infty)^r, \Omega(\mu,\varepsilon) \otimes_{M_\mu} \mathbb{C})$, the complex vector space of locally constant compactly supported functions on $(\mathbf{V}^\infty)^r$ valued in $\Omega(\mu,\varepsilon) \otimes_{M_\mu} \mathbb{C}$. We now review.

Denote by

- $\mathrm{P}_{r,\varepsilon}$ the parabolic subgroup of $\mathrm{G}_{r,\varepsilon}$ stabilizing the subspace $\bigoplus_{i=1}^r Ee_i$,
- $\mathrm{M}_{r,\varepsilon}$ the subgroup of $\mathrm{P}_{r,\varepsilon}$ stabilizing both subspaces $\bigoplus_{i=1}^r Ee_i$ and $\bigoplus_{i=1}^r Ee_{-i}$, which is a Levi subgroup,
- $\mathrm{N}_{r,\varepsilon}$ the unipotent radical of $\mathrm{P}_{r,\varepsilon}$, and
- $\mathrm{U}_{r,\varepsilon}$ the center of $\mathrm{N}_{r,\varepsilon}$.

If we use the basis $\{e_1,\dots,e_r,e_0,e_{-1},\dots,e_{-r}\}$, then the skew-Hermitian form is given by the matrix

$$\mathrm{w}_{r,\varepsilon} := \begin{pmatrix} & & \mathbf{1}_r \\ & \varepsilon & \\ -\mathbf{1}_r & & \end{pmatrix};$$

and we have

$$\mathrm{M}_{r,\varepsilon} = \left\{ \mathrm{m}(a,\alpha) := \begin{pmatrix} a & & \\ & \alpha & \\ & & {}^{\mathrm{t}}a^{\mathrm{c},-1} \end{pmatrix} \middle| a \in \operatorname{Res}_{E/F}\mathrm{GL}_r, \alpha \in \mathrm{G}_{0,\varepsilon} \right\},$$

$$\mathrm{N}_{r,\varepsilon} = \left\{ \mathrm{n}(b,\beta) := \begin{pmatrix} \mathbf{1}_r & \varepsilon b & \frac{\varepsilon b\, {}^{\mathrm{t}}b^{\mathrm{c}}}{2} + \beta \\ & 1 & {}^{\mathrm{t}}b \\ & & \mathbf{1}_r \end{pmatrix} \middle| b \in \operatorname{Res}_{E/F}\mathrm{Mat}_{r,1}, \beta \in \mathrm{Herm}_r \right\},$$

$$\mathrm{U}_{r,\varepsilon} = \{\mathrm{n}(0,\beta) \mid \beta \in \mathrm{Herm}_r\} \subseteq \mathrm{N}_{r,\varepsilon}.$$

For $\phi \in \mathscr{S}((\mathbf{V}^\infty)^r, \Omega(\mu,\varepsilon) \otimes_{M_\mu} \mathbb{C})$, we have

- for every $(a,\alpha) \in \mathrm{GL}_r(\mathbf{A}_E^\infty) \times \mathrm{G}_{0,\varepsilon}(\mathbf{A}_F^\infty)$ that

$$(\omega_{r,\varepsilon;\mu,\nu}(\mathrm{m}(a,\alpha))\phi)(x) = \nu(\det a) \cdot |\det a|_E^{n/2} \cdot \phi(xa),$$

- for every $(b,\beta) \in \mathrm{Mat}_{r,1}(\mathbf{A}_E^\infty) \times \mathrm{Herm}_r(\mathbf{A}_F^\infty)$ that

$$(\omega_{r,\varepsilon;\mu,\nu}(\mathrm{n}(b,\beta))\phi)(x) = \psi_F(\operatorname{tr}\beta T(x)) \cdot ([xb,0].\phi(x)),$$

  where $T(x) \in \mathrm{Herm}_r(\mathbf{A}_F^\infty)$ denotes the moment matrix of $x \in (\mathbf{V}^\infty)^r$, and $[xb,0].$ denotes the action of the element $[xb,0] \in \mathrm{H}_\varepsilon(\mathbf{V})(\mathbf{A}_F^\infty)$ on $\Omega(\mu,\varepsilon)\otimes_{M_\mu}\mathbb{C}$ (see Remark 3.5).

The action of $\mathrm{w}_{r,\varepsilon}$ on ϕ is given by some Fourier transform, whose formula we omit.

Now for every $x \in (\mathbf{V}^\infty)^r$ and every $\phi \in \Omega(A_\mu, i_\mu, \varepsilon)^K$, we put

$$\widetilde{Z}(x,\phi)_K := q^*\widetilde{Z}(x)_K \cdot \widetilde{Z}(\phi)_K \in \mathrm{CH}^{r+\dim A_\mu}(\widetilde{\mathrm{Sh}}(\mathbf{V})_K \times A_\mu)_{M_\mu},$$

where $q\colon \widetilde{\mathrm{Sh}}(\mathbf{V})_K \times A_\mu \to \widetilde{\mathrm{Sh}}(\mathbf{V})_K$ is the projection. As before, we denote by $\widetilde{Z}(x,\phi)_K^\natural$ the image of $\widetilde{Z}(x,\phi)_K$ in $\mathrm{CH}^{r+\dim A_\mu}(\widetilde{\mathrm{Sh}}(\mathbf{V})_K \times A_\mu)_{M_\mu}^\natural$. By linearity, we obtain a complex linear map

$$\widetilde{Z}(x,-)_K^\natural \colon \Omega(\mu,\varepsilon)^K \to \mathrm{CH}^r_\mu(\widetilde{\mathrm{Sh}}(\mathbf{V})_K)_{\mathbb{C}}^\natural.$$

However, unlike $\widetilde{Z}(\phi)_K$, the elements $\{\widetilde{Z}(x,\phi)_K^\natural\}_K$ are not necessarily compatible with respect to K. To remedy the issue, we need to take average.

Definition 4.1. We say that an element $\phi \in \mathscr{S}((\mathbf{V}^{\infty})^r, \Omega(\mu, \varepsilon) \otimes_{M_\mu} \mathbb{C})$ is *strongly K-invariant* for some (sufficiently small) open compact subgroup $K \subseteq \mathrm{U}(\mathbf{V})(\mathbb{A}_F^{\infty})$ if

- $\phi(x) \in \Omega(\mu, \varepsilon)^K \otimes_{M_\mu} \mathbb{C}$ for every $x \in (\mathbf{V}^{\infty})^r$; and
- $\phi(hx) = \phi(x)$ for every $x \in (\mathbf{V}^{\infty})^r$ and $h \in K$.

It is clear that every element $\phi \in \mathscr{S}((\mathbf{V}^{\infty})^r, \Omega(\mu, \varepsilon) \otimes_{M_\mu} \mathbb{C})$ is strongly K-invariant for some K.

Let $\mathrm{Herm}_r(F)_+$ be the subset of $\mathrm{Herm}_r(F)$ consisting of totally semi-positive definite Hermitian matrices. Take $\phi \in \mathscr{S}((\mathbf{V}^{\infty})^r, \Omega(\mu, \varepsilon) \otimes_{M_\mu} \mathbb{C})$ and $T \in \mathrm{Herm}_r(F)_+$. For every K such that ϕ is strongly K-invariant, we put

$$\widetilde{Z}(T, \phi)_K^{\natural} := \sum_{x \in K \backslash (\mathbf{V}^{\infty})^r, T(x) = T} \widetilde{Z}(x, \phi(x))_K^{\natural},$$

which is a finite sum in $\mathrm{CH}^r_\mu(\widetilde{\mathrm{Sh}}(\mathbf{V})_K)^{\natural}_{\mathbb{C}}$. It is straightforward to show that now $\{\widetilde{Z}(T, \phi)_K^{\natural}\}_K$ are compatible with respect to K, hence induce an element $\widetilde{Z}(T, \phi)^{\natural} \in \mathrm{CH}^r_\mu(\widetilde{\mathrm{Sh}}(\mathbf{V}))^{\natural}_{\mathbb{C}}$. Then we put

$$\widetilde{Z}(\phi)^{\natural} := \sum_{T \in \mathrm{Herm}_r(F)_+} e^{-\pi \sum_{\tau \in \Phi_F} \tau(\mathrm{tr}\, T)} \cdot \widetilde{Z}(T, \phi)^{\natural}$$

as a formal series of $\mathrm{CH}^r_\mu(\widetilde{\mathrm{Sh}}(\mathbf{V}))^{\natural}_{\mathbb{C}}$.

Remark 4.2. If $\phi \in \mathscr{S}((\mathbf{V}^{\infty})^r, \Omega(\mu, \varepsilon) \otimes_{M_\mu} \mathbb{C})$ takes values in $\Omega(A_\mu, i_\mu, \varepsilon)$, then we can define $\widetilde{Z}(T, \phi)$ (resp. $\widetilde{Z}(\phi)$) as an element of $\mathrm{CH}^{r+\dim A_\mu}(\widetilde{\mathrm{Sh}}(\mathbf{V}) \times A_\mu)_{M_\mu}$ (resp. a formal series of $\mathrm{CH}^{r+\dim A_\mu}(\widetilde{\mathrm{Sh}}(\mathbf{V}) \times A_\mu)_{\mathbb{C}}$), which lifts $\widetilde{Z}(T, \phi)^{\natural}$ (resp. $\widetilde{Z}(\phi)^{\natural}$).

Definition 4.3. For every $\phi \in \mathscr{S}((\mathbf{V}^{\infty})^r, \Omega(\mu, \varepsilon) \otimes_{M_\mu} \mathbb{C})$, we define the *(mixed) arithmetic generating function* associated to ϕ to be the function $\widetilde{Z}^{\natural, \phi}_{r,\varepsilon;\mu,\nu}$ on $\mathrm{G}_{r,\varepsilon}(\mathbb{A}_F^{\infty})$ valued in (formal series of) $\mathrm{CH}^r_\mu(\widetilde{\mathrm{Sh}}(\mathbf{V}))^{\natural}_{\mathbb{C}}$ such that

$$\widetilde{Z}^{\natural, \phi}_{r,\varepsilon;\mu,\nu}(g) = \widetilde{Z}(\omega_{r,\varepsilon;\mu,\nu}(g)\phi)^{\natural}.$$

Remark 4.4. Note that one can specialize Definition 4.3 to the case $r = 0$. Then it coincides with Definition 3.7 once we identify $\mathscr{S}((\mathbf{V}^{\infty})^0, \Omega(\mu, \varepsilon) \otimes_{M_\mu} \mathbb{C})$ with $\Omega(\mu, \varepsilon) \otimes_{M_\mu} \mathbb{C}$.

5 Product Relation for Arithmetic Generating Functions

Let $\mathbf{V}$, (μ, ε), $r \geq 0$, and ν be as in the previous discussion. We have constructed

- for every $\phi \in \mathscr{S}((\mathbf{V}^\infty)^r, \Omega(\mu,\varepsilon) \otimes_{M_\mu} \mathbb{C})$, a function $\widetilde{Z}^{\natural,\phi}_{r,\varepsilon;\mu,\nu}$ on $\mathrm{G}_{r,\varepsilon}(\mathbf{A}^\infty_F)$ valued in (formal series of) $\mathrm{CH}^r_\mu(\widetilde{\mathrm{Sh}}(\mathbf{V}))^\natural_\mathbb{C}$, from Sects. 3 and 4; and
- for every $\phi \in \mathscr{S}((\mathbf{V}^\infty)^r, \mathbb{C})$, a function $\widetilde{Z}^\phi_{r;\nu}$ on $\mathrm{G}_r(\mathbf{A}^\infty_F)$ valued in (formal series of) $\mathrm{CH}^r(\widetilde{\mathrm{Sh}}(\mathbf{V}))_\mathbb{C}$, from [Liu11a] by taking the splitting character χ in [Liu11a] to be ν.

For readers convenience, we briefly review the construction of $\widetilde{Z}^\phi_{r;\nu}$. For every K under which ϕ is invariant, we put

$$\widetilde{Z}(T,\phi)_K := \sum_{x \in K\backslash(\mathbf{V}^\infty)^r, T(x)=T} \phi(x)\cdot \widetilde{Z}(x)_K;$$

they are compatible with respect to K, hence induce an element $\widetilde{Z}(T,\phi) \in \mathrm{CH}^r(\widetilde{\mathrm{Sh}}(\mathbf{V}))_\mathbb{C}$. Then we put

$$\widetilde{Z}(\phi) := \sum_{T\in \mathrm{Herm}_r(F)_+} e^{-\pi\sum_{\tau\in\Phi_F}\tau(\mathrm{tr}\,T)}\cdot \widetilde{Z}(T,\phi)$$

as a formal series of $\mathrm{CH}^r(\widetilde{\mathrm{Sh}}(\mathbf{V}))_\mathbb{C}$. Finally, we put

$$\widetilde{Z}^\phi_{r;\nu}(g) := \widetilde{Z}(\omega_{r;\nu}(g)\phi),$$

where $\omega_{r;\nu}$ is the relevant Weil representation of $\mathrm{G}_r(\mathbf{A}^\infty_F)$ on $\mathscr{S}((\mathbf{V}^\infty)^r,\mathbb{C})$ with respect to the splitting character ν.

We also

- denote by $\widetilde{Z}^{\natural,\phi}_{r;\nu}$ the function on $\mathrm{G}_r(\mathbf{A}^\infty_F)$ valued in (formal series of) $\mathrm{CH}^r(\widetilde{\mathrm{Sh}}(\mathbf{V}))^\natural_\mathbb{C}$ induced from $\widetilde{Z}^\phi_{r;\nu}$; and
- change $\widetilde{Z}$ to Z if we restrict from $\widetilde{\mathrm{Sh}}(\mathbf{V})$ to $\mathrm{Sh}(\mathbf{V})$ for every meaningful notation.

It is known that $\{Z^\phi_{r;\nu}\}_{r\geq 0}$ satisfy certain product relation, which we recall as follows. For $r, r' \geq 0$, we have a natural isomorphism $\mathrm{W}_r \oplus \mathrm{W}_{r'} \simeq \mathrm{W}_{r+r'}$ which induces a homomorphism

$$\diamond\colon \mathrm{G}_r \times \mathrm{G}_{r'} \to \mathrm{G}_{r+r'}.$$

On the other hand, we have a natural map

$$\diamond\colon \mathscr{S}((\mathbf{V}^\infty)^r,\mathbb{C}) \times \mathscr{S}((\mathbf{V}^\infty)^{r'},\mathbb{C}) \to \mathscr{S}((\mathbf{V}^\infty)^{r+r'},\mathbb{C})$$

by taking exterior product, and a natural map

$$\diamond\colon \mathrm{CH}^r(\mathrm{Sh}(\mathbf{V}))_{\mathbb{C}} \times \mathrm{CH}^{r'}(\mathrm{Sh}(\mathbf{V}))_{\mathbb{C}} \to \mathrm{CH}^{r+r'}(\mathrm{Sh}(\mathbf{V}))_{\mathbb{C}}$$

by taking intersection product. Then the product relation is stated as in the following proposition.

Proposition 5.1. *For $\phi \in \mathscr{S}((\mathbf{V}^\infty)^r, \mathbb{C})$ and $\phi' \in \mathscr{S}((\mathbf{V}^\infty)^{r'}, \mathbb{C})$, we have*

$$Z^{\phi}_{r;\nu}(g) \diamond Z^{\phi'}_{r';\nu}(g') = Z^{\phi\diamond\phi'}_{r+r';\nu}(g \diamond g').$$

Proof. This follows by the same argument for [YZZ09, Theorem 1.1]. □

We would like to extend the above product relation to involve $\{Z^{\natural,\phi}_{r,\varepsilon;\mu,\nu}\}$ as well.

We first consider the product of $Z^{\natural,\phi}_{r,\varepsilon;\mu,\nu}$ and $Z^{\natural,\phi'}_{r';\nu}$. For $r, r' \geq 0$, we have a natural isomorphism $\mathrm{W}_{r,\varepsilon} \oplus \mathrm{W}_{r'} \simeq \mathrm{W}_{r+r',\varepsilon}$ which induces a homomorphism

$$\diamond\colon \mathrm{G}_{r,\varepsilon} \times \mathrm{G}_{r'} \to \mathrm{G}_{r+r',\varepsilon}.$$

On the other hand, we have a natural map

$$\diamond\colon \mathscr{S}((\mathbf{V}^\infty)^r, \Omega(\mu,\varepsilon)\otimes_{M_\mu}\mathbb{C}) \times \mathscr{S}((\mathbf{V}^\infty)^{r'}, \mathbb{C}) \to \mathscr{S}((\mathbf{V}^\infty)^{r+r'}, \Omega(\mu,\varepsilon)\otimes_{M_\mu}\mathbb{C})$$

by taking exterior product, and a natural map

$$\diamond\colon \mathrm{CH}^r_\mu(\mathrm{Sh}(\mathbf{V}))^\natural_{\mathbb{C}} \times \mathrm{CH}^{r'}(\mathrm{Sh}(\mathbf{V}))^\natural_{\mathbb{C}} \to \mathrm{CH}^{r+r'}_\mu(\mathrm{Sh}(\mathbf{V}))^\natural_{\mathbb{C}}$$

by first pulling back along the projection $\mathrm{Sh}(\mathbf{V})_K \times A_\mu \to \mathrm{Sh}(\mathbf{V})_K$ and then taking intersection product.

Proposition 5.2. *For $\phi \in \mathscr{S}((\mathbf{V}^\infty)^r, \Omega(\mu,\varepsilon)\otimes_{M_\mu}\mathbb{C})$ and $\phi' \in \mathscr{S}((\mathbf{V}^\infty)^{r'}, \mathbb{C})$, we have*

$$Z^{\natural,\phi}_{r,\varepsilon;\mu,\nu}(g) \diamond Z^{\natural,\phi'}_{r';\nu}(g') = Z^{\natural,\phi\diamond\phi'}_{r+r',\varepsilon;\mu,\nu}(g \diamond g').$$

Proof. By definition, we have the relation

$$\omega_{r,\varepsilon;\mu,\nu}(g)\phi \diamond \omega_{r';\nu}(g')\phi' = \omega_{r+r',\varepsilon;\mu,\nu}(g \diamond g')(\phi \diamond \phi')$$

for $g \in \mathrm{G}_{r,\varepsilon}(\mathbf{A}^\infty_F)$ and $g' \in \mathrm{G}_{r'}(\mathbf{A}^\infty_F)$. Thus, we may assume $g = g' = 1$ in the proposition. The case $r = 0$ follows directly from the construction. In general, by linearity, we may assume $\phi = \phi_1 \otimes \phi_2$ for $\phi_1 \in \Omega(\mu,\varepsilon)\otimes_{M_\mu}\mathbb{C}$ and $\phi_2 \in \mathscr{S}((\mathbf{V}^\infty)^r, \mathbb{C})$. Then we have

$$\mathrm{Z}(\phi)^\natural = \mathrm{Z}(\phi_1)^\natural \diamond \mathrm{Z}(\phi_2)^\natural$$

by the case $r = 0$. Thus, we have

$$\begin{aligned} Z(\phi)^{\natural} \diamond Z(\phi')^{\natural} &= (Z(\phi_1)^{\natural} \diamond Z(\phi_2)^{\natural}) \diamond Z(\phi')^{\natural} \\ &= Z(\phi_1)^{\natural} \diamond (Z(\phi_2)^{\natural} \diamond Z(\phi')^{\natural}) \\ &= Z(\phi_1)^{\natural} \diamond Z(\phi_2 \diamond \phi')^{\natural} \end{aligned}$$

by Proposition 5.1. By the case $r = 0$ again, the above expression equals

$$Z(\phi_1 \diamond \phi_2 \diamond \phi')^{\natural} = Z(\phi \diamond \phi')^{\natural}.$$

The proposition is proved. □

Remark 5.3. We may define similarly $Z^{\natural,\phi'}_{r';\nu}(g') \diamond Z^{\natural,\phi}_{r,\varepsilon;\mu,\nu}(g)$. Then we clearly have

$$Z^{\natural,\phi'}_{r';\nu}(g') \diamond Z^{\natural,\phi}_{r,\varepsilon;\mu,\nu}(g) = Z^{\natural,\phi}_{r,\varepsilon;\mu,\nu}(g) \diamond Z^{\natural,\phi'}_{r';\nu}(g').$$

Then we consider the product of $Z^{\natural,\phi}_{r,\varepsilon;\mu,\nu}$ and $Z^{\natural,\phi'}_{r',-\varepsilon;\mu^{c},\nu}$. Note that we have $M_{\mu^{c}} = M_{\mu}$; that Φ_{μ} and $\Phi_{\mu^{c}}$ are complementary; and that $-\varepsilon$ is μ^{c}-admissible. For $r, r' \geq 0$, we have a natural isomorphism $\mathrm{W}_{r,\varepsilon} \oplus \mathrm{W}_{r',-\varepsilon} \simeq \mathrm{W}_{r+r'+1}$ as follows: Let $\{e'_1, \ldots, e'_r, e'_0, e'_{-1}, \ldots, e'_{-r}\}$, $\{e''_1, \ldots, e''_{r'}, e''_0, e''_{-1}, \ldots, e''_{-r'}\}$, and $\{e_1, \ldots, e_{r+r'+1}, e_{-1}, \ldots, e_{-r+r'+1}\}$ be standard bases of $\mathrm{W}_{r,\varepsilon}$, $\mathrm{W}_{r',-\varepsilon}$, and $\mathrm{W}_{r+r'+1}$, respectively. Then under this isomorphism, $e_{\pm i} = e'_{\pm i}$ for $1 \leq i \leq r$, $e_{\pm(r+i)} = e''_{\pm i}$ for $1 \leq i \leq r'$, $e_{r+r'+1} = e'_0 + e''_0$, and $e_{-(r+r'+1)} = \frac{e'_0 - e''_0}{2\varepsilon}$. Such isomorphism induces a homomorphism

$$\diamond\colon \mathrm{G}_{r,\varepsilon} \times \mathrm{G}_{r',\varepsilon} \to \mathrm{G}_{r+r'+1}.$$

On the other hand, there is a natural isomorphism

$$(\Omega(\mu, \varepsilon) \otimes_{M_{\mu}} \mathbb{C}) \otimes_{\mathbb{C}} (\Omega(\mu^{c}, -\varepsilon) \otimes_{M_{\mu}} \mathbb{C}) \simeq \mathscr{S}(\mathbf{V}^{\infty}, \mathbb{C})$$

of Weil representations of $\mathrm{U}(\mathbf{V})(\mathbb{A}_F^{\infty})$ [Liu, Lemma 5.5], which induces a map

$$\diamond\colon \mathscr{S}((\mathbf{V}^{\infty})^r, \Omega \otimes_{M_{\mu}} \mathbb{C}) \times \mathscr{S}((\mathbf{V}^{\infty})^{r'}, \Omega(\mu^{c}, -\varepsilon) \otimes_{M_{\mu}} \mathbb{C}) \to \mathscr{S}((\mathbf{V}^{\infty})^{r+r'+1}, \mathbb{C}).$$

We then have the relation

$$(\mu, \varepsilon)\omega_{r,\varepsilon;\mu,\nu}(g)\phi \diamond \omega_{r',-\varepsilon;\mu^{c},\nu}(g')\phi' = \omega_{r+r'+1;\nu}(g \diamond g')(\phi \diamond \phi') \tag{5.1}$$

for $g \in \mathrm{G}_{r,\varepsilon}(\mathbb{A}_F^{\infty})$ and $g' \in \mathrm{G}_{r',-\varepsilon}(\mathbb{A}_F^{\infty})$.

It remains to define the operation $\diamond$ on Chow groups. Take an object $(A_{\mu}, i_{\mu}) \in \mathcal{A}(\mu)$. Then we have $(A^{\vee}_{\mu}, i^{\vee}_{\mu}) \in \mathcal{A}(\mu^{c})$ by [Liu, Proposition 4.6]. For every K, we define a map

$$\diamond_K\colon \mathrm{CH}^{r+\dim A_\mu}(\mathrm{Sh}(\mathbf{V})_K\times A_\mu)\times \mathrm{CH}^{r'+\dim A_\mu}(\mathrm{Sh}(\mathbf{V})_K\times A_\mu^\vee)\to \mathrm{CH}^{r+r'+1}(\mathrm{Sh}(\mathbf{V})_K)$$

by the formula

$$Z\diamond_K Z' = q_*(p^*Z\cdot p'^*Z'\cdot(\mathrm{Sh}(\mathbf{V})_K\times\mathcal{P}_\mu)),$$

where $p\colon \mathrm{Sh}(\mathbf{V})_K\times A_\mu\times A_\mu^\vee\to\mathrm{Sh}(\mathbf{V})_K\times A_\mu$, $p'\colon \mathrm{Sh}(\mathbf{V})_K\times A_\mu\times A_\mu^\vee\to \mathrm{Sh}(\mathbf{V})_K\times A_\mu^\vee$, and $q\colon \mathrm{Sh}(\mathbf{V})_K\times A_\mu\times A_\mu^\vee\to\mathrm{Sh}(\mathbf{V})_K$ are natural projections; and $\mathcal{P}_\mu$ is the Poincaré divisor on $A_\mu\times A_\mu^\vee$. It is clear that the maps $\{\diamond_K\}_K$ are compatible with respect to K under pullbacks, hence induce a map

$$\diamond\colon \mathrm{CH}^r_\mu(\mathrm{Sh}(\mathbf{V}))^\natural_\mathbb{C}\times\mathrm{CH}^{r'}_{\mu^c}(\mathrm{Sh}(\mathbf{V}))^\natural_\mathbb{C}\to\mathrm{CH}^{r+r'+1}(\mathrm{Sh}(\mathbf{V}))^\natural_\mathbb{C}.$$

Theorem 5.4. *For $\phi\in\mathscr{S}((\mathbf{V}^\infty)^r,\Omega(\mu,\varepsilon)\otimes_{M_\mu}\mathbb{C})$ and $\phi'\in\mathscr{S}((\mathbf{V}^\infty)^{r'},\Omega(\mu^c,-\varepsilon)\otimes_{M_\mu}\mathbb{C})$, we have*

$$Z^{\natural,\phi}_{r,\varepsilon;\mu,\nu}(g)\diamond Z^{\natural,\phi'}_{r',-\varepsilon;\mu^c,\nu}(g') = Z^{\natural,\phi\diamond\phi'}_{r+r'+1;\nu}(g\diamond g'). \tag{5.2}$$

Remark 5.5. Theorem 5.4 has an amazing consequence. Suppose that $\{\mathrm{Sh}(\mathbf{V})_K\}_K$ are proper. Then $\mathrm{CH}^1(\mathrm{Sh}(\mathbf{V})_K)^\natural_\mathbb{C}=\mathrm{CH}^1(\mathrm{Sh}(\mathbf{V})_K)_\mathbb{C}$ by the injectivity of the Kummer map. We have, for elements $\phi\in\Omega(A_\mu,i_\mu,\varepsilon)^K$ and $\phi'\in\Omega(A_\mu^\vee,i_\mu^\vee,-\varepsilon)^K$, that the "product"

$$Z(\phi)_K\diamond_K Z(\phi')_K\in\mathrm{CH}^1(\mathrm{Sh}(\mathbf{V})_K)_\mathbb{C},$$

which equals $Z(\phi\diamond\phi')_K$ by Theorem 5.4, is a generating series of divisors! This strengthens [Liu, Lemma 5.6].

Remark 5.6. In fact, the "product" $Z(\phi)_K\diamond_K Z(\phi')_K$ is defined for every $\phi\in\Omega(A_\mu,i_\mu,\varepsilon)^K$ and $\phi'\in\Omega(A_\mu^\vee,i_\mu^\vee,\varepsilon')^K$ where $\varepsilon'\in E^{\times-}_{\mu^c}$ might not be $-\varepsilon$. Then in view of Remark 5.5, $Z(\phi)_K\diamond_K Z(\phi')_K$ can be regarded as a generating series of divisors associated to the section $\phi\diamond\phi'$ in a Weil representation of $\mathrm{U}(\mathrm{W}_{0,\varepsilon}\oplus\mathrm{W}_{0,\varepsilon'})(\mathbf{A}_F^\infty)\times\mathrm{U}(\mathbf{V})(\mathbf{A}_F^\infty)$. In this way, we may define arithmetic generating functions on $\mathrm{U}(\mathrm{W})(\mathbf{A}_F^\infty)$ valued in $\mathrm{CH}^1(\mathrm{Sh}(\mathbf{V}))_\mathbb{C}$ for an arbitrary skew-Hermitian space W over E of dimension 2 of signature (1, 1), not necessarily the split one W_1. More generally, one may define arithmetic generating functions on $\mathrm{U}(\mathrm{W})(\mathbf{A}_F^\infty)$ for an arbitrary skew-Hermitian space W over E as long as $\mathrm{W}\otimes_\mathbb{Q}\mathbb{R}$ is split.

To prove Theorem 5.4, we need some preparation. The following lemma reduces the theorem to the case $r=r'=0$.

Lemma 5.7. *Fix E/F, $\mathbf{V}$, (μ,ε), and ν. If Theorem 5.4 holds for $r=r'=0$, then it holds for every r and r'.*

Proof. By (5.1), we may assume $g = g' = 1$ in (5.2). By linearity, we may assume $\phi = \phi_1 \diamond \phi_2$ with $\phi_1 \in \mathscr{S}((\mathbf{V}^\infty)^r, \mathbb{C})$ and $\phi_2 \in \Omega(\mu, \varepsilon) \otimes_{M_\mu} \mathbb{C}$, and $\phi' = \phi'_1 \diamond \phi'_2$ with $\phi'_1 \in \mathscr{S}((\mathbf{V}^\infty)^{r'}, \mathbb{C})$ and $\phi'_2 \in \Omega(\mu^c, -\varepsilon) \otimes_{M_\mu} \mathbb{C}$. Then by Proposition 5.2, we have

$$Z(\phi)^\natural = Z(\phi_1)^\natural \diamond Z(\phi_2)^\natural, \quad Z(\phi')^\natural = Z(\phi'_1)^\natural \diamond Z(\phi'_2)^\natural.$$

By the associativity and commutativity of various "products" $\diamond$, we have

$$\begin{aligned} Z(\phi)^\natural \diamond Z(\phi')^\natural &= (Z(\phi_1)^\natural \diamond Z(\phi_2)^\natural) \diamond (Z(\phi'_1)^\natural \diamond Z(\phi'_2)^\natural) \\ &= Z(\phi_1)^\natural \diamond (Z(\phi_2)^\natural \diamond Z(\phi'_1)^\natural) \diamond Z(\phi'_2)^\natural \\ &= Z(\phi_1)^\natural \diamond (Z(\phi'_1)^\natural \diamond Z(\phi_2)^\natural) \diamond Z(\phi'_2)^\natural \\ &= (Z(\phi_1)^\natural \diamond Z(\phi'_1)^\natural) \diamond (Z(\phi_2)^\natural \diamond Z(\phi'_2)^\natural), \end{aligned}$$

which, by Proposition 5.1 and the case for $r = r' = 0$, equals

$$Z(\phi_1 \diamond \phi'_1)^\natural \diamond Z(\phi_2 \diamond \phi'_2)^\natural,$$

which equals $Z(\phi \diamond \phi')^\natural$ by Proposition 5.2 again. The lemma follows. □

For the case $r = r' = 0$, we need a pullback formula.

Lemma 5.8. *Suppose* $\mathbf{V} = \mathbf{V}_\bullet \oplus \mathrm{V}_\circ \otimes_E \mathbf{A}_E$ *for some totally definite Hermitian space* $\mathrm{V}_\circ$ *over* E. *Take* $K \subseteq \mathrm{U}(\mathbf{V})(\mathbf{A}_F^\infty)$ *and put* $K_\bullet := K \cap \mathrm{U}(\mathbf{V}_\bullet)(\mathbf{A}_F^\infty)$. *For two elements* $\phi \in \Omega(A_\mu, i_\mu, \varepsilon)_{\mathbf{V}}^K$ *and* $\phi' \in \Omega(A_\mu^\vee, i_\mu^\vee, -\varepsilon)_{\mathbf{V}}^K$,[3] *write*

$$\phi \diamond \phi' = \sum_\alpha (\phi_{\bullet\alpha} \diamond \phi'_{\bullet\alpha}) \diamond \phi_{\circ\alpha}$$

as a finite sum, where $\phi_{\bullet\alpha} \in \Omega(A_\mu, i_\mu, \varepsilon)_{\mathbf{V}_\bullet}^{K_\bullet}$, $\phi'_{\bullet\alpha} \in \Omega(A_\mu^\vee, i_\mu^\vee, -\varepsilon)_{\mathbf{V}_\bullet}^{K_\bullet}$, *and* $\phi_{\circ\alpha} \in \mathscr{S}(\mathrm{V}_\circ \otimes_E \mathbf{A}_E^\infty, \mathbb{C})$. *Then we have*

$$\varsigma_{\mathrm{V}_\circ}^*(Z(\phi)_K \diamond_K Z(\phi')_K) = \sum_\alpha \theta(\phi_{\circ\alpha}) \cdot (Z(\phi_{\bullet\alpha})_{K_\bullet} \diamond_{K_\bullet} Z(\phi'_{\bullet\alpha})_{K_\bullet})$$

in $\mathrm{CH}^1(\mathrm{Sh}(\mathbf{V}_\bullet)_{K_\bullet})_\mathbb{C}$, *where* $\varsigma_{\mathrm{V}_\circ} \colon \mathrm{Sh}(\mathbf{V}_\bullet)_{K_\bullet} \to \mathrm{Sh}(\mathbf{V})_K$ *is the induced morphism* (2.1); *and*

$$\theta(\phi_{\circ\alpha}) := \sum_{x \in \mathrm{V}_\circ} e^{-\sum_{\tau \in \Phi_F} \tau((x,x)_\mathrm{V})} \cdot \phi_{\circ\alpha}(x)$$

are theta sums.

[3] Here, we use the subscript $\mathbf{V}$ to indicate that they are associated to $\mathbf{V}$.

Proof. Let $\xi_{\mathbf{V}_\circ}\colon A(\mathbf{V}_\bullet)_{K_\bullet} \to A(\mathbf{V})_K$ be the homomorphism induced from $\varsigma_{\mathbf{V}_\circ}\colon \mathrm{Sh}(\mathbf{V}_\bullet)_{K_\bullet} \to \mathrm{Sh}(\mathbf{V})_K$. Let $\Delta_\bullet\colon A(\mathbf{V}_\bullet)_{K_\bullet} \to A(\mathbf{V}_\bullet)_{K_\bullet} \times A(\mathbf{V}_\bullet)_{K_\bullet}$ and $\Delta\colon A(\mathbf{V})_K \to A(\mathbf{V})_K \times A(\mathbf{V})_K$ be diagonal homomorphisms. Put $\mathcal{Q}_\mu := (\mathrm{T}_\mu^{\mathrm{can},\mathrm{t}} \otimes \mathrm{T}_{\mu^c}^{\mathrm{can}})^*\mathcal{P}_\mu \in \mathrm{CH}^1(A_\mu \times A_{\mu^c})_{\mathbb{C}}$. Then it suffices to show that

$$\xi_{\mathbf{V}_\circ}^*\Delta^*(\phi \times \phi')^*\mathcal{Q}_\mu = \sum_\alpha \theta(\phi_{\circ\alpha}) \cdot \Delta_\bullet^*(\phi_{\bullet\alpha} \times \phi'_{\bullet\alpha})^*\mathcal{Q}_\mu \tag{5.3}$$

holds in $\mathrm{CH}^1(A(\mathbf{V}_\bullet)_{K_\bullet})_{\mathbb{C}}$. Since $\mathcal{Q}_\mu$ is a symmetric line bundle, it suffices to show that (5.3) holds in cohomology. For this, we choose an embedding $\hat{\tau}\colon E \hookrightarrow \mathbb{C}$ which induces $\tau \in \Phi_F$, a τ-nearby Hermitian space $\mathrm{V}_\bullet$ of $\mathbf{V}_\bullet$; and put $\mathrm{V} := \mathrm{V}_\bullet \oplus \mathrm{V}_\circ$ which is a τ-nearby Hermitian space of $\mathbf{V}$. Then the (de Rham) Chern class of $\Delta^*(\phi \times \phi')^*\mathcal{Q}_\mu$ is represented by the Kudla–Milson $(1,1)$-form $\omega(\phi \diamond \phi')$ on $\mathrm{Sh}(\mathrm{V}, \mathrm{h}_{\mathrm{V},\hat{\tau}})_K(\mathbb{C})$ associated to $\phi \diamond \phi'$ (see the discussion in [Liu, Section 5.1]). By the pullback formula

$$\varsigma_{\mathrm{V}_\circ}^*\omega(\phi \diamond \phi') = \sum_\alpha \theta(\phi_{\circ\alpha}) \cdot \omega(\phi_{\bullet\alpha} \diamond \phi'_{\bullet\alpha})$$

of Kudla–Milson forms, the two sides of (5.3) have the same (de Rham) Chern class. Thus, the lemma follows. □

Lemma 5.9. *Suppose $n \geq 4$. Then Theorem 5.4 holds for $r = r' = 0$.*

Proof. By (5.1), we may assume $g = g' = 1$ in (5.2). Note that both sides of (5.2) give a linear functional from $\mathscr{S}(\mathbf{V}^\infty)$ to $\mathrm{CH}^1(\mathrm{Sh}(\mathbf{V}))_{\mathbb{C}}^\natural$ that is $\mathrm{U}(\mathbf{V})(\mathbf{A}_F^\infty)$-equivariant. By [Liu, Lemma 5.6],[4] (5.2) holds up to homological equivalence. Therefore, the difference of the two sides of (5.2) defines a $\mathrm{U}(\mathbf{V})(\mathbf{A}_F^\infty)$-equivariant linear functional from $\mathscr{S}(\mathbf{V}^\infty)$ to $\mathrm{CH}^1(\mathrm{Sh}(\mathbf{V}))_{\mathbb{C}}^{\natural 0}$. For every rational prime ℓ, we choose an embedding $\mathbb{Q}_\ell \hookrightarrow \mathbb{C}$. Then we have an $\mathrm{U}(\mathbf{V})(\mathbf{A}_F^\infty)$-equivariant embedding

$$\begin{aligned}\mathrm{CH}^1(\mathrm{Sh}(\mathbf{V}))_{\mathbb{C}}^{\natural 0} &\hookrightarrow \prod_\ell \mathrm{H}^1(E, \mathrm{H}^1_{\text{ét}}(\mathrm{Sh}(\mathbf{V})_{E^{\mathrm{ac}}}, \mathbb{Q}_\ell)) \otimes_{\mathbb{Q}_\ell} \mathbb{C} \\ &\simeq \prod_\ell \mathrm{H}^1(E, \mathrm{H}^1_{\text{ét}}(\widetilde{\mathrm{Sh}}(\mathbf{V})_{E^{\mathrm{ac}}}, \mathbb{Q}_\ell)) \otimes_{\mathbb{Q}_\ell} \mathbb{C}\end{aligned}$$

by Definition 3.1 and [MR92, Lemma 1] as $n \geq 4$. Thus, if (5.2) fails, then we obtain a nonzero $\mathrm{U}(\mathbf{V})(\mathbf{A}_F^\infty)$-equivariant linear functional from $\mathscr{S}(\mathbf{V}^\infty)$ to some $\omega(\mu', \varepsilon', \chi')$ by [Liu, Proposition 4.13]. However, this is not possible as theta lifts

[4] Although it is assumed that $\mathrm{Sh}(\mathbf{V})$ is proper in [Liu, Section 5], actually the proof of [Liu, Lemma 5.6] works without this assumption.

from $\mathrm{G}_{0,\varepsilon}(\mathbf{A}_F^\infty)$ and $\mathrm{G}_1(\mathbf{A}_F^\infty)$ to $\mathrm{U}(\mathbf{V})(\mathbf{A}_F^\infty)$ are disjoint when $n \geq 4$. Thus, the lemma follows.

□

Proof of Theorem 5.4. By Lemma 5.7, we may assume $r = r' = 0$. By Lemma 5.8 and a similar pullback formula for $Z(\phi \diamond \phi')$ which can be proved by the same argument for [YZZ09, Proposition 3.1], we may assume $n \geq 4$. Then the theorem follows from Lemma 5.9.

□

6 Mixed Arithmetic Theta Lifting

Let $\mathbf{V}$, (μ, ε), $r \geq 0$, and ν be as in the previous discussion. Then for every $\tau \in \Phi_F$, there exists a unique integer $w(\nu_\tau \mu_\tau^{-n})$ such that

$$(\nu_\tau \mu_\tau^{-n})(z) = \left(\frac{z}{\overline{z}}\right)^{w(\nu_\tau \mu_\tau^{-n})}$$

for $z \in \mathbb{C} = E \otimes_{E,\hat{\tau}} \mathbb{C}$ where $\hat{\tau} \in \Phi_\mu$ is the unique element above τ.

Conjecture 6.1 (Modularity of Mixed Arithmetic Generating Functions). *For every element $\phi \in \mathscr{S}((\mathbf{V}^\infty)^r, \Omega(\mu, \varepsilon) \otimes_{M_\mu} \mathbb{C})$, there exists a finite dimensional complex subspace C_ϕ of $\mathrm{CH}^r_\mu(\mathrm{Sh}(\mathbf{V}))^\natural_\mathbb{C}$ such that $Z^{\natural,\phi}_{r,\varepsilon;\mu,\nu}$ takes values in formal series of C_ϕ, is absolutely convergent, and extends to a (C_ϕ-valued) automorphic form on $\mathrm{G}_{r,\varepsilon}(\mathbf{A}_F)$ whose archimedean component is a lowest weight vector in the holomorphic discrete series representation with the Harish-Chandra parameter $\{w(\nu_\tau\mu_\tau^{-n}), w(\nu_\tau\mu_\tau^{-n}) - 1, \ldots, w(\nu_\tau\mu_\tau^{-n}) - 2r \mid \tau \in \Phi_F\}$ with respect to the CM type Φ_μ.*

Remark 6.2. We have the following remarks concerning Conjecture 6.1.

(1) In the conjecture, if the extension to an automorphic form exists, then it must be unique as $\mathrm{G}_{r,\varepsilon}(F)\mathrm{G}_{r,\varepsilon}(\mathbf{A}_F^\infty)$ is dense in $\mathrm{G}_{r,\varepsilon}(\mathbf{A}_F)$.
(2) The conjecture is known when $r = 0$ by Remark 3.8.
(3) We expect that the similar modularity property in the conjecture holds for $\widetilde{Z}^{\natural,\phi}_{r,\varepsilon;\mu,\nu}$ at least when $0 \leq r \leq n-2$.

From now on, we assume Conjecture 6.1 and regard $Z^{\natural,\phi}_{r,\varepsilon;\mu,\nu}$ as an automorphic form, in particular, a function on $\mathrm{G}_{r,\varepsilon}(F)\backslash\mathrm{G}_{r,\varepsilon}(\mathbf{A}_F)$ valued in a finite dimensional subspace of $\mathrm{CH}^r_\mu(\mathrm{Sh}(\mathbf{V}))^\natural_\mathbb{C}$.

Let $\mathscr{A}_{\mathrm{cusp}}(\mathrm{G}_{r,\varepsilon})$ be the space of cuspidal automorphic forms on $\mathrm{G}_{r,\varepsilon}(\mathbf{A}_F)$. For every $f \in \mathscr{A}_{\mathrm{cusp}}(\mathrm{G}_{r,\varepsilon})$ and every $\phi \in \mathscr{S}((\mathbf{V}^\infty)^r, \Omega(\mu, \varepsilon) \otimes_{M_\mu} \mathbb{C})$, we put

$$\Theta^{\natural,\phi}_{\mu,\nu}(f) := \int_{\mathrm{G}_{r,\varepsilon}(F)\backslash\mathrm{G}_{r,\varepsilon}(\mathbf{A}_F)} f(g) \cdot \overline{Z^{\natural,\phi}_{r,\varepsilon;\mu,\nu}(g)}\, dg,$$

where $\mathrm{d}g$ is the Tamagawa measure on $\mathrm{G}_{r,\varepsilon}(\mathbf{A}_F)$, as an element of $\mathrm{CH}^r_\mu(\mathrm{Sh}(\mathbf{V}))^\natural_\mathbb{C}$.

Definition 6.3 (Mixed Arithmetic Theta Lifting). For an irreducible cuspidal automorphic representation π of $\mathrm{G}_{r,\varepsilon}$ realized on $V_\pi \subseteq \mathscr{A}_{\mathrm{cusp}}(\mathrm{G}_{r,\varepsilon})$, we defined its *(mixed) arithmetic theta lifting* to $\mathbf{V}$ to be the complex subspace

$$\Theta^\natural_{\mu,\nu}(V_\pi, \mathbf{V}) := \{\Theta^{\natural,\phi}_{\mu,\nu}(f) \mid f \in V_\pi, \phi \in \mathscr{S}((\mathbf{V}^\infty)^r, \Omega(\mu,\varepsilon) \otimes_{M_\mu} \mathbb{C})\}$$

of $\mathrm{CH}^r_\mu(\mathrm{Sh}(\mathbf{V}))^\natural_\mathbb{C}$.

7 Conjecture on Arithmetic Inner Product Formula for Odd Ranks

Now we assume that $n = 2r + 1$ is odd (with $r \geq 1$). We also assume $F \neq \mathbb{Q}$ for simplicity. In particular, $\{\mathrm{Sh}(\mathbf{V})_K\}_K$ are projective of dimension $2r$. We assume Conjecture 6.1.

Take an irreducible representation π of $\mathrm{G}_{r,\varepsilon}(\mathbf{A}_F)$ realized on $V_\pi \subseteq \mathscr{A}_{\mathrm{cusp}}(\mathrm{G}_{r,\varepsilon})$ such that π_τ is the holomorphic discrete series representation with the Harish-Chandra parameter $\{w(\nu_\tau\mu_\tau^{-n}), w(\nu_\tau\mu_\tau^{-n}) - 1, \ldots, w(\nu_\tau\mu_\tau^{-n}) - 2r\}$ for every $\tau \in \Phi_F$ with respect to the CM type Φ_μ. By the same proof of [Liu11a, Proposition 3.9], we know that $\Theta^\natural_{\mu,\nu}(V_\pi)$ is contained in $\mathrm{CH}^r_\mu(\mathrm{Sh}(\mathbf{V}))^{\natural 0}_\mathbb{C}$.

For every K, there is a (bilinear) Beilinson–Bloch–Poincaré height pairing

$$\langle\, ,\, \rangle^{\mathrm{BBP}}_{\mathrm{Sh}(\mathbf{V})_K, A_\mu} \colon \mathrm{CH}^{r+\dim A_\mu}(\mathrm{Sh}(\mathbf{V})_K \times A_\mu)^0_\mathbb{C} \times \mathrm{CH}^{r+\dim A_\mu}(\mathrm{Sh}(\mathbf{V})_K \times A^\vee_\mu)^0_\mathbb{C} \to \mathbb{C}$$

constructed in [Liu, Section 3.2], conditional on the existence of the Beilinson–Bloch height pairing. Conjecturally, such pairing descends to a pairing

$$\langle\, ,\, \rangle^{\natural,\mathrm{BBP}}_{\mathrm{Sh}(\mathbf{V})_K, A_\mu} \colon \mathrm{CH}^{r+\dim A_\mu}(\mathrm{Sh}(\mathbf{V})_K \times A_\mu)^{\natural 0}_\mathbb{C} \times \mathrm{CH}^{r+\dim A_\mu}(\mathrm{Sh}(\mathbf{V})_K \times A^\vee_\mu)^{\natural 0}_\mathbb{C} \to \mathbb{C}.$$

Assuming this, then by restriction, we obtain a bilinear pairing

$$\langle\, ,\, \rangle^{\natural,\mathrm{BBP}}_{\mathrm{Sh}(\mathbf{V})_K, \mu} \colon \mathrm{CH}^r_\mu(\mathrm{Sh}(\mathbf{V})_K)^{\natural 0}_\mathbb{C} \times \mathrm{CH}^r_{\mu^\mathtt{c}}(\mathrm{Sh}(\mathbf{V})_K)^{\natural 0}_\mathbb{C} \to \mathbb{C},$$

which is independent of the choice of $(A_\mu, i_\mu) \in \mathcal{A}(\mu)$.

Let $\pi^\vee$ be the contragredient representation of π, realized on $V_{\pi^\vee} \subseteq \mathscr{A}_{\mathrm{cusp}}(\mathrm{G}_{r,\varepsilon})$, the complex conjugation space of V_π. Take a quadruple $(f, f^\vee; \phi, \phi^\vee)$ where $f \in V_\pi$, $f^\vee \in V_{\pi^\vee}$, $\phi \in \mathscr{S}((\mathbf{V}^\infty)^r, \Omega(\mu,\varepsilon) \otimes_{M_\mu} \mathbb{C})$ and $\phi^\vee \in \mathscr{S}((\mathbf{V}^\infty)^r, \Omega(\mu^\mathtt{c}, -\varepsilon) \otimes_{M_\mu} \mathbb{C})$. For every K such that both f and $f^\vee$ are K-invariant and both ϕ and $\phi^\vee$ are strongly K-invariant (Definition 4.1), we then obtain a complex number

$$\mathrm{vol}(K)\cdot\langle\Theta^{\natural,\phi}_{\mu,\nu}(f),\Theta^{\natural,\phi^\vee}_{\mu^c,\nu^c}(f^\vee)\rangle^{\natural,\mathrm{BBP}}_{\mathrm{Sh}(\mathbf{V})_K,\mu},$$

where $\mathrm{vol}(K)$ is the canonical volume of K [Liu, Definition 4.22]. It is easy to see that they are independent of the choice of K. Thus, it makes sense to denote the common value by

$$\langle\Theta^{\natural,\phi}_{\mu,\nu}(f),\Theta^{\natural,\phi^\vee}_{\mu^c,\nu^c}(f^\vee)\rangle^{\natural,\mathrm{BBP}}_{\mathrm{Sh}(\mathbf{V}),\mu}.$$

Conjecture 7.1 (Arithmetic Inner Product Formula). *Let notation be as above. In particular, $F\neq\mathbb{Q}$ and $n=2r+1\geq 3$ is odd. Suppose that $(f=\otimes_v f_v, f^\vee=\otimes_v f_v^\vee;\phi=\otimes_v\phi_v,\phi^\vee=\otimes_v\phi_v^\vee)$ is a decomposable quadruple. Then we have*

$$\langle\Theta^{\natural,\phi}_{\mu,\nu}(f),\Theta^{\natural,\phi^\vee}_{\mu^c,\nu^c}(f^\vee)\rangle^{\natural,\mathrm{BBP}}_{\mathrm{Sh}(\mathbf{V}),\mu}=\frac{L'(\frac{1}{2},\mathrm{BC}(\pi)\otimes\nu)}{\prod_{i=1}^n L(i,\mu^i_{E/F})}\prod_v Z^*(0,\nu_v,f_v,f_v^\vee,\Phi_v).$$

Here, $\mathrm{BC}(\pi)$ denotes the standard base change to $\mathrm{Res}_{E/F}\mathrm{GL}_n$; and $Z^*(0,\nu_v,f_v,f_v^\vee,\Phi_v)$ is the normalized local zeta integral (with respect to a decomposition of the Tamagawa measure) appearing in the usual Rallis inner product formula for which we take the Schwartz function Φ_v to be $\phi_v\diamond\phi_v^\vee$ (resp. the standard Gaussian function on $\mathbb{C}^n$) for v nonarchimedean (resp. archimedean), which equals 1 for all but finitely many v.

Remark 7.2. We have the following remarks concerning Conjecture 7.1.

(1) Since n is odd, ν has to be a conjugate symplectic character in the terminology of [GGP12]. Thus, $\mathrm{BC}(\pi)\otimes\nu$ is of symplectic type.
(2) We may always take $\nu=\mu$ after replacing π by $\pi\otimes(\nu\mu^{-1})^\flat$. When $\nu=\mu$, the Harish-Chandra parameter of π_τ should be $\{r,r-1,\dots,-r\}$ for every $\tau\in\Phi_F$.
(3) By the local theta dichotomy (see [HKS96, GG11] and [SZ15]), there exists a totally positive definite Hermitian space $\mathbf{V}$ over $\mathbf{A}_E$ of rank n, unique up to isomorphism, such that $\prod_v Z^*(0,\nu_v,f_v,f_v^\vee,\Phi_v)\neq 0$ for some choice of $(f,f^\vee;\phi,\phi^\vee)$. Moreover, such Hermitian space $\mathbf{V}$ is coherent (resp. incoherent) if $\epsilon(\frac{1}{2},\mathrm{BC}(\pi)\otimes\nu)=1$ (resp. $\epsilon(\frac{1}{2},\mathrm{BC}(\pi)\otimes\nu)=-1$). Thus the conjecture implies that, under the condition $\epsilon(\frac{1}{2},\mathrm{BC}(\pi)\otimes\nu)=-1$, if $L'(\frac{1}{2},\mathrm{BC}(\pi)\otimes\nu)\neq 0$, then $\Theta^\natural_{\mu,\nu}(V_\pi,\mathbf{V})$ is nontrivial for exactly one $\mathbf{V}$; the converse is also expected to hold.
(4) In fact, we further conjecture that

$$\sum_{\mathbf{V}}\dim\Theta^\natural_{\mu,\nu}(V_\pi,\mathbf{V})=\begin{cases}1, & \text{if } \mathrm{ord}_{s=\frac{1}{2}}L(s,\mathrm{BC}(\pi)\otimes\nu)=1,\\ 0, & \text{otherwise,}\end{cases}$$

where the sum is taken over all isomorphism classes of totally positive definite incoherent Hermitian spaces over $\mathbf{A}_E$ of rank n.

(5) If one believes Remark 6.2(3), then one can define $\widetilde{\Theta}^{\natural,\phi}_{\mu,\nu}(f)$ and $\widetilde{\Theta}^{\natural}_{\mu,\nu}(V_\pi, \mathbf{V})$, and formulate Conjecture 7.1 without assuming $F \neq \mathbb{Q}$.
(6) The (conjectural) arithmetic inner product formula when $n = 2r$ is even has been formulated in [Liu11a] using generating functions $Z^{\phi}_{r;\nu}$, and was confirmed when $n = 2$ in [Liu11b].

Acknowledgments The author would like to thank the Simons Foundation and the organizers for inviting him to the Simons Symposium *Relative Trace Formulas* in April 2018 in Germany, from which the work has been benefited. He also thanks the anonymous referee for valuable suggestions to improve the article. The research of the author is partially supported by NSF grant DMS–1702019 and a Sloan Research Fellowship.

References

[DMOS82] P. Deligne, J.S. Milne, A. Ogus, K.-Y Shih, *Hodge Cycles, Motives, and Shimura Varieties*. Lecture Notes in Mathematics, vol. 900 (Springer, Berlin, 1982). MR654325 ↑3

[GGP12] W.T. Gan, B.H. Gross, D. Prasad, Symplectic local root numbers, central critical L values, and restriction problems in the representation theory of classical groups. Astérisque **346**, 1–109 (2012) (English, with English and French summaries). Sur les conjectures de Gross et Prasad. I. MR3202556 ↑16

[GG11] Z. Gong, L. Grenié, An inequality for local unitary theta correspondence. Ann. Fac. Sci. Toulouse Math. (6) **20**(1), 167–202 (2011) (English, with English and French summaries). MR2830396 ↑16

[Gro] B.H. Gross, Incoherent definite spaces and Shimura varieties, in *Relative Trace Formulas* (Springer, Cham, 2021). https://doi.org/10.1007/978-3-030-68506-5. ↑3

[HKS96] M. Harris, S.S. Kudla, W.J. Sweet, Theta dichotomy for unitary groups. J. Am. Math. Soc. **9**(4), 941–1004 (1996). MR1327161 ↑16

[Kud97] S.S. Kudla, Algebraic cycles on Shimura varieties of orthogonal type. Duke Math. J. **86**(1), 39–78 (1997). https://doi.org/10.1215/S0012-7094-97-08602-6. MR1427845 ↑1

[Kud02] S.S. Kudla, Derivatives of Eisenstein series and arithmetic geometry, in *Proceedings of the International Congress of Mathematicians*, vol. II (Higher Ed. Press, Beijing, 2002), pp. 173–183. MR1957031 ↑1

[Kud04] S.S. Kudla, *Special Cycles and Derivatives of Eisenstein Series*. Heegner Points and Rankin L-Series, Math. Sci. Res. Inst. Publ., vol. 49 (Cambridge Univ. Press, Cambridge, 2004), pp. 243–270. https://doi.org/10.1017/CBO9780511756375.009. MR2083214 ↑1

[KR94] S.S. Kudla, S. Rallis, A regularized Siegel-Weil formula: the first term identity. Ann. Math. (2) **140**(1), 1–80 (1994). https://doi.org/10.2307/2118540. MR1289491 ↑1

[Liu11a] Y. Liu, Arithmetic theta lifting and L-derivatives for unitary groups, I. Algebra Number Theory **5**(7), 849–921 (2011). MR2928563 ↑1, 2, 9, 15, 16

[Liu11b] Y. Liu, Arithmetic theta lifting and L-derivatives for unitary groups, II. Algebra Number Theory **5**(7), 923–1000 (2011). MR2928564 ↑2, 16

[Liu] Y. Liu, Fourier–Jacobi cycles and arithmetic relative trace formula (with an appendix by Chao Li and Yihang Zhu) (Preprint). https://arxiv.org/abs/2102.11518. ↑2, 4, 5, 6, 7, 11, 12, 13, 14, 15

[MR92] V.K. Murty, D. Ramakrishnan, The Albanese of unitary Shimura varieties, in *The Zeta Functions of Picard Modular Surfaces* (Univ. Montréal, Montreal, 1992), pp. 445–464. MR1155237 ↑14

[Ral87] S. Rallis, *L-functions and the Oscillator Representation*. Lecture Notes in Mathematics, vol. 1245 (Springer, Berlin, 1987). MR887329 ↑1

[SZ15] B. Sun, C.-B. Zhu, Conservation relations for local theta correspondence. J. Am. Math. Soc. **28**(4), 939–983 (2015). MR3369906 ↑16

[YZZ09] X. Yuan, S.-W. Zhang, W. Zhang, The Gross-Kohnen-Zagier theorem over totally real fields. Compos. Math. **145**(5), 1147–1162 (2009). https://doi.org/10.1112/S0010437X08003734. MR2551992 ↑10, 14

Twists of $GL(3)$ L-functions

Ritabrata Munshi

Abstract Let π be a $SL(3,\mathbb{Z})$ Hecke-Maass cusp form, and let χ be a primitive Dirichlet character modulo M, which we assume to be prime. We establish the following unconditional subconvex bound for the twisted L-function:

$$L\left(\tfrac{1}{2},\pi\otimes\chi\right) \ll_{\pi,\varepsilon} M^{3/4-1/308+\varepsilon}.$$

2010 *Mathematics Subject Classification* 11F66

1 Introduction

In this paper we return to the subconvexity problem addressed in [Mu15b], with the aim of presenting an argument, a variation of the $GL(2)$ delta method introduced in [Mu15b], which is more transparent and technically much simpler. As an advantage we are now able to write down an explicit subconvex exponent. More importantly the present argument, unlike [Mu15b], does not rely on the Ramanujan-Selberg conjecture for $GL(3)$ automorphic forms. We will prove the following unconditional subconvexity result.

Theorem 1. *Suppose π is a $SL(3,\mathbb{Z})$ Hecke-Maass cusp form, and χ is a primitive Dirichlet character modulo M (which we assume to be prime). Then we have*

$$L\left(\tfrac{1}{2},\pi\otimes\chi\right) \ll M^{3/4-1/308+\varepsilon}. \tag{1}$$

Recall that the exponent $3/4+\varepsilon$ corresponds to the convexity bound. In general we will stick to the notations used in [Mu15b]. The reader may refer to that paper for a broader introduction to the problem and for basic definitions (also see [Go]). The analogues of our result in the case of degree one L-function was obtained by

R. Munshi (✉)
Statistics and Mathematics Unit, Indian Statistical Institute, Kolkata, India

W. Müller et al. (eds.), *Relative Trace Formulas*, Simons Symposia,
https://doi.org/10.1007/978-3-030-68506-5_11

Burgess [Bu], and the degree two case (for modular forms) was first tackled by Duke, Friedlander and Iwaniec [DFI]. This paper unconditionally settles the degree three case.

Our analysis should work even when M is not a prime, but for simplicity we only supply details for M prime. From the technical point of view this paper puts the $GL(2)$ delta method in a better perspective. We explain how one can introduce the convenient 'congruence-equation trick' (see [BrMu, Mu15a, Mu15c]) in this set up. Also we introduce a trick for splitting the moduli, post reciprocity, which could be useful elsewhere. These two tricks allow us to avoid intertwining two applications of $GL(3)$ Voronoi summation with reciprocity, which was a major technicality in the earlier paper [Mu15b].

2 Petersson Trace Formula and $GL(2)$ δ-Method

We start by recalling the $GL(2)$ delta method, which is an upshot of the Petersson trace formula. The delta method of [DFI], in contrast, can be termed as the $GL(1)$ delta method, as it only utilizes the harmonics of the circle group, i.e. the trigonometric functions $e^{2\pi ix}$, to get a Fourier expansion of the delta function $\delta : \mathbb{Z} \mapsto \{0, 1\}$,

$$\delta(n) = \begin{cases} 1 & \text{if } n = 0 \\ 0 & \text{otherwise.} \end{cases}$$

As such the delta method of [DFI] can be termed as a refinement of the classical circle method of Ramanujan-Hardy-Littlewood. Whereas the $GL(1)$ delta method has been extensively used to study problems in analytic theory of L-functions (e.g. to tackle the shifted convolution sum problem), it can hardly be used to count rational points on varieties beyond degree three (see [H-B]) due to the natural constraint—the modulus of harmonics have to be as large as the square root of the size of the equation. One way to overcome this constraint will be to utilize harmonics from higher rank groups like $GL(2)$. Indeed the relative trace formula of Petersson (and Kuznetsov) can be employed to derive such an expansion of the delta function, as we will see below. (In general, one may use the Kuznetsov trace formula for $GL(n)$ to derive expansions of the delta function in terms of higher rank harmonics. To this end one has to first make the Kuznetsov formula 'explicit'. Recent works of Blomer [Bl] and Buttcane [But], at least give such an explicit formula for $GL(3)$.)

Let p be a prime number and let $k \equiv 3 \bmod 4$ be a positive integer (which will be of the size $1/\varepsilon$). Let ψ be a character of $\mathbb{F}_p^\times$ satisfying $\psi(-1) = -1 = (-1)^k$. We consider ψ as a character modulo pM. The main novelty in this paper is the use of the space $S_k(pM, \psi)$ (of cusp forms of level pM, nebentypus ψ and weight k) in place of $S_k(p, \psi)$, for the $GL(2)$ δ-method. The inclusion of M, which is the conductor of χ, in the level is an analogue of the 'congruence-equation trick' which was used in [Mu15a] in the context of the usual delta method. (This trick has turned out to be useful in other problems as well, e.g. [BrMu, Mu15c].) Let $H^\star(pM,\psi)$ be the set of Hecke-normalized newforms and we extend it to $H_k(pM,\psi)$—an orthogonal Hecke basis of the space of cusp forms $S_k(pM,\psi)$. Let P be a parameter, which will be optimally chosen later, and let

$$P^\star = \sum_{\substack{P<p<2P\\ p \text{ prime}}} \sum_{\psi \bmod p} (1-\psi(-1)),$$

which is of size P^2. The Petersson trace formula for the space $S_k(pM,\psi)$ states

$$\sum_{f\in H_k(pM,\psi)} \omega_f^{-1}\lambda_f(n)\overline{\lambda_f(r)}=\delta(n,r)+2\pi i^{-k}\sum_{c=1}^{\infty}\frac{S_\psi(r,n;cpM)}{cpM}J_{k-1}\left(\frac{4\pi\sqrt{nr}}{cpM}\right).$$

Here $\delta(n,r)=\delta(n-r)$, $\lambda_f(n)$ are the normalized Fourier coefficients of the form f, ω_f^{-1} is the spectral weight, $S_\psi(a,b;c)$ is the generalized Kloosterman sum and $J_{k-1}(x)$ is the Bessel function (see [IwKo]). Using the Petersson formula we derive

$$\delta(n,r)=\frac{1}{P^\star}\sum_{\substack{P<p<2P\\ p \text{ prime}}}\sum_{\psi \bmod p}(1-\psi(-1))\sum_{f\in H_k(pM,\psi)}\omega_f^{-1}\lambda_f(n)\overline{\lambda_f(r)} \tag{1}$$

$$-\frac{2\pi i}{P^\star}\sum_{\substack{P<p<2P\\ p \text{ prime}}}\sum_{c=1}^{\infty}\frac{1}{cpM}\sum_{\psi \bmod p}(1-\psi(-1))\,S_\psi(r,n;cpM)J_{k-1}\left(\frac{4\pi\sqrt{nr}}{cpM}\right),$$

i.e. the right hand side is 1 if $n=r$, and is equal to 0 otherwise.

Let $\mathcal{L}$ be a set of primes in the range $[L,2L]$, with $L\ll\min\{P^{1-\varepsilon},M^{1-\varepsilon}\}$ and $|\mathcal{L}|=:L^\star\gg L^{1-\varepsilon}$. Consider the sum

$$\mathcal{F}=\frac{1}{L^\star P^\star}\sum_{\substack{P<p<2P\\ p \text{ prime}}}\sum_{\psi \bmod p}(1-\psi(-1))\sum_{f\in H_k(pM,\psi)}\omega_f^{-1} \tag{2}$$

$$\times\sum_{\ell\in\mathcal{L}}\bar{\chi}(\ell)\sum\sum_{m,n=1}^{\infty}\lambda(m,n)\lambda_f(n\ell)W\left(\frac{nm^2}{N}\right)\sum_{r=1}^{\infty}\overline{\lambda_f(r)}\chi(r)V\left(\frac{r}{N\ell}\right).$$

Here W is a smooth bump function with support $[1,2]$ and V is a smooth function supported in $[M^{-4\theta},4]$, with $V(x)=1$ for $x\in[2M^{-4\theta},2]$, and satisfying $y^jV^{(j)}(y)\ll_j 1$. (Here $0<\theta<1/2$ is a parameter to be chosen optimally later.) Using the Petersson formula we see that the diagonal term is exactly the sum of interest

$$S^\star(N)=\sum_{m,n=1}^{\infty}\sum\lambda(m,n)\chi(n)W\left(\frac{nm^2}{N}\right)V\left(\frac{n}{N}\right).$$

In Lemma 6 of Section 2 of [Mu15b] we showed that

$$L(\tfrac{1}{2},\pi\otimes\chi)\ll M^{\varepsilon}\sup_{N}\frac{|S^\star(N)|}{\sqrt{N}}+M^{3/4-\theta/2+\varepsilon},\tag{3}$$

where the supremum is taken over N in the range $M^{3/2-\theta}<N<M^{3/2+\theta}$. The weight function W appears after taking a smooth dyadic subdivision of the approximate functional equation. In fact from (1) it follows that

$$S^\star(N)=\mathcal{F}-2\pi i\,\mathcal{O}\tag{4}$$

where the off-diagonal is given by

$$\mathcal{O}=\frac{1}{L^\star P^\star}\sum_{\substack{P<p<2P\\ p\text{ prime}}}\sum_{\psi\bmod p}(1-\psi(-1))\sum_{m,n=1}^{\infty}\sum\lambda(m,n)W\left(\frac{nm^2}{N}\right)\tag{5}$$

$$\times\sum_{\ell\in\mathcal{L}}\sum_{r=1}^{\infty}\chi(r\bar{\ell})V\left(\frac{r}{N\ell}\right)\sum_{c=1}^{\infty}\frac{S_{\psi}(r,n\ell;cpM)}{cpM}J_{k-1}\left(\frac{4\pi\sqrt{n\ell r}}{cpM}\right).$$

The set of primes $\mathcal{L}$ plays a subtle role. This has been introduced to split the modulus, at one stage, as a product of two numbers. This is not completely apparent in the beginning as it takes up the role of a modulus only after an application of reciprocity. The splitting of the modulus is used at the last application of Cauchy inequality followed by the Poisson summation. One will notice that putting the whole modulus inside makes the modulus for Poisson too large for any saving, whereas putting the whole modulus outside makes the diagonal too small. The situation is similar to one that we faced in [Mu14], where we employed Jutila's version of the circle method to split the modulus.

Theorem 1 will follow from the following two propositions.

Proposition 1. *Let $\mathcal{O}$ be as defined in* (5). *Suppose $L<P$ and $\theta<1/2$. We have*

$$\mathcal{O}\ll N^{1/2}M^{\varepsilon}\left\{\frac{M^{3/4+3\theta/2}L}{P}+\frac{M^{1+\theta}}{P}\right\}.$$

Proposition 2. *If $M^{2\theta+\varepsilon}\ll L\ll M^{-\varepsilon}\min\{M^2/N,\sqrt{P}\}$, we have*

$$\mathcal{F}\ll M^{\varepsilon}N^{1/2}\left[P^{1/2}M^{1/2+9\theta}+\frac{P^{1/4}M^{5/8+17\theta/4}}{L^{1/2}}\left(L^{3/4}+\frac{M^{11\theta/4}P^{3/4}}{M^{1/8}}\right)\right]$$

Proof of Theorem 1. Plugging the bounds from the above propositions into (4) and (3) we obtain

$$L(\tfrac{1}{2}, \pi \otimes \chi) \ll M^{\varepsilon} \left\{ P^{1/2} M^{1/2+9\theta} + \frac{P^{1/4} M^{5/8+17\theta/4}}{L^{1/2}} \left(L^{3/4} + \frac{M^{11\theta/4} P^{3/4}}{M^{1/8}} \right) \right\}$$
$$+ M^{\varepsilon} \left\{ \frac{M^{3/4+3\theta/2} L}{P} + \frac{M^{1+\theta}}{P} \right\} + M^{3/4-\theta/2+\varepsilon}.$$

Then we choose the three parameters—P, L and θ—to minimize

$$\frac{P^{1/4} M^{5/8+17\theta/4+\varepsilon}}{L^{1/2}} \left(L^{3/4} + \frac{M^{11\theta/4} P^{3/4}}{M^{1/8}} \right) + \frac{M^{1+\theta}}{P} + M^{3/4-\theta/2+\varepsilon}. \tag{6}$$

The optimum choice of L is obtained by equating the first two terms. This gives $L = PM^{-1/6+11\theta/3}$, and reduces the above bound to

$$M^{7/12+31\theta/6+\varepsilon} P^{1/2} + \frac{M^{1+\theta}}{P} + M^{3/4-\theta/2+\varepsilon}.$$

Equating the first two terms we now get the optimum choice for P, which turns put to be $P = M^{5/18-25\theta/9}$. Ultimately we find that the optimum choice of θ is given by $\theta = 1/154$. It follows that the expression in (6) is bounded by $O(M^{3/4-1/308+\varepsilon})$. Then we check that this also dominates the remaining two terms in the bound for $L(1/2, \pi \otimes \chi)$ above. This completes the proof of the theorem. □

Notation. Suppose $\mathcal{A} \ll M^{\varepsilon} \sum_{b \in \mathcal{F}} |\mathcal{B}_b| + M^{-2016}$ where $|\mathcal{F}| \ll M^{\varepsilon}$ and the implied constants depend only on ε. Then we write

$$\mathcal{A} \lhd_{\mathcal{F}} \mathcal{B}_b, \quad \text{or simply as} \quad \mathcal{A} \lhd \mathcal{B},$$

where there is no scope of confusion.

3 Outline of the Proof

In this section we give a brief outline of the proof. Let $N = M^{3/2}$ and temporarily assume the Ramanujan conjecture $|\lambda(m,n)| \ll (mn)^{\varepsilon}$. First consider the off-diagonal term as given in (5). For convenience assume that $m = 1$, and that c is in the transition range, i.e. $c \sim C = NL/PM = M^{1/2}L/P$. Consider the generic case $p \nmid c$. The character sum

$$\sum_{\psi \bmod p} (1 - \psi(-1))\, S_{\psi}(r, n\ell; cpM)$$

can be partially evaluated, and one gets

$$pS(\bar{p}r, \bar{p}n\ell; cM)e\left(\pm\frac{\overline{cM}(r+n\ell)}{p}\right).$$

Applying the reciprocity relation we see that the sum (5) is essentially given by

$$\frac{1}{NL^2P}\sum_{\substack{P<p<2P\\ p\text{ prime}}}\sum_{n\sim N}\lambda(1,n)\sum_{\ell\in\mathcal{L}}\sum_{r\sim NL}\chi(r\bar{\ell})\sum_{c\sim C}S(\bar{p}r, \bar{p}n\ell; cM)e\left(\frac{\overline{p}(r+n\ell)}{cM}\right).$$

Notice the presence of M in the modulus cM. This acts as a conductor lowering trick as in [Mu15a].

Assuming the Weil bound, we see that the Petersson formula gives a saving of size $\sqrt{PM}/\sqrt{C}$, and in addition we have saved $\sqrt{P}$ in the sum over ψ. Next we apply the Poisson summation formula on the r sum. Since the length of the sum $r \sim NL$ is larger than the modulus $cM \sim NL/P$, we are only left with the zero frequency. Hence from Poisson we save $\sqrt{CM}$. Our initial target was to save NL, and so far we have saved PM. So now it remains to save $M^{1/2}L/P$, in the sum

$$\sum_{\substack{P<p<2P\\ p\text{ prime}}}\sum_{n\sim N}\lambda(1,n)\sum_{\ell\in\mathcal{L}}\sum_{c\sim C}\chi(pc\bar{\ell})\mathfrak{D}(\bar{p}\bar{c}n\ell; M)$$

where the character sum $\mathfrak{D}$ is as given in (1). We apply the Cauchy inequality and reduce the problem to that of saving ML^2/P^2 in the sum

$$\sum_{n\sim N}\Bigl|\sum_{\substack{P<p<2P\\ p\text{ prime}}}\sum_{\ell\in\mathcal{L}}\sum_{c\sim C}\chi(pc\bar{\ell})\mathfrak{D}(\bar{p}\bar{c}n\ell; M)\Bigr|^2.$$

Next we open the absolute square (after smoothing) and apply the Poisson summation on the sum over n with modulus M. Only the zero frequency makes a sizeable contribution, which we need to estimate. In the diagonal we at most save $PLC \sim M^{1/2}L^2$ (which will be smaller than the modulus M) and in the off-diagonal we save M. Crucially the structure of the character sum $\mathfrak{D}$ is such that for the zero-th frequency the saving is the full modulus M and not just the square-root of the modulus. It turns out that the off-diagonal $\mathcal{O}$ is fine if $P \gg \max\{L, M^{1/4}\}$. This is the content of Proposition 1. Note that we do not require to utilize the oscillation in the Fourier coefficients $\lambda(1,n)$ (as in [Mu15b]).

Next we consider the sum $\mathcal{F}$ as given in (2). This is essentially given by

$$\frac{1}{LP^2}\sum_{\substack{P<p<2P\\ p \text{ prime}}}\sum_{\psi \bmod p}(1-\psi(-1))\sum_{f\in H_k(pM,\psi)}\omega_f^{-1}$$
$$\times\sum_{\ell\in\mathcal{L}}\bar{\chi}(\ell)\lambda_f(\ell)\sum_{n\sim N}\lambda(1,n)\lambda_f(n)\sum_{r\sim NL}\overline{\lambda_f(r)}\chi(r).$$

We apply $GL(3)\times GL(2)$ Rankin-Selberg functional equation to the n sum and $GL(2)$ Voronoi summation to the r sum. We save $N/(P^3M^3)^{1/2}$ in the n-sum and save $NL/MP^{1/2}$ in the Voronoi summation. As initially we needed to save NL, it follows that we now need to save MP^2 in the sum

$$\sum_{\substack{P<p<2P\\ p \text{ prime}}}\sum_{\psi \bmod p}\chi(p)\,(1-\psi(-1))\,\psi(M)g_\psi^2\sum_{f\in H_k(pM,\psi)}\omega_f^{-1}$$
$$\times\sum_{n\sim NP^3}\lambda(1,n)\bar{\lambda}_f(p^2Mn)\sum_{\ell\in\mathcal{L}}\sum_{r\sim M^{1/2}P/L}\lambda_f(r\ell)\bar{\chi}(r\ell).$$

Observe that r and ℓ occur together and it appears as if we have split the dual variable into a product of two variables, whose individual sizes can be regulated by choosing L appropriately. We apply the Petersson formula to arrive at

$$\sum_{\substack{P<p<2P\\ p \text{ prime}}}\sum_{\psi \bmod p}\chi(p)\,(1-\psi(-1))\,\psi(M)g_\psi^2$$
$$\times\sum_{n\sim NP^3}\lambda(1,n)\sum_{\ell\in\mathcal{L}}\sum_{r\sim M^{1/2}P/L}\bar{\chi}(r\ell)\sum_{c\sim C}S_\psi(p^2Mn,r\ell;cpM),$$

where the new transition range is given by $C = M^{1/2}P^2$. We have saved $\sqrt{PM}/\sqrt{C}$ (assuming Weil) from Petersson and we now need to save $M^{3/4}P^{5/2}$. The Kloosterman sum splits as

$$S(Mn,r\ell;cM)S_\psi(0,r\ell\overline{cM};p).$$

The ψ sum now gives a saving of size $P^{1/2}$ and the $GL(3)$ Voronoi summation gives a saving of size $M^{3/4}$. (Notice that the modulus is just c.) Also the $GL(3)$ Voronoi transforms the Kloosterman sum into an additive character. Our job reduces to saving P^2 in the sum

$$\sum_{\substack{P<p<2P\\ p \text{ prime}}}\chi(p)\sum_{\ell\in\mathcal{L}}\sum_{r\sim M^{1/2}P/L}\sum_{c\sim C}\bar{\chi}(r\ell)e\left(\frac{c\overline{r\ell}}{p}\right)\sum_{n\sim P^3}\lambda(n,1)e\left(-\frac{\overline{r\ell}Mn}{c}\right).$$

Applying reciprocity we obtain

$$\sum_{\substack{P<p<2P\\ p\text{ prime}}} \chi(p) \sum_{\ell\in\mathcal{L}} \sum_{r\sim M^{1/2}P/L} \sum_{c\sim C} \bar{\chi}(r\ell) e\left(-\frac{c\bar{p}}{r\ell}\right) \sum_{n\sim P^3} \lambda(n,1) e\left(\frac{\bar{c}Mn}{r\ell}\right).$$

We can now apply the Poisson summation on the sum over c. The length of the sum is $M^{1/2}P^2$ and the modulus is $r\ell \sim M^{1/2}P$. So we are just left with the zero frequency and get a saving of $M^{1/4}P^{1/2}$. It remains to save $P^{3/2}/M^{1/4}$ in the sum

$$\sum_{\substack{P<p<2P\\ p\text{ prime}}} \chi(p) \sum_{\ell\in\mathcal{L}} \sum_{r\sim M^{1/2}P/L} \bar{\chi}(r\ell) \sum_{n\sim P^3} \lambda(n,1) S(Mn,-\bar{p};r\ell).$$

Applying Cauchy we see that we need to save $P^3/M^{1/2}$ in the sum

$$\sum_{n\sim P^3} \sum_{r\sim M^{1/2}P/L} \Big| \sum_{\substack{P<p<2P\\ p\text{ prime}}} \sum_{\ell\in\mathcal{L}} \chi(p\bar{\ell})\, \lambda(n,1) S(Mn,-\bar{p};r\ell)\Big|^2.$$

The diagonal is fine if $PL > P^3/M^{1/2}$ or $L > P^2/M^{1/2}$. We save $P^3/M^{1/4}(PL)^{1/2}$ in the off-diagonal, which is fine if $L < M^{1/2}/P$. This is the content of Proposition 2. In particular we have a choice for L as long as $P < M^{1/3}$. Recall that the off-diagonal $\mathcal{O}$ was fine if $P > M^{1/4}$. Consequently we have a choice for the parameters P and L such that we have subconvex bounds for both $\mathcal{F}$ and $\mathcal{O}$.

4 The Off-Diagonal

The off-diagonal contribution $\mathcal{O}$, is analyzed in the same spirit as Section 4 of [Mu15b]. For notational convenience we will only consider the sub-sum of (5) where $p \nmid c$, which will be denoted by $\mathcal{O}_0$. The other case can be tackled in the same fashion (by extracting the highest power p dividing c), and we at the end get a stronger bound for that sum. Let

$$\mathfrak{D}(u;M) = \sum_{\substack{b \bmod M\\ (b(b-1),M)=1}} \bar{\chi}(b-1) e\left(\frac{(\bar{b}-1)u}{M}\right), \tag{1}$$

and

$$\mathfrak{I}(n,\ell;v) = \int_{\mathbb{R}} e\left(\frac{N\ell y + n\ell}{v}\right) J_{k-1}\left(\frac{4\pi\sqrt{Nn\ell^2 y}}{v}\right) V(y)\mathrm{d}y.$$

Let

$$\mathcal{C} = NLM^{\varepsilon}/PM.$$

Lemma 1. *We have*

$$\mathcal{O}_0 \lhd \frac{N}{P^2M^{3/2}} \sum_{\substack{P<p<2P \\ p \text{ prime}}} \sum\sum_{m,n=1}^{\infty} \lambda(m,n) W\left(\frac{nm^2}{N}\right) \sum_{\ell \in \mathcal{L}} \chi(\bar{\ell}) \tag{2}$$
$$\times \sum_{c \ll \mathcal{C}} \chi(pc) \frac{\mathfrak{D}(\overline{pc}n\ell; M)}{c} \mathfrak{I}(n, \ell; cpM).$$

Proof. Consider the sum in (5) with $p \nmid c$. (In the non-generic case where $p|c$, one needs to pull out the highest power of p dividing c, before running the argument given below.) The Bessel function is negligibly small if $c \gg \mathcal{C}$. Hence we only consider $c \ll \mathcal{C}$. (So $(c, M) = 1$.) As $p \nmid c$, the character sum

$$\sum_{\psi \bmod p} (1 - \psi(-1))\, S_{\psi}(r, n\ell; cpM)$$

can be replaced by a linear combination of

$$pS(\bar{p}r, \bar{p}n\ell; cM)e\left(\pm\frac{\overline{cM}(r + n\ell)}{p}\right).$$

Consider the plus term. We apply the reciprocity relation

$$e\left(\frac{\overline{cM}(r + n\ell)}{p}\right) = e\left(-\frac{\overline{p}(r + n\ell)}{cM}\right) e\left(\frac{(r + n\ell)}{cpM}\right),$$

and then push the last term to the weight function.

Next we apply the Poisson summation formula on the sum over r with modulus cM. It turns out that the nonzero frequencies make a negligible contribution. (Indeed the total conductor of the sum is of size $cM(1+NL/cpM) \ll M^{\varepsilon}NL/P$, and hence the length of the dual sum is $M^{\varepsilon}/P \ll M^{-\varepsilon}$.) Hence the sum

$$\sum_{r=1}^{\infty} \chi(r)S(\bar{p}r, \bar{p}n\ell; cM)e\left(-\frac{\overline{p}(r + n\ell)}{cM}\right) e\left(\frac{(r + n\ell)}{cpM}\right) V\left(\frac{r}{N\ell}\right) J_{k-1}\left(\frac{4\pi\sqrt{n\ell r}}{cpM}\right),$$

up to a negligible error term, reduces to

$$\frac{N\ell}{cM}\mathfrak{C}(n, p, \ell; cM)\, \mathfrak{I}(n, \ell; cpM)$$

where the character sum is given by

$$\mathfrak{C}(n,p,\ell;cM)=\sum_{a \bmod cM}\chi(a)S(\bar{p}a,\bar{p}n\ell;cM)e\left(-\frac{\overline{p}(a+n\ell)}{cM}\right)$$

and $\mathfrak{I}(n,\ell;cpM)$ is as above. Using the standard bound for the Bessel function and the second derivative bound we get that $\mathfrak{I}(\dots)\ll cPMm/NL$. It follows that we have

$$\mathcal{O}_0\lhd\frac{N}{P^2M^2}\sum_{\substack{P<p<2P\\ p\text{ prime}}}\sum\sum_{m,n=1}^{\infty}\lambda(m,n)W\left(\frac{nm^2}{N}\right)\sum_{\ell\in\mathcal{L}}\chi(\bar{\ell})$$
$$\times\sum_{c\ll C}\frac{\mathfrak{C}(n,p,\ell;cM)}{c^2}\mathfrak{I}(n,\ell;cpM).$$

The character sum can be partially evaluated. First it decomposes as a product of two character sums. The one with modulus c is given by

$$\mathfrak{C}_1=\sum_{a \bmod c}S(\bar{p}\bar{M}a,\bar{p}\bar{M}n\ell;c)e\left(-\frac{\overline{pM}(a+n\ell)}{c}\right)=c.$$

The other one with modulus M is given by

$$\mathfrak{C}_2=\sum_{a \bmod M}\chi(a)S(\bar{p}\bar{c}a,\bar{p}\bar{c}n\ell;M)e\left(-\frac{\overline{pc}(a+n\ell)}{M}\right)$$
$$=\sum_{b \bmod M}^{\star}e\left(\frac{(\bar{b}-1)\overline{pc}n\ell}{M}\right)\sum_{a \bmod M}\chi(a)e\left(\frac{(b-1)\overline{pc}a}{M}\right).$$

The inner sum vanishes unless $(b-1,M)=1$, in which case it is given by $\chi(pc)\bar{\chi}(b-1)g_\chi$. One gets

$$\mathfrak{C}(\dots)=cg_\chi\chi(pc)\sum_{\substack{b \bmod M\\(b(b-1),M)=1}}\bar{\chi}(b-1)e\left(\frac{(\bar{b}-1)\overline{pc}n\ell}{M}\right).$$

The lemma follows. □

Applying the Cauchy inequality to the right hand side of (2), we arrive at

$$\mathcal{O}_0\lhd\frac{N}{P^2M^{3/2}}\mathfrak{V}^{1/2}\,\mathfrak{U}^{1/2},\tag{3}$$

where

$$\mathfrak{V} = \sum\sum_{nm^2 \sim N} m\ |\lambda(m,n)|^2$$

and

$$\mathfrak{U} = \sum\sum_{nm^2 \sim N} \frac{1}{m} \Big| \sum_{\substack{P<p<2P \\ p \text{ prime}}} \sum_{\ell \in \mathcal{L}} \sum_{c \ll \mathcal{C}} \frac{\chi(pc\bar{\ell})}{c} \mathfrak{D}(\overline{pc}n\ell; M)\mathfrak{I}(n,\ell; cpM)\Big|^2.$$

Lemma 2. *We have*

$$\mathfrak{V} \ll N^{1+\varepsilon}$$

and

$$\mathfrak{U} \ll NL^2P^2 + M^2P^2.$$

Proof. The Hecke relation implies that

$$|\lambda(m,n)|^2 \ll N^{\varepsilon} \sum_{d|(m,n)} |\lambda(m/d,1)|^2 |\lambda(1,n/d)|^2,$$

and we have the Ramanujan bound on average

$$\sum_{e \ll E} |\lambda(1,e)|^2 \ll E^{1+\varepsilon}. \tag{4}$$

It follows that

$$\begin{aligned}\mathfrak{V} &\ll N^{\varepsilon} \sum_{m \ll N^{1/2}} m \sum_{d|m} |\lambda(m/d,1)|^2 \sum_{n \ll N/m^2 d} |\lambda(1,n)|^2 \\ &\ll N^{1+\varepsilon} \sum_{d \ll N^{1/2}} \frac{1}{d^2} \sum_{m \ll N^{1/2}/d} \frac{|\lambda(m,1)|^2}{m} \ll N^{1+\varepsilon}.\end{aligned}$$

Next we consider $\mathfrak{U}$. The Bessel function in the integral $\mathfrak{I}$ is negligibly small if $m \gg NLM^{\varepsilon}/PM =: \mathcal{M}$. For $m \ll \mathcal{M}$, and $C \ll \mathcal{C}/m$ we consider

$$\mathfrak{U}_m = \sum_{n \in \mathbb{Z}} W\left(\frac{n}{N}\right) \Big| \sum_{\substack{P<p<2P \\ p \text{ prime}}} \sum_{\ell \in \mathcal{L}} \sum_{c \sim C} \frac{\chi(pc\bar{\ell})}{c} \mathfrak{D}(\overline{pc}n\ell; M)\mathfrak{I}(n,\ell; cpM)\Big|^2,$$

where

$$\mathcal{N}=\max\left\{\frac{N}{m^2},\frac{NL}{CPm}\right\}M^{\varepsilon},$$

and W is a smooth function supported in $[-1,1]$. Opening the absolute square we apply the Poisson summation on the outer sum with modulus M. Taking into account the possible oscillation in the weight function, we see that our choice of the extended length of the sum implies that the nonzero frequencies make a negligible contribution.[1] It follows that

$$\mathfrak{U}_m=\frac{\mathcal{N}}{M}\mathop{\sum\sum}_{\substack{P<p,p'<2P\\ p,p'\text{ prime}}}\mathop{\sum\sum}_{\ell,\ell'\in\mathcal{L}}\mathop{\sum\sum}_{c,c'\sim C}\frac{\chi(pc\overline{p'c'}\bar{\ell}\ell')}{cc'}\,\mathfrak{C}_3\,\mathfrak{J}+O(M^{-2016})$$

where the character sum is given by

$$\mathfrak{C}_3=\sum_{a\bmod M}\mathfrak{D}(\overline{pc}a\ell;M)\overline{\mathfrak{D}(\overline{p'c'}a\ell';M)}$$

and the integral is given by

$$\mathfrak{J}=\int_{\mathbb{R}}\mathfrak{I}(\mathcal{N}y,\ell;cpM)\overline{\mathfrak{I}(\mathcal{N}y,\ell';c'p'M)}W(y)dy.$$

We use the trivial bound $\mathfrak{J}\ll(CPMm)^2/(NL)^2$. Now consider the character sum $\mathfrak{C}_3$ which is given by

$$\mathop{\sum\sum}_{\substack{b,b'\bmod M\\ (bb'(b-1)(b'-1),M)=1}}\bar{\chi}(b-1)\chi(b'-1)\sum_{a\bmod M}e\left(\frac{(\bar{b}-1)\overline{pc}a\ell-(\bar{b}'-1)\overline{p'c'}a\ell'}{M}\right).$$

We get

[1]Referee's comment: '...if one continued this investigation/path and considered the t-aspect problem (in such case the oscillatory phase of $\mathfrak{I}(n,\ell;cpM)$ is of size t), then one should be able to establish a t-aspect subconvexity for $L(s,\pi)$; in fact I would guess one major variant of Yongxiao Lin's preprint entitled 'Bounds for twists of GL(3) L-functions' which takes care the t-aspect case, was influenced by this remark of the author. In that work (as pointed out in Remark 5.2 in Lin's paper) Lin proved an analog of the bound of $\mathfrak{C}_3$ sum..., which is sufficient and is also indispensable (using this line of method) to imply a t-aspect subconvexity.'

$$\mathfrak{C}_3 = M \sum\sum_{\substack{b,b' \bmod M \\ (bb'(b-1)(b'-1),M)=1 \\ \bar{b}' \equiv 1+(\bar{b}-1)v \bmod M}} \bar{\chi}(b-1)\chi(b'-1)$$

where $v \equiv \overline{pc\ell p'c'\bar{\ell}'} \bmod M$. Now

$$\overline{b-1}(\overline{1+(\bar{b}-1)v}-1) \equiv \overline{1+b(\bar{v}-1)}$$

and consequently

$$\mathfrak{C}_3 = M \sum_{\substack{b \bmod M \\ (b(b-1),M)=1}} \bar{\chi}(1+b(\bar{v}-1)) = \begin{cases} M(M-2) & \text{if } v \equiv 1 \bmod M \\ O(M) & \text{otherwise.} \end{cases}$$

It follows that

$$\mathfrak{U}_m \ll \mathcal{N}\left(\frac{P^4M^2C^2m^2}{N^2} + \frac{P^3M^3Cm^2}{N^2L}\right),$$

and we have

$$\mathfrak{U} \ll \sum_{m \ll \mathcal{M}} \frac{\mathcal{N}}{m}\left(\frac{P^4M^2C^2m^2}{N^2} + \frac{P^3M^3Cm^2}{N^2L}\right) \ll NL^2P^2 + M^2P^2.$$

In the last inequality we assumed that $L < P$ and $M < N$ (say $\theta < 1/2$). □

Plugging in the bounds from Lemma 2 to (3) we get the bound of Proposition 1 for the sum $\mathcal{O}_0$. In the case $p|c$, we obtain a stronger bound by employing the above analysis.

5 Treating the Old Forms

In the rest of the paper we will prove Proposition 2. To analyse the sum $\mathcal{F}$ we use the functional equation for the $GL(3) \times GL(2)$ Rankin-Selberg convolution, as in [Mu15b]. There are two new issues. First we need to split $\lambda_f(n\ell)$ using the Hecke relation and secondly we need to take care of the oldforms. To this end let $\mathcal{F}^\sharp$ be same as the expression in (2) with $\lambda_f(n\ell)$ replaced by $\lambda_f(n)\lambda_f(\ell)$.

Lemma 3. *Suppose $L \gg M^{2\theta+\varepsilon}$. We have*

$$\mathcal{F} = \mathcal{F}^\sharp + O(M^{-2016}).$$

Proof. On using the Hecke relation in (2) we get two terms, one of them being $\mathcal{F}^\sharp$. To tackle the other term, consider

$$\sum_{f\in H_k(pM,\psi)}\omega_f^{-1}\sum_{\ell\in\mathcal{L}}\bar{\chi}(\ell)\sum\sum_{m,n=1}^{\infty}\lambda(m,n\ell)\lambda_f(n)W\left(\frac{n\ell m^2}{N}\right) \tag{1}$$
$$\times\sum_{r=1}^{\infty}\overline{\lambda_f(r)}\chi(r)V\left(\frac{r}{N\ell}\right).$$

To this we apply the Petersson formula. Observe that the diagonal does not exist and the off-diagonal is given by

$$\sum_{c=1}^{\infty}\sum_{\ell\in\mathcal{L}}\bar{\chi}(\ell)\sum\sum_{m,n=1}^{\infty}\lambda(m,n\ell)W\left(\frac{n\ell m^2}{N}\right)$$
$$\times\sum_{r=1}^{\infty}\chi(r)\frac{S_\psi(r,n;cpM)}{cpM}J_{k-1}\left(\frac{4\pi\sqrt{rn}}{cpM}\right)V\left(\frac{r}{N\ell}\right).$$

For $c\gg NM^\varepsilon/pM$ the Bessel function is negligibly small. For smaller values of c, we apply the Poisson summation formula on the sum over r. As $L\gg M^{2\theta+\varepsilon}$, it follows that the nonzero frequencies make a negligible contribution. For the zero frequency we have the character sum

$$\sum_{a \bmod cpM}\chi(a)S_\psi(a,n;cpM)$$

which vanishes. More precisely the part modulo p-power vanishes. Hence the above sum (1) is negligibly small. Consequently in (2) we can replace $\lambda_f(n\ell)$ by $\lambda_f(n)\lambda_f(\ell)$ at a cost of a negligible error term. □

Next we take into account the contribution of the old forms.

Lemma 4. *We have*

$$\mathcal{F}=\mathcal{F}^\star+O(P^2M^\varepsilon),$$

where

$$\mathcal{F}^\star=\frac{1}{L^\star P^\star}\sum_{\substack{P<p<2P\\ p\ prime}}\sum_{\psi \bmod p}(1-\psi(-1))\sum_{f\in H_k^\star(pM,\psi)}\omega_f^{-1} \tag{2}$$
$$\times\sum_{\ell\in\mathcal{L}}\lambda_f(\ell)\bar{\chi}(\ell)\sum\sum_{m,n=1}^{\infty}\lambda(m,n)\lambda_f(n)W\left(\frac{nm^2}{N}\right)\sum_{r=1}^{\infty}\overline{\lambda_f(r)}\chi(r)V\left(\frac{r}{N\ell}\right).$$

Proof. For $g \in S_k(p, \psi)$, we set $g|_M(z) = M^{k/2}g(Mz)$ which lies in $S_k(pM, \psi)$. Define $g^\star = g|_M - \langle g|_M, g\rangle \langle g, g\rangle^{-1} g$. Then $\{g, g^\star : g \in H_k(p, \psi)\}$ gives an orthogonal Hecke basis of the space of oldforms. For $f = g$, the functional equations yield the bound

$$\sum\sum_{m,n=1}^{\infty} \lambda(m,n)\lambda_f(n)W\left(\frac{nm^2}{N}\right)\sum_{r=1}^{\infty}\overline{\lambda_f(r)}\chi(r)V\left(\frac{r}{N\ell}\right) \ll p^{3/2}\, p^{1/2}M^{1+\varepsilon},$$

as $\pi \otimes f$ has conductor p^3 and $f \otimes \chi$ has conductor pM^2. Now to treat $f = g^\star$, note that $\lambda_{g|_M}(r) = 0$ unless $M|r$. But in this case we have $\chi(r) = 0$. Also $\langle g|_M, g|_M\rangle = \langle g, g\rangle$, hence by Bessel inequality we have $| \langle g|_M, g\rangle \langle g, g\rangle^{-1} | \leq 1$. Consequently for $f = g^\star$, the above sum reduces exactly to the case $f = g$ with a multiplier of size 1. Hence the above bound also holds for $f = g^\star$. Consequently the contribution of the oldforms is bounded by p^2M^ε. (We have not mentioned the size of ω_f^{-1} for $f = g, g^\star$. For details see [S-PY].) □

6 Applying Functional Equations

Next we will apply the functional equation for $GL(3) \times GL(2)$ Rankin-Selberg convolution, and $GL(2)$ Voronoi summation formula to the sums over (m, n) and r respectively. This will lead us to the family of dual sums

$$\begin{aligned}\mathcal{D}^\star =&\mathcal{D}^\star(\tilde{N}, \tilde{R}; W, V) = \frac{N^2}{M^2P^5} \sum_{\substack{P<p<2P \\ p \text{ prime}}} \chi(p) \\ &\times \sum_{\psi \bmod p} (1-\psi(-1))\,\psi(-M)g_\psi^2 \sum_{f\in H_k^\star(pM,\psi)} \omega_f^{-1}\overline{\lambda_f(p^2M)}\sum_{\ell\in\mathcal{L}}\lambda_f(\ell)\bar{\chi}(\ell) \\ &\times \sum\sum_{m,n=1}^{\infty}\lambda(n,m)\overline{\lambda_f(n)}W\left(\frac{nm^2}{\tilde{N}}\right)\sum_{r=1}^{\infty}\lambda_f(r)\bar{\chi}(r)V\left(\frac{r}{\tilde{R}}\right)\end{aligned} \tag{1}$$

where, g_ψ is the Gauss sum associated with the character ψ, W is a bump function with support $[1, 2]$, V has the same support but $V^{(j)} \ll_j M^{2j\theta}$, and the lengths of the sums are determined by the following restrictions

$$\frac{P^3M^3}{NM^\varepsilon} \ll \tilde{N} \ll \frac{P^3M^{3+\varepsilon}}{N} \quad\text{and}\quad \frac{M^2P}{NLM^\varepsilon} \ll \tilde{R} \ll \frac{M^{2+4\theta+\varepsilon}P}{NL}. \tag{2}$$

We can take $\tilde{R}$, $\tilde{N}$ to be dyadic, so that the size of the family is $O(M^\varepsilon)$.

Lemma 5. *We have*

$$\mathcal{F}^\star \lhd \mathcal{D}^\star,$$

where we are using the shorthand notation introduced in the Sect. 1.

Proof. We use the functional equation of $L(s, \pi \otimes f)$ to derive the following summation formula. Let $\mathcal{U}$ be a partition of unity (see Lemma 1 of [Mu15b]) and let $\mathcal{U}^\dagger$ be the subset of $\mathcal{U}$ consisting of those pairs $(U, \tilde{N})$ which have $\tilde{N}$ in the range $[P^3M^{3-\varepsilon}/N, P^3M^{3+\varepsilon}/N]$. We have

$$\sum\sum_{m,n=1}^{\infty} \lambda(m,n)\lambda_f(n)W\left(\frac{m^2n}{N}\right)$$

$$= i^{3k}\psi(-M^2)g_\psi^3 M^2\overline{\lambda_f(p^3M)} \sum_{\mathcal{U}^\dagger}\sum\sum_{m,n=1}^{\infty} \frac{\lambda(n,m)\overline{\lambda_f(n)}}{m^2n} U\left(\frac{m^2n}{\tilde{N}}\right)$$

$$\times \frac{1}{2\pi i}\int_{(0)} \tilde{W}(s)\left(\frac{m^2nN}{p^3M^3}\right)^s \frac{\gamma(1-s)}{\gamma(s)}\mathrm{d}s + O(M^{-2016}).$$

The epsilon factor of the functional equation is derived from the epsilon factor ε_f of $L(s, f)$ where $f \in H_k^\star(pM, \psi)$. This was originally worked out in [AtLi] and [Li]. Appendix A of [KMV] nicely encapsulates the essentials. By Proposition A.1 we have $\varepsilon_f = \psi(M)\eta_f(p)\eta_f(M)$. Now $\eta_f(p) = g_\psi/\sqrt{p}\lambda_f(p) = g_\psi\overline{\lambda_f(p)}/\sqrt{p}$ where g_ψ stands for the Gauss sum for the character ψ. Also $\eta_f(M) = -1/\sqrt{M}\lambda_f(M)$. Now using $(\sqrt{M}\lambda_f(M))^2 = \psi(M)$ (in particular $|\sqrt{M}\lambda_f(M)| = 1$), we get $\eta_f(M) = -\sqrt{M}\,\overline{\lambda_f(M)}$. Hence $\varepsilon_f = -\sqrt{M}\psi(M)g_\psi\overline{\lambda_f(pM)}/\sqrt{p}$ and so $\varepsilon_f^3 = -M^{1/2}\psi(M^2)g_\psi^3\overline{\lambda_f(p^3M)}/p^{3/2}$. Thus we obtain the above epsilon factor. To the other sum

$$\sum_{r=1}^{\infty}\overline{\lambda_f(r)}\chi(r)V\left(\frac{r}{N\ell}\right) = g_{\bar{\chi}}^{-1}\sum_{a \bmod M}\bar{\chi}(a)\sum_{r=1}^{\infty}\overline{\lambda_f(r)}e\left(\frac{ar}{M}\right)V\left(\frac{r}{N\ell}\right)$$

we apply the Voronoi summation formula (from Appendix A of [KMV]). This transforms the above sum into

$$2\pi i^k\chi(-p)\bar{\psi}(-M)\frac{g_\chi g_{\bar{\psi}}}{g_{\bar{\chi}}}\frac{N\ell}{Mp}\sum_{r=1}^{\infty}\lambda_f(rp)\bar{\chi}(r)\int_0^\infty V(x)J_{k-1}\left(\frac{4\pi\sqrt{N\ell r x}}{M\sqrt{p}}\right)\mathrm{d}x.$$

In the last integral V is supported in $[M^{-4\theta}, 4]$, and satisfies $y^jV^{(j)}(y) \ll 1$. As k is large the Bessel function is negligibly small if $r \ll M^2P/NLM^\varepsilon$. On the other hand making the change of variables $y^2 = x$, pulling out the oscillation of the Bessel function and integrating by parts we get that the integral is negligibly small if $r \gg M^{2+4\theta+\varepsilon}P/NL$. This reduces the analyses of the sum in (2) to that of the sums of the type $\mathcal{D}^{\star}$. The lemma follows. (More details can be found in Section 5 of [Mu15b].) □

Next we extend the sum over f in (1) to a full orthogonal basis by introducing in the sum $g, g^{\star}$ with $g \in S_k^{\star}(p, \psi)$. Let the full sum be denoted by $\mathcal{D}^{\sharp}$.

Lemma 6. *Suppose $L \ll P^{1/2}M^{-\varepsilon}$. We have*

$$\mathcal{F}^{\star} \lhd \mathcal{D}^{\sharp} + N^{1/2}\frac{P^{1/2}M^{1/2+4\theta}}{L^{1/2}}.$$

Proof. If $f = g$ is an oldform coming from level p, the sub sum over (m, n) in (1) is negligibly small, as the length of the sum $\tilde{N} \gg P^3M^{\varepsilon}$ is larger than the size of the conductor P^3. To tackle the case of $f = g^{\star}$ with g a newform of level p, we note that $\lambda_{g^{\star}}(h)\bar{\chi}(h) = -\langle g|_M, g\rangle \langle g, g\rangle^{-1} \lambda_g(h)\bar{\chi}(h)$ where $\langle g|_M, g\rangle \langle g, g\rangle^{-1} \ll M^{-1/2}$ (see Theorem 3.2 of [S-PY]). Also $\lambda_{g^{\star}}(p^2M) = \sqrt{M}\lambda_g(p^2) - \langle g|_M, g\rangle \langle g, g\rangle^{-1} \lambda_g(p^2M) \ll \sqrt{M}$. Moreover, as $\langle g^{\star}, g^{\star}\rangle = \langle g|_M, g|_M\rangle - |\langle g|_M, g\rangle|^2 \langle g, g\rangle^{-1} = \langle g, g\rangle + O(\langle g, g\rangle / M) \asymp PM$, we get $\omega_{g^{\star}}^{-1} \asymp PM$. So the total contribution of the forms $g^{\star}$ is dominated by

$$\frac{N^2}{M^{7/2}P^5} \sum_{\substack{P<p<2P \\ p \text{ prime}}} \sum_{\substack{\psi \bmod p \\ \psi(-1)=-1}} \sum_{g \in H_k^{\star}(p,\psi)} \tag{3}$$

$$\times \Big| \sum_{\ell \in \mathcal{L}} \lambda_g(\ell)\bar{\chi}(\ell) \sum\sum_{m,n=1}^{\infty} \lambda(n, m)\overline{\lambda_{g|M}(n)} W\left(\frac{nm^2}{\tilde{N}}\right)\Big| \Big| \sum_{r=1}^{\infty} \lambda_g(r)\bar{\chi}(r) V\left(\frac{r}{\tilde{R}}\right)\Big|.$$

To the sum over (m, n) we apply the inverse Mellin transform to get

$$\sqrt{M}\frac{1}{2\pi i}\int_{(1/2)} \tilde{W}(s)\left(\frac{\tilde{N}}{M}\right)^s L_M(s, \pi \otimes g) \sum\sum_{i,j=0}^{\infty} \frac{\lambda(M^{i+1}, M^j)\lambda_g(M^i)}{M^{(i+2j)s}} ds,$$

where L_M stands for the L-function without the Euler factor at the prime M. Using the trivial bound for the $GL(3)$ Fourier coefficients, and using the approximate functional equation we replace the above expression by sums of the form

$$M^{1/2}\tilde{N}^{1/2} \Big| \sum\sum_{nm^2 \ll P^{3/2+\varepsilon}} \frac{\lambda(n, m)\lambda_g(n)}{(nm^2)^{1/2}} \Big|.$$

As such we are able to dominate (3) by

$$\frac{N^2(L\tilde{N})^{1/2}}{M^3P^5}\sum_{\substack{P<p<2P\\ p \text{ prime}}}\ \sum_{\substack{\psi \bmod p\\ \psi(-1)=-1}}\ \sum_{g\in H_k^\star(p,\psi)}$$

$$\times\Big|\sum\sum_{\substack{nm^2\ll P^{3/2+\varepsilon}\\ \ell\in\mathcal{L}}}\sum\frac{\lambda(n,m)\bar{\chi}(\ell)\lambda_g(n)\lambda_g(\ell)}{(\ell nm^2)^{1/2}}\Big|\Big|\sum_{r=1}^{\infty}\lambda_g(r)\bar{\chi}(r)V\left(\frac{r}{\tilde{R}}\right)\Big|.$$

We apply the Cauchy inequality. In the sum over (m, n, ℓ) we use the Hecke relation. Then applying the large sieve inequality for $GL(2)$ cuspforms (see [Dj]) and using the Ramanujan bound on average we get that (assuming that $L \ll P^{1/2}M^{-\varepsilon}$)

$$\sum_{\substack{P<p<2P\\ p \text{ prime}}}\ \sum_{\substack{\psi \bmod p\\ \psi(-1)=-1}}\ \sum_{g\in H_k^\star(p,\psi)}\Big|\sum\sum_{\substack{nm^2\ll P^{3/2+\varepsilon}\\ \ell\in\mathcal{L}}}\sum\frac{\lambda(n,m)\bar{\chi}(\ell)\lambda_g(n)\lambda_g(\ell)}{(\ell nm^2)^{1/2}}\Big|^2 \ll P^3M^{\varepsilon}.$$

For the other sum we use the trivial bound. Hence we get that (3) is dominated by

$$\frac{N^2(L\tilde{N})^{1/2}}{M^3P^5}\,P^3\tilde{R} \ll N^{1/2}\frac{P^{1/2}M^{1/2+4\theta+\varepsilon}}{L^{1/2}},$$

and the lemma follows. □

So the sum over f can be extended to a complete Hecke basis at a cost of a manageable error term. Notice that using the Ramanujan conjecture we can get a much better error term. Next we use the Hecke relation. We analyse the generic term. The other term can be analysed in the same fashion and at the end we get a stronger bound for it. Consider (1) with $\lambda_f(\ell)\lambda_f(r)$ (resp. $\bar{\lambda}_f(p^2M)\bar{\lambda}_f(n)$) replaced by $\lambda_f(\ell r)$ (resp. $\bar{\lambda}_f(np^2M)$) and the sum over f is extended to a full Hecke basis. We will denote this sum by $\mathcal{D}$. Using the Petersson formula this reduces to

$$\mathcal{D}=\frac{N^2}{M^2P^5}\sum_{\substack{P<p<2P\\ p \text{ prime}}}\chi(p)\sum_{\psi \bmod p}(1-\psi(-1))\,\psi(-M)g_{\psi}^2 \tag{4}$$

$$\times\sum_{c=1}^{\infty}\frac{1}{cpM}\sum_{\ell\in\mathcal{L}}\sum_{r=1}^{\infty}\bar{\chi}(r\ell)V\left(\frac{r}{\tilde{R}}\right)$$

$$\times\sum\sum_{m,n=1}^{\infty}\lambda(m,n)S_{\psi}(np^2M, r\ell; cpM)J_{k-1}\left(\frac{4\pi\sqrt{nr\ell}}{c\sqrt{M}}\right)W\left(\frac{nm^2}{\tilde{N}}\right).$$

(The diagonal vanishes as $\chi(M) = 0$.) The Kloosterman sum vanishes if $M|c$. On the other hand if $p|c$ then the sum vanishes unless $p|r$. The contribution of these terms can be shown to be much smaller compared to the generic terms $p \nmid cr$. For the generic term where $(r, p) = 1$, we take dyadic subdivision of the m sum and a smooth dyadic subdivision of the c sum. Let the contribution of the block with $m \sim \mathfrak{m}$ and $c \sim C$ be denoted by $\mathcal{D}(C, \mathfrak{m})$. This sum is negligibly small if

$$C \gg \mathcal{C} = (\tilde{N}\tilde{R}L)^{1/2} M^{\varepsilon} / M^{1/2}\mathfrak{m}. \tag{5}$$

Lemma 7. *For $C \ll \mathcal{C}$, we have*

$$\mathcal{D}(C, \mathfrak{m}) = \frac{N^2}{CM^3P^4} \sum_{\substack{P<p<2P \\ p \text{ prime}}} \chi(p) \sum_{\ell \in \mathcal{L}} \sum_{\substack{c,r=1 \\ (p,r)=1}}^{\infty} \bar{\chi}(r\ell) e\left(\frac{c\overline{r\ell}}{p}\right)$$

$$\times \sum_{m \sim \mathfrak{m}} \sum_{n=1}^{\infty} \lambda(n, m) S(n, r\ell\bar{M}; c) W\left(\frac{c}{C}, \frac{n}{\tilde{N}_0}, \frac{r}{\tilde{R}}\right) + O(M^{-2016})$$

where

$$W(x, y, z) = J_{k-1}\left(\frac{4\pi\sqrt{\ell \tilde{N}_0 \tilde{R} yz}}{\sqrt{M}Cx}\right) x^{-1} W(x) W(y) V(z).$$

Here W and V are as in (1).

Proof. Observe that the Kloosterman sum in (4) vanishes if $M|c$ as $(M, r\ell) = 1$. Hence we get

$$S_{\psi}(np^2M, r\ell; cpM) = -S_{\psi}(np^2, r\ell\bar{M}; cp),$$

and for $p \nmid c$ it further reduces to $\psi(r\ell)\bar{\psi}(cM) g_{\bar{\psi}} S(n, r\ell\bar{M}; c)$. Then executing the sum over ψ, and taking a dyadic subdivision of the m sum, smooth dyadic partition for the c-sum, we arrive at sums of the type given in the statement of the lemma. □

7 Intertwining Voronoi and Poisson Summations with Reciprocity

The next step involves an application of the Voronoi summation formula (see [L, MiSc]) on the sum over n. Let

$$\tilde{W}(x, s, z) = \int_0^{\infty} W(x, y, z)\, y^{s-1} dy,$$

and for $\ell = 0, 1$ define

$$\gamma_\ell(s) = \frac{1}{2\pi^{3(s+\frac{1}{2})}} \prod_{i=1}^{3} \frac{\Gamma\left(\frac{1+s+\alpha_i+\ell}{2}\right)}{\Gamma\left(\frac{-s-\alpha_i+\ell}{2}\right)} \tag{1}$$

and set $\gamma_\pm(s) = \gamma_0(s) \mp i\gamma_1(s)$. We define the integral transforms

$$W_\pm^\star(x, y, z) = \frac{1}{2\pi i} \int_{(\sigma)} y^{-s} \gamma_\pm(s) \tilde{W}(x, -s, z) ds \tag{2}$$

where $\sigma > -1 + \max\{-\mathrm{Re}(\alpha_1), -\mathrm{Re}(\alpha_2), -\mathrm{Re}(\alpha_3)\}$. Here α_i are the Langlands parameter of π.

Lemma 8. *We can write $\mathcal{D}(C, \mathfrak{m})$ as a sum of two similar sums $\mathcal{D}^\pm(C, \mathfrak{m})$, where*

$$\begin{aligned}\mathcal{D}^\pm(C, \mathfrak{m}) =& \frac{N^2}{CM^3P^4} \sum_{\substack{P<p<2P \\ p \text{ prime}}} \chi(p) \sum_{\ell\in\mathcal{L}} \sum_{\substack{c,r=1 \\ (p,r)=1}}^{\infty} c^2 \bar{\chi}(r\ell) e\left(\frac{\overline{cr\ell}}{p}\right) \\ &\times \sum_{m\sim\mathfrak{m}} \sum_{m'|cm} \sum_{n=1}^{\infty} \frac{\lambda(m', n)}{m'n} \sum_{d|c} \frac{\mu(d)}{d} \\ &\times \sum_{\substack{\beta \bmod mc/m' \\ r\ell\bar{M}+\beta m'\equiv 0 \bmod c/d}}^{\star} e\left(\pm\frac{\bar{\beta}n}{mc/m'}\right) W_\pm^\star\left(\frac{c}{C}, \frac{m'^2 n\tilde{N}_0}{c^3 m}, \frac{r}{\tilde{R}}\right).\end{aligned} \tag{3}$$

Proof. The lemma follows from the $GL(3)$ Voronoi summation formula (see Lemmas 19 and 20 of [Mu15b]).

The character sum can be evaluated quite easily. We write cd in place of c, and then set $c_1 = (m', c)$. Let $c = c_1c_2$, $m' = c_1m''$. We will only provide the details for the generic case $\ell \nmid c_1$. In the case $\ell | c_1$, one needs to pull out the top power of ℓ dividing c_1 and run the following argument with suitable modifications. At the end the bound that we obtain in this case is stronger.

Lemma 9. *Suppose $md/m'' = q_1q_2$ with $(q_1, c_2) = 1$ and $q_2|c_2^\infty$. Suppose $\ell \nmid c_1$. Then the character sum in (3) vanishes unless $c_1|r$ and $q_2|n$, in which case we have*

$$\sum_{\substack{\beta \bmod mc/m' \\ r\ell\bar{M}+\beta m'\equiv 0 \bmod c/d}}^{\star} e\left(\frac{\bar{\beta}n}{mc/m'}\right) = q_2\mathfrak{c}_{q_1}(n) e\left(-\frac{\overline{r'\ell m'' M}n}{qc_2}\right),$$

where $\mathfrak{c}_{q_1}(n)$ is the Ramanujan sum with modulus q_1 and $r' = r/c_1$.

Proof. From the congruence condition it follows that c_1 necessarily divides $r\ell$. Since we are taking ℓ to be prime, and dealing with the generic case $\ell \nmid c_1$, it follows that $c_1|r$, and we write $r = c_1 r'$. The congruence condition now yields

$$\bar{\beta} \equiv -\overline{r'\ell m'' M} \bmod c_2.$$

Hence

$$\sum_{\substack{\beta \bmod mc/m' \\ r\ell\bar{M}+\beta m'\equiv 0 \bmod c/d}}^{\star} e\left(\frac{\bar{\beta}n}{mc/m'}\right) = e\left(-\frac{\overline{r'\ell m''Mn}}{qc_2}\right) \sum_{\delta \bmod q}^{\dagger} e\left(\frac{\delta n}{q}\right)$$

where $q = md/m''$. Write $q = q_1 q_2$ with $(q_1, c_2) = 1$ and $q_2|c_2^{\infty}$. The dagger on the sum means that $(\delta, q_1) = 1$. It follows that the sum vanishes unless $q_2|n$, in which case it reduces to $q_2 \mathfrak{c}_{q_1}(n)$. The lemma follows. □

We set

$$\mathfrak{D} = \frac{N^2 C}{M^2 P^4 (L\tilde{R})^2} \sum\sum_{\substack{m\sim\mathfrak{m} \\ d\sim D}} \sum\sum_{c_1 m''|dm} \sum\sum_{\substack{q_1q_2=dm/m'' \\ \tilde{q}_2\sim Q}} \frac{c_1 m'' q_2}{d^2 m \tilde{q}_2} \Omega(\dots) \tag{4}$$

where

$$\Omega(\dots) = \sum\sum_{\substack{r'\sim\tilde{R}/c_1 \\ n\ll\mathfrak{N}}} |\mathfrak{c}_{q_1}(n)|\, |\lambda(c_1 m'', q_2 n)| \times\Big| \sum_{\substack{P<p<2P \\ p \text{ prime}}} \sum_{\ell\in\mathcal{L}} \sum_{c_2\in\mathcal{S}} \chi(p\bar{\ell})\, S(c_2 - \tilde{q}_2\bar{p},\ \bar{q}_1\bar{\tilde{q}}_2 m'' Mn; r'\ell)\Big|. \tag{5}$$

Here $\tilde{q}_2$ stands for the product of the prime factors of q_2, the set

$$\mathcal{S} = \left\{c_2 \in \mathbb{Z} : \left|c_2 - \frac{d\tilde{q}_2}{p}\right| \ll \frac{DQ\tilde{R}LM^{\varepsilon}}{C} =: C_2\right\}$$

and

$$\mathfrak{N} = \frac{(L\tilde{R})^{3/2}\tilde{N}^{1/2}M^{\varepsilon}}{q_2 c_1^2 m''^2 M^{3/2}}. \tag{6}$$

Lemma 10. *We have*

$$\mathcal{D}^{+}(C, \mathfrak{m}) \ \lhd\ \mathfrak{D}. \tag{7}$$

Proof. We use the reciprocity relation

$$e\left(-\frac{\overline{r'\ell}m''Mn}{qc_2}\right)=e\left(\frac{\overline{qc_2}m''Mn}{r'\ell}\right)e\left(-\frac{m''Mn}{qc_2r'\ell}\right)$$

and push the last term to the weight function. In fact the function

$$e\left(-\frac{m''Mn}{qc_2r'\ell}\right)W_+^\star\left(\frac{c_1c_2d}{C},\frac{m''^2n\tilde{N}_0}{c_1c_2^3d^3m},\frac{r}{\tilde{R}}\right)$$

is 'nice'. More precisely, this may be replaced by

$$\frac{m'^2nM}{L\tilde{R}C\mathfrak{m}}V\left(\frac{c_1c_2d}{C}\right)V\left(\frac{m'^2n\tilde{N}_0}{C^3m}\right)V\left(\frac{r}{\tilde{R}}\right)$$

where V are bump functions, at a cost of introducing a family of size M^ε. The proof of this fact follows in the same line as the analysis given in Section 4 of [L]. (A detailed proof of this fact will appear in an upcoming joint work with R. Holowinsky and Z. Qi.) Also observe that

$$e\left(\frac{\overline{cr\ell}}{p}\right)=e\left(\frac{c_2d\overline{r'\ell}}{p}\right)=e\left(-\frac{c_2d\overline{p}}{r'\ell}\right)e\left(\frac{c_2d}{pr'\ell}\right)$$

where the last factor is only mildly oscillating and we are going to absorb it in the weight function.

We are now ready to apply the Poisson summation formula on the sum over c_2. Observe that since $q_2|c_2^\infty$ we have $\tilde{q}_2|c_2$ (where $\tilde{q}_2$ stands for the product of the prime factors of q_2). We write nq_2 in place of n, and $\tilde{q}_2c_2$ in place of c_2. The Poisson summation transforms the sum

$$\sum_{c_2\in\mathbb{Z}}e\left(\frac{\overline{q_1\tilde{q}_2c_2m''Mn}-\tilde{q}_2c_2\bar{p}}{r'\ell}\right)V\left(\frac{c_1c_2d\tilde{q}_2}{C}\right)e\left(\frac{c_2d\tilde{q}_2}{pr'\ell}\right)$$

to

$$\frac{C}{c_1d\tilde{q}_2r'\ell}\sum_{c_2\in\mathbb{Z}}S(c_2-\tilde{q}_2\bar{p},\bar{q}_1\bar{\tilde{q}}_2m''Mn;r'\ell)\int_{\mathbb{R}}V(y)e\left(\frac{Cy}{c_1pr'\ell}-\frac{c_2Cy}{c_1d\tilde{q}_2r'\ell}\right)\mathrm{d}y.$$

The sum over c_2 can be truncated as

$$\left|c_2-\frac{d\tilde{q}_2}{p}\right|\ll\frac{c_1d\tilde{q}_2r'\ell M^\varepsilon}{C}$$

at a cost of a negligible error term. The lemma now follows by getting rid of the weight function by introducing a family of sums of the type (5). □

8 Applying Cauchy Inequality Followed by Poisson Summation

It now remains to analyse the sums (4) and (5). The next lemma is a consequence of the Weil bound for the Kloosterman sum.

Lemma 11. *Suppose $C_2 \gg M^{-\varepsilon}$ then we have*

$$\mathfrak{D} \ll (NP)^{1/2} M^{1/2+9\theta+\varepsilon}.$$

Proof. Using the Weil bound we get

$$\Omega(\dots) \ll \sum_{\substack{r' \sim \tilde{R}/c_1 \\ n \ll \mathfrak{N}}}\sum |\mathfrak{c}_{q_1}(n)|\,|\lambda(c_1 m'', q_2 n)|$$
$$\times \sum_{\substack{P<p<2P \\ p \text{ prime}}} \sum_{\ell \in \mathcal{L}} \sum_{c_2 \in \mathcal{S}} (pc_2 - \tilde{q}_2,\ m''n; r'\ell)(r'\ell)^{1/2}.$$

(Note that $\tilde{R} < M$, and we are assuming that M is a prime.) Summing over r' and ℓ we get

$$\Omega(\dots) \ll \frac{(\tilde{R}L)^{3/2} P C_2}{c_1^{3/2}} \sum_{n \ll \mathfrak{N}} |\mathfrak{c}_{q_1}(n)|\,|\lambda(c_1 m'', q_2 n)|.$$

Consequently we get

$$\mathfrak{D} \ll \frac{N^2 (\tilde{R}L)^{1/2}}{M^2 P^3} \sum_{\substack{m \sim \mathfrak{m} \\ d \sim D}}\sum \sum_{c_1 m'' | dm}\sum \sum_{\substack{q_1 q_2 = dm/m'' \\ \tilde{q}_2 \sim Q}}\sum \frac{m'' q_2}{c_1^{1/2} dm}$$
$$\times \sum_{n \ll \mathfrak{N}} |\mathfrak{c}_{q_1}(n)|\,|\lambda(c_1 m'', q_2 n)|.$$

Now $|\mathfrak{c}_{q_1}(n)| \le (q_1, n) = h$, and we write $q_1 = hq_3$, and replace n by nh. Using Hecke relation (see Theorem 6.4.11 of [Go]) and the trivial bound $|\lambda(1, \ell)| \ll \ell^{1/2}$, we get

$$|\lambda(c_1 m'', q_2 nh)| \ll \sum_{\nu | (c_1 m'', q_2 nh)} (c_1 m''/\nu)^{1/2} |\lambda(1, q_2 nh/\nu)|.$$

It follows that

$$\mathfrak{D} \ll \frac{N^2(\tilde{R}L)^{1/2}}{M^2P^3} \sum\sum\sum\sum_{hq_2q_3m''\sim \mathfrak{m}D} \sum\sum_{\substack{c_1|hq_2q_3\\ \nu|c_1m''}} \frac{m''^{1/2}}{q_3} \sum_{\substack{n\ll \mathfrak{N}/h\\ q_2nh\equiv 0 \bmod \nu}} |\lambda(1, q_2nh/\nu)|.$$

Now taking dyadic subdivision for each variables ($h \sim H$, $q_2 \sim Q_2$), and then gluing q_2nh we get

$$\mathfrak{D} \lhd \frac{N^2(\tilde{R}L)^{1/2}}{M^2P^3} \sum\sum_{q_3m''\sim \mathfrak{m}D/HQ_2} \frac{m''^{1/2}}{q_3} \sum_{u\ll U} |\lambda(1,u)|.$$

where

$$U = \frac{(L\tilde{R})^{3/2}\tilde{N}^{1/2}M^{\varepsilon}}{m''^2M^{3/2}}.$$

Using the Ramanujan bound on average (4) we get

$$\mathfrak{D} \ll M^{\varepsilon}\frac{N^2(\tilde{R}L)^2\tilde{N}^{1/2}}{M^{7/2}P^3}.$$

The lemma follows. □

On the other hand if $C_2 \leq 1$ the Weil bound yields that the expression in (4) is dominated by

$$O\left(\frac{N^{3/2}P^{3/2}}{M}\right).$$

So one still needs to save $P^{3/2}/M^{1/4}$. (Recall that to break the convexity bound in the off-diagonal we needed to take $P > M^{1/4}$.) This can be achieved by using the Cauchy inequality and then applying the Poisson summation formula on the sum over n.

Lemma 12. *If $C_2 \ll M^{-\varepsilon}$, so that the set $\mathcal{S}$ is at most singleton, then we have*

$$\mathfrak{D} \ll M^{\varepsilon}\frac{(NP)^{1/4}M^{1+4\theta}}{L^{1/2}}\left(L^{3/4} + \frac{(\tilde{R}L)^{1/2}P^{1/4}}{N^{1/4}}\right).$$

Proof. If $\mathcal{S}$ is empty there is nothing to prove. Suppose $\mathcal{S}$ is singleton with one element c_2. Using the bound $|\mathfrak{c}_{q_1}(n)| \leq (q_1, n) = h$ and applying the Cauchy inequality we obtain

$$\Omega(\dots) \ll \sum_{h|q_1} h\, \mathfrak{V}^{1/2}\, \mathfrak{U}^{1/2}$$

where

$$\mathfrak{V}=\sum\sum_{\substack{r'\sim\tilde{R}/c_1 \\ n\ll\mathfrak{N}/h}} |\lambda(c_1m'', hq_2n)|^2$$

and

$$\mathfrak{U}=\sum\sum_{\substack{r'\sim\tilde{R}/c_1 \\ n\ll\mathfrak{N}/h}} \Big| \sum_{\substack{P<p<2P \\ p \text{ prime}}} \sum_{\ell\in\mathcal{L}} \chi(p\bar{\ell})\, S(c_2-\tilde{q}_2\bar{p},\, \bar{q}_1\bar{\tilde{q}}_2m''Mhn; r'\ell)\Big|^2.$$

To $\mathfrak{U}$, utilizing positivity, we first introduce a smooth weight of the form $W(nh/\mathfrak{N})$ with W a smooth function supported in $[-\gamma,\gamma]$ for some constant γ. Then we open the absolute square and apply the Poisson summation formula on the sum over n with modulus $r'\ell\ell'$. This yields

$$\mathfrak{U}\ll\frac{c_1\mathfrak{N}}{hL^2\tilde{R}}\sum_{r'\sim\tilde{R}/c_1}\sum\sum_{\substack{P<p,p'<2P \\ p,p' \text{ prime}}}\sum\sum_{\ell,\ell'\in\mathcal{L}}\sum_{|n|\ll hL^2\tilde{R}/c_1\mathfrak{N}} |\mathfrak{C}_4| + M^{-2016}$$

where the character sum $\mathfrak{C}_4$ is given by

$$\sum_{a \bmod r'\ell\ell'} S(c_2-\tilde{q}_2\bar{p},\, \bar{q}_1\bar{\tilde{q}}_2m''Mha; r'\ell) S(c_2-\tilde{q}_2\bar{p}',\, \bar{q}_1\bar{\tilde{q}}_2m''Mha; r'\ell')e\left(\frac{an}{r'\ell\ell'}\right).$$

This character sum has also appeared in [Mu15a], where we proved square root cancellation in the sum using Deligne's result.

It follows that

$$\begin{aligned}\mathfrak{U}&\ll\frac{\tilde{R}^{5/2}P^2L^4}{c_1^{5/2}}+\frac{\mathfrak{N}\tilde{R}^2L^2}{hc_1^2}\left(P+\frac{P^2c_1}{\tilde{R}}\right)\\ &\ll M^{\varepsilon}\,\frac{(\tilde{R}L)^{5/2}P^2}{c_1^{5/2}}\left(L^{3/2}+\frac{\tilde{R}LP^{1/2}}{N^{1/2}}\right)\end{aligned}$$

where in the last inequality we assumed that $L\ll M^2/N$. Consequently we get the bound

$$\Omega(\dots)\ll M^{\varepsilon}\,\frac{(\tilde{R}L)^{5/4}P}{c_1^{5/4}}\left(L^{3/4}+\frac{(\tilde{R}L)^{1/2}P^{1/4}}{N^{1/4}}\right)\sum_{h|q_1} h\,\mathfrak{V}^{1/2}$$

and hence

$$\mathfrak{D} \ll \frac{N^2 C M^\varepsilon}{M^2 P^3 (L\tilde{R})^{3/4}} \left(L^{3/4} + \frac{(\tilde{R}L)^{1/2} P^{1/4}}{N^{1/4}} \right)$$
$$\times \sum_{\substack{m \sim \mathfrak{m} \\ d \sim D}} \sum \sum_{c_1 m'' | dm} \sum \sum_{\substack{q_1 q_2 = dm/m'' \\ \tilde{q}_2 \sim Q}} \sum \frac{m'' q_2}{c_1^{1/4} d^2 m \tilde{q}_2} \sum_{h | q_1} h\, \mathfrak{Y}^{1/2}.$$

Consider the second line of the above expression. We glue dm into a single variable u of size $D\mathfrak{m}$, and write $hh' = q_1$, so that $hh' q_2 m'' = dm = u$. It follows that the second line is dominated by

$$\frac{1}{D} \sum_{u \sim D\mathfrak{m}} \sum \sum_{c_1 m'' | u} \sum \sum \sum_{hh' q_2 = u/m''} \frac{1}{c_1^{1/4} h'} \mathfrak{Y}^{1/2}.$$

Applying the Cauchy inequality we get that this is bounded by

$$\frac{\mathfrak{m}^{1/2}}{D^{1/2}} \left[\sum_{u \sim D\mathfrak{m}} \sum \sum_{c_1 m'' | u} \sum \sum_{h q_2 | u/m''} \sum \sum_{\substack{r' \sim \tilde{R}/c_1 \\ n \ll \mathfrak{N}/h}} |\lambda(c_1 m'', h q_2 n)|^2 \right]^{1/2}. \tag{1}$$

Using Hecke relation as in the proof of the previous lemma we get that the expression within the braces is dominated by

$$\sum_{u \sim D\mathfrak{m}} \sum \sum \sum_{\substack{c_1 m'' | u \\ \nu | c_1 m''}} \sum \sum_{h q_2 | u/m''} \sum \sum_{\substack{r' \sim \tilde{R}/c_1 \\ n \ll \mathfrak{N}/h \\ h q_2 n \equiv 0 \bmod \nu}} c_1 m'' \, |\lambda(1, h q_2 n/\nu)|^2.$$

Here the sum over ν is over square-free integers. Next we sum over r'. Writing $hq_2 m'' \upsilon = u$, and then restricting all the variables in dyadic intervals $h \sim H$, $q_2 \sim Q_2$, $m'' \sim M''$, $\upsilon \sim V$, with $H Q_2 M'' V \ll D\mathfrak{m}$, we reduce the analysis of the above expression to that of

$$\tilde{R} M'' \sum \sum \sum \sum_{\substack{h \sim H, q_2 \sim Q_2 \\ m'' \sim M'', \upsilon \sim V}} \sum_{c_1 | h q_2 \upsilon} \sum_{\nu | c_1 m''} \sum_{\substack{n \ll \mathfrak{N}/h \\ h q_2 n \equiv 0 \bmod \nu}} |\lambda(1, h q_2 n/\nu)|^2.$$

This is dominated by

$$\tilde{R} M'' M^\varepsilon \sum \sum \sum \sum_{\substack{h \sim H, q_2 \sim Q_2 \\ m'' \sim M'', \upsilon \sim V}} \sum_{\nu | h q_2 \upsilon m''} \sum_{\substack{n \ll \mathfrak{N}^\star / Q_2 H \\ h q_2 n \equiv 0 \bmod \nu}} |\lambda(1, h q_2 n/\nu)|^2,$$

where (recall the definition of $\mathfrak{N}$ in (6))

$$\mathfrak{N}^\star = \frac{(L\tilde{R})^{3/2}\tilde{N}^{1/2}M^\varepsilon}{M''^2 M^{3/2}}.$$

Since for any $\xi \ll \mathfrak{N}^\star$ there are $O(M^\varepsilon)$ many (h, q_2, n, ν) such that $hq_2n/\nu = \xi$, we see that the above sum is dominated by

$$\tilde{R}M''^2 V M^\varepsilon \sum_{\xi \ll \mathfrak{N}} |\lambda(1,\xi)|^2.$$

By applying the Ramanujan bound on average (4) we see that the last expression is dominated by $O(\tilde{R}M''2V\mathfrak{N}^\star M^\varepsilon)$. Substituting in (1) we see that the expression in (1) is dominated by $O(M^\varepsilon \mathfrak{m}\tilde{R}^{5/4}L^{3/4}\tilde{N}^{1/4}M^{-3/4})$. The lemma follows. □

From Lemmas 11 and 12 we get that

$$\mathfrak{D} \ll M^\varepsilon N^{1/2}\left[P^{1/2}M^{1/2+9\theta} + \frac{P^{1/4}M^{5/8+17\theta/4}}{L^{1/2}}\left(L^{3/4} + \frac{M^{11\theta/4}P^{3/4}}{M^{1/8}}\right)\right].$$

From this and taking into account the error term from Lemma 6, which is dominated by the bound in Lemma 11, we conclude Proposition 2.

Acknowledgments I wish to thank Roman Holowinsky and Zhi Qi for several helpful discussions related to the method presented in this paper, especially I thank Qi for pointing out the cancellation of the oscillatory factor of the Bessel function, which is used in the proof of Lemma 10. This paper was written as a supplement to my paper [Mu15b] almost 5 years ago and a preliminary version of it was put in the arxiv in 2016. In the meantime Holowinsky and Nelson [HoNe] have simplified the proof of the main result of [Mu15b] tremendously. However I believe that this paper contains some useful ideas, e.g. the idea of 'mass transfer', which has recently been used by my student Sharma [Sh] to establish subconvexity for character twists of $GL(2) \times GL(3)$ Rankin-Selberg L-functions. Let me thank Nicolas Templier for his patience and for encouraging me to submit this article to this volume. Finally I thank the referees for many helpful comments which greatly improved the quality of exposition.

References

[AtLi] A.O.L. Atkin, W.C.W. Li, Twists of newforms and pseudo-eigenvalues of W-operators. Invent. Math. **48**, 221–243 (1978)

[Bl] V. Blomer, Applications of the Kuznetsov formula on $GL(3)$. Invent. Math. **194**, 673–729 (2013)

[BrMu] T.D. Browning, R. Munshi, Rational points on singular intersections of quadrics. Compositio Math., **149**, 1457–1494 (2013)

[Bu] D.A. Burgess, On character sums and primitive roots. Proc. London Math. Soc. **12**, 179–192 (1962)

[But] J. Buttcane, The spectral Kuznetsov formula on $SL(3)$. Trans. Am. Math. Soc. **368**, 6683–6714 (2016)

[Dj] G. Djankovic, A larger GL_2 large sieve in the level aspect. Cent. Eur. J. Math. **10**, 748–760 (2012)

[DFI] W. Duke, J.B. Friedlander, H. Iwaniec, Bounds for automorphic L-functions. Invent. Math. **112**, 1–8 (1993)

[Go] D. Goldfeld: Automorphic Forms and L-Functions for the Group $GL(n, \mathbb{R})$. Cambridge Univ. Press, (2006), vol. **99**, Cambridge.

[H-B] D.R. Heath-Brown, A new form of the circle method, and its application to quadratic forms. J. Reine Angew. Math. **481**, 149–206 (1996)

[HoNe] R. Holowinsky, P. Nelson, Subconvex bounds on GL_3 via degeneration to frequency zero. Math. Ann. **372**, 299–319 (2018)

[IwKo] H. Iwaniec, E. Kowalski, Analytic number theory. Am. Math. Soc. Coll. Publ. **53** (2004). American Mathematical Society, Providence, RI .

[KMV] E. Kowalski, P. Michel, J. Vanderkam, Rankin-Selberg L-functions in the level aspect. Duke Math. J. **114**, 123–191 (2002)

[Li] W.C.W. Li, Newforms and functional equations, Math. Ann. **212**, 285–315 (1975)

[L] X. Li, Bounds for $GL(3) \times GL(2)$ L-functions and $GL(3)$ L-functions. Ann. Math. **173**, 301–336 (2011)

[MiSc] S.D. Miller, W. Schmid, Automorphic distributions, L-functions, and Voronoi summation for $GL(3)$. Ann. Math. **164**, 423–488 (2006)

[Mu14] R. Munshi, The circle method and bounds for L-functions—I. Math. Annalen, **358**, 389–401 (2014)

[Mu15a] R. Munshi, The circle method and bounds for L-functions—II. Subconvexity for twists of $GL(3)$ L-functions. Am. J. Math. **137**, 791–812 (2015)

[Mu15b] R. Munshi, The circle method and bounds for L-functions—IV. Subconvexity for twists of $GL(3)$ L-functions. Ann. Math. **182**, 617–672 (2015)

[Mu15c] R. Munshi, Pairs of quadrics in 11 variables. Compos. Math. **151**, 1189–1214 (2015)

[S-PY] R. Schulze-Pillot, A. Yenirce, Petersson products of bases of spaces of cusp forms and estimates for Fourier coefficients. Int. J. Number Theory **14**, 2277–2290 (2018)

[Sh] P. Sharma, Subconvexity for $GL(3) \times GL(2)$ twists in level aspect. arXiv:1906.09493 [math.NT] (2019)

Modular forms on G_2 and their Standard L-Function

Aaron Pollack

Abstract The purpose of this partly expository paper is to give an introduction to modular forms on G_2. We do this by focusing on two aspects of G_2 modular forms. First, we discuss the Fourier expansion of modular forms, following work of Gan-Gross-Savin and the author. Then, following Gurevich-Segal and Segal, we discuss a Rankin-Selberg integral yielding the standard L-function of modular forms on G_2. As a corollary of the analysis of this Rankin-Selberg integral, one obtains a Dirichlet series for the standard L-function of G_2 modular forms; this involves the arithmetic invariant theory of cubic rings. We end by analyzing the archimedean zeta integral that arises from the Rankin-Selberg integral when the cusp form is an even weight modular form.

1 Introduction

If a reductive group G over $\mathbf{Q}$ has a Hermitian symmetric space $G(\mathbf{R})/K$, then there is a notion of *modular forms* on G. Namely, one can consider the automorphic forms on G that give rise to holomorphic functions on $G(\mathbf{R})/K$. However, if G does not have a Hermitian symmetric space, then G has no obvious notion of modular forms, or very special automorphic forms. The split exceptional group G_2 is an example: its symmetric space $G_2(\mathbf{R})/\left(\mathrm{SU}(2)\times\mathrm{SU}(2)\right)$ has no complex structure. Nevertheless, Gan-Gross-Savin [GGS02] defined and studied modular forms on G_2 by building off of work of Gross-Wallach [GW96] on quaternionic discrete series representations.

The purpose of this paper is to give an introduction to modular forms on G_2, with the hope that we might interest others in this barely-developed area of mathematics. Indeed, very little is known about G_2-modular forms. For example, as pointed out

A. Pollack (✉)
Department of Mathematics, Institute for Advanced Study, Princeton, NJ, USA

Department of Mathematics, Duke University, Durham, NC, USA
e-mail: aaronjp@math.ias.edu; apollack@math.duke.edu

W. Müller et al. (eds.), *Relative Trace Formulas*, Simons Symposia,
https://doi.org/10.1007/978-3-030-68506-5_12

in [GGS02], one does not even know the smallest positive integer k for which there is a nonzero level-one cuspidal modular form on G_2 of weight k.

Besides recalling for the reader some basic facts about the group G_2 and the definition of its modular forms, we focus on two aspects of G_2-modular forms. The first thing we focus on is their Fourier expansion. Using a certain multiplicity one result of Wallach [Wal03] on "generalized Whittaker vectors", Gan-Gross-Savin defined the (generic) Fourier coefficients of a G_2 modular form. This involves Wallach's multiplicity one result and the arithmetic invariant theory of cubic rings. Using their definition of Fourier coefficients, Gan-Gross-Savin computed the Fourier coefficients of an explicit theta function and (using [JR97]) some Eisenstein series, and also computed how Hecke operators act on these Fourier coefficients.

The paper [Pol18] makes explicit the Fourier *expansion* of modular forms, i.e., it involves functions on $G_2(\mathbf{R})$ and not just the Fourier coefficients, which are complex numbers. Furthermore, as a consequence of *loc cit* one can also define the non-generic Fourier coefficients of G_2-modular forms. The first result we discuss is this Fourier expansion of G_2-modular forms. This is the G_2-special case of the main result of [Pol18]. Because the constructions and arguments in *loc cit* cover all the exceptional Dynkin types, they rely on sometimes lengthy Jordan algebra computations. For G_2 one does not need Jordan algebras, so we give a treatment that is simpler, and that we expect is easier for the non-expert to follow.

The second aspect of G_2-modular forms that we discuss is their standard L-function. Here we rely on the exceptional papers [GS15] and [Seg17] of Gurevich-Segal and Segal. These works give a Rankin-Selberg integral yielding the degree 7 standard L-function of automorphic cusp forms on G_2. Moreover, the integrals of *loc cit* are non-vanishing for G_2-modular forms, and in fact rely on the same type of Fourier coefficients mentioned above. (The older Rankin-Selberg integral [Gin93] of Ginzburg relies on a Whittaker coefficient and vanishes identically for G_2-modular forms.)

The beautiful results of the papers [Seg17] and [GS15] have somewhat involved proofs. We give a streamlined account of the integrals of *loc cit* by simplifying a bit some of the computations. We then use this analysis of the Rankin-Selberg integral to a give a Dirichlet series for the standard L-function of modular forms on G_2. The Dirichlet series has a nice expression in terms of cubic rings and the Fourier coefficients of modular forms.

The Dirichlet series comes from analyzing the local zeta integrals at the good finite primes that arise from the global Rankin-Selberg convolution. One can also ask for an analysis of the archimedean zeta integral, when the cusp form involved is a modular form. Given our computation of the Fourier expansion of modular forms, this question becomes tractable. In the final section, we analyze this archimedean zeta integral and relate it to products of gamma functions.

We now give an outline of the rest of the paper. In Sect. 2, we review the octonions and basic facts about the group G_2. Next, in Sect. 3, we recall the definition from [GGS02] of modular forms on G_2 and their Fourier coefficients. We also state the results about the Fourier expansion (Theorem 3.4) and Dirichlet series (Theorem 3.5) of G_2-modular forms. In Sect. 4, we prove Theorem 3.4, after

[Pol18]. In Sect. 5 we discuss the Rankin-Selberg integral of [GS15] and [Seg17], and also the Dirichlet series for the standard L-function. In particular, we prove Theorem 3.5. Finally, in Sect. 6, we apply our results on the Fourier expansion of modular forms to analyze the archimedean zeta integral that arises from the Rankin-Selberg integral of *loc cit* when the cusp form is a modular form.

2 Octonions and the Group G_2

In this section, we define the group G_2 as the automorphisms of the split octonions, and review basic structural information about G_2: its Lie algebra, Heisenberg parabolic subgroup, Cartan involution, and maximal compact subgroup.

2.1 The Octonions

In this subsection we very briefly review the octonions, following the discussion in [PWZ19, Section 1.1]. The reader might see [SV00] for a more thorough treatment.

Thus suppose F is field of characteristic 0. We denote by Θ the split octonions over F. Let V_3 denote the standard representation of the group SL_3, and $V_3^\vee$ the dual representation. As a vector space,

$$\Theta = \left\{ \begin{pmatrix} a & v \\ \phi & d \end{pmatrix} : a, d \in F, v \in V_3, \phi \in V_3^\vee \right\}.$$

This is the so-called Zorn model of the octonions.

As an F-algebra, Θ is non-commutative and non-associative. Because V_3 is considered as a representation of SL_3, we assume given an identification $\wedge^3 V_3 \simeq F$, which induces $\wedge^2 V_3 \simeq V_3^\vee$ and $\wedge^2 V_3^\vee \simeq V_3$. The multiplication in Θ is given as

$$\begin{pmatrix} a & v \\ \phi & d \end{pmatrix} \begin{pmatrix} a' & v' \\ \phi' & d' \end{pmatrix} = \begin{pmatrix} aa' + \phi'(v) & av' + d'v - \phi \wedge \phi' \\ a'\phi + d\phi' + v \wedge v' & \phi(v') + dd' \end{pmatrix}.$$

The octonions also come equipped with a quadratic norm $N : \Theta \to F$ defined by $N\left(\begin{smallmatrix} a & v \\ \phi & d \end{smallmatrix}\right) = ad - \phi(v)$. If $x, x' \in \Theta$, then $N(xx') = N(x)N(x')$. Associated to the norm N is a symmetric bilinear form defined by $(x, y) = N(x+y) - N(x) - N(y)$. For $x = \begin{pmatrix} a & v \\ \phi & d \end{pmatrix}$ and $x' = \begin{pmatrix} a' & v' \\ \phi' & d' \end{pmatrix}$, one has

$$(x, x') = ad' + a'd - \phi(v') - \phi'(v).$$

This bilinear form is non-degenerate, and is split. I.e., there are isotropic subspaces of Θ of dimension four.

Like a ring of quaternions over F, the octonions come equipped with a conjugation $*: \Theta \to \Theta$, which is an order-reversing involution on Θ. The conjugate x^* of $x = \begin{pmatrix} a & v \\ \phi & d \end{pmatrix}$ is $x^* = \begin{pmatrix} d & -v \\ -\phi & a \end{pmatrix}$. One has $xx^* = N(x)1$ inside Θ. Define the trace of $x \in \Theta$ as $\mathrm{tr}(\left(\begin{smallmatrix} a & v \\ \phi & d \end{smallmatrix}\right)) = a + d$. For $x \in \Theta$, one has the quadratic equation $x^2 = \mathrm{tr}(x)x - N(x)$.

Write V_7 for the elements of Θ with trace 0. Thus $\left(\begin{smallmatrix} a & v \\ \phi & d \end{smallmatrix}\right) \in V_7$ if and only if $a + d = 0$. The trace 0 elements are the perpendicular space to the line generated by the element $1 \in \Theta$. For $x \in \Theta$, we write $Im(x) = \frac{1}{2}(x - x^*)$, so that $x = \mathrm{tr}(x)/2 \cdot 1 + Im(x)$. This is the orthogonal decomposition of x into $F \cdot 1 \oplus V_7$.

The split algebraic group G_2 over F is defined to be the automorphisms of Θ:

$$G_2(F) = \{g \in \mathrm{GL}(\Theta) : g(x \cdot y) = gx \cdot gy \text{ for all } x, y \in \Theta\}.$$

One has $g1 = 1$, and $g(x^*) = (gx)^*$. Since G_2 stabilizes the element 1, it acts on V_7, preserving the induced quadratic form. This representation $G_2 \to \mathrm{GL}(V_7)$ factors through $\mathrm{SO}(V_7)$.

The group G_2 has two conjugacy classes of maximal parabolic subgroups. Their associated flag varieties can be neatly described in terms of subspaces of Θ with particular properties under the multiplication on Θ. See, for example, [PWZ19, Proposition 1.3].

2.2 *The Lie Algebra* $\mathfrak{g}_2$

One can write down the Lie algebra of G_2 relatively easily inside the Lie algebra of $\mathrm{SO}(V_7)$. To do this, first recall that if V is a non-degenerate quadratic space, then $\mathfrak{so}(V) \simeq \wedge^2 V$. In this identification, an element $w \wedge x$ acts (on the left) of V as

$$(w \wedge x) \cdot v = (x, v)w - (w, v)x.$$

The Lie bracket in this notation is

$$[w \wedge x, y \wedge z] = (x, y)w \wedge z - (x, z)w \wedge y - (w, y)x \wedge z + (w, z)x \wedge y. \quad (1)$$

The multiplication on Θ defines a map $V_7 \otimes V_7 \to V_7$ via $w \otimes x \mapsto Im(wx)$. This map is alternating, i.e., $Im(xw) = -Im(wx)$ for $x, w \in V_7$, and thus induces a map $\wedge^2 V_7 \to V_7$. This map is surjective and G_2-equivariant. Denote by $\mathfrak{g}$ the kernel of $\wedge^2 V_7 \to V_7$. Then $\mathfrak{g}$ is closed under the Lie bracket induced from (1), and is the Lie algebra of G_2.

We now give an explicit basis for $\mathfrak{g}$, and make some special notation which we will use throughout. First, denote by e_1, e_2, e_3 a fixed basis of V_3, and write e_1^*, e_2^*, e_3^* for the basis of $V_3^\vee$ dual to the e_i. Set $u_0 = \begin{pmatrix} 1 & \\ & -1 \end{pmatrix} \in V_7$. We will abuse notation and also let e_i, e_j^* denote elements in V_7. Thus $(e_i, e_j^*) = -\delta_{ij}$, $(u_0, u_0) = -2$, and $(u_0, e_i) = (u_0, e_j^*) = 0$ for all i, j.

We set $E_{kj} = e_j^* \wedge e_k$, $v_j = u_0 \wedge e_j + e_{j+1}^* \wedge e_{j+2}^*$, and $\delta_j = u_0 \wedge e_j^* + e_{j+1} \wedge e_{j+2}$ (indices taken modulo three). One checks immediately from the definition of multiplication in Θ that the elements v_j and δ_j are in the kernel of $\wedge^2 V_7 \to V_7$, and thus in $\mathfrak{g}$. The same goes for E_{kj} so long as $j \neq k$. A sum $\alpha_1 E_{11} + \alpha_2 E_{22} + \alpha_3 E_{33}$ is in $\mathfrak{g}$ if and only if $\alpha_1 + \alpha_2 + \alpha_3 = 0$.

The above elements span $\mathfrak{g}$. We write

$$\mathfrak{h} = \{\alpha_1 E_{11} + \alpha_2 E_{22} + \alpha_3 E_{33} : \alpha_1 + \alpha_2 + \alpha_3 = 0\}.$$

This is a Cartan subalgebra of $\mathfrak{g}$. The Cartan subalgebra $\mathfrak{h}$ acts on E_{kj} by $\alpha_k - \alpha_j$, i.e.,

$$[\sum_i \alpha_i E_{ii}, E_{kj}] = (\alpha_k - \alpha_j) E_{kj}.$$

These are the long roots for $\mathfrak{h}$. Together with $\mathfrak{h}$, the E_{jk} span the Lie algebra $\mathfrak{sl}_3$. The Cartan $\mathfrak{h}$ acts on v_j via α_j and δ_j via $-\alpha_j$. These are the short roots.

One has the following Lie bracket relations, which can be checked easily.

- $[\delta_{j-1}, v_j] = 3E_{j,j-1}$
- $[v_{j-1}, \delta_j] = -3E_{j-1,j}$
- $[\delta_{j-1}, \delta_j] = 2v_{j+1}$
- $[v_{j-1}, v_j] = 2\delta_{j+1}$
- $[\delta_j, v_j] = 3E_{jj} - (E_{11} + E_{22} + E_{33})$.

All indices here are taken modulo 3. We will choose a positive system on $\mathfrak{g}$ by letting E_{12} and v_2 be the positive simple roots.

Abstractly, $\mathfrak{g} = \mathfrak{g}_2 = \mathfrak{sl}_3 \oplus V_3 \oplus V_3^\vee$, and this is a $\mathbf{Z}/3$-grading. See, e.g., [FH91, chapter 22]. In fact, all the (split) exceptional Lie algebras have $\mathbf{Z}/3$-grading that generalizes this one. See [Rum97] or [Sav99]. In this $\mathbf{Z}/3$-model, the Lie bracket is given as follows:

(1) the commutators $[\mathfrak{sl}_3, V_3]$ and $[\mathfrak{sl}_3, V_3^\vee]$ are given by the action of $\mathfrak{sl}_3$ on V_3 and $V_3^\vee$;
(2) the commutators $[V_3, V_3]$ and $[V_3^\vee, V_3^\vee]$ are given by[1] $[x, y] = 2x \wedge y \in V_3^\vee$ for $x, y \in V_3$ and $[\gamma, \delta] = 2\gamma \wedge \delta \in V_3$ for $\gamma, \delta \in V_3^\vee$;

[1]The 2 here is $1_J \times 1_J = 2 1_J$.

(3) if $x \in V_3$ and $\gamma \in V_3^\vee$ then[2] $[\gamma, x] = 3x \otimes \gamma - (x, \gamma)I_3$, which is in $\mathfrak{sl}_3$. Here $x \otimes \gamma \in V_3 \otimes V_3^\vee \simeq End(V_3)$ is considered as an element of $End(V_3)$, i.e., $x \otimes \gamma$ is the linear map $V_3 \to V_3$ sending $y \in V_3$ to $(\gamma, y)x$.

The elements E_{ij}, v_j, δ_k of $\mathfrak{g}$ defined above are simply the standard basis vectors for $\mathfrak{sl}_3$, V_3 and $V_3^\vee$ in the decomposition $\mathfrak{g} = \mathfrak{sl}_3 \oplus V_3 \oplus V_3^\vee$.

2.3 The Heisenberg Parabolic Subgroup

We now describe the Heisenberg parabolic subgroup $P = MN$ of G_2. This is the subgroup that stabilizes the two-dimensional subspace of V_7 spanned by e_3^* and e_1. The unipotent radical N is two-step, $N \supseteq N_0 \supseteq 1$, with N/N_0-four-dimensional abelian, and $N_0 = [N, N]$ one-dimensional. The Lie algebra $\mathfrak{n}$ of N is spanned by v_1, δ_3, and the strictly upper triangular matrices in $\mathfrak{sl}_3$. The element E_{13} of $\mathfrak{sl}_3$ spans N_0.

We chose the Levi subgroup M of P by requiring it to have Lie algebra spanned by v_2, δ_2, and the Cartan $\mathfrak{h}$. This Levi subgroup M of the Heisenberg parabolic is isomorphic to GL_2. The isomorphism $M \simeq \mathrm{GL}_2$ is chosen so that $Ad(m)E_{13} = \det(m)E_{13}$. The conjugation action of M on N/N_0 is the representation $Sym^3(V_2) \otimes \det(V_2)^{-1}$ of GL_2, where V_2 denotes the standard representation of GL_2. On the Lie algebra level, the isomorphism $M_2(F) \simeq \mathfrak{m}$ is given by

$$\left(\begin{smallmatrix} a & b \\ c & d \end{smallmatrix}\right) \mapsto dE_{11} + (a-d)E_{22} - aE_{33} + bv_2 - c\delta_2. \qquad (2)$$

2.4 Cartan Involution

Before defining a Cartan involution Θ on $\mathfrak{g}$ and a maximal compact subgroup K of $G_2(\mathbf{R})$, we briefly discuss the Killing form on $\mathfrak{g}$. If V is a quadratic space, then the Killing form on $\mathfrak{so}(V) \simeq \wedge^2 V$ is proportional to the quadratic form given by $(w \wedge x, y \wedge z) = (w, z)(x, y) - (w, y)(x, z)$. This pairing on $\wedge^2 V_7$ induces one on $\mathfrak{g}$ which is proportional to the Killing form on $\mathfrak{g}$.

Denote $\iota : \Theta \to \Theta$ the involution given by $\iota\left(\begin{pmatrix} a & v \\ \phi & d \end{pmatrix}\right) = \begin{pmatrix} d & -\tilde{\phi} \\ -\tilde{v} & a \end{pmatrix}$. Here $\tilde{e_i} = e_i^*$, $\tilde{e_j^*} = e_j$ and $\tilde{\cdot}$ is extended by linearity. The involution ι has ± 1 eigenspaces on Θ each with dimension 4, and the quadratic form $(\cdot, \cdot)$ on Θ is positive-definite on the $+1$-eigenspace, and negative-definite on the -1 eigenspace.

The element $1 \in \Theta$ is fixed by ι, and abusing notation we denote by ι the restriction of this involution to V_7. Then ι induces an involution θ on $\wedge^2 V_7$ via

[2]The 3 here is $\mathrm{tr}_J(1_J)$.

$\theta(w \wedge x) = \iota(w) \wedge \iota(x)$. Since $\iota(u_0) = -u_0$, $\iota(e_i) = -e_i^*$, and $\iota(e_i^*) = -e_i$, it is clear from our basis above that θ preserves $\mathfrak{g} \subseteq \wedge^2 V_7$.

Claim 2.1. *The involution θ is a Cartan involution on $\mathfrak{g}$.*

Proof. Because $\iota \in O(V_7)$, θ preserves the bracket on $\mathfrak{so}(V_7)$, and thus also on $\mathfrak{g}$. It is clear as well that θ is an involution, because ι is. That θ is a Cartan involution now follows easily from the definition: I.e., $B_\theta(x, y) := -B(x, \theta(y))$ is positive-definite on $\mathfrak{g}$, where B is the Killing form. □

The maximal compact subgroup of $G_2(\mathbf{R})$ is $(\mathrm{SU}(2) \times \mathrm{SU}(2))/\mu_2$. Below we will see this explicitly on the Lie algebra level.

3 Modular forms on G_2

In this section we define modular forms on G_2, after [GGS02]. We then discuss their "abstract" Fourier expansion in terms of cubic rings, which follows from a multiplicity one result of Wallach at the archimedean place and the arithmetic invariant theory of $\mathrm{GL}_2(\mathbf{Z})$ acting on $Sym^3(\mathbf{Z}^2) \otimes \det^{-1}$ (the orbits parametrize cubic rings). We then state the two main theorems that we will prove:

(1) An explicit, or refined Fourier expansion. This is a purely archimedean result. It is the G_2-special case of the main result of [Pol18].
(2) A Dirichlet series for the standard L-function of a modular form on G_2, away from finitely many bad primes. This is really a p-adic result (for p finite). With a small amount of work, it can be extracted from the papers [GS15] and [Seg17]. Namely, these papers give Rankin-Selberg integrals for cusp forms on G_2, and the Dirichlet series is essentially equivalent to the unramified computation in these two papers.

We hope people will be interested in this beautiful Rankin-Selberg integral of Gurevich-Segal and Segal. However, the computations in *loc cit* are rather difficult (at least for this author!). Thus we give a new presentation of these Rankin integrals, following the same outline as [GS15, Seg17] but with a few technical simplifications. We explain below the manner in which we accomplish these simplifications.

3.1 Quaternionic Discrete Series

The definition of modular forms on G_2, from [GGS02], relies on that of the quaternionic discrete series representations [GW96]. These are a family of certain discrete series representations π_n of $G_2(\mathbf{R})$, for integers $n \geq 2$, singled out by

Gross and Wallach [GW94, GW96]. In this subsection, we briefly discuss these representations, and the definition of modular forms on G_2.

By way of analogy, first consider the group Sp_{2g}, and the Siegel modular forms. The group $\mathrm{Sp}_{2g}(\mathbf{R})$ has holomorphic discrete series representations, and the automorphic representations of $\mathrm{Sp}_{2g}(\mathbf{A})$ that have these as their infinite components give rise to the Siegel modular forms. More precisely, the maximal compact subgroup of $\mathrm{Sp}_{2g}(\mathbf{R})$ is the unitary group $U(g)$. Consider the holomorphic discrete series representation σ_n of $\mathrm{Sp}_{2g}(\mathbf{R})$, whose minimal K-type is the one-dimensional representation $\det^{\otimes n}$ of $U(g)$. The representations σ_n on $\mathrm{Sp}_{2n}(\mathbf{R})$ have the smallest Gelfand-Kirillov dimension of the discrete series on Sp_{2n} [Wal15], and simplest minimal K-type. These representations are responsible for the Siegel modular forms on Sp_{2g} of weight n.

The discrete series representations π_n on $G_2(\mathbf{R})$ are analogues of the representations σ_n; they too have the smallest Gelfand-Kirollov dimension among the discrete series on $G_2(\mathbf{R})$ [Wal15], and have the simplest minimal K-type. More precisely, set $K \simeq (\mathrm{SU}(2) \times \mathrm{SU}(2))/\mu_2$ the maximal compact subgroup of $G_2(\mathbf{R})$, that corresponds to the Cartan involution θ defined in Sect. 2.4. From [GW96], one has the K-type decomposition

$$\pi_n|_K = \bigoplus_{k\geq 0} Sym^{2n+k}(\mathbf{C}^2) \boxtimes Sym^k(W_{\mathbf{C}}). \tag{1}$$

Here $\mathbf{C}^2$ is the defining two-dimensional representation of the "long root" SU(2), which is the first SU(2)-factor in $K = (\mathrm{SU}(2) \times \mathrm{SU}(2))/\mu_2$, and $W_{\mathbf{C}}$ is the four-dimensional irreducible representation of the "short root" SU(2), which is the second SU(2)-factor in K. (The summands in (1) are not generally irreducible.) Thus the minimal K-type of π_n is $\mathbb{V}_n := Sym^{2n}(\mathbf{C}^2) \boxtimes \mathbf{1}$. Note that K has no nontrivial one-dimensional representations, and thus there are no discrete series of $G_2(\mathbf{R})$ with one-dimensional minimal K-types.

Roughly, the modular forms on G_2 of weight n correspond to certain automorphic forms in automorphic representations $\pi = \pi_f \otimes \pi_\infty$ with $\pi_\infty = \pi_n$. More precisely, one defines[3] modular forms on G_2 as follows. Denote by $\mathcal{A}(G_2)$ the space of automorphic forms on $G_2(\mathbf{A})$. Then, following [GGS02], one sets

$$\mathrm{MF}(G_2)_n = \mathrm{Hom}_{G_2(\mathbf{R})}(\pi_n, \mathcal{A}(G_2))$$

the space of modular forms of weight n on G_2. More concretely, associated to a homomorphism $\varphi \in \mathrm{Hom}_{G_2(\mathbf{R})}(\pi_n, \mathcal{A}(G_2))$, one gets a "vector-valued" function F_φ on $G_2(\mathbf{A})$ by restricting φ to the minimal K-type $\mathbb{V}_n$ of π_n. That is, one defines $F_\varphi : G_2(\mathbf{A}) \to \mathbb{V}_n^\vee$ as $F_\varphi(g)(v) = \varphi(v)(g)$. We will think about $\mathrm{MF}(G_2)_n$, the

[3] In the paper [Pol18] we generalize slightly the definition of modular forms on G_2; for example, one can also define modular forms of weight 1.

modular forms on G_2 of weight n, concretely through their associated vector-valued functions F_φ.

3.2 *Wallach's Result and Fourier Coefficients*

Suppose φ is a modular form of weight n on G_2. We now discuss the definition of the Fourier coefficients of φ, or equivalently the Fourier coefficients of F_φ. We will consider characters of $N(\mathbf{Q})\backslash N(\mathbf{A})$, where recall N denotes the unipotent radical of the Heisenberg parabolic. Thus suppose $\chi : N(\mathbf{Q})\backslash N(\mathbf{A}) \to \mathbf{C}^\times$ denotes a unitary character of N. (In particular, χ is trivial on $N_0(\mathbf{A})$, because $N_0 = [N, N]$ is the commutator subgroup of N.) If ξ is an automorphic form on G_2, then $\xi_\chi(g) := \int_{N(\mathbf{Q})\backslash N(\mathbf{A})} \chi^{-1}(n)\xi(ng)\,dn$ is a function $G_2(\mathbf{A}) \to \mathbf{C}$ satisfying $\xi(ng) = \chi(n)\xi(g)$ for all $n \in N(\mathbf{A})$. Similarly, we define $F_{\varphi,\chi} : G_2(\mathbf{A}) \to \mathbb{V}_n^\vee$ as

$$F_{\varphi,\chi}(g) = \int_{N(\mathbf{Q})\backslash N(\mathbf{A})} \chi^{-1}(n)F_\varphi(ng)\,dn.$$

Of course, for general automorphic forms ξ on $G_2(\mathbf{A})$, not much can be said about the function ξ_χ on G_2. However, when φ is a modular form, it turns out that the function $F_{\varphi,\chi}$ restricted to the real points of the Levi subgroup $M(\mathbf{R}) = \mathrm{GL}_2(\mathbf{R})$ is a particular nice function, and that much arithmetical information about φ can be extracted from $F_{\varphi,\chi}$. Elaborating on these claims is what this paper is about.

The first nontrivial fact in this direction is the following multiplicity one result of Wallach, which is a special case of the main result of [Wal03]. In the following theorem, a character $\chi : N(\mathbf{R}) \to \mathbf{C}^\times$ is called generic if it is in the open orbit for action of $M(\mathbf{C})$ on $N^{ab}(\mathbf{C})^\vee$.

Theorem 3.1 (Wallach). *Suppose that $\chi : N(\mathbf{R}) \to \mathbf{C}^\times$ is a generic character. Then*

$$\dim \mathrm{Hom}_{N(\mathbf{R})}(\pi_n, \chi) \leq 1,$$

i.e., there is at most a one-dimensional space of χ-equivariant moderate growth linear functionals on π_n.

In fact, the single open orbit of $M(\mathbf{C})$ on $N^{ab}(\mathbf{C})^\vee$ breaks up into two orbits of $M(\mathbf{R})$ on $N^{ab}(\mathbf{R})^\vee$, a "positive" one and a "negative" one. Wallach proves that $\mathrm{Hom}_{N(\mathbf{R})}(\pi_n, \chi) = 0$ if χ is in the negative orbit, $\mathrm{Hom}_{N(\mathbf{R})}(\pi_n, \chi)$ is one-dimensional if χ is in the positive orbit, and furthermore determines $\mathrm{Hom}_{N(\mathbf{R})}(\pi_n, \chi)$ as a representation of the stabilizer of χ in M.

From now on we denote $W = N^{ab}$, so that[4] $W = Sym^3(V_2) \otimes \det^{-1}$ as a representation of $M \simeq \mathrm{GL}_2$. Using the above results of Wallach, Gan-Gross-Savin [GGS02] defined a notion of Fourier coefficient for modular forms φ on G_2, as follows.

Fix $\chi_0 : N(\mathbf{R}) \to \mathbf{C}^\times$ a non-degenerate character in the "positive" orbit of $M(\mathbf{R})$, and a basis ℓ_0 of $\mathrm{Hom}_{N(\mathbf{R})}(\pi_n, \chi_0)$. Suppose that $\chi : N(\mathbf{Q})\backslash N(\mathbf{A}) \to \mathbf{C}^\times$ is a non-degenerate character, and that $\chi|_{N(\mathbf{R})}$ is in the positive orbit. Then using χ_0 and Wallach's result on the stabilizer, one can make a canonical basis element $\ell_\chi \in \mathrm{Hom}_{N(\mathbf{R})}(\pi_n, \chi)$ (that depends on χ_0 and ℓ_0.) See [GGS02, pg. 124], where our ℓ_χ is denoted $\lambda_k(g) \cdot (g \cdot \ell_0)$ in *loc cit.*

Now, suppose φ is a modular form on G_2 of weight n. Then integration

$$v \in \pi_n \mapsto \int_{N(\mathbf{Q})\backslash N(\mathbf{A})} \chi^{-1}(n)\varphi(v)(n)\, dn$$

defines an element $I_{\varphi,\chi}$ of $\mathrm{Hom}_{N(\mathbf{R})}(\pi_n, \chi)$. Thus, there is $a_\varphi(\chi) \in \mathbf{C}$ such that $I_{\varphi,\chi} = a_\varphi(\chi)\ell_\chi$ as elements of $\mathrm{Hom}_{N(\mathbf{R})}(\pi_n, \chi)$.

Definition 3.2 (Gan-Gross-Savin). Suppose φ is a modular form on G_2 of weight n, and χ_0, ℓ_0 have been fixed as above. Then the complex number $a_\varphi(\chi)$ is called the χ-Fourier coefficient of φ. If $\chi|_{N(\mathbf{R})}$ is in the negative orbit, then the χ-Fourier coefficient of φ is 0.

It is a consequence of the first main result that we will prove that the definition of Fourier coefficients of φ can also be extended to the characters χ on $N(\mathbf{Q})\backslash N(\mathbf{A})$ that are not-necessarily generic, so long as they are not the trivial character. Cusp forms, however, only have generic Fourier coefficients.

3.3 *Cubic Rings and Fourier Coefficients*

We have now briefly explained the analytic result of Wallach that goes into defining the Fourier coefficients of modular forms on G_2. Using the arithmetic invariant theory of cubic rings, which parametrize the orbits of $\mathrm{GL}_2(\mathbf{Z})$ on $W(\mathbf{Z}) = Sym^3(\mathbf{Z}^2) \otimes \det^{-1}$, Gan-Gross-Savin refined and enriched the above notion of Fourier coefficients of modular forms.

To understand this, first fix the standard additive character $\psi : \mathbf{Q}\backslash\mathbf{A} \to \mathbf{C}^\times$. With ψ fixed, characters χ on $N(\mathbf{Q})\backslash N(\mathbf{A})$ correspond to elements $w \in W(\mathbf{Q})$. Namely, W has a symplectic form $\langle\ ,\ \rangle$ satisfying $\langle gw_1, gw_2\rangle = \det(g)\langle w_1, w_2\rangle$ for all $w_1, w_2 \in W$ and $g \in \mathrm{GL}_2$. Then, associated to $w \in W(\mathbf{Q})$ one defines

[4] The Levi subgroup M of P is conjugate in $G_2(\mathbf{C})$ to the short root SU(2), and through this conjugation, $W \otimes \mathbf{C}$ becomes identified with the $W_{\mathbf{C}}$ appearing in the K-type decomposition of π_n. This is why we use the same letter for both spaces.

$\chi_w : N(\mathbf{Q})\backslash N(\mathbf{A}) \to \mathbf{C}^\times$ as $\chi_w(n) = \psi(\langle w, \overline{n}\rangle)$, where $\overline{n}$ is the image of n in $N^{ab} \simeq W$. For $w \in W(\mathbf{Q})$ generic and φ a modular form, we set $a_\varphi(w) := a_\varphi(\chi_w)$.

Now, suppose that φ is a modular form of weight n and *level one* for G_2. That φ is level one means that $F_\varphi(gk) = F_\varphi(g)$ for all $k \in G_2(\widehat{\mathbf{Z}})$. It then follows easily that $a_\varphi(w) \neq 0$ implies $w \in W(\mathbf{Z})$. Furthermore, if φ is of weight n, then $a_\varphi(k \cdot w) = \det(k)^n a_\varphi(w)$ for all $k \in \mathrm{GL}_2(\mathbf{Z})$ [GGS02, pg 125]. Consequently, if the weight n of φ is even, the Fourier coefficients of φ are invariants of the orbits of $\mathrm{GL}_2(\mathbf{Z})$ acting on $W(\mathbf{Z})$.

As mentioned, the orbits of $\mathrm{GL}_2(\mathbf{Z})$ on $W(\mathbf{Z})$ parametrize cubic rings over $\mathbf{Z}$. A cubic ring T over $\mathbf{Z}$ is a commutative ring that is isomorphic to $\mathbf{Z}^3$ as a $\mathbf{Z}$-module. The parametrization is as follows [DF64, GGS02, GL09]. An element of $W(\mathbf{Z})$ is binary cubic form $f(x, y) = ax^3 + bx^2y + cxy^2 + dy^3$, with $a, b, c, d \in \mathbf{Z}$. Associated to f, one sets $T = \mathbf{Z} \oplus \mathbf{Z}\omega \oplus \mathbf{Z}\theta$, with multiplication table

- $\omega\theta = -ad$
- $\omega^2 = -ac + a\theta - b\omega$
- $\theta^2 = -bd + c\theta - d\omega$.

The basis $1, \omega, \theta$ of T has the special property that $\omega\theta \in \mathbf{Z}$. One calls bases $1, \omega, \theta$ of cubic rings T with this property *good bases*. It is easily seen that every cubic ring has a good basis. The above association $f(x, y) \mapsto (T, \{1, \omega, \theta\})$ induces a bijection between binary cubic forms and isomorphism classes of cubic rings with a good basis. There is an action of $\mathrm{GL}_2(\mathbf{Z})$ on the set of good bases of a fixed cubic ring T, and the association $f(x, y) \mapsto (T, \{1, \omega, \theta\})$ is equivariant for the action of $\mathrm{GL}_2(\mathbf{Z})$ on both sides. It induces a bijection between orbits of $\mathrm{GL}_2(\mathbf{Z})$ on $W(\mathbf{Z})$ and isomorphism classes of cubic rings.

One can also see [Pola, Section 2.3] for a discussion of the parametrization of cubic rings, from the point of view of "lifting laws". The lifting law will appear when we consider the Rankin-Selberg integral in Sect. 5.

As a consequence of the above discussion, one can parametrize the Fourier coefficients of a modular form on G_2 in terms of cubic rings.

Definition 3.3 (Gan-Gross-Savin). Suppose that φ is a modular form on G_2 of level one and even weight, and that T is a non-degenerate cubic ring over $\mathbf{Z}$, i.e., assume $T \otimes_{\mathbf{Z}} \mathbf{Q}$ is an étale $\mathbf{Q}$-algebra. Define $a_\varphi(T) := a_\varphi(w)$, where $w \in W(\mathbf{Z})$ is any element of the $\mathrm{GL}_2(\mathbf{Z})$-orbit on $W(\mathbf{Z})$ that corresponds to the ring T.

Over $\mathbf{R}$, there are two types of non-degenerate cubic rings: $\mathbf{R}\times\mathbf{R}\times\mathbf{R}$ and $\mathbf{R}\times\mathbf{C}$. The first type corresponds to the "positive" orbit of $\mathrm{GL}_2(\mathbf{R})$ on $W(\mathbf{R})$, and the second to the "negative" orbit. Thus, if T is a cubic ring and $T \otimes \mathbf{R} = \mathbf{R} \times \mathbf{C}$, then $a_\varphi(T) = 0$.

3.4 The Theorems

We now state the two main results that we will prove below.

3.4.1 The Fourier Expansion

The first theorem is the analytic description of the Fourier expansion of modular forms of weight n on G_2. To setup the result, if $F = F_\varphi : G_2(\mathbf{A}) \to \mathbb{V}_n^\vee$ is associated to a modular form φ on G_2 of weight n, set $F_0(g) = \int_{N_0(\mathbf{Q})\backslash N_0(\mathbf{A})} F(ng)\,dn$. Then we can Fourier expand F_0 along $W = N/N_0$.

To do this, first recall that the representation W of GL_2 has an invariant symplectic form $\langle\ ,\ \rangle$, mentioned above. If $w \in W$ corresponds to the binary cubic form $f(x, y) = ax^3 + bx^2y + cxy^2 + dy^3$, and $w' \in W$ corresponds to the binary cubic $f'(x, y) = a'x^3 + b'x^2y + c'xy^2 + d'y^3$, then $\langle w, w'\rangle = ad' - \frac{1}{3}bc' + \frac{1}{3}cb' - da'$. Instead of writing out binary cubic forms, from now on we write elements of W as four-tuples $w = (a, \frac{b}{3}, \frac{c}{3}, d)$; this element corresponds to the binary cubic $ax^3 + bx^2y + cxy^2 + dy^3$.

Associated to an element $w \in W(\mathbf{R})$, we have the polynomial $h_w(z) = az^3 + bz^2 + cz + d$ on $\mathbf{C}$. We write $w \geq 0$ if $h_w(z)$ has no roots on the upper (equivalently) lower half-plane $\mathcal{H}_1^\pm$. This notion plays an important role in the statement of Theorem 3.4, i.e., in the Fourier expansion of modular forms on G_2.

Here are some facts, examples and non-examples:

(1) Suppose $w = \frac{1}{3}(0, 1, -1, 0)$. Then $h_w(z) = z^2 - z = z(z-1)$, which has only real roots. Thus $w \geq 0$.
(2) Suppose $w = (1, 0, 0, d)$. Then $h_w(z) = x^3 + d$. Hence if $d \neq 0$, w is not ≥ 0, but if $d = 0$, then $w \geq 0$.
(3) The condition $w \geq 0$ is equivalent to the binary cubic $f_w(x, y) = ax^3 + bx^2y + cxy^2 + dy^3$ factoring into three linear factors over $\mathbf{R}$, and thus is an invariant of the orbit $\mathrm{GL}_2(\mathbf{R})w$.
(4) If w is in the open orbit for $\mathrm{GL}_2(\mathbf{C})$ on $W(\mathbf{C})$, so that associated to w is a cubic etale $\mathbf{Q}$-algebra E_w, then $w \geq 0$ if and only if $E_w \otimes \mathbf{R} = \mathbf{R} \times \mathbf{R} \times \mathbf{R}$.
(5) Note that *the degree* of the polynomial h_w can change in the $\mathrm{GL}_2(\mathbf{R})$-orbit, even though the condition $w \geq 0$ is invariant. The reason we define the condition $w \geq 0$ in terms of the single variable polynomial h_w not having 0's, as opposed to the binary cubic f_w factoring, is that the condition h_w not having 0's on $\mathcal{H}_1^\pm$ generalizes nicely to the larger quaternionic exceptional groups.

Just like the Fourier expansion of Siegel modular forms on $\mathrm{Sp}_{2n}(\mathbf{R})$ involves the special functions $e^{2\pi i\,\mathrm{tr}(TZ)}$ on the Siegel upper half-space, the Fourier expansion of modular forms on G_2 involves some special functions. In preparation for the statement of Theorem 3.4, we now define these functions.

Fix a positive integer n. For $w \in W(\mathbf{R})$ with $w \geq 0$, and an integer v with $-n \leq v \leq n$, define the function $\mathcal{W}_w^v : \mathrm{GL}_2(\mathbf{R}) \to \mathbf{C}$ as

$$\mathcal{W}_w^v(m) = \left(\frac{|j(m,i)h_w(z)|}{j(m,i)h_w(z)}\right)^v \det(m)^n |\det(m)| K_v(|j(m,i)h_w(z)|). \tag{2}$$

Here, K_v denotes the vth K-Bessel function. Furthermore, for $m \in \mathrm{GL}_2(\mathbf{R})$, $z = m \cdot i$ in $\mathcal{H}_1^{\pm}$ and $j(m, i) = (ci + d)^3 \det(m)^{-1}$ if $m = \left(\begin{smallmatrix} a & b \\ c & d \end{smallmatrix}\right)$. Note that it is important that $h_w(z)$ has no zeros on $\mathcal{H}_1$ for this function to be defined, because the Bessel function K_v is not defined at 0.

Now, fix the standard basis x, y of weight vectors of the standard representation of the long-root SU(2), so that $\{x^{n+v}y^{n-v}\}_{-n \leq v \leq n}$ are a basis of $\mathbb{V}_n = Sym^{2n}(V_2) \boxtimes \mathbf{1}$. Define $\mathcal{W}_w : \mathrm{GL}_2(\mathbf{R}) \to \mathbb{V}_n^\vee$ as

$$\mathcal{W}_w(m) = \sum_{-n \leq v \leq n} \mathcal{W}_w^v \frac{x^{n+v}y^{n-v}}{(n+v)!(n-v)!}.$$

The function $\mathcal{W}_w$ is the special function that controls the Fourier expansion of modular forms on G_2, as stated in Theorem 3.4.

Finally, denote $n : W \to N/N_0$ the identification of the $\mathbf{Q}$-vector space W with the unipotent group N/N_0. With this preparation, we have the following result, from [Pol18]:

Theorem 3.4. *Suppose $F = F_\varphi$ corresponds to a modular form φ on G_2 of weight n. Denote by F_{00} the constant term of F along N. Then there are complex numbers $a_F(w)$ for $w \in 2\pi W(\mathbf{Q})$, such that for all $x \in W(\mathbf{R})$ and $m \in M(\mathbf{R}) \simeq \mathrm{GL}_2(\mathbf{R})$ one has*

$$F_0(n(x)m) = F_{00}(m) + \sum_{w \in 2\pi W(\mathbf{Q}), w \geq 0} a_F(w) e^{-i\langle w, x \rangle} \mathcal{W}_w(m).$$

Furthermore, the constant term F_{00} is of the form

$$F_{00}(m) = \Phi(m)x^{2n} + \beta x^n y^n + \Phi(mw_0)y^{2n}$$

with $\beta \in \mathbf{C}$, $w_0 = \left(\begin{smallmatrix} & -1 \\ 1 & \end{smallmatrix}\right) \in \mathrm{GL}_2$, and Φ associated to a holomorphic modular form of weight $3n$ on GL_2.

Theorem 3.4 gives the explicit Fourier expansion of modular forms φ on G_2 of weight n. This result is the G_2-special case of the main result of [Pol18]. Note that the sum over w in the theorem is only over those w with $w \geq 0$, just like, for example, the fact that Siegel modular forms on $\mathrm{Sp}_{2n}(\mathbf{R})$ only have Fourier coefficients corresponding to positive semi-definite matrices T.

3.4.2 The Dirichlet Series

The second main result we discuss is a Dirichlet series for the standard L-function of cuspidal representations π associated to modular forms on G_2. This Dirichlet series follows from [GS15] and [Seg17] without too much work; it is more-or-less equivalent to the unramified computation in these two papers.

Theorem 3.5. *Suppose $\pi = \pi_f \otimes \pi_k$ is a cuspidal automorphic representation of $G_2(\mathbf{A})$, unramified at all finite places, and φ is the level one modular form of weight k on G_2 associated to π. Suppose that k is even. Furthermore, suppose that E is a cubic étale $\mathbf{Q}$-algebra with $E \otimes \mathbf{R} \simeq \mathbf{R} \times \mathbf{R} \times \mathbf{R}$, and $\mathcal{O}_E$ is the maximal order in E. Finally, denote by S the set of places of $\mathbf{Q}$ consisting of 2, 3 and all the primes that ramify in E. Then*

$$\sum_{T \subseteq \mathcal{O}_E, n \geq 1, \text{ both prime to } S} \frac{a_\varphi(\mathbf{Z} + nT)}{[\mathcal{O}_E : T]^{s-k+1} n^s}$$
$$= a_\varphi(\mathcal{O}_E) \frac{L^S(\pi, Std, s - 2k + 1)}{L^S(E, s - 2k + 2)\zeta^S_{\mathbf{Q}}(2s - 4k + 2)}. \qquad (3)$$

Here $L(E, s) = \zeta_E(s)/\zeta_{\mathbf{Q}}(s)$ is the ratio of the Dedekind zeta functions of E and $\mathbf{Q}$; that T and n are both prime to S means that $[\mathcal{O}_E : T]$ and n are not divisible by primes in S; and the superscript S on the L-functions on the right-hand side of (3) means that the local factor at primes in S have been removed.

Note that if T is a cubic ring, then $\mathbf{Z}+nT$ is again a cubic ring, so the expression on the left-hand side of (3) makes sense. Also, it would of course be desirable to improve this result so that the places in S are accounted for.

It is a fact (see [Pol17, Corollary 4.2]) that the Dirichlet series for the Spin L-function of Siegel modular forms on PGSp_6 has a remarkably similar expression. In the case of Siegel modular forms on GSp_6, the Fourier coefficients turn out to be parametrized by orders in quaternion algebras that are ramified at infinity (see, e.g., [GL09]), and the Dirichlet series for the Spin L-function turns out to be a very similar sum as appears in the left-hand side of (3).

4 The Generalized Whittaker Function

In this section we sketch the proof of Theorem 3.4. This section is completely independent of Sect. 5; thus the reader interested in the contents of that section may now skip there.

Recall that $\varphi : \pi_n \to \mathcal{A}(G_2(\mathbf{A}))$ is a $G_2(\mathbf{R})$-equivariant homomorphism, and

$$F = F_\varphi : G_2(\mathbf{Q})\backslash G_2(\mathbf{A}) \to \mathbb{V}_n^\vee$$

is defined as $F_\varphi(g)(v) = \varphi(v)(g)$. Because F is defined through the minimal K-type $\mathbb{V}_n$ of π_n, it is annihilated by a certain first-order linear differential operator $\mathcal{D}_n$. We will shortly describe this operator, which is commonly referred to as the Schmid operator.

Suppose $\chi : N(\mathbf{R}) \to \mathbf{C}^\times$ is a character. The Fourier expansion of F along N/N_0 is controlled by functions $\mathcal{W}^\chi$ satisfying $\mathcal{W}^\chi(ng) = \chi(n)\mathcal{W}^\chi(g)$ for all $n \in$

$N(\mathbf{R})$ and $\mathcal{W}^{\chi}(gk) = k^{-1} \cdot \mathcal{W}^{\chi}(g)$ for all $k \in K \subseteq G_2(\mathbf{R})$ the maximal compact. In order to prove Theorem 3.4, we find all these functions $\mathcal{W}^{\chi}$ that are annihilated by $\mathcal{D}_n$ and have moderate growth. It turns out that so long as χ is nontrivial, there is at most a one-dimensional space of such functions, and that the differential equations $\mathcal{D}_n \mathcal{W}^{\chi} = 0$ allow one to solve for this function explicitly. The outcome of this calculation immediately results in Theorem 3.4.

Let us now describe the differential operator $\mathcal{D}_n$ more precisely. Recall the Cartan involution θ, and let $\mathfrak{g} = \mathfrak{k} \oplus \mathfrak{p}$ denote the decomposition into $\theta = 1$ and $\theta = -1$ parts. Suppose $\{X_i\}_i$ is a basis of $\mathfrak{p}$, and $\{X_i^*\}_i$ the dual basis of $\mathfrak{p}^{\vee}$. Set

$$\widetilde{D}F(g) = \sum_i X_i F(g) \otimes X_i^*,$$

a function $G_2(\mathbf{A}) \to \mathbb{V}_n^{\vee} \otimes \mathfrak{p}^{\vee}$. It is independent of the choice of basis X_i.

Now, as a representation of $\mathrm{SU}(2) \times \mathrm{SU}(2)$, $\mathfrak{p} \simeq V_2 \boxtimes W$, where W is the symmetric cube of the defining two-dimensional representation of the short-root $\mathrm{SU}(2)$. Thus, $\mathbb{V}_n^{\vee} \otimes \mathfrak{p}^{\vee}$ surjects onto $Sym^{2n-1}(V_2)^{\vee} \boxtimes W^{\vee}$. The composition of $\widetilde{D}$ with this projection is the operator $\mathcal{D}_n$. It induces a map

$$\left(C^{\infty}(G(\mathbf{R})) \otimes Sym^{2n}(V_2)^{\vee} \boxtimes \mathbf{1}\right)^K \to \left(C^{\infty}(G(\mathbf{R})) \otimes Sym^{2n-1}(V_2)^{\vee} \boxtimes W^{\vee}\right)^K.$$

Because the K-type $Sym^{2n-1}(V_2) \boxtimes W$ does not appear in $\pi_n|_K$, see (1), one sees that $\mathcal{D}_n F_{\varphi} = 0$. A good bit of the work to prove Theorem 3.4 will be to make explicit the equations $\mathcal{D}_n \mathcal{W}^{\chi} = 0$. See [KO95, Yam90, Yam91], and [YY96a, YY96b, Yos99] for some related work.

4.1 The Cartan Decomposition

In this subsection, we make some explicit computations with the Cartan decomposition $\mathfrak{g} = \mathfrak{k} \oplus \mathfrak{p}$. As mentioned, $\mathfrak{k} \simeq \mathfrak{sl}_2 \oplus \mathfrak{sl}_2$, and $\mathfrak{p} \simeq V_2 \boxtimes W$ as a representation of $\mathfrak{k}$. The purpose of this subsection is to record these isomorphisms explicitly.

First, observe that $\theta(v_j) = \delta_j$, and $\theta(E_{i,j}) = -E_{j,i}$. Thus $\mathfrak{p} = \mathfrak{g}^{\theta=-1}$ is spanned by the Cartan $\mathfrak{h}$, $E_{i,j} + E_{j,i}$ for $i \neq j$, and $v_j - \delta_j$. One finds $\mathfrak{k} = \mathfrak{g}^{\theta=1}$ is spanned by $v_j + \delta_j$ and $E_{i,j} - E_{j,i}$. To make explicit the equation $\mathcal{D}_n \mathcal{W}^{\chi} = 0$, we will need bases of $\mathfrak{k}$ and $\mathfrak{p}$ that make clear the $\mathfrak{k}$-action.

4.1.1 Basis of $\mathfrak{k}$

To do this, for $j \in \{1, 2, 3\}$, and indices taken modulo 3, set

$$u_j = \frac{1}{4}\left(E_{j+2,j+1} - E_{j+1,j+2} + v_j + \delta_j\right)$$

$$r_j = \frac{1}{4}\left(3E_{j+2,j+1} - 3E_{j+1,j+2} - v_j - \delta_j\right).$$

The elements u_j and r_j for $j = 1, 2, 3$ span $\mathfrak{k}$. In fact, the u_j span the long-root $\mathfrak{sl}_2$ and the r_j span the short-root $\mathfrak{sl}_2$. As the reader may check by explicit computation, they satisfy the following commutation relations:

- $[u_i, r_j] = 0$ for all i, j;
- $[u_j, u_{j+1}] = u_{j+2}$
- $[r_j, r_{j+1}] = r_{j+2}$.

Write $i = \sqrt{-1}$. For later use, we set

- $h_u = 2iu_2$
- $e_u = u_1 - iu_3$
- $f_u = -u_1 - iu_3$
- $h_r = 2ir_2$
- $e_r = r_1 - ir_3$
- $f_r = -r_1 - ir_3$.

These are $\mathfrak{sl}_2$-triples. I.e., $[h_u, e_u] = 2e_u$, $[h_u, f_u] = -2f_u$, $[e_u, f_u] = h_u$, and similarly for the r's.

4.1.2 Basis of $\mathfrak{p}$

In this paragraph, we write down a good basis of $\mathfrak{p}$. For $j = 1, 2, 3$, set $y_j = v_j - \delta_j$, and $f_j = E_{j+1,j+2} + E_{j+2,j+1}$, with indices taken modulo 3. Together with the split cartan $\mathfrak{h}$, these elements span $\mathfrak{p}$. We will now explicitly describe a basis of $\mathfrak{p}$ that is well-suited to our $\mathfrak{sl}_2$-triples above. Here is that basis.

$$d_3 := f_1 - y_1 + i(f_3 + y_3) \qquad h_3 := 2(E_{33} - E_{11}) + 2if_2$$

$$d_1 := -\frac{2}{3}(E_{11} - 2E_{22} + E_{33}) + \frac{2}{3}iy_2 \qquad h_1 := \frac{3f_1 + y_1}{3} + i\left(\frac{3f_3 - y_3}{3}\right)$$

$$d_{-1} := -\frac{3f_1 + y_1}{3} + i\left(\frac{3f_3 - y_3}{3}\right) \qquad h_{-1} := -\frac{2}{3}(E_{11} - 2E_{22} + E_{33}) - \frac{2}{3}iy_2$$

$$d_{-3} := 2(E_{33} - E_{11}) - 2if_2 \qquad h_{-3} := y_1 - f_1 + i(f_3 + y_3).$$

Denote by x_ℓ, y_ℓ a standard basis for the standard 2-dimensional representation of the $\mathfrak{sl}_2$ spanned by e_u, h_u, f_u, and denote by x_s, y_s a standard basis for the standard 2-dimensional representation of the $\mathfrak{sl}_2$ spanned by e_r, h_r, f_r. (The "ℓ" and "s" are for *long* and *short*.) As a representation of $\mathfrak{sl}_2$, the triple $\{e_u, h_u, f_u\}$ spans a copy of $Sym^2(V_2)$. We normalize the choice of x_ℓ, y_ℓ by the correspondence $e_u \mapsto x_\ell^2$, $h_u \mapsto -2x_\ell y_\ell$, and $f_u \mapsto -y_\ell^2$.

Proposition 4.1. *The map $\mathfrak{p} \to V_2 \boxtimes S^3 V_2$ sending $h_{3-2i} \mapsto x_\ell \boxtimes x_s^{3-i} y_s^i$, and $d_{3-2i} \mapsto y_\ell \boxtimes x_s^{3-i} y_s^i$ is an isomorphism of $\mathfrak{sl}_2 \oplus \mathfrak{sl}_2$ modules. In particular, h_u acts on h_* by the eigenvalue 1 and d_* by the eigenvalue -1. The element h_r acts on d_k and h_k by the eigenvalue k.*

Proof. Again, this is just explicit computation. □

We will also need to know how the Killing form restricts to the h_i and d_j's. This is recorded in the following lemma. Recall from Sect. 2.4 the pairing $(\cdot,\cdot)$ on $\mathfrak{g}_2$, which is restricted from $\mathfrak{so}(V_7) \simeq \wedge^2 V_7$.

Lemma 4.2. *One has $d_3 = -\overline{h}_{-3}$, $d_1 = \overline{h}_{-1}$, $d_{-1} = -\overline{h}_1$, and $d_{-3} = \overline{h}_3$. Futhermore, one has the following pairings:*

- $(\overline{h}_{-3}, h_{-3}) = 16$
- $(\overline{h}_{-1}, h_{-1}) = 16/3$
- $(\overline{h}_1, h_1) = 16/3$
- $(\overline{h}_3, h_3) = 16.$

Finally, $(h_i, h_j) = (\overline{h}_i, \overline{h}_j) = 0$ for all i, j and $(\overline{h}_i . h_j) = 0$ if $i \neq j$.

Remark 4.3. In [Pol18], we can replace some of the above explicit computation, as follows. As described above, $\mathfrak{g} = \mathfrak{g}_2$ has a $\mathbf{Z}/3$-grading, and one can define the Lie algebra structure through this $\mathbf{Z}/3$-model. The Lie algebra $\mathfrak{g}$ also has a $\mathbf{Z}/2$-grading, $\mathfrak{g} = \mathfrak{g}^0 \oplus \mathfrak{g}^1$ and one can define the structure through this $\mathbf{Z}/2$-model. All of the larger quaternionic exceptional Lie algebras also have both a $\mathbf{Z}/3$ and $\mathbf{Z}/2$ model. Denote by G the adjoint group associated to the Lie algebra $\mathfrak{g}$. In [Pol18, section 5] we write down an explicit element $\mathcal{C} \in G(\mathbf{C})$ with the property that $\mathcal{C}(\mathfrak{g}^0 \otimes \mathbf{C}) = \mathfrak{k}$ and $\mathcal{C}(\mathfrak{g}^1 \otimes \mathbf{C}) = \mathfrak{p}$. One gets good bases for $\mathfrak{k}$ and $\mathfrak{p}$ and all their associated structure from the structure of $\mathfrak{g}^0$ and $\mathfrak{g}^1$ and the fact that $\mathcal{C}$ is an automorphism of $\mathfrak{g} \otimes \mathbf{C}$.

In order to define $\mathcal{C}$, we use both the $\mathbf{Z}/3$-model, the $\mathbf{Z}/2$-model, and an explicit isomorphism between them. This element $\mathcal{C}$ is a sort of explicit, exceptional "Cayley" transform. The reason for the name is that classical Cayley transform has an analogous property: On $\mathfrak{sp}_{2n}$, the Cayley transform is an element $C \in \mathrm{Sp}_{2n}(\mathbf{C})$ that conjugates the decomposition $\mathfrak{n} \oplus \mathfrak{m} \oplus \overline{\mathfrak{n}}$ to $\mathfrak{p}^+ \oplus \mathfrak{k} \oplus \mathfrak{p}^-$, where $\mathfrak{m}$ is the Lie algebra of the Levi of the Siegel parabolic, $\mathfrak{n}$ is its unipotent radical, and $\overline{\mathfrak{n}}$ is the opposite unipotent. A more abstract Cayley transform for the quaternionic Lie algebras can be found in [GW96]. To keep things as elementary as possible, we have defined $\mathfrak{g}_2$ through the trace 0 octonions V_7 and used the $\mathbf{Z}/3$-model of $\mathfrak{g}$, and thus not introduced the Cayley transform $\mathcal{C}$.

4.2 Iwasawa Decomposition

To write down the equations $\mathcal{D}_n \mathcal{W}^\chi = 0$, and to utilize the equivariance properties of the functions $\mathcal{W}^\chi$, we will make use of the Iwasawa decomposition on $\mathfrak{g}$. That is, if $X \in \mathfrak{g}$, one can write $X = n + m + k$ for $n \in \mathfrak{n}$ the Lie algebra of $N(\mathbf{R})$, $m \in \mathfrak{gl}_2$ the Lie algebra of the Levi $M(\mathbf{R})$ of the Heisenberg parabolic P, and $k \in \mathfrak{k}$.

Because the abelianization of N is $Sym^3(V_2)$, we have a map $\mathfrak{n} \to Sym^3(V_2)$ sending an element $X \in \mathfrak{n}$ to a binary cubic polynomial p_X in the two variables u, v. Specifically, one sends

$$X = aE_{12} + \frac{b}{3}v_1 + \frac{c}{3}\delta_3 + dE_{23} + \mu E_{13} \mapsto p_X = au^3 + bu^2v + cuv^2 + dv^3 = (a, \frac{b}{3}, \frac{c}{3}, d).$$

We now record the Iwasawa decomposition of the elements h_i using this polynomial map. (Because $\mathcal{W}^\chi(\exp(\mu E_{13})g) = \mathcal{W}^\chi(g)$, the loss of the information about E_{13} spanning $N_0 = [N, N]$ will not cause any problem.)

Set $\epsilon_1 = E_{22} - E_{33}$ and $\epsilon_2 = E_{11} - E_{22}$. These are a basis of $\mathfrak{h}$. In fact, under the map $\mathfrak{gl}_2 \to \mathfrak{g}_2$ from (2), ϵ_1 is the image of $\left(\begin{smallmatrix}1&0\\0&0\end{smallmatrix}\right)$ and ϵ_2 the image of $\left(\begin{smallmatrix}0&0\\0&1\end{smallmatrix}\right)$.

We have

$$\begin{aligned}
h_3 &\equiv -2(\epsilon_1 + \epsilon_2) - (h_u + h_r)\\
h_1 &\equiv 2(v + iu)^2(v - iu) - \frac{4}{3}f_r\\
h_{-1} &\equiv \frac{2}{3}(\epsilon_1 - \epsilon_2 - 2iv_2) + \frac{1}{3}(3h_u - h_r)\\
h_{-3} &\equiv -2(v + iu)^3 - 4e_u.
\end{aligned}$$

We are here employing the map $\mathfrak{n} \to Sym^3(V_2)$ just described; the $\equiv$ means that the sides differ by an element of $\mathfrak{n}_0$. One obtains the Iwasawa decomposition of the $\overline{h}_j$'s by taking the complex conjugate of the above expressions.

4.3 Actions of Lie Algebra Elements

To write down the equations $\mathcal{D}_n \mathcal{W}^\chi = 0$ concretely, we coordinatize the Levi subgroup $M \simeq \mathrm{GL}_2$ of the Heisenberg parabolic, and make explicit how the Lie algebra elements appearing in the Iwasawa decomposition above act in these coordinates.

First, because $\mathcal{W}^\chi$ has equivariance properties with respect to $N(\mathbf{R})$ and the maximal compact subgroup K of $G_2(\mathbf{R})$, it suffices to choose coordinates for $B(\mathbf{R})^0$, the connected component of the identity of the real points of the uppertriangular Borel subgroup of $M \simeq \mathrm{GL}_2$. We use the coordinates $g = \begin{pmatrix} 1 & x \\ & 1 \end{pmatrix}\begin{pmatrix} y^{1/2} & \\ & y^{-1/2} \end{pmatrix}\begin{pmatrix} w & \\ & w \end{pmatrix}$ for $x \in \mathbf{R}$, and $y, w \in \mathbf{R}^{\times}_{>0}$.

More precisely, suppose $\mathcal{W}^{\chi}$ is a generalized Whittaker function for χ annihilated by $\mathcal{D}_n$. That is, assume that $\mathcal{W}^{\chi} : G_2(\mathbf{R}) \to \mathbb{V}_n^{\vee}$ is a smooth function of moderate growth, with $\mathcal{D}_n\mathcal{W}^{\chi} = 0$, and that $\mathcal{W}^{\chi}(ngk) = \chi(n)k^{-1} \cdot \mathcal{W}^{\chi}(g)$ for all $n \in N(\mathbf{R})$, $k \in K$ and $g \in G_2(\mathbf{R})$. We define $\phi : \mathbf{R} \times \mathbf{R}^{\times}_{>0} \times \mathbf{R}^{\times}_{>0} \to \mathbb{V}_n^{\vee}$ as

$$\phi(x, y, w) = \mathcal{W}^{\chi}\left(\begin{pmatrix} 1 & x \\ & 1 \end{pmatrix}\begin{pmatrix} y^{1/2} & \\ & y^{-1/2} \end{pmatrix}\begin{pmatrix} w & \\ & w \end{pmatrix}\right).$$

We will translate the condition $\mathcal{D}_n\mathcal{W}^{\chi} = 0$ into explicit equations for ϕ.

Now, because $\mathcal{W}^{\chi}(gk) = k^{-1} \cdot \mathcal{W}^{\chi}(g)$ for all $k \in K$, if $X \in \mathfrak{k}$ then $(X\mathcal{W}^{\chi})(g) = -X \cdot \mathcal{W}^{\chi}(g)$. The character χ on $N(\mathbf{R})$ is of the form $\chi(X) = e^{i\langle \omega, p_X \rangle}$ for a unique $\omega \in W$. It follows that if $X \in \mathfrak{n}$ and $m \in M(\mathbf{R})$, one has $(X\mathcal{W}^{\chi})(m) = i\langle \omega, m \cdot p_X \rangle \mathcal{W}^{\chi}(m)$. Finally, by definition, one has

$$\begin{aligned}
(\epsilon_1 + \epsilon_2)\phi(x, y, w) &= w\partial_w\phi(x, y, w) \\
(\epsilon_1 - \epsilon_2)\phi(x, y, w) &= 2y\partial_y\phi(x, y, w) \\
v_2\phi(x, y, w) &= y\partial_x\phi(x, y, w). \qquad (1)
\end{aligned}$$

4.4 Schmid Equation

In this subsection we write down the equation $\mathcal{D}_n\mathcal{W}^{\chi} = 0$ explicitly in coordinates. First, recall that $\mathcal{D}_n = pr_- \circ \widetilde{D}$, where

$$pr_- : \left(Sym^{2n}(V_2) \boxtimes \mathbf{1}\right) \otimes V_2 \boxtimes W \to Sym^{2n-1}(V_2) \boxtimes W$$

is the non-trivial K-equivariant map (unique up to scalar multiple). We fix the map $Sym^{2n}(V_2) \otimes V_2 \to Sym^{2n-1}(V_2)$ as

$$x_\ell^a y_\ell^b \otimes v \mapsto a\langle x_\ell, v \rangle x_\ell^{a-1} y_\ell^b + b\langle y_\ell, v \rangle x_\ell^a y_\ell^{b-1}$$

where $\langle \cdot, \cdot \rangle$ is the invariant symplectic form on V_2 with $\langle x_\ell, y_\ell \rangle = 1$.

Thus, for any function $F : G_2(\mathbf{R}) \to \mathbb{V}_n^{\vee}$ we have

$$\begin{aligned}
\widetilde{D}F = &-d_3F \otimes h_{-3} + 3d_1F \otimes h_{-1} - 3d_{-1}F \otimes h_1 + d_{-3}F \otimes h_3 \\
&+ h_3F \otimes d_{-3} - 3h_1F \otimes d_{-1} + 3h_{-1}F \otimes d_1 - h_{-3}F \otimes d_3.
\end{aligned}$$

Applying the Iwasawa decomposition of the elements $h_i, \overline{h_j}$ computed above, and the equivariance properties of $\mathcal{W}^\chi$ we obtain the following for $\widetilde{D}\mathcal{W}^\chi$:

$$\begin{aligned}\widetilde{D}\mathcal{W}^\chi &= -2i\langle \omega, m\cdot(v-iu)^3\rangle \mathcal{W}^\chi \otimes h_{-3} - 4f_u\cdot(\mathcal{W}^\chi \otimes h_{-3}) + 4\mathcal{W}^\chi \otimes d_{-3} \\ &- 2(\epsilon_2 - \epsilon_1 - 2iv_2)\mathcal{W}^\chi \otimes h_{-1} + (3h_u - h_r - 4)\cdot(\mathcal{W}^\chi \otimes h_{-1}) \\ &+ 6i\langle \omega, m\cdot(v-iu)^2(v+iu)\rangle \mathcal{W}^\chi \otimes h_1 - 4e_r\cdot(\mathcal{W}^\chi \otimes h_1) + 4\mathcal{W}^\chi \otimes h_3 \\ &- 2(\epsilon_1+\epsilon_2)\mathcal{W}^\chi \otimes h_3 - (h_u + h_r - 4)\cdot(\mathcal{W}^\chi \otimes h_3) \\ &- 2(\epsilon_1+\epsilon_2)\mathcal{W}^\chi \otimes d_{-3} + (h_u + h_r + 4)\cdot(\mathcal{W}^\chi \otimes d_{-3}) \\ &- 6i\langle \omega, m\cdot(v+iu)^2(v-iu)\rangle \mathcal{W}^\chi \otimes d_{-1} - 4f_r\cdot(\mathcal{W}^\chi \otimes d_{-1}) + 4\mathcal{W}^\chi \otimes d_{-3} \\ &- 2(\epsilon_2 - \epsilon_1 + 2iv_2)\mathcal{W}^\chi \otimes d_1 + (h_r - 3h_u - 4)\cdot(\mathcal{W}^\chi \otimes d_1) \\ &+ 2i\langle \omega, m\cdot(v+iu)^3\rangle \mathcal{W}^\chi \otimes d_3 - 4e_u\cdot(\mathcal{W}^\chi \otimes d_3) + 4\mathcal{W}^\chi \otimes h_3.\end{aligned}$$

Write $[x_\ell^r]$ to mean $\frac{x_\ell^r}{r!}$, and similarly for y. For $k \in \mathbf{Z}$ with $-n \le k \le n$, define $\mathcal{W}^{\chi,k} : G_2(\mathbf{R}) \to \mathbf{C}$ via $\mathcal{W}^\chi(g) = \sum_{-n\le k\le n} \mathcal{W}^{\chi,k}[x_\ell^{n+k}][y_\ell^{n-k}]$. Then applying the contraction pr_-, we get

$$\begin{aligned}\mathcal{D}_n\mathcal{W}^\chi &= \sum_k \mathcal{D}_{3,0}^{n,k}\mathcal{W}^{\chi,k}[x_\ell^{n+k}][y_\ell^{n-k-1}] \boxtimes x_s^3 + \sum_k \mathcal{D}_{2,1}^{n,k}\mathcal{W}^{\chi,k}[x_\ell^{n+k}][y_\ell^{n-k-1}] \boxtimes x_s^2 y_s \\ &+ \sum_k \mathcal{D}_{1,2}^{n,k}\mathcal{W}^{\chi,k}[x_\ell^{n+k}][y_\ell^{n-k-1}] \boxtimes x_s^1 y_s^2 + \sum_k \mathcal{D}_{0,3}^{n,k}\mathcal{W}^{\chi,k}[x_\ell^{n+k}][y_\ell^{n-k-1}] \boxtimes y_s^3\end{aligned}$$

for functions $\mathcal{D}_{i,j}^{n,k}\mathcal{W}^{\chi,k}$ that are as follows:

$$\begin{aligned}\frac{1}{2}\mathcal{D}_{3,0}^{n,k}\mathcal{W}^{\chi,k} &= (\epsilon_1+\epsilon_2-(2n+2)-k)\mathcal{W}^{\chi,k} + i\langle \omega, m\cdot(v+iu)^3\rangle \mathcal{W}^{\chi,k+1} \\ \frac{1}{2}\mathcal{D}_{0,3}^{n,k}\mathcal{W}^{\chi,k} &= -(\epsilon_1+\epsilon_2-(2n+2)+k+1)\mathcal{W}^{\chi,k+1} + i\langle \omega, m\cdot(v-iu)^3\rangle \mathcal{W}^{\chi,k} \\ \frac{1}{2}\mathcal{D}_{2,1}^{n,k}\mathcal{W}^{\chi,k} &= -3i\langle \omega, m\cdot(v-iu)^2(v+iu)\rangle \mathcal{W}^{\chi,k} - (\epsilon_2-\epsilon_1+2iv_2+3(k+1))\mathcal{W}^{\chi,k+1} \\ \frac{1}{2}\mathcal{D}_{1,2}^{n,k}\mathcal{W}^{\chi,k} &= -3i\langle \omega, m\cdot(v+iu)^2(v-iu)\rangle \mathcal{W}^{\chi,k+1} + (\epsilon_2-\epsilon_1-2iv_2-3k)\mathcal{W}^{\chi,k}.\end{aligned}$$

Since $\mathcal{D}_n\mathcal{W}^\chi = 0$, the above expressions yield differential-difference equations satisfied by the $\mathcal{W}^{\chi,k}$. Recall that we defined $\phi(x,y,w) = \mathcal{W}^\chi\left(\begin{pmatrix}1 & x\\ & 1\end{pmatrix}\begin{pmatrix}y^{1/2} & \\ & y^{-1/2}\end{pmatrix}\begin{pmatrix}w & \\ & w\end{pmatrix}\right)$, and similarly define $\phi_k(x,y,w) : \mathbf{R}\times\mathbf{R}_{>0}^\times\times\mathbf{R}_{>0}^\times \to \mathbf{C}$ as

$$\phi_k(x,y,w) = \mathcal{W}^{\chi,k}\left(\begin{pmatrix}1 & x\\ & 1\end{pmatrix}\begin{pmatrix}y^{1/2} & \\ & y^{-1/2}\end{pmatrix}\begin{pmatrix}w & \\ & w\end{pmatrix}\right).$$

Applying the identities (1), we obtain the following differential equations for the ϕ_k.

Proposition 4.4. *Write $m \in \mathrm{GL}_2(\mathbf{R})$ for the element* $\begin{pmatrix} 1 & x \\ & 1 \end{pmatrix} \begin{pmatrix} y^{1/2} & \\ & y^{-1/2} \end{pmatrix} \begin{pmatrix} w & \\ & w \end{pmatrix}$. *The ϕ_k satisfy the following differential-difference equations:*

(1) $(w\partial_w - (2n+2) - k)\phi_k + i\langle \omega, m \cdot (v+iu)^3 \rangle \phi_{k+1} = 0$
(2) $-(w\partial_w - (2n+2) + k)\phi_{k+1} + i\langle \omega, m \cdot (v-iu)^3 \rangle \phi_k = 0$
(3) $-3i\langle \omega, m \cdot (v-iu)^2(v+iu) \rangle \phi_k - (-2y\partial_y + 2iy\partial_x + 3(k+1))\phi_{k+1} = 0$
(4) $-3i\langle \omega, m \cdot (v+iu)^2(v-iu) \rangle \phi_{k+1} + (-2y\partial_y - 2iy\partial_x - 3k)\phi_k = 0.$

4.5 The Formula

In this subsection we analyze the equations of Proposition 4.4. We find that if $\omega \neq 0$ then there is at most a one-dimensional space of solutions to these equations that are of moderate growth, and that the solutions are given as in (2). In particular, we prove Thereom 3.4.

First, set $z = x + iy$, $\partial_z = \frac{1}{2}\left(\partial_x - i\partial_y\right)$ and $\partial_{z^*} = \frac{1}{2}(\partial_x + i\partial_y)$. As is clear from Proposition 4.4, this variable change will help solve the above differential equations. Furthermore, we note that if f is a polynomial in u, v, and $m \in \mathrm{GL}_2$, as we have in Proposition 4.4, then $(m \cdot f)(u, v) = \det(m)^2 f\left(m^{-1}\begin{pmatrix} u \\ v \end{pmatrix}\right)$. To go further, we give expressions for the terms in Proposition 4.4 of the form $\langle \omega, m \cdot f \rangle$ for the binary cubics f appearing in this proposition.

The following lemma may be checked by direct computation.

Lemma 4.5. *Let the notation be as above. Then $m \cdot (u - iv)^3 = wy^{-3/2}(u - zv)^3$. Furthermore, suppose $\omega = -(a, \frac{b}{3}, \frac{c}{3}, d)$, and define $p_\chi(t) = at^3 + bt^2 + ct + d$, a cubic polynomial. Then*

(1) $\langle \omega, m \cdot (-v+iu)^3 \rangle = -iwy^{-3/2}p_\chi(z^*)$
(2) $\langle \omega, m \cdot (-v-iu)^3 \rangle = iwy^{-3/2}p_\chi(z)$
(3) $\langle \omega, m \cdot 3(-v-iu)^2(-v+iu) \rangle = -wy^{-3/2}(2yp'_\chi(z) + 3ip_\chi(z)) = -2wy^{-5/2}\partial_z\left(p_\chi(z)y^{-3}\right)$
(4) $\langle \omega, m \cdot 3(-v+iu)^2(-v-iu) \rangle = -2wy^{-5/2}\partial_{z^*}\left(p_\chi(z^*)y^{-3}\right).$

Applying the variable change and the lemma, we obtain the following simplified differential equations. Define a function G_k by the equality $\phi_k = w^{2n+2}G_k$.

Corollary 4.6. *The G_k satisfy the following differential-difference equations.*

(1) $(w\partial_w + k + 1)G_{k+1} = -wy^{-3/2}p_\chi(z^*)G_k$
(2) $(w\partial_w - k)G_k = -wy^{-3/2}p_\chi(z)G_{k+1}$
(3) $(4iy\partial_z + 3k)G_k = -2iwy^{5/2}\partial_z(p_\chi(z)y^{-3})G_{k+1}$
(4) $(4iy\partial z^* + 3k)G_k = -2iwy^{5/2}\partial_{z^*}(p_\chi(z^*)y^{-3})G_{k-1}.$

To solve the equations of Corollary 4.6, we require a couple preliminaries. First, set $u = |p_\chi(z)|y^{-3/2}w$. Note that, for a function f,

(1) $(4iy\partial_z + 3v)(y^{-3v/2}f) = y^{-3v/2}(4iy\partial_z f)$
(2) $(4iy\partial_{z^*} + 3v)(y^{3v/2}f) = y^{3v/2}(4iy\partial_{z^*} f)$
(3) $\partial_{z^*}(u) = \frac{1}{2}wy^{3/2}\frac{p_\chi(z)}{|p_\chi(z)|}\partial_{z^*}(p_\chi(z^*)y^{-3})$
(4) $\partial_z(u) = \frac{1}{2}wy^{3/2}\frac{p_\chi(z^*)}{|p_\chi(z)|}\partial_z(p_\chi(z)y^{-3})$.

The solution to the equations of Corollary 4.6 will involve K-Bessel functions. We recall the following standard facts regarding these functions:

Theorem 4.7. *The K-Bessel functions satisfy the following identities:*

(1) $((z\partial_z)^2 - v^2)K_v(z) = z^2K_v(z)$
(2) $-z^{-v}\partial_z(z^vK_v) = K_{v-1}$
(3) $-z^v\partial_z(z^{-v}K_v) = K_{v+1}$
(4) $-(z\partial_z - v)K_v = zK_{v+1}$
(5) $-(z\partial_z + v)K_v = zK_{v-1}$.

We now come to the proof of the explicit formula.

Proof of Theorem 3.4. Applying the differential equations in w, one gets

$$\begin{aligned}(w\partial_w + v)(w\partial_w - v)G_v &= (w\partial_w + v)\left(-wy^{-3/2}p_\chi(z)G_{v+1}\right)\\ &= -wy^{-3/2}p_\chi(z)(w\partial_w + v + 1)\left(G_{v+1}\right)\\ &= (wy^{-3/2}p_\chi(z))wy^{-3/2}p_\chi(z^*)G_v = u^2G_v.\end{aligned}$$

It follows that G_v is a Bessel function in w. More precisely, since G_v cannot grow exponentially as $w \to \infty$, $G_v = K_v(u)Y_v(z, z^*)$ for some function Y_v of z, z^*.

Now, using this, one gets

$$\begin{aligned}wp_\chi(z)y^{-3/2}Y_{v+1}(z, z^*)K_{v+1}(u) &= -(w\partial_w - v)G_v\\ &= |p_\chi(z)|y^{-3/2}wY_v(z, z^*)K_{v+1}(u).\end{aligned}$$

Hence $Y_{v+1} = \frac{|p_\chi(z)|}{p_\chi(z)}Y_v(z, z^*) = \frac{p_\chi(z)^*}{|p_\chi(z)|}Y_v(z, z^*)$. One concludes that, on an open set where $p_\chi(z) \neq 0$, $Y_v(z, z^*) = \left(\frac{|p_\chi(z)|}{p_\chi(z)}\right)^v Y_0(z, z^*)$.

It remains to understand the function Y_0. To do so, one applies the differential equations with ∂_z and ∂_{z^*} to show $\partial_z Y_0 = \partial_{z^*}Y_0 = 0$. Thus, Y_0 is constant. It follows that on an open set where $p_\chi(z) \neq 0$, one obtains that ϕ_k is proportional to $\left(\frac{|p_\chi(z)|}{p_\chi(z)}\right)^v w^{2n+2}K_v(|p_\chi(z)|y^{-3/2}w)$. Note that if w stays constant but $p_\chi(z) \to 0$, this function diverges. Hence, in order for there to be any nonzero solutions to the equations $\mathcal{D}_n\mathcal{W}^\chi = 0$, we require that $p_\chi(z)$ is never 0 on the upper half-plane $\mathcal{H}_1$.

One obtains the exact expression for the functions $\mathcal{W}_w(m)$ by imposing the K-equivariance $\mathcal{W}_w(gk) = k^{-1}\mathcal{W}_w(g)$. Finally, one can check that the solutions found really do satisfy all the differential equations. □

Remark 4.8. The reader wishing to analyze the constant term of modular forms on G_2 must simply take $\omega = 0$ in Proposition 4.4. Substantial simplification occurs immediately.

5 The Rankin-Selberg Integral

In this section we prove Theorem 3.5. In fact, this Dirichlet series is more-or-less equivalent to the unramified computation of the Rankin-Selberg integrals in [GS15, Seg17], so we also recall these integrals.

5.1 *The Global Integral*

We now recall the global Rankin-Selberg integrals of [GS15] and [Seg17]. Fix F a number field, and set $G = G_2$ the split exceptional group over F. The global integral depends on an auxiliary cubic étale extension E of F. Out of E, one can define a reductive F-group G_E which is of type D_4, and for which there is a natural inclusion $G \subseteq G_E$.

5.1.1 The Group G_E

More precisely, first suppose $E = F \times F \times F$ is split, and Θ denotes the split octonions over F. Denote by $(\cdot, \cdot, \cdot)$ the trilinear form on Θ given by $(x_1, x_2, x_3) = \mathrm{tr}((x_1x_2)x_3) = \mathrm{tr}(x_1(x_2x_3))$. Then $G_E^{sc} = \mathrm{Spin}(8)$ is the group of triples of automorphisms $g = (g_1, g_2, g_3) \in \mathrm{SO}(\Theta)^3$ satisfying $(g_1x_1, g_2x_2, g_3x_3) = (x_1, x_2, x_3)$ for all $x_1, x_2, x_3 \in \Theta$. The group $G = G_2$ embeds in diagonally G_E^{sc} as $g \mapsto (g, g, g)$.

For a general cubic étale extension E of F, one defines G_E^{sc} as follows. The F-vector space $E \otimes_F \Theta$ has a unique linear form ℓ that is the descent of the linear form $(\cdot, \cdot, \cdot) : (F \times F \times F) \otimes_F \Theta \to F$ defined using the trace on Θ. The form ℓ satisfies $\ell(xv) = N_{E/F}(x)\ell(v)$ for all $x \in E$ and $v \in E \otimes \Theta$. Then one defines G_E to consist of the E-linear automorphisms g of $E \otimes \Theta$ in $\mathrm{SO}_E(E \otimes \Theta)$ satisfying $\ell(gv) = \ell(v)$ for all $v \in E \otimes \Theta$. The group $G = G_2$ embeds in G_E^{sc} as the E-linear extension of F-automorphisms of Θ.

Alternatively, instead of the using a simply-connected group of type D_4, one could use the adjoint group of type D_4 to define the Rankin-Selberg integral. For this, one considers $\mathfrak{g}_E = \mathfrak{sl}_3 \oplus E^{\mathrm{tr}=0} \oplus (V_3 \otimes E) \oplus (V_3 \otimes E)^\vee$. One can define the structure of a Lie algebra on $\mathfrak{g}_E$, for which $\mathfrak{sl}_3 \oplus E^{\mathrm{tr}=0}$, $(V_3 \otimes E)$, and $(V_3 \otimes E)^\vee$ are the pieces of a $\mathbf{Z}/3$-grading. See [Rum97, section 1], or also [Pol18, section 4]. One then defines G_E^{ad} as the adjoint group associated to this Lie algebra. The Lie algebra of $\mathfrak{g}$ of $G = G_2$ embeds in $\mathfrak{g}_E$ as $\mathfrak{sl}_3 \oplus V_3 \oplus V_3^\vee$. This embedding yields an action of $\mathfrak{g}$ on $\mathfrak{g}_E$, which gives rise to a map $G \to G_E^{ad}$. (The group G_2 is simply-connected and adjoint.) In fact, denote by E^0 the elements of E with trace equal to 0. Then one has an isomorphism of $\mathfrak{g}_2$-modules $\mathfrak{g}_E = \mathfrak{g}_2 \oplus V_7 \otimes E^0$ with E^0 a trivial $\mathfrak{g}_2$-module. Here the map $V_7 \otimes E^0 \to \mathfrak{g}_E$ is given as follows: for $j \in \{1, 2, 3\}$ and $\epsilon \in E^0$, $e_j \otimes \epsilon \mapsto -v_j \otimes \epsilon$, $e_j^* \otimes \epsilon \mapsto \delta_j \otimes \epsilon$, and $u_0 \otimes \epsilon \mapsto \Phi_{\epsilon,1}$ in the notation of [Pola, section 3].

Now set $G_E = G_E^{sc}$ or G_E^{ad}. The Rankin-Selberg integral is formed by defining a degenerate Eisenstein series on G_E for the Heisenberg parabolic, and pulling it back to G. Because we have $G \to G_E^{sc} \to G_E^{ad}$, and this Eisenstein series on G_E^{sc} is the pullback of the degenerate Eisenstein series on G_E^{ad}, one can use either G_E^{sc} or G_E^{ad} to define the global integral. We have described both the simply-connected and the adjoint form of the group G_E, because both are convenient for calculating with this integral: The simply-connected group is convenient for unfolding the integral and doing orbit calculations, while the adjoint group is useful for doing certain calculations once the integral is unfolded.

5.1.2 The Global Integral

Let us now define the global integral precisely. Define P_E to be the subgroup of G_E preserving the line spanned by $E_{13} \in \mathfrak{sl}_3 \subseteq \mathfrak{g}_E$. Then P_E is the Heisenberg parabolic subgroup of G_E, and the action of P_E on E_{13} gives the similitude character $\nu : P_E \to \mathrm{GL}_1$, $Ad(p)E_{13} = \nu(p)E_{13}$. The modulus character of P_E is $\delta_{P_E}(p) = |\nu(p)|^5$.

The Eisenstein series on G_E is defined as follows. Suppose that Φ is a Schwartz-Bruhat function on $\mathfrak{g}_E \otimes \mathbf{A}$. Define

$$f(g, \Phi, s) = \int_{\mathrm{GL}_1(\mathbf{A})} \Phi(tAd(g)^{-1}E_{13})|t|^s\, dt.$$

If $p \in P_E$, then $f(pg, \Phi, s) = |\nu(p)|^s f(g, \Phi, s)$. The Eisenstein series is

$$E(g, \Phi, s) = \sum_{\gamma \in P_E(F)\backslash G_E(F)} f(\gamma g, \Phi, s).$$

Defining $f(g, \Phi, s)$ as an integral of a Schwartz-Bruhat function Φ on $\mathfrak{g}_E$ is one of the modifications we make to the computation in [GS15, Seg17]. The use of Φ will lead to simplifications later.

Suppose that π is a cuspidal automorphic representation on $G(\mathbf{A})$, and φ is a cusp form in the space of π. The global integral is

$$I(\varphi, \Phi, s) = \int_{G(F)\backslash G(\mathbf{A})} \varphi(g) E(g, \Phi, s)\, dg.$$

With our normalizations, the main results of [GS15] and [Seg17] imply that this integral represents the ratio of partial L-functions $\frac{L^S(\pi, Std, s-2)}{L^S(E, s-1)\zeta_F^S(2s-4)}$.

5.2 *Embedding in* SO(7)

Before unfolding the global integral $I(\varphi, \Phi, s)$, we dispense with a preliminary that we will need below. Namely, we describe explicitly how the Heisenberg parabolic subgroup $P = MN$ of G_2 embeds in SO(7). Of course, the SO(7) is SO(V_7), with quadratic form coming from the norm on the octonions.

Fix the ordered basis

$$e_1, e_3^*, e_2^*, u_0, -e_2, -e_1^*, -e_3$$

of V_7. With this ordered basis, the Gram matrix for the bilinear form $(\cdot, \cdot)$ on V_7 is

$$\begin{pmatrix} & & I_2 \\ & S & \\ I_2 & & \end{pmatrix} \text{ with } S = \begin{pmatrix} & & 1 \\ & -2 & \\ 1 & & \end{pmatrix}.$$

Denote by V_2 the standard representation of $\mathrm{GL}_2 \simeq M$. Then, as a representation of GL_2, we have $V_7 \simeq V_2 \oplus M_2^{\mathrm{tr}=0} \oplus V_2^\vee$. Here $M_2^{\mathrm{tr}=0}$ denotes the 2×2 matrices with trace 0, on which GL_2 acts by conjugation. The quadratic form S on $M_2^{\mathrm{tr}=0}$ is given by the determinant. Put another way, one embeds $\mathrm{GL}_2 \simeq M$ into SO(V_7) as $g \mapsto \mathrm{diag}(g, Ad^0(g), {}^t g^{-1})$. Here $Ad^0(g)$ is the map $M_2^{\mathrm{tr}=0} \to M_2^{\mathrm{tr}=0}$ given by $x \mapsto gxg^{-1}$. In this setup, the identification Span(e_2^*, u_0, e_2) with $M_2^{\mathrm{tr}=0}$ is given by $e_2^* \mapsto \left(\begin{smallmatrix} 0 & -1 \\ 0 & 0 \end{smallmatrix}\right)$, $u_0 \mapsto \left(\begin{smallmatrix} 1 & 0 \\ 0 & -1 \end{smallmatrix}\right)$, and $e_2 \mapsto \left(\begin{smallmatrix} 0 & 0 \\ -1 & 0 \end{smallmatrix}\right)$.

Caution: Our identification $M \simeq \mathrm{GL}_2$ is **not** the same identification that was described in Sects. 2 or 4. The identification $M \simeq \mathrm{GL}_2$ just described differs from the previous ones by a conjugation by the matrix $\left(\begin{smallmatrix} & 1 \\ -1 & \end{smallmatrix}\right)$. We will use this new identification of M with GL_2 for the rest of this section.

Now, if one associates to the element $X = u_1 E_{12} + \frac{u_2}{3} v_1 + \frac{u_3}{3} \delta_3 + u_4 E_{23}$ the polynomial $f_X(x, y) = u_1 x^3 + u_2 x^2 y + u_3 x y^2 + u_4 y^3$, and if $g \in \mathrm{GL}_2 \simeq M$, then $f_{g \cdot X}(x, y) = \det(g)^{-1} f((x, y)g)$.

Claim 5.1. *Write* $(\alpha, \beta, \gamma, \delta)$ *for the element* $\alpha E_{12} + \beta v_1 + \gamma \delta_3 + \delta E_{23}$ *of* $\mathfrak{g}$. *The Heisenberg Lie algebra* $\mathfrak{n}$ *is embedded as*

$$\mu E_{13} + (\alpha, \beta, \gamma, \delta) \mapsto \begin{pmatrix} 0 & h' & -\mu J_2 \\ 0 & 0 & h \\ 0 & 0 & 0 \end{pmatrix} \text{ with } h' = -{}^t h S$$

and $h = \begin{pmatrix} \alpha & \beta \\ \beta & \gamma \\ \gamma & \delta \end{pmatrix}$. *Here* $J_2 = \begin{pmatrix} & 1 \\ -1 & \end{pmatrix}$.

Proof. Recall that $v_1 = u_0 \wedge e_1 + e_2^* \wedge e_3^*$ and $\delta_3 = u_0 \wedge e_3^* + e_1 \wedge e_2$. Thus

$$\alpha E_{12} + \beta v_1 + \gamma \delta_3 + \delta E_{23} = \alpha e_2^* \wedge e_1 + \beta(u_0 \wedge e_1 + e_2^* \wedge e_3^*) + \gamma(u_0 \wedge e_3^* + e_1 \wedge e_2) + \delta e_3^* \wedge e_2.$$

Evaluating this Lie algebra element on $-e_1^*$ and $-e_3$ we get

$$(\alpha E_{12} + \beta v_1 + \gamma \delta_3 + \delta E_{23})(-e_1^*) = \alpha e_2^* + \beta u_0 + \gamma(-e_2)$$

and

$$(\alpha E_{12} + \beta v_1 + \gamma \delta_3 + \delta E_{23})(-e_3) = \beta e_2^* + \gamma u_0 + \delta(-e_2).$$

The claim follows. □

The Heisenberg subgroup N consists of elements of the form

$$\begin{pmatrix} I_2 & h' & x \\ & I_3 & h \\ & & I_2 \end{pmatrix} \text{ with } {}^t h S h + x + {}^t x = 0.$$

5.2.1 Unfolding

For the detailed unfolding of the integral $I(\varphi, \Phi, s)$, we refer the reader to [Seg17, Theorem 3.2]. We will content ourselves with just indicating the lines of the proof, because we state the unfolding a bit differently than it appears in *loc cit.*

As preparation, denote by $\mathcal{O}_E$ the maximal order in E, and recall from Sect. 3.3 the notion of a good basis of $\mathcal{O}_E$. Suppose $1, \omega, \theta$ is a good basis of $\mathcal{O}_E$, with multiplication table as in Sect. 3.3 with the constants a, b, c, d. Set

$$v_E = aE_{12} + \frac{b}{3}v_1 + \frac{c}{3}\delta_3 + dE_{23} \in \mathfrak{g} = \mathfrak{sl}_3 \oplus V_3 \oplus V_3^\vee$$

and

$$\widetilde{v_E} = aE_{12} - v_1 \otimes \omega + \delta_3 \otimes \theta + dE_{23} \in \mathfrak{g}_E = (\mathfrak{sl}_3 \oplus E^{\mathrm{tr}=0}) \oplus (V_3 \otimes E) \oplus (V_3^\vee \otimes E).$$

The element $\widetilde{v_E}$ is the "rank one lift" of the element v_E in the sense of [Pola, Section 2.3].

One can show that $\widetilde{v_E}$ is in the $G_E(F)$-orbit of E_{13}; say $\gamma_0^{-1}E_{13} = \widetilde{v_E}$. In fact, one has

$$\exp\left(ad\left(-aE_{32} + v_3 \otimes \omega + \delta_1 \otimes \theta + dE_{21}\right)\right)(E_{13}) = E_{13} + \widetilde{v_E}$$

and from this the existence of γ_0 follows easily. Finally, define $N^{0,E} \subseteq N \subseteq G_2 = G$ the subgroup of the unipotent radical of the Heisenberg parabolic N consisting of those n with $\langle v_E, \overline{n}\rangle = 0$. Here $\overline{n}$ is the image of n in $N^{ab} \simeq W$ and $\langle \cdot, \cdot \rangle$ is the symplectic form on W.

The global integral unfolds as follows. For a cusp form φ on G, and a character χ on $N(\mathbf{Q})\backslash N(\mathbf{A})$, define

$$\varphi_\chi(g) = \int_{N(\mathbf{Q})\backslash N(\mathbf{A})} \chi^{-1}(n)\varphi(ng)\, dn.$$

Theorem 5.2 (Gurevich-Segal, Segal). *Fix an additive character $\psi : F\backslash\mathbf{A} \to \mathbf{C}^\times$, and define a character $\chi = \chi_E$ on N as $\chi(n) = \psi(\langle v_E, \overline{n}\rangle)$. One has*

$$\begin{aligned} I(\varphi, \Phi, s) &= \int_{N^{0,E}(\mathbf{A})\backslash G(\mathbf{A})} f(\gamma_0 g, \Phi, s)\varphi_\chi(g)\, dg \\ &= \int_{N^{0,E}(\mathbf{A})\backslash \mathrm{GL}_1(\mathbf{A}) \times G(\mathbf{A})} |t|^s \Phi(tg^{-1}\widetilde{v_E})\varphi_\chi(g)\, dg \\ &= \int_{N(\mathbf{A})\backslash \mathrm{GL}_1(\mathbf{A}) \times G(\mathbf{A})} |t|^s \Phi_\chi(t, g)\varphi_\chi(g)\, dg \end{aligned}$$

where

$$\Phi_\chi(t, g) = \int_{N^{0,E}(\mathbf{A})\backslash N(\mathbf{A})} \chi(n)\Phi(tg^{-1}n^{-1}\widetilde{v_E})\, dn.$$

Proof. Denote by N_E the unipotent radical of the Heisenberg parabolic P_E. The element $\widetilde{v_E}$ is "rank one" (see [Pola, section 2.3]) in the abelianization N_E^{ab} of N_E. As mentioned above, this fact is equivalent to the fact that $\widetilde{v_E}$ is in the $G_E(F)$-orbit of E_{13}, which is the minimal orbit of G_E on its Lie algebra $\mathfrak{g}_E$. The line $F\widetilde{v_E}$ gives rise to a double coset in $P_E(F)\backslash G_E(F)/G_2(F)$, represented by γ_0. One obtains a contribution to $I(\varphi, \Phi, s)$ of the form

$$\int_{\left(G_2(F)\cap\gamma_0^{-1}P_E(F)\gamma_0\right)\backslash G_2(\mathbf{A})} \varphi(g) f(\gamma_0 g, \Phi, s)\, dg.$$

The group $G_2(F) \cap \gamma_0^{-1} P_E(F)\gamma_0$, which is the stabilizer of the line $F\widetilde{v_E}$ in $G_2(F)$, can be computed without much difficulty, using the explicit decomposition of $\mathfrak{g}_E$ as a G_2-module described above. One obtains the integral as in the statement of the theorem.

The other double cosets in $P_E(F)\backslash G_E(F)/G_2(F)$ give rise to integrals that vanish because φ is a cusp form; see [GS15] and [Seg17] for this. □

We will begin the unramified computation of this integral in the next subsection. By defining the Eisenstein section $f(g, \Phi, s)$ through the Schwartz-Bruhat function Φ, we were able to write the unfolding integral in terms of Φ and the element $\widetilde{v_E}$. This small tweak on [GS15, Seg17] will enable us to simplify some the computations of *loc cit.*

Remark 5.3. The Dirichlet series of Theorem 3.5, the arithmetic invariant theory of cubic rings, the Rankin-Selberg integral of Theorem 5.2, and the lifting law of [Pola, section 2.3] are all tied together. Namely, out of the cubic ring $\mathcal{O}_E$, one constructs the group G_E and the Rankin-Selberg integral. This Rankin integral yields the standard L-function on G_2, whose Dirichlet series, given in Theorem 3.5, is most neatly expressed in terms of the arithmetic invariant theory of cubic rings and binary cubic forms. Finally, the lifting law $v_E \rightsquigarrow \widetilde{v_E}$ controls the Rankin integral, relating the Eisenstein series on G_E to the cusp form on G. It also controls the arithmetic invariant theory of cubic rings; see [Pola].

5.3 *The Unramified Computation: Overview*

In this subsection we give an overview of the proof that the integral of Theorem 5.2 represents the partial standard L-function $L^S(\pi, Std, s)$. The technique to evaluate $I(\varphi, \Phi, s)$, following [GS15, Seg17], is that of "non-unique models", due to Piatetski-Shapiro and Rallis [PSR88]. Ultimately, it boils down to expressing the Fourier coefficients of automorphic forms in terms of Hecke eigenvalues.

For the rest of this section, everything is local at a finite prime unless explicitly mentioned to the contrary. Thus F denotes a p-adic local field, $\mathcal{O}$ its ring of integers with uniformizer p, and E is a cubic étale extension of F. Furthermore, we assume that p is prime to 2, 3 and that $\mathcal{O}_E$ is an unramified extension of $\mathcal{O}$. Here is the statement of the unramified computation. Define the lattice

$$\mathfrak{g}_E(\mathcal{O}_E) = (\mathfrak{sl}_3(\mathcal{O}_E) \oplus \mathcal{O}_E^{\mathrm{tr}=0}) \oplus (V_3(\mathbf{Z}) \otimes \mathcal{O}_E) \oplus (V_3(\mathbf{Z})^\vee \otimes \mathcal{O}_E).$$

Theorem 5.4. *Suppose the representation π_p is unramified, and denote by V_p the space of π_p. Suppose that v_0 is a spherical vector in V_p, and $L : V_p \to \mathbf{C}$ is a linear functional satisfying $L(nv) = \chi(n)L(v)$ for all $n \in N(F)$ and $v \in V_p$. Furthermore, suppose that Φ is the characteristic function of $\mathfrak{g}_E(\mathcal{O}_E)$. Then*

$$\int_{N(F)\backslash \mathrm{GL}_1(F)\times G(F)} |t|^s \Phi_\chi(t,g)L(gv_0)\,dg = L(v_0)\frac{L(\pi, Std, s-2)}{L(E, s-1)\zeta_F(2s-4)}. \tag{1}$$

The reader can no-doubt easily convince themselves that combining Theorem 5.4 with Theorem 5.2 shows that the global integral $I(\varphi, \Phi, s)$ represents the ratio of global partial L-functions $\frac{L^S(\pi,Std,s-2)}{L^S(E,s-1)\zeta_F^S(2s-4)}$. The formal argument that one makes to conclude this from Theorem 5.4 is spelled out in many places, e.g. [BFG95, GS15, PS17].

The first step in applying the technique of "non-unique models" is to write down what can be called an approximate basic function. A basic function is, roughly speaking, the precise way one encodes a local L-function into Hecke operators, and an approximate basic function is a Hecke operator that encodes an approximation to a local L-function. In practice, making computations with basic functions is very difficult, but approximate basic functions can be much easier to manipulate, which is why they are useful.

In [GS15], the authors construct an approximate basic function on G_2 by pulling back to G_2 (a well-known) approximate basic function $\Delta_{\mathrm{SO}(7)}$ on $\mathrm{SO}(V_7)$. We make a small modification to this construction, as follows: We define an approximate basic function for $\mathrm{GL}_1 \times G_2$, as opposed to just G_2. Namely, for $t \in \mathrm{GL}_1$ and $h \in G_2$, define $\Delta(t,h) = \mathrm{char}(t \cdot r_7(h) \in End(V_7(\mathcal{O})))$, where r_7 denotes the representation of G_2 on the trace 0 elements V_7 of Θ. That is, $\Delta(t,h) = 1$ if the endomorphism $t \cdot r_7(h)$ preserves the integer lattice $V_7(\mathcal{O})$ and is 0 otherwise. The approximate basis function we will use going forward is the function $\Delta(t,h)|t|^s$.

Because $\mathrm{GL}_1 \times G_2$ has a natural character $\mathrm{GL}_1 \times G_2 \to \mathrm{GL}_1$, this modification aligns with the philosophy of [BK00, Ngo14] and the theme of the paper [Polb]. That $\Delta(t,h)|t|^s$ is an approximate basic function is made precise in the following proposition. The key point is that if one integrates out the GL_1-factor to obtain a function $\Delta_s(h) = \int_{\mathrm{GL}_1} |t|^s \Delta(t,h)\,dt$ on G_2, then one recovers the approximate basic function of [GS15].

Denote by q the order of the residue field $\mathcal{O}/p$. Define the Hecke operator T on G_2 to be q^{-3} times the characteristic function of those $g \in G_2(F)$ with $p \cdot r_7(g) \in M_7(\mathcal{O}) = End(V_7(\mathcal{O}))$. For ease of notation, set $z = q^{-s}$.

Proposition 5.5. *Recall that $v_0 \in V_p$ is a spherical vector for π_p, and suppose that $\ell : V_p \to \mathbf{C}$ is a linear functional. Then*

$$\int_{\mathrm{GL}_1(F)\times G(F)} |t|^{s+3}\Delta(t,g)\ell(gv_0)\,dg\,dt$$

$$= (1-z)(1-q^{-1}z)L(\pi_p, Std, s)\ell(N_0 * v_0)$$
$$= (1-z)(1-q^{-1}z)N_0(\pi_p, s)L(\pi_p, Std, s)\ell(v_0).$$

Here

$$N_0 = 1 + (q^{-1}+1)z + \frac{z^2}{q} + (q^{-2}+q^{-1})z^3 + \frac{z^4}{q^2} - \frac{z^2}{q}T$$

and $N_0(\pi_p, s)$ *is the meromorphic function of* s *defined by the equality* $N_0 * v_0 = N_0(\pi_p, s)v_0$.

Proof. This follows from [GS15, Proposition 7.1] by integrating over the $t \in \mathrm{GL}_1$. □

The proposition says that applying the Hecke operator $\Delta(t,h)|t|^{s+3}$ to the function $\ell(hv_0)$ yields the standard L-function of π_p, up to a correction factor that is a polynomial in q^{-s}. We will apply Proposition 5.5 to the case $\ell = L$ where L is as in Theorem 5.4.

It follows from Proposition 5.5 that in order to prove Theorem 5.4, it suffices to prove that

$$\int_{\mathrm{GL}_1(F)\times G(F)} |t|^{s+2}\Delta(t,g)L(gv_0)\,dt\,dg$$
$$= M(\pi_p, s)\int_{N(F)\backslash \mathrm{GL}_1(F)\times G(F)} |t|^{s+1}\Phi_\chi(t,g)L(gv_0)\,dt\,dg \quad (2)$$

where

$$M(\pi_p, s) = (1-qz)(1-z)N_0(\pi_p, s-1)L(E,s)\zeta_F(2s-2).$$

One does this by explicitly manipulating the two-sides of this expression. This is the technique of non-unique models, in this particular case. The rest of the paper is concerned with doing this manipulation.

5.4 Local Unramifed Integral

To begin to analyze the left-hand side of (1), one needs to compute Φ_χ explicitly. Denote by $m : \mathrm{GL}_2 \to G_2$ our chosen identification of the Levi subgroup of the Heisenberg parabolic with GL_2. Furthermore, for $h \in \mathrm{GL}_2$, set $\widetilde{h} = \det(h)h^{-1}$.

Lemma 5.6. *Let the assumptions and notation be as in Theorem 5.4. For* $t \in \mathrm{GL}_1$ *and* $h \in \mathrm{GL}_2$, *define* $\lambda = \lambda(t,h) = \det(h)/t$. *Then*

$$\Phi_\chi(t, m(h)) = |\lambda|\,\mathrm{char}(\lambda \in \mathcal{O}, \lambda^{-1}h \in M_2(\mathcal{O}), \lambda^{-1}\widetilde{h}\widetilde{v_E} \in \mathfrak{g}_E(\mathcal{O}_E)).$$

Here and below, by char(P) for a property P we mean the function that is 1 if P is satisfied and 0 otherwise.

Proof. Suppose $n \in N$, and denote by $\overline{n}$ the image of n in $N^{ab} \simeq W$. Then

$$th^{-1}n^{-1}\widetilde{v_E} = th^{-1}\widetilde{v_E} + t\det(h)^{-1}\langle v_E, \overline{n}\rangle E_{13}. \tag{3}$$

The lemma now follows easily from (3). □

For $\lambda \in \mathrm{GL}_1$ and $h \in \mathrm{GL}_2$, define

$$A_0(\lambda, h) = \mathrm{char}(\lambda \in \mathcal{O}, \lambda^{-1}h \in M_2(\mathcal{O}), \lambda^{-1}\widetilde{h}\widetilde{v_E} \in \mathfrak{g}_E(\mathcal{O}_E)).$$

This function A_0 is related to cubic rings over $\mathcal{O}$ that are subrings of $\mathcal{O}_E$. To make this precise, we setup some notation and review elements of the arithmetic invariant theory of binary cubic forms and cubic rings.

First, for a cubic ring T over $\mathcal{O}$, we denote by $c(T)$ the p-adic content of T. This is the largest integer c so that $T = \mathcal{O} + p^c T_0$ for a cubic ring T_0 over $\mathcal{O}$. If T is a free rank three $\mathcal{O}$-module in E of the form $T = \mathcal{O} \oplus \mathcal{O}\omega' \oplus \mathcal{O}\theta'$, then $c(T)$ is also defined, but could be negative. We will use this more general notion moving forward.

Next, for an element $x = \begin{pmatrix} \alpha & \beta \\ \gamma & \delta \end{pmatrix}$ of $\mathrm{GL}_2(F)$, denote by $T(x)$ the $\mathcal{O}$-module spanned by 1, $\delta\omega - \beta\theta$ and $-\gamma\omega + \alpha\theta$. Here $1, \omega, \theta$ is the good basis of the maximal order $\mathcal{O}_E$ fixed above. Then $T(x)$ is a ring, i.e., $T(x)$ is closed under multiplication, if and only if $\widetilde{x}\widetilde{v_E} \in \mathfrak{g}_E(\mathcal{O}_E)$. This is closely connected to [Pola, section 2.3]. As a consequence of this fact, one has

$$A_0(\lambda, h) = \mathrm{char}(\lambda \in \mathcal{O}, T(\lambda^{-1}h) \text{ a ring}) = \mathrm{char}(\lambda \in \mathcal{O}, \lambda | p^{c(T(h))}).$$

For ease of notation, we now set $c(x) := c(T(x))$ for $x \in \mathrm{GL}_2(F)$. Denote by $I(s)$ the left-hand side of (1), and recall that $z = q^{-s}$. The local unramifed integral to be computed is as follows. Applying the Iwasawa decomposition, one obtains

$$\begin{aligned} I(s+1) &= \int_{\mathrm{GL}_1 \times \mathrm{GL}_2} \delta_P^{-1}(m(h))|\lambda||\det(h)/\lambda|^{s+1} A_0(\lambda, h) L(m(h)v)\, d\lambda\, dh \\ &= \int_{\mathrm{GL}_2} |\det(h)|^{s-2} L(m(h)v) \left(\int_{\mathrm{GL}_1} |\lambda|^{-s} A_0(\lambda, h)\, d\lambda\right) dh. \end{aligned}$$

We convert this integral into a sum as follows. The integral over λ becomes

$$\int_{\mathrm{GL}_1} |\lambda|^{-s} A_0(\lambda, h)\, d\lambda = \sum_{0 \le j \le c(h)} z^{-j} = z^{-c(h)} \left(\frac{1 - z^{c(h)+1}}{1 - z}\right).$$

Recall from, e.g., [GGS02, section 4] that

(1) every cubic ring T over $\mathcal{O}$ contained in E is of the form $T(h)$ for some $h \in \mathrm{GL}_2(F)$;
(2) the rings $T(h_1)$ and $T(h_2)$ are isomorphic as rings if and only if $h_1 = h_2 k$ for some $k \in \mathrm{GL}_2(\mathcal{O})$.

For an element $h \in \mathrm{GL}_2(F)$, write $[h]$ for the coset $h\,\mathrm{GL}_2(\mathcal{O})$. Thus

$$I(s+1) = \frac{1}{1-z}\left(\sum_{[h] \text{ cubic ring}} L(m(h)v)|\det(h)|^{-2} z^{\mathrm{val}(\det(h))-c(h)}(1-z^{c(h)+1})\right)$$

$$= \frac{1}{1-z}\left(\sum_{[h] \text{ cubic ring}} L(m(h)v)|\det(h)|^{-2} P_h(z)\right) \tag{4}$$

with

$$P_h(z) = z^{\mathrm{val}(\det(h))-c(h)}(1-z^{c(h)+1}),$$

a polynomial in z. The sum is over all cubic subrings of $\mathcal{O}_E$. In order to prove the equality in (2), we will further manipulate the expression (4) and the left-hand side of (2).

5.5 The Fourier Coefficient of the Approximate Basic Function

In this subsection, we manipulate the left-hand side of (2). More precisely, for $t \in \mathrm{GL}_1(F)$ and $h \in \mathrm{GL}_2(F)$ set

$$D_\chi(t,h) = \int_{N(F)} \chi(n)\Delta(t, nm(h))\,dn.$$

Then the left-hand side of (2) is

$$D(s) := \int_{\mathrm{GL}_1 \times \mathrm{GL}_2} \delta_P^{-1}(m(h))|t|^{s+2} D_\chi(t,h) L(m(h)v)\,dh\,dt.$$

One of the main computations is the expression for $D_\chi(t,h)$. Set $f_{max}(x,y) = ax^3 + bx^2y + cxy^2 + dy^3$, the binary cubic form corresponding to the maximal order $\mathcal{O}_E$, together with its good basis $1, \omega, \theta$. For a binary cubic form V with coefficients in $\mathcal{O}$, define $N(V)$ to be the number of 0's of V in $\mathbb{P}^1(\mathbb{F}_q)$. Also, for an element $h \in \mathrm{GL}_2(F)$, define $\mathrm{val}(h) \in \mathbf{Z}$ to be the largest integer n so that $p^{-n}h \in M_2(\mathcal{O})$.

Proposition 5.7. *Define $x_0(h)$ by $h = p^{\mathrm{val}(h)}x_0(h)$, and as before $\lambda = \det(h)/t$. Write $D'_\chi(\lambda, h) = D_\chi(t, h)$, i.e., D'_χ is the same function as D_χ, except expressed in terms of the new variables λ, h. Finally, define*

$$\epsilon(x_0(h)) = \begin{cases} 1 & \text{if } x_0(h) \in \mathrm{GL}_2(\mathcal{O}) \\ 2 & \text{if } x_0(h) \notin \mathrm{GL}_2(\mathcal{O}). \end{cases}$$

Then

$$D'_\chi(\lambda, h) = |\det(\lambda^{-1}h)|^{-1}\,\mathrm{char}(h \in M_2(\mathcal{O}), \mathrm{val}(\lambda^{-1}h) \in \{0, 1\}, T(x_0(h)) \text{ a ring})$$
$$\times \begin{cases} 1 & \text{if } \mathrm{val}(\lambda^{-1}h) = 0 \\ N(f_{max}) - \epsilon(x_0(h)) & \text{if } \mathrm{val}(\lambda^{-1}h) = 1. \end{cases}$$

Proof. We sketch the proof. From Sect. 5.2, we have that $r_7(nm(h))$ is of the form

$$\begin{pmatrix} h & y'Ad^0(h) & z\,{}^th^{-1} \\ & Ad^0(h) & y\,{}^th^{-1} \\ & & {}^th^{-1} \end{pmatrix}$$

with y of the form $\begin{pmatrix} \alpha & \beta \\ \beta & \gamma \\ \gamma & \delta \end{pmatrix}$, $y' = -{}^tyS$ with S given in Sect. 5.2, and z satisfying $z + z^t + {}^tySy = 0$.

Because the character χ corresponds to the maximal order $\mathcal{O}_E$ of E, one can show that in order for $D_\chi(t, h)$ to be nonzero, one needs $h \in M_2(\mathcal{O})$. Thus assume for the rest of the proof that $h \in M_2(\mathcal{O})$. Set $x = t\,{}^th^{-1}$. Thus, $\Delta(t, nm(h)) \neq 0$ if and only if

(1) $x \in M_2(\mathcal{O})$;
(2) $yx \in M_{3,2}(\mathcal{O})$;
(3) $zx \in M_2(\mathcal{O})$.

The element z above is of the form $\mu \begin{pmatrix} & 1 \\ -1 & \end{pmatrix} - \frac{1}{2}{}^tySy$ for some $\mu \in F$. Note that for y as above,

$$\mu \begin{pmatrix} & 1 \\ -1 & \end{pmatrix} - \frac{1}{2}{}^tySy = \begin{pmatrix} \beta^2 - \alpha\gamma & \frac{1}{2}(\beta\gamma - \alpha\delta) + \mu \\ \frac{1}{2}(\beta\gamma - \alpha\delta) - \mu & \gamma^2 - \beta\delta \end{pmatrix}.$$

Denote by W' the set of 3×2 matrices of the form $\begin{pmatrix} a' & b' \\ b' & c' \\ c' & d' \end{pmatrix}$. Now, for $x \in \mathrm{GL}_2(F) \cap M_2(\mathcal{O})$, define

$$V(x) = \left\{ y \in W' : yx \in M_{3,2}(\mathcal{O}) \text{ and } \exists \mu \in F \text{ s.t. } \left(\mu \begin{pmatrix} & 1 \\ -1 & \end{pmatrix} - \frac{1}{2} {}^t y S y \right) x \in M_2(\mathcal{O}) \right\}.$$

We thus have that

$$D_\chi(t, h) = ||x||^{-1} \int_{W'} \chi(y) \operatorname{char}(y \in V(x))\, dy \tag{5}$$

where $x = t\,{}^t h^{-1}$ and $||x|| = |p|^{\mathrm{val}(x)}$.

Suppose that $x = k_1 \begin{pmatrix} p^j & \\ & p^k \end{pmatrix} k_2$ with $j \geq k$ and $k_1, k_2 \in \mathrm{GL}_2(\mathcal{O})$. Note that $V(x) = V(xk_2^{-1})$. Furthermore, by changing variables in the integral in (5) to eliminate k_1 and by replacing χ by an equivalent character, we can assume that $x = \operatorname{diag}(p^j, p^k)$ with $j \geq k \geq 0$. We therefore obtain

$$\begin{aligned} D_\chi(t, h) = q^k \int_{w=(\alpha,\beta,\gamma,\delta) \in W(F)} & \psi(\langle \omega, w \rangle) \operatorname{char}(p^j \alpha, p^k(\beta, \gamma, \delta)) \\ & \times \operatorname{char}(p^j(\beta^2 - \alpha\gamma) \in \mathcal{O}, p^j(\alpha\delta - \beta\gamma) \in \mathcal{O}, p^k(\gamma^2 - \beta\delta) \in \mathcal{O}))\, dw. \end{aligned}$$

Here $\omega = (a, \frac{b}{3}, \frac{c}{3}, d) \in W(\mathcal{O})$ is a binary cubic form representing the maximal order $\mathcal{O}_E$ of E. Set $r = j - k$, so that $x = p^k \operatorname{diag}(p^r, 1)$. For any $\alpha_0 \in \mathcal{O}$, making a change of variable $\alpha \mapsto \alpha + p^{-r}\alpha_0$ on the one hand fixes this integral, but on the other hand multiplies it by $\psi(-p^{-r} d\alpha_0)$. It follows that for $D_\chi(t, h)$ to be nonzero, we must have $p^r | d$. This is equivalent to $T(x_0(h))$ is a ring, so we have checked this condition of the statement of the proposition.

From now assume that $p^r | d$. Making a change of variables, we obtain

$$D_\chi(t, h) = q^j \sum_{w=(\alpha,\beta,\gamma,\delta) \in W(\mathcal{O})/p^k, **} \psi\left(\frac{\langle \omega, (p^{-r}\alpha, \beta, \gamma, \delta) \rangle}{p^k} \right) \tag{6}$$

where the conditions $**$ mean that

$$\alpha\gamma - p^r\beta^2 \in p^k\mathcal{O}, \alpha\delta - p^r\beta\gamma \in p^k\mathcal{O}, \gamma^2 - \beta\delta \in p^k\mathcal{O}.$$

In case $k = 0$ or $k = 1$, this sum can be directly evaluated by hand, and one obtains values as in the statement of the proposition. One is left to consider the case $k \geq 2$, i.e., the when $p^2 | x$. In the following lemma, we show that $\int_{V(x)} \chi(y)\, dy$ vanishes if $p^2 | x$, completing the proof of the proposition. □

Lemma 5.8. *The sum on the right-hand side of (6) vanishes if* $k \geq 2$.

Proof. The sum breaks up into two pieces, consisting of those w with $p | w$ and those w with $p \nmid w$. Consider first the subsum consisting of those w with $p | w$:

$$D_\chi^p := \sum_{w=(\alpha,\beta,\gamma,\delta)\in W(\mathcal{O})/p^k,\, p|w \text{ and } **} \psi\left(\frac{\langle\omega, (p^{-r}\alpha, \beta, \gamma, \delta)\rangle}{p^k}\right).$$

If $w_0 = (0, b_0, c_0, d_0) \in W(\mathcal{O})$ is fixed, then making the variable change $w \mapsto w + p^{k-1}w_0$ preserves D_χ^p. However, this changes the sum by $\psi(\langle\omega, w_0\rangle/p)$. Because ω is non-degenerate modulo p, there is w_0 with $\psi(\langle\omega, w_0\rangle/p) \neq 1$. Hence $D_\chi^p = 0$, and it suffices to evaluate

$$D_\chi^1 := \sum_{w=(\alpha,\beta,\gamma,\delta)\in W(\mathcal{O})/p^k,\, p\nmid w \text{ and } **} \psi\left(\frac{\langle\omega, (p^{-r}\alpha, \beta, \gamma, \delta)\rangle}{p^k}\right).$$

In case $r = 0$, the evaluation of the sum D_χ^1 now becomes a fun exercise, which we leave to the reader; it can be analyzed by the same method that we will use to handle the case $r \geq 1$. Thus assume $r \geq 1$. In this case, if $w = (\alpha, \beta, \gamma, \delta)$ satisfies $**$, one finds that there are three disjoint cases:

(1) $\alpha \in (\mathcal{O}/p^k)^\times$. In this case, $w \equiv \alpha(1, \beta, p^r\beta^2, p^{2r}\beta^3)$ for $\beta \in \mathcal{O}/p^k$.
(2) $\delta \in (\mathcal{O}/p^k)^\times$. In this case, $w \equiv \delta(p^r\gamma^3, \gamma^2, \gamma, 1)$ for $\gamma \in \mathcal{O}/p^k$.
(3) Neither α nor δ are units. In this case, $w \equiv \beta(\alpha, 1, \gamma, \gamma^2)$ with $\alpha\gamma \equiv p^r \pmod{p^k}$ and both α and γ divisible by p.

Consider first the case where $\delta \in (\mathcal{O}/p^k)^\times$. Define the polynomial $g(x) = a - bx + cx^2 - dx^3$. Then the subsum of D_χ^1 corresponding to $\delta \in (\mathcal{O}/p^k)^\times$ becomes

$$\sum_{\delta\in(\mathcal{O}/p^k)^\times,\gamma\in\mathcal{O}/p^k} \psi\left(\frac{\delta g(\gamma)}{p^k}\right).$$

If γ is such that $g(\gamma)$ is a unit, then the sum over δ with γ fixed is 0, because $k \geq 2$. Thus one need only sum over those γ with $g(\gamma) \equiv 0 \pmod{p}$. Now, fix a root γ_0 of g modulo p, and consider the sum

$$\sum_{\delta\in(\mathcal{O}/p^k)^\times,\gamma\equiv\gamma_0 \pmod{p}} \psi\left(\frac{\delta g(\gamma)}{p^k}\right).$$

The domain of this sum is preserved under $\gamma \mapsto \gamma + p^{k-1}\gamma'$ for some fixed γ'. But

$$\frac{g(\gamma + p^{k-1}\gamma')}{p^k} \equiv \frac{g(\gamma)}{p^k} + \frac{g'(\gamma_0)}{p}\gamma'.$$

Thus the sum over $\gamma \equiv \gamma_0$ vanishes unless $g'(\gamma_0) \equiv 0 \pmod{p}$. But then $g(x)$ modulo p has a double root at γ_0, which implies that the binary cubic $f(x, y) \equiv \mu\ell^2 x \pmod{p}$ for some $\mu \in (\mathcal{O}/p)^\times$ and line ℓ. But this contradicts the assumption that f is non-degenerate modulo p, so there are no such γ_0 with $g(\gamma_0) \equiv g'(\gamma_0) \equiv 0 \pmod{p}$. It follows that the subsum consisting of those w with δ a unit vanishes.

Next consider the case where $\alpha \in (\mathcal{O}/p^k)^\times$. Set

$$\omega' = (p^{2r}a, \frac{p^r b}{3}, \frac{c}{3}, p^{-r}d) = (a', \frac{b'}{3}, \frac{c'}{3}, d')$$

and define the polynomial $h(x) = a'x^3 - b'x^2 + c'x - d'$. Note that ω' is integral. Furthermore, because f is non-degenerate but $p|d$, we must have $p \nmid c$ and thus $c = c'$ is a unit. Now, the subsum of D^1_χ corresponding to $\alpha \in (\mathcal{O}/p^k)^\times$ becomes

$$\sum_{\alpha \in (\mathcal{O}/p^k)^\times, \beta \in \mathcal{O}/p^k} \psi\left(\frac{\alpha h(\beta)}{p^k}\right).$$

Just like in the previous case, this sum immediately reduces to the subsum consisting of those β equivalent to some β_0 modulo p with $h(\beta_0) \equiv h'(\beta_0) \equiv 0 \pmod{p}$. But since p divides a' and b', but c' is prime to p, there are no double roots β_0. Hence this subsum of D^1_χ vanishes as well.

We are left with the final case, in which neither α nor δ is unit. Set $d' = p^{-r}d$, which is in $\mathcal{O}$. In this case, we must evaluate

$$\sum_{\beta \in (\mathcal{O}/p^k)^\times, \alpha, \gamma \in \mathcal{O}/p^k ***} \psi\left(\frac{\beta(a\gamma^2 - b\gamma + c - d'\alpha)}{p^k}\right)$$

where the conditions $***$ are that $\alpha\gamma \equiv p^r \pmod{p^k}$ and both α and γ are divisible by p. But note that, as mentioned above, c is a unit. Thus the fact that $\alpha \equiv \gamma \equiv 0 \pmod{p}$ imply that $a\gamma^2 - b\gamma + c - d'\alpha$ is a unit mod p. Hence the sum over β, with α and γ fixed vanishes. Thus this subsum of D^1_χ vanishes, and the proof of the lemma is complete. □

We now reconsider the condition "$T(x_0(h))$ a ring" from the above proposition. We claim that this condition is equivalent to $\mathrm{val}(h) = c(h)$. Clearly, $T(x_0(h))$ a ring means that $c(h) \geq \mathrm{val}(h)$. Recall $f_{max}(x, y) = ax^3+bx^2y+cxy^2+dy^3$, the binary cubic form corresponding to the maximal order $\mathcal{O}_E$, together with its good basis $1, \omega, \theta$. By assumption $f_{max}(x, y)$ is non-degenerate modulo p. It follows from this that $c(x_0(h))$ must be 0. Indeed, suppose $x_0(h) = k\begin{pmatrix} p^r & \\ & 1 \end{pmatrix}k'$ with $k, k' \in \mathrm{GL}_2(\mathcal{O})$, and $f' = f \cdot k = a'x^3 + b'x^2y + c'xy^2 + d'y^3$. Then $T(x_0(h))$ is represented by the binary cubic

$$\frac{1}{p^r} f_{max}((x, y)\widetilde{\mathrm{diag}(p^r, 1)k}) = \frac{1}{p^r}a'x^3 + b'x^2y + p^r c'xy^2 + p^{2r}d'y^3.$$

Thus, if $c(x_0(h)) \geq 1$, then $p^{r+1}|a'$ and $p|b'$, so that f' and then f_{max} is degenerate modulo p.

We therefore obtain

$$D(s) = \sum_{[h] \text{ cubic ring}} L(hv)|\det(h)|^{-2}$$

$$\mathrm{char}(\mathrm{val}(h) = c(h)) z^{\mathrm{val}(\det(h))-c(h)}(1 + (N(f_{max}) - \epsilon(h_0))z).$$

This is the expression that we will have to compare to the Hecke-manipulated local integral. Define $N'(f_{max}, h_0) = N(f_{max}) - \epsilon(h_0) + 1$. We thus have

$$D(s) = \sum_{[h] \text{ cubic ring}} L(hv)|\det(h)|^{-2}$$

$$\mathrm{char}(\mathrm{val}(h) = c(h)) z^{\mathrm{val}(\det(h))-c(h)}(1 - z + N'(f_{max}, h_0)z).$$

For later use, we set

$$B_0(z) = 1 + (q+1)z + qz^2 + (q^2+q)z^3 + q^2z^4$$

so that $N_0(s-1) = B_0(z) - qz^2T$.

5.6 The Hecke Operator S

Define the Hecke operator S on $G_2(F)$ to be the bi-$G_2(\mathcal{O})$ invariant function on $G_2(F)$ that is the characteristic function of those $g \in G_2(F)$ with $p \cdot r_7(g) \in M_7(\mathcal{O}) \setminus pM_7(\mathcal{O})$. Thus $T = q^{-3}(S + \mathbf{1}_{G_2(\mathcal{O})})$, where $\mathbf{1}_{G_2(\mathcal{O})}$ denotes the characteristic function of $G_2(\mathcal{O})$. In order to evaluate the right-hand side of (2), we will need a coset decomposition of S. This was given in [GGS02, sections 13 and 14].

We require some notation for Hecke operators on the Levi GL_2 of the Heisenberg parabolic. Define

$$T(p) = \mathrm{GL}_2(\mathcal{O}) \begin{pmatrix} p & \\ & 1 \end{pmatrix} \mathrm{GL}_2(\mathcal{O})$$

and

$$T(p^{-1}) := \mathrm{GL}_2(\mathcal{O}) \begin{pmatrix} p^{-1} & \\ & 1 \end{pmatrix} \mathrm{GL}_2(\mathcal{O}).$$

One has the following proposition regarding $L(m(h)S * v_0)$, which is a conglomeration of [GGS02, Corollary 13.3, Proposition 14.2 and Proposition 14.5], or essentially equivalent to [GGS02, Proposition 15.6].

Proposition 5.9 (Gan-Gross-Savin). *Suppose* $h \in \mathrm{GL}_2(F)$ *and* $L(m(h)v_0) \neq 0$. *Then*

$$L(m(h)S * v_0) = L(m(hp^{-1})v_0) + qL(m(h * T(p^{-1}))v_0) + q^4 L(m(h * T(p))v_0) + q^6 L(m(hp)v_0) + \left(-1 + q^2(N(f_{max} \cdot h) - 1)\right) L(hv_0).$$

5.7 Hecke Operator Applied to Local Integral

We can now apply $N_0(s-1)$ to the local integral $I(s+1)$. First, note that for $g \in \mathrm{GL}_2(F)$,

$$\sum_{[h]} L(m(hg)v)|\det(h)|^{-2} P_h(z) = \sum_{[h]} L(m(h)v)|\det(h)|^{-2}|\det(g)|^2 P_{hg^{-1}}(z).$$

We thus have that

$$q^{-2} \sum_{[h]} L\left(m(h)(S+1) * v\right) |\det(h)|^{-2} P_h(z) = \sum_{[h]} L(m(h)v)|\det(h)|^{-2} M_h(z)$$

with

$$M_h(z) = q^2 P_{hp}(z) + q P_{h*T(p)}(z) + P_{h*T(p^{-1})}(z) + P_{hp^{-1}}(z) + (N(f_{max} \cdot h) - 1) P_h(z).$$

Here, for a Hecke operator Y on GL_2 with coset decomposition $Y = \sum_i a_i [y_i \mathrm{GL}_2(\mathcal{O})]$, $P_{h*Y}(z) = \sum_i a_i P_{hy_i}(z)$. Hence

$$N_0(s-1)I(s+1) = \frac{1}{1-z} \left(\sum_{[h]} L(m(h)v)|\det(h)|^{-2}(B_0(z)P_h(z) - z^2 M_h(z)) \right),$$

and thus

$$\begin{aligned}&(1-qz)(1-z)N_0(s-1)I(s+1) \\ &\quad = (1-qz) \sum_{[h]} L(m(h)v)|\det(h)|^{-2}(B_0(z)P_h(z) - z^2 M_h(z)).\end{aligned}$$

5.8 The Cubic Ring Identity

Thus, to prove the equality of Eq. (2) and thus that global integral represents $L^S(\pi, Std, s-2)$, we need to verify the identity

$$\zeta_F(2s-2)L(E,s)(1-qz)(B_0(z)P_h(z) - z^2M_h(z)) = \text{char}(\text{val}(h) = c(h))z^{\text{val}(\det(h))-c(h)} \times (1 - z + N'(f_{max}, h_0)z). \tag{7}$$

There is one immediate simplification here: $\zeta(2s-2)(1-qz) = (1+qz)^{-1}$.

To verify (7), we need to know how the content of cubic rings change under the action of the Hecke operator $T(p)$. That is, we must understand how the multiset $\{c(hy_i)\}_i$ is related to $c(h)$, where $T(p) = \sqcup y_i \, \mathrm{GL}_2(\mathcal{O})$ is the coset decomposition of $T(p)$.

Slightly more canonically, one can reformulate this in terms of $\mathcal{O}$-lattices Λ in F^2. Namely, fix the lattice $\Lambda_0 = \mathcal{O}^2 \subseteq F^2$, with F^2 thought of as row vectors. Now, for any $y \in \mathrm{GL}_2(F)$ and lattice Λ in F^2, we get a new lattice Λy. Identifying Λ_0 with the rank two $\mathcal{O}$-module $\mathcal{O}_E/\mathcal{O}$, one can consider the rank three $\mathcal{O}$-module $T(\Lambda)$ corresponding to Λ, defined by $T(\Lambda) = T(h)$ if $\Lambda = \Lambda_0 h$. We also write $c(\Lambda) = c(T(\Lambda))$ for the content of $T(\Lambda)$. In this setup, the Hecke operator $T(p)$ replaces a lattice Λ with the $q+1$-sublattices of Λ of index q. We thus need to understand the multiset $\{c(\Lambda')\}$ where $\Lambda' \subseteq \Lambda$ is an index q sublattice.

Suppose $h = p^c h_0$, with $f_{max} \cdot h_0$ integral and not divisible by p, so that the content of h is c, i.e., $c(h) = c$. Set $f_0 = f_{max} \cdot h_0$. Then f_0 is one of five types modulo p:

(1) f_0 is irreducible modulo p;
(2) $f_0 = \ell q$ with ℓ a line and q irreducible modulo p;
(3) $f_0 = \ell_1\ell_2\ell_3$ with ℓ_i distinct lines modulo p;
(4) $f_0 = \ell_1^2\ell_2$ with ℓ_1, ℓ_2 distinct lines modulo p;
(5) $f_0 = \alpha\ell^3$, with ℓ a line modulo p and $\alpha \in (\mathcal{O}/p)^\times$

One has the following lemma.

Lemma 5.10. *Suppose $f = p^c f_0$, with f_0 in each of the cases enumerated above. Denote by Λ_f the rank two $\mathcal{O}$-lattice corresponding to f. The content $c(\Lambda')$ of the index q sublattices of Λ_f are given as follows, in each of the various cases:*

(1) *f_0 irreducible mod p: all $q+1$ index q sublattices Λ' have $c(\Lambda') = c-1$;*
(2) *$f_0 = \ell q$ mod p: 1 sublattice $\Lambda' = \Lambda_\ell$ has $c(\Lambda_\ell) = c$; the other q sublattices have $c(\Lambda') = c-1$;*
(3) *$f_0 = \ell_1\ell_2\ell_3$ mod p: Three lattices $\Lambda' = \Lambda_{\ell_i}$, $i = 1,2,3$, have $c(\Lambda_{\ell_i}) = c$; the other $q-2$ sublattices have $c(\Lambda') = c-1$.*
(4) *$f_0 = \ell_1^2\ell_2$ mod p: 1 lattice $\Lambda' = \Lambda_{\ell_2}$ has $c(\Lambda_{\ell_2}) = c$; one lattice Λ_{ℓ_1} has $c(\Lambda_{\ell_1}) = c+1$; the other $q-1$ lattices have $c(\Lambda') = c-1$.*

(5) $f_0 = \alpha\ell^3$ *mod p*: 1 *lattice* Λ_ℓ *has* $c(\Lambda_\ell) = c+2$; *the other q lattices have* $c(\Lambda') = c-1$.

Proof. This is straightforward. □

From the lemma, and the fact that $T(p^{-1}) = p^{-1}T(p)$, it is straightforward to evaluate $M_h(z)$. For example, if $c = c(h) \geq 2$, then there are no "boundary conditions", and one obtains the following for each of the 5 possible cases. Set $v = \mathrm{val}(\det(h))$ and $g(z) = q^2 z^{v-c+1}(1-z^{c+2}) + z^{v-c-1}(1-z^c) + qz^{v-c}(1-z^{c+1})$. Then

$$f_0 \text{ irred} : g(z) + q(q+1)z^{v-c+2}(1-z^c) + (q+1)z^{v-c+1}(1-z^{c-1})$$

$$f_0 = \ell q : g(z) + q\left(z^{v-c+1}(1-z^{c+1}) + qz^{v-c+2}(1-z^c)\right) + \left(z^{v-c}(1-z^c) + qz^{v-c+1}(1-z^{c-1})\right)$$

$$f_0 = \ell_1\ell_2\ell_3 : g(z) + q\left(3z^{v-c+1}(1-z^{c+1}) + (q-2)z^{v-c+2}(1-z^c)\right) + \left(3z^{v-c}(1-z^c) + (q-2)z^{v-c+1}(1-z^{c-1})\right)$$

$$f_0 = \ell_1^2\ell_2 : g(z) + q\left(z^{v-c+1}(1-z^{c+1}) + z^{v-c}(1-z^{c+2}) + (q-1)z^{v-c+2}(1-z^c)\right) + \left(z^{v-c}(1-z^c) + z^{v-c-1}(1-z^{c+1}) + (q-1)z^{v-c+1}(1-z^{c-1})\right)$$

$$f_0 = \alpha\ell^3 : g(z) + q\left(z^{v-c-1}(1-z^{c+3}) + qz^{v-c+2}(1-z^c)\right) + \left(z^{v-c+2}(1-z^{c+2}) + qz^{v-c+1}(1-z^{c-1})\right).$$

From this table, one obtains that $B_0(z)P_h(z) - z^2 M_h(z)$ is as follows, in each of the above cases, when $c \geq 2$:

f_0 irred	$z^{v-c}\left(1+qz-z^3-qz^4\right) = z^{v-c}(1+qz)(1-z^3)$
$f_0 = \ell q$	$z^{v-c}\left(1+qz-z^2-qz^3\right) = z^{v-c}(1+qz)(1-z^2)$
$f_0 = \ell_1\ell_2\ell_3$	$z^{v-c}\left(1+qz-3z^2+(2-3q)z^3+2qz^4\right) = z^{v-c}(1+qz)(1-z)^2(1+2z)$.
$f_0 = \ell_1^2\ell_2$	$z^{v-c}\left(1+(q-1)z-(q+1)z^2-(q-1)z^3+qz^4\right) = z^{v-c}(1+qz)(1-z)(1-z^2)$
$f_0 = \alpha\ell^3$	0.

We also require the following lemma.

Lemma 5.11. *One has* $\mathrm{val}(h) = c(h)$ *if and only if* $f_0 \neq \ell^3$ *modulo p.*

Proof. First suppose that $\mathrm{val}(h) = c(h)$. This means $p \nmid h_0$, and thus we can suppose (making an appropriate change of variables) that $h_0 = \mathrm{diag}(p^r, 1)$ with $r \geq 0$. And if $r = 0$, $f_0 = f_{max}$, so we are done. Thus we can suppose $r \geq 1$. Then $f_0(x, y) = \frac{1}{p^r} f_{max}(x, p^r y) = \frac{a}{p^r}x^3 + bx^2y + cp^r xy^2 + p^{2r}dy^3$. Hence $p^r|a$; set $a' = a/p^r$. Then $f_0 = a'x^3 + bx^2y = x^2(a'x+by)$ modulo p. To check this direction of the lemma, we thus need to see that b is not 0 modulo p. But if $b = 0$ mod p, then $f_{max} = cxy^2 + dy^3$ mod p, which cannot happen by assumption. Hence this direction of the lemma is verified.

Conversely, suppose $v(h) \neq c(h)$, so that we may assume $h_0 = p^t \operatorname{diag}(p^r, 1)$ with $t \geq 1$. Now, one always has that $f_{max} \cdot (p^r \operatorname{diag}(p^r, 1))$ is integral, thus $1 \leq t \leq r$. Hence

$$\begin{aligned} f_0 = f_{max} \cdot h_0 &= \frac{1}{p^{r-t}}(ax^3 + bp^r x^2 y + cp^{2r} xy^2 + dp^{3r} y^3) \\ &= a' x^3 + bp^t x^2 y + cp^{r+t} xy^2 + dp^{2r+t} y^3. \end{aligned}$$

Since $t \geq 1$, $f_0 = a'x^3 = \alpha\ell^3$ modulo p, as desired. □

Combining the previous calculations, we obtain the following proposition. Note that $1 - z + N'(T, h_0)z = 1 + (N(T_{max}) - \epsilon(h_0))z$.

Proposition 5.12. *Let the notation be as above. Then*

$$\begin{aligned} &(1 + qz)^{-1} L(E, s)\left(B_0(z)P_h(z) - z^2 M_h(z)\right) \\ &\quad = \operatorname{char}(\operatorname{val}(h) = c(h)) z^{v-c}(1 + (N(f_{max}) - \epsilon(h_0))z). \end{aligned}$$

Proof. We leave the proof of this identity in the cases $c(h) = 0$ and $c(h) = 1$ to reader. Thus, suppose $c = c(h) \geq 2$.

Now, in all but the case that $f_0 = \ell_1^2\ell_2$ mod p, the proposition follows immediately from the above calculations. In case $f_0 = \ell_1^2\ell_2$ mod p, then we can have either $E = F \times K$ or $E = F \times F \times F$. In the first case, the left-hand side becomes $z^{v-c}(1 - z)$. In the second case, the left-hand side becomes $z^{v-c}(1+z)$. In both cases, because $\ell_1^2\ell_2$ does not correspond to the maximal ring, $\epsilon(h_0) = 2$. Thus, when $E = F \times K$, $1 + (N(T_{max}) - \epsilon(h_0))z = 1 + (1 - 2)z = 1 - z$, and when $E = F \times F \times F$, $1 + (N(f_{max}) - \epsilon(h_0))z = 1 + (3 - 2)z = 1 + z$. Thus we get agreement in both cases. This completes the proof of the proposition. □

We have now proved (7), and with it, Theorem 5.4.

5.9 *The Dirichlet Series*

By Proposition 5.12, one obtains that

$$\frac{L(\pi_p, Std, s - 1)}{L(E, s)\zeta_F(2s - 2)} = \frac{1}{1 - z}\left(\sum_{[h]} L(m(h)v)|\det(h)|^{-2} P_h(z)\right).$$

In other words, one obtains

$$\frac{L(\pi_p, Std, s-1)}{L(E,s)\zeta_F(2s-2)} = \sum_{[h],[\lambda]} L(m(h)v)|\det(h)|^{s-2}|\lambda|^{-s} A_0(\lambda, h).$$

Make the change of variables $h = \lambda x$. Then we finally obtain

$$\frac{L(\pi_p, Std, s-1)}{L(E,s)\zeta_F(2s-2)} = \sum_{[x] \text{ cubic ring}, [\lambda]} L(m(\lambda x)v)|\lambda^2 \det(x)|^{-2}|\lambda \det(x)|^{s}.$$

This is the desired Dirichlet series for the standard L-function on G_2. Converting this local expression to the "classical" notation of Sect. 3 proves Theorem 3.5.

6 Archimedean Zeta Integral

In this final section, we combine our work from the previous two sections, and analyze the archimedean zeta integral that arises from the global integral $I(\varphi, \Phi, s)$ when φ is a modular form of even weight n. Throughout this section, the ground field F is the field of real numbers $\mathbf{R}$. We are able to relate this integral to products of gamma functions $\Gamma(s)$ and a certain explicit integral involving binary cubic forms, of the type considered in [Shi72].

Suppose $\Phi : \mathfrak{g}_E \to \mathbb{V}_n^\vee$ is a Schwartz function satisfying $\Phi(rv) = r \cdot \Phi(v)$ for all $v \in \mathfrak{g}_E$ and $r \in K_E$, the maximal compact subgroup of $G_E(\mathbf{R})$. Denote by $\{\cdot, \cdot\}_K$ the symmetric K_E-invariant pairing on $\mathbb{V}_n$ or $\mathbb{V}_n^\vee$. In this section, we analyze the integral

$$I(s, \Phi) = \int_{N^{0,E}(\mathbf{R})\backslash \mathrm{GL}_1(\mathbf{R}) \times G_2(\mathbf{R})} |t|^s \{\Phi(tg^{-1}\widetilde{v_E}), \mathcal{W}_\chi(g)\}_K \, dg$$

for a particular choice of Φ.

6.1 Setup

We now record the setup in greater detail. Denote by $pr_K : \mathfrak{g}_E \otimes \mathbf{C} \to \mathfrak{sl}_2$ the K_E-equivariant surjection to the Lie algebra of the long root $\mathrm{SU}(2) \subseteq K_E$. As in Sect. 4, we can identify this $\mathfrak{sl}_2$ with $Sym^2(V_2)$, and thus there is a nth power map $\mathfrak{sl}_2 \to \mathbb{V}_n \simeq \mathbb{V}_n^\vee$, which we write as $X \mapsto X^n$ for $X \in \mathfrak{sl}_2$. We write $\{e_\ell, h_\ell, f_\ell\}$ for the $\mathfrak{sl}_2$-triple of this $\mathfrak{sl}_2$ from [Pol18]. For $X \in \mathfrak{g}_E$, we write $||X|| = B_\theta(X, X)^{1/2}$. Here $B_\theta(X, Y) = -B(X, \theta(Y))$ is a positive-definite symmetric pairing on $\mathfrak{g}_E$. See [Pol18, section 4] for the Cartan involution θ and invariant form $B(X, Y)$ on $\mathfrak{g}_E$. With these preliminaries, we define $\Phi(v) = pr_K(v)^n e^{-||v||^2}$.

From this definition, one obtains

$$
\begin{aligned}
f(\gamma_0 g, \Phi, s) &= \int_{\mathrm{GL}_1(\mathbf{R})} |t|^s \Phi(tg^{-1}\widetilde{v_E})\, dt \\
&= \int_{\mathrm{GL}_1(\mathbf{R})} |t|^s pr_K(tg^{-1}\widetilde{v_E})^n e^{-t^2||g^{-1}\widetilde{v_E}||^2}\, dt \\
&= \Gamma\left(\frac{s+n}{2}\right) pr_K(g^{-1}\widetilde{v_E})^n ||g^{-1}\widetilde{v_E}||^{-(s+n)}.
\end{aligned}
$$

Here we have used that n is even.

Denote by $\chi : N(\mathbf{R}) \to \mathbf{C}^\times$ the character given by $\chi(n) = e^{i\langle v_E, n\rangle}$. For $z \in \mathcal{H}_1^\pm$, set

$$
r_0(z) = (1, -z, z^2, -z^3) = E_{12} - zv_1 + z^2\delta_3 - z^3 E_{23}
$$

in $N^{ab} \otimes \mathbf{C}$. From Theorem 3.4, we have $\mathcal{W}_\chi(m) = \sum_{-n\leq v\leq n} \mathcal{W}_\chi^v(m)\frac{x^{n+v}y^{n-v}}{(n+v)!(n-v)!}$ with

$$
\mathcal{W}_\chi^v(m) = \det(m)^n|\det(m)| \left(\frac{|\langle v_E, mr_0(i)\rangle|}{\langle v_E, mr_0(i)\rangle}\right)^v K_v\left(|\langle v_E, mr_0(i)\rangle|\right). \tag{1}
$$

For $m \in M(\mathbf{R})$ and $n \in N(\mathbf{R})$ in the Levi and unipotent radical of the Heisenberg parabolic of $G_2(\mathbf{R})$, we set

$$
x(n, m) = (nm)^{-1}\widetilde{v_E} = m^{-1}\widetilde{v_E} + \det(m)^{-1}\langle v_E, \overline{n}\rangle E_{13} \in \mathfrak{g}_E.
$$

Lemma 6.1. *With the notation above, one has the following:*

(1) $pr_K(x(n, m)) = \frac{1}{4}\det(m)^{-1}\left(\langle\widetilde{v_E}, mr_0(-i)\rangle e_\ell - \langle\widetilde{v_E}, mr_0(i)\rangle f_\ell - i\langle\widetilde{v_E}, \overline{n}\rangle h_\ell\right)$;
(2) $||x(n, m)||^2 = |\det(m)|^{-2}\left(|\langle\widetilde{v_E}, mr_0(i)\rangle|^2 + |\langle\widetilde{v_E}, \overline{n}\rangle|^2\right)$.

Proof. Denote by B the scalar multiple of the Killing form on $\mathfrak{g}_E$ normalized so that $B(e_\ell, f_\ell) = 1$. Then for $x \in \mathfrak{g}_E$, one has

$$
pr_K(x) = B(x, f_\ell)e_\ell + \frac{1}{2}B(x, h_\ell)h_\ell + B(x, e_\ell)f_\ell.
$$

From this, one obtains the first statement of the lemma. For the second statement, one uses the identity $||m^{-1}\widetilde{v_E}||^2 = |\langle m^{-1}\widetilde{v_E}, r_0(i)\rangle|^2$ that holds because $m^{-1}\widetilde{v_E}$ is rank one. □

Putting the various pieces together, and applying the Iwasawa decomposition, we arrive at

$$
I(s, \Phi) = \Gamma\left(\frac{s+n}{2}\right)\int_{\mathrm{GL}_2(\mathbf{R})}\int_{(N^{0,E}\backslash N)(\mathbf{R})} \frac{|det(m)|^{-3}e^{i\langle v_E, \overline{n}\rangle}}{||x(n, m)||^{(s+n)}}\{pr_K(x(n, m))^n, \mathcal{W}_\chi(m)\}_K\, dn\, dm.
$$

In the rest of this section we will compute this integral.

6.2 The Computation

To go further, we need an expression for $\{pr_K(x(n,m))^n, \mathcal{W}_\chi(m)\}_K$. To ease notation, set $\alpha = \langle \widetilde{v_E}, mr_0(i)\rangle$ and $\beta = \langle \widetilde{v_E}, \overline{n}\rangle$. The Lie algebra $\mathfrak{sl}_2$ of the complexified long root SU(2) $\subseteq K$ is identified with $Sym^2(V_2)$ via the map $e_\ell \mapsto x^2, h_\ell \mapsto -2xy$ and $f_\ell \mapsto -y^2$. Thus $x(n,m) = \frac{1}{4}\det(m)^{-1}(\alpha^* x^2 + 2i\beta xy + \alpha y^2)$. We normalize the pairing $\{\cdot,\cdot\}_K$ so that $\{x^{n-v}y^{n+v}, x^{n+v}y^{n-v}\}_K = (-1)^{n+v}(n+v)!(n-v)!$. Plugging in (1), with these normalizations one obtains

$$\{pr_K(x(n,m))^n, \mathcal{W}_\chi(m)\}_K = \frac{n!}{2^{2n}}|\det(m)| \sum_{j_1+j_2+j_3=n} (-1)^{j_3-j_1} \frac{(\alpha^*)^{j_1}}{(j_1)!}\frac{(2i\beta)^{j_2}}{(j_2)!}\frac{\alpha^{j_3}}{(j_3)!} \times \left(\frac{|\alpha|}{\alpha}\right)^{j_3-j_1} K_{j_3-j_1}(|\alpha|)$$

$$= \frac{|\det(m)|}{2^{2n}} \sum_{j_1+j_2+j_3=n} (-1)^{j_3-j_1} \frac{n!}{(j_1)!(j_2)!(j_3)!} (2i\beta)^{j_2}|\alpha|^{n-j_2} K_{j_3-j_1}(|\alpha|).$$

Iterating the identity $K_{n+1}(x) + K_{n-1}(x) = -2K_n'(x)$, one obtains

$$\sum_{j_1,j_3\geq 0, j_1+j_3=n-j_2} \frac{(n-j_2)!}{(j_1)!(j_3)!} K_{j_3-j_1}(|\alpha|) = (-2)^{n-j_2} K_0^{(n-j_2)}(|\alpha|).$$

Thus

$$\{pr_K(x(n,m))^n, \mathcal{W}_\chi(m)\}_K = \frac{|\det(m)|}{2^n} \sum_j \binom{n}{j} (i\beta)^j |\alpha|^{n-j} K_0^{(n-j)}(|\alpha|).$$

Therefore

$$I(s,\Phi) = 2^{-n}\Gamma\left(\frac{s+n}{2}\right) \int_{\mathrm{GL}_2(\mathbf{R})\times(N_{0,E}\backslash N)(\mathbf{R})} \frac{|\det(m)|^{s+n-2}e^{i\beta}}{(|\alpha|^2+|\beta|^2)^{\frac{s+n}{2}}} \times \left(\sum_j \binom{n}{j}(i\beta)^{n-j}|\alpha|^j K_0^{(j)}(|\alpha|)\right) dn\, dm.$$

For $r, y > 0$, one has

$$\frac{\Gamma(s)}{2\Gamma(1/2)} \int_{\mathbf{R}} \frac{e^{irx}}{(x^2+y^2)^s} dx = \left(\frac{r}{2y}\right)^{s-1/2} K_{s-1/2}(ry)$$

and thus

$$\frac{\Gamma(s)}{2\Gamma(1/2)}\int_{\mathbf{R}}\frac{(ix)^k e^{irx}}{(x^2+y^2)^s}\,dx=\partial_r^k\left(\left(\frac{r}{2y}\right)^{s-1/2}K_{s-1/2}(ry)\right).$$

We obtain that

$$\Gamma\left(\frac{s+n}{2}\right)\left(\int_{\mathbf{R}}\frac{e^{i\beta}(i\beta)^j}{(|\alpha|^2+|\beta|^2)^{(s+n)/2}}\,d\beta\right)|\alpha|^{n-j}K_0^{(n-j)}(|\alpha|)$$

differs from

$$\left\{\partial_r^j\left(\left(\frac{r}{2|\alpha|}\right)^{(s+n-1)/2}K_{(s+n-1)/2}(r|\alpha|)\right)\partial_r^{n-j}\left(K_0(r|\alpha|)\right)\right\}|_{r=1}$$

by a nonzero absolute constant. Equivalently,

$$\Gamma\left(\frac{s+n}{2}\right)\left(\int_{\mathbf{R}}\frac{e^{i\beta}(i\beta)^j}{(|\alpha|^2+|\beta|^2)^{(s+n)/2}}\,d\beta\right)|\alpha|^{n-j}K_0^{(n-j)}(|\alpha|)$$

is equal to

$$2^{-s/2}|\alpha|^{1-s}\left\{\partial_y^{n-j}\left(K_0(y)\right)\partial_y^j\left(y^{(s+n-1)/2}K_{(s+n-1)/2}(y)\right)\right\}|_{y=|\alpha|}$$

up to a nonzero absolute constant.

We have therefore proved

$$I(s,\Phi)\doteq 2^{-s/2}\int_{\mathrm{GL}_2(\mathbf{R})}|\det(m)|^{s+n-2}|\alpha|^{1-s}\partial_y^n \left(y^{(s+n-1)/2}K_{(s+n-1)/2}(y)K_0(y)\right)|_{y=|\alpha|}\,dm \qquad (2)$$

where $\doteq$ means that the two sides are equal up to an element of $\mathbf{C}^\times$, and $\alpha=\langle\widetilde{v_E},mr_0(i)\rangle$. Recall $w_0=\begin{pmatrix} & -1\\ 1 & \end{pmatrix}$. Because $w_0r_0(i)=r_0(-i)=r_0(i)^*$ and the integrand in (2) only depends on the absolute value of $|\alpha|$, this integral can be replaced by an integral over $\mathrm{GL}_2(\mathbf{R})^0$. Thus we can split the integral in (2) into an integral over $\mathrm{SL}_2(\mathbf{R})$ and $\mathbf{R}^\times_{>0}$.

For $m=gw\in\mathrm{GL}_2(\mathbf{R})$, with $g\in\mathrm{SL}_2(\mathbf{R})$ and $w\in\mathbf{R}^\times_{>0}$, set $\alpha_0=\langle\widetilde{v_E},gr_0(i)\rangle$ so that $|\alpha|=w|\alpha_0|$. Thus

$$I(s,\Phi)$$

$$\doteq 2^{-s/2}\int_{\mathrm{SL}_2(\mathbf{R})\times\mathbf{R}^\times_{>0}} w^{s+2n-3}|\alpha_0|^{1-s}\partial_y^n\left(y^{(s+n-1)/2}K_{(s+n-1)/2}(y)K_0(y)\right)|_{y=|\alpha|}\,dg\,d^\times w$$

$$= 2^{-s/2} \int_{\mathrm{SL}_2(\mathbf{R}) \times \mathbf{R}^\times_{>0}} w^{s+2n-3} |\alpha_0|^{-(2s+2n-4)} \partial_y^n \left(y^{(s+n-1)/2} K_{(s+n-1)/2}(y) K_0(y) \right) |_{y=w} \, dg \, d^\times w$$

$$= 2^{-s/2} \left(\int_{\mathrm{SL}_2(\mathbf{R})} |\alpha_0|^{-(2s+2n-4)} \, dg \right)$$

$$\times \left(\int_0^\infty w^{s+2n-3} \partial_y^n \left(y^{(s+n-1)/2} K_{(s+n-1)/2}(y) K_0(y) \right) |_{y=w} \frac{dw}{w} \right).$$

Integrating by parts on the second integral, we obtain

$$I(s, \Phi) \doteq 2^{-s/2} \frac{\Gamma(s+2n-3)}{\Gamma(s+n-3)} J(s+n-2) \int_0^\infty w^{(3s+3n-7)/2} K_{(s+n-1)/2}(w) K_0(w) \frac{dw}{w}$$

where $J(\nu) = \int_{\mathrm{SL}_2(\mathbf{R})} |\alpha_0|^{-2\nu} \, dg$. From the Mellin transform of the K-Bessel functions and an application of Barnes' lemma one has the general identity

$$\int_0^\infty K_\mu(y) K_\nu(y) y^s \frac{dy}{y} = 2^{s-3} \frac{\Gamma\left(\frac{s+\mu+\nu}{2}\right) \Gamma\left(\frac{s+\mu-\nu}{2}\right) \Gamma\left(\frac{s-\mu+\nu}{2}\right) \Gamma\left(\frac{s-\mu-\nu}{2}\right)}{\Gamma(s)},$$

valid for $Re(s)$ large compared with $Re(\nu)$, $Re(\mu)$. Applying this, we get

$$I(s, \Phi) \doteq 2^s J(s+n-2) \frac{\Gamma(s+2n-3)\Gamma(s+n-2)^2 \Gamma((s+n-3)/2)^2}{\Gamma(s+n-3)\Gamma((3s+3n-7)/2)}.$$

It remains to evaluate the function $J(\nu)$. In coordinates, one has

$$J(\nu) = \int_{\mathcal{H}_1} |p_E(z)|^{-2\nu} y^{3\nu} \frac{dx \, dy}{y^2}$$

where p_E is the cubic polynomial defining the cubic extension $E/\mathbf{Q}$. One can also manipulate as follows:

$$\Gamma(\nu) J(\nu) = \Gamma(\nu) \int_{\mathrm{SL}_2(\mathbf{R})} |\alpha_0|^{-2\nu} \, dg = \int_{\mathrm{SL}_2(\mathbf{R})} \left(\int_0^\infty t^\nu e^{-|\alpha_0|^2 t} \frac{dt}{t} \right) dg$$

$$= \int_{\mathrm{GL}_2(\mathbf{R})^+} |\det(g)|^\nu e^{-|\alpha|^2} \, dg$$

$$\doteq \int_{V^*} |q(v)|^{\nu/2} e^{-|\langle v, r_0(i) \rangle|^2} \frac{dV}{|q(v)|}.$$

Here q is the invariant quartic form on $W = Sym^3(V_2) \otimes \det^{-1}$; V^* is the $\mathrm{GL}_2(\mathbf{R})$-orbit that consists of the binary cubics with 3 distinct roots; and dV is the additive Haar measure on W, so that $\frac{dV}{|q(v)|}$ is $\mathrm{GL}_2(\mathbf{R})$-invariant.

One can show that the function $v \mapsto e^{-|\langle v, r_0(i) \rangle|^2}$ is a Schwartz function on V^*. It follows that this final integral if of the form considered in [Shi72]. From [Shi72, page 162], one obtains

$$\Gamma(\nu)J(\nu) = h(\nu)\Gamma\left(\frac{\nu}{2} - \frac{1}{6}\right)\Gamma\left(\frac{\nu}{2}\right)^2\Gamma\left(\frac{\nu}{2} + \frac{1}{6}\right)$$

for some function $h(\nu)$ that is analytic in ν.

Summing up, we have proved the following result.

Theorem 6.2. *Let the notation be as above.*

(1) *One has*

$$I(s, \Phi) \doteq 2^s \left(\int_{V^*} |q(v)|^{(s+n-2)/2} e^{-|\langle v, r_0(i)\rangle|^2} \frac{dV}{|q(v)|} \right) \times \frac{\Gamma(s+2n-3)\Gamma(s+n-2)\Gamma((s+n-3)/2)^2}{\Gamma(s+n-3)\Gamma((3s+3n-7)/2)}.$$

(2) *There is an entire function $R_n(s)$ so that*

$$I(s, \Phi) = R_n(s) \frac{\Gamma(s+2n-3)\Gamma(s+n-2)\Gamma(s+n-3)}{\Gamma((s+n-1)/2)}.$$

Proof. The first part we have already for proved. The second follows from the first, [Shi72, page 162], and a "cubic" application of the Gauss multiplication formula for the gamma function. □

Evaluating the integral $J(\nu)$ exactly remains an interesting problem.

Acknowledgments We thank Takuya Yamauchi for helpful comments on an earlier version of this manuscript. We also thank the anonymous referee for his or her comments and corrections, which have improved the readability of this paper.

This work supported by the Schmidt fund at IAS. We thank the IAS for its hospitality and for providing an excellent working environment. We also thank the Simons Foundation for its support of the symposium on the Relative Trace Formula, where some aspects of this work were discussed.

References

[BFG95] D. Bump, M. Furusawa, D. Ginzburg, Non-unique models in the Rankin-Selberg method. J. Reine Angew. Math. **468**, 77–111 (1995). MR 1361787

[BK00] A. Braverman, D. Kazhdan, γ-functions of representations and lifting. Geom. Funct. Anal. **Special Volume**(Part I), 237–278 (2000). With an appendix by V. Vologodsky, GAFA 2000 (Tel Aviv, 1999)

[DF64] B.N. Delone, D.K. Faddeev, The theory of irrationalities of the third degree, in *Translations of Mathematical Monographs*, vol. 10 (American Mathematical Society, Providence, 1964). MR 0160744

[FH91] W. Fulton, J. Harris, Representation theory, in *Graduate Texts in Mathematics*, vol. 129 (Springer, New York, 1991). A first course, Readings in Mathematics. MR 1153249

[GGS02] W.T. Gan, B. Gross, G. Savin, Fourier coefficients of modular forms on G_2. Duke Math. J. **115**(1), 105–169 (2002). MR 1932327

[Gin93] D. Ginzburg, On the standard L-function for G_2. Duke Math. J. **69**(2), 315–333 (1993). MR 1203229

[GL09] B.H. Gross, M.W. Lucianovic, On cubic rings and quaternion rings. J. Number Theory **129**(6), 1468–1478 (2009). MR 2521487

[GS15] N. Gurevich, A. Segal, The Rankin-Selberg integral with a non-unique model for the standard L-function of G_2. J. Inst. Math. Jussieu **14**(1), 149–184 (2015). MR 3284482

[GW94] B.H. Gross, N.R. Wallach, A distinguished family of unitary representations for the exceptional groups of real rank $= 4$, in *Lie theory and geometry, Program Mathematical*, vol. 123 (Birkhäuser, Boston, MA, 1994), pp. 289–304. MR 1327538

[GW96] B.H. Gross, N.R. Wallach, On quaternionic discrete series representations, and their continuations. J. Reine Angew. Math. **481**, 73–123 (1996). MR 1421947

[JR97] D. Jiang, S. Rallis, Fourier coefficients of Eisenstein series of the exceptional group of type G_2. Pacific J. Math. **181**(2), 281–314 (1997). MR 1486533

[KO95] H. Koseki, T. Oda, Whittaker functions for the large discrete series representations of $\mathrm{SU}(2, 1)$ and related zeta integrals. Publ. Res. Inst. Math. Sci. **31**(6), 959–999 (1995). MR 1382562

[Ngo14] B.C. Ngo, On a certain sum of automorphic L-functions, in *Automorphic forms and Related Geometry: Assessing the Legacy of I.* I. Piatetski-Shapiro, Contemp. Math., vol. 614 (American Mathematical Society, Providence, 2014), pp. 337–343

[Pola] A. Pollack, Lifting laws and arithmetic invariant theory. Camb. J. Math. (2016). arXiv preprint arXiv:1609.08273

[Polb] A. Pollack, Unramified Godement-Jacquet theory for the spin similitude group. J. Raman. Math. Soc. (to appear)

[Pol17] A. Pollack, The spin L-function on GSp_6 for Siegel modular forms. Compos. Math. **153**(7), 1391–1432 (2017). MR 3705262

[Pol18] A. Pollack, The Fourier expansion of modular forms on quaternionic exceptional groups. Duke Math. J. **169**(7), 1209–1280 (2020)

[PS17] A. Pollack, S. Shah, On the Rankin-Selberg integral of Kohnen and Skoruppa. Math. Res. Lett. **24**(1), 173–222 (2017). MR 3662279

[PSR88] I. Piatetski-Shapiro, S. Rallis, A new way to get Euler products. J. Reine Angew. Math. **392**, 110–124 (1988). MR 965059

[PWZ19] A. Pollack, C. Wan, M. Zydor, A G_2-period of a Fourier coefficient of an Eisenstein series on E_6. Isr. J. Math. **234**(1), 229–279 (2019)

[Rum97] K.E. Rumelhart, Minimal representations of exceptional p-adic groups. Represent. Theory **1**, 133–181 (1997). MR 1455128

[Sav99] G. Savin, A class of supercuspidal representations of $G_2(k)$. Canad. Math. Bull. **42**(3), 393–400 (1999). MR 1703700

[Seg17] A. Segal, A family of new-way integrals for the standard L-function of cuspidal representations of the exceptional group of type G_2. Int. Math. Res. Not. IMRN **7**, 2014–2099 (2017). MR 3658191

[Shi72] T. Shintani, On Dirichlet series whose coefficients are class numbers of integral binary cubic forms. J. Math. Soc. Japan **24**, 132–188 (1972). MR 0289428

[SV00] T.A. Springer, F.D. Veldkamp, Octonions, Jordan algebras and exceptional groups, in *Springer Monographs in Mathematics* (Springer, Berlin, 2000). MR 1763974

[Wal03] N.R. Wallach, Generalized Whittaker vectors for holomorphic and quaternionic representations. Comment. Math. Helv. **78**(2), 266–307 (2003). MR 1988198

[Wal15] N.R. Wallach, On the Gelfand-Kirillov dimension of a discrete series representation, in *Representations of Reductive Groups*. Program Mathematical, vol. 312 (Birkhäuser/Springer, Cham, 2015), pp. 505–516. MR 3495808

[Yam90] H. Yamashita, Embeddings of discrete series into induced representations of semisimple Lie groups. I. General theory and the case of $\mathrm{SU}(2, 2)$. Jpn. J. Math. (N.S.) **16**(1), 31–95 (1990). MR 1064445

[Yam91] H. Yamashita, Embeddings of discrete series into induced representations of semisimple Lie groups. II. Generalized Whittaker models for SU(2, 2). J. Math. Kyoto Univ. **31**(2), 543–571 (1991). MR 1121177

[Yos99] T. Yoshinaga, Embeddings of discrete series into some induced representations of a group of G_2 type, in *Sūrikaisekikenkyūsho Kōkyūroku* (1094), 29–43 (1999). Automorphic forms on $Sp(2; \mathbf{R})$ and $SU(2, 2)$, II (Kyoto, 1998). MR 1751054

[YY96a] T. Yoshinaga, H. Yamashita, The embeddings of discrete series into principal series for an exceptional real simple Lie group of type G_2. J. Math. Kyoto Univ. **36**(3), 557–595 (1996). MR 1417827

[YY96b] T. Yoshinaga, H. Yamashita, The embeddings of discrete series into principal series for an exceptional real simple Lie group of type G_2. Proc. Jpn. Acad. Ser. A Math. Sci. **72**(4), 78–81 (1996). MR 1391903